ELECTRIC CIRCUITS AND NETWORK ANALYSIS

Other CBS book by the same author

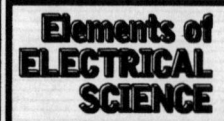

SEVENTH EDITION

ELECTRIC CIRCUITS AND NETWORK ANALYSIS

P. M. Chandrashekharaiah
ME, FIE, MISTE

Vice-Principal and
Professor and Head
Department of Electrical Engineering and Electronics
Shridevi Institute of Engineering and Technology
Sira Road, Tumkur 572106

Former
Professor and Head,
Department of Electrical Engineering and Electronics
Bapuji Institute of Engineering and Technology, Davangere
and
Dean, Engineering Faculty, Kuvempu University

CBS

CBS Publishers & Distributors Pvt. Ltd.

New Delhi • Bengaluru • Chennai • Kochi • Kolkata • Mumbai
Hyderabad • Uttarakhand • Nagpur • Patna • Pune • Jharkhand

ISBN: 81-239-1410-5

First Combined Edition: 2007
Reprint: 2010, 2012, 2015, 2019

Published by **Satish Kumar Jain** and produced by **Varun Jain** for
CBS Publishers & Distributors Pvt. Ltd.,
4819/XI Prahlad Street, 24 Ansari Road, Daryaganj, New Delhi - 110002
delhi@cbspd.com, cbspubs@airtelmail.in • www.cbspd.com
Ph.: 23289259, 23266861, 23266867 • Fax: 011-23243014

Corporate Office: 204 FIE, Industrial Area, Patparganj, Delhi - 110 092
Ph: 49344934 • Fax: 011-49344935
E-mail: publishing@cbspd.com • publicity@cbspd.com

Branches:
- *Bengaluru:* 2975, 17th Cross, K.R. Road, Bansankari 2nd Stage,
  Bengaluru - 70 • Ph: +91-80-26771678/79 • Fax: +91-80-26771680
  E-mail: cbsbng@gmail.com, bangalore@cbspd.com
- *Chennai:* No. 7, Subbaraya Street, Shenoy Nagar, Chennai - 600030
  Ph: +91-44-26681266, 26680620 • Fax: +91-44-42032115
  E-mail: chennai@cbspd.com
- *Kochi:* Ashana House, 39/1904, A.M. Thomas Road, Valanjambalam,
  Ernakulum, Kochi • Ph: +91-484-4059061-65
  Fax: +91-484-4059065 • E-mail: cochin@cbspd.com
- *Kolkata:* 6-B, Ground Floor, Rameshwar Shaw Road, Kolkata - 700014
  Ph: +91-33-22891126/7/8 • E-mail: kolkata@cbspd.com
- *Mumbai:* 83-C, Dr. E. Moses Road, Worli, Mumbai - 400018
  Ph: +91-9833017933, 022-24902340/41 • E-mail: mumbai@cbspd.com

Representatives:

- Hyderabad: 0-9885175004
- Patna: 0-9334159340
- Jharkhand: 0-9811541605
- Nagpur: 0-9021734563
- Pune: 0-9623451994
- Uttarakhand: 0-9716462459

Printed at:
India Binding House, Noida, UP (India)

Foreword

It gives me immense pleasure in introducing the textbook *Electric Circuits and Network Analysis* by Prof. P.M. Chandrashekharaiah, Vice-Principal and Head of the Department of Electrical Engineering and Electronics, SIET, Tumkur. This book is written, keeping in view the subject "Network Analysis" prescribed for III Sem. (E & E) students of Visvesvaraya Technological University, Karnataka. I have known Prof. P.M. Chandrashekharaiah as my student and also as a teacher for the last 35 years and he has the potential to write textbooks of high quality and this book is an example as to how the fundamental concepts have to be presented.

The circuit concepts have been clearly explained so that, on self-reading, the students can understand the fundamentals easily. I am very much impressed by the way in which the chapter "Network Topology" has been written. The textbook contains a large number of worked problems and a large number of numerical problems for which the answers are given. The students can really feel the pulse of the type of problems that may appear in the examinations, by going through worked problems and by solving the numerical problems.

Prof. P.M. Chandrashekharaiah is already the author of the book *Elements of Electrical Science* which is well written and well received by both the students and staff of engineering colleges and polytechnics.

I hope and wish that this book also will be well received. I congratulate Prof. P.M. Chandrashekharaiah for writing this book and wish him good luck in all his future endeavours.

Dr. T. Basavaraju
Chairman
Department of Electrical Sciences
University Visvesvaraya College of Engineering
Bangalore 560001

to
my teacher
and benefactor

Sri B.L. Siddaiah
MA,BEd

Preface

A sound knowledge of the fundamentals of electric circuits and network analysis is essential for the understanding of any branch of science such as mechanical, hydraulic, thermal, nuclear, traffic flow, weather prediction, because all such systems can be simulated by an electric circuit. The performance of any electrical device or machine is always studied by drawing its electrical equivalent circuit.

Out of my nearly 35 years of teaching experience, I have observed that a single textbook never gives all that a student requires for the study of a particular subject. Keeping this in view, an attempt is made in this book to give a comprehensive material required by students of engineering studying in the Indian universities to study the subject *Electric Circuits and Network Analysis* with special reference to students studying electrical science branches.

The textbook contains 15 chapters: 9 chapters are covered under the section **Electric Circuits** and 6 chapters under **Network Analysis**. The demarcation between 'electric circuits' and 'network analysis' is only arbitrary and hypothetical, as any chapter given in the textbook can be studied under either of the above two concepts.

In some universities, the contents of the textbook are studied as one course at the undergraduate level and in some other universities, it is studied as two different courses.

A large number of problems taken from the question papers of various universities are worked out for the benefit of the students. A large number of unsolved numerical problems are given at the end of each chapter for which answers are given at the end pages of the book. The students are advised to solve these problems which will give them confidence about their understanding of the concepts clearly.

I would like to thank M/s CBS Publishers & Distributors, New Delhi, for the efforts they have taken to publish this book. My special thanks to Mr. Y.N. Arjuna, Publishing Director, CBS P&D, for his constant interaction and support.

I hope that this book will be well received by the students and staff of all the engineering colleges. Any suggestions for the improvement of the book will be thankfully acknowledged.

P.M. Chandrashekharaiah

Acknowledgement

My interaction with the students, both inside and outside the classroom, during the course of my 35 years of teaching experience, has enabled me to understand the way in which the fundamental concepts of electric circuits have to be presented so that they will be well received and perceived by them. I would like to thank all my students for their interaction with me.

This book is dedicated to Sri B.L. Siddaiah, my teacher, when I was studying in high school, at Koratagere, Tumkur district, who was not only a teacher but also the guardian, throughout the three years of my stay in that school. I will never forget the love and affection showered on me by Sri B.L. Siddaiah, his wife Smt. Geervani Siddaiah, and all the family members. I shall always remain indebted and grateful to this family for all the help I received, both moral and material.

Dr. T. Basavaraju, Chairman, Department of Electrical Sciences, University Visvesvaraya College of Engineering, Bangalore, who has also taught me during my engineering studies, was kind enough to write the Foreword to this book. I would like to thank him for his gesture, constant help and guidance he has rendered during the course of my studies and also afterwards.

This textbook is written, composed and got printed within a short period of 3 months of time. This would not have been possible but for the help and cooperation of my colleagues in the department. I wish to thank Sri B.S. Thirtharaju, Lecturer in the Department, who interacted with me constantly throughout the preparation of this book, during composing, identifying problems, both worked and numerical, verifying the answers, going through proofreading, but for whose efforts, this book would not have been brought out in such a very short time. I would like to thank him with affection for his constant cooperation and efforts he put in during the preparation of this book. I would like to specially thank Sri. S. Chidanandappa, Assistant Professor in the Department, who gave very useful suggestions, and in writing the chapter on Network Topology. I would like to thank my colleagues, Sri M.S. Nagaraj, Sri A.H. Thejaswi and Sri S.R. Basavarajappa, who actively interacted with me in one way or the other during the preparation of this book.

I would like to thank my wife Smt. K.M. Rajeswari, and sons P.C. Sangameswara, P.C. Naveen and P.C. Praveen, who were always a constant source of inspiration during the process of writing this book.

Finally, I would like to thank all those, who are directly or indirectly responsible for making this book a possibility.

P.M. Chandrashekharaiah

Contents

═══ ELECTRIC CIRCUITS ═══

NETWORK ANALYSIS

two port networks • *T* section representation of a two port
network • π section representation of a two port network.

Chapter 16
NETWORK SYNTHESIS

ELECTRIC CIRCUITS

Basic Circuit Concepts: D.C. Circuits

1.1 INTRODUCTION

An electric circuit is an interconnection of various elements in which there is at least one closed path in which current can flow. Any engineering system uses in one way or the other, an electric circuit as a component. The performance of any electrical device or machine is always studied by drawing its electrical equivalent circuit. The fundamental requirement for the understanding of any branch of science is the sound knowledge of electric circuits. Any type of system such as mechanical, hydraulic, thermal, nuclear, traffic flow, weather prediction, etc. can always be simulated by an electric circuit. All control systems are studied by representing them in the form of electric circuits. When the techniques of circuit theory are mastered, we effectively also learn the analysis of any system.

1.2 ELEMENTS OF AN ELECTRIC CIRCUIT

An electric circuit consists of two types of elements (i) *active elements* or *sources* and (ii) *passive elements* or *sinks*.

Active elements are the elements of a circuit which possess energy of their own and can impart it to other elements of the circuit. There are two types of active elements (i) *voltage source* and (ii) *current source*.

An *ideal voltage source* is one which delivers energy to the load at a constant terminal voltage, irrespective of the current drawn by the load.

An *ideal current source* is one, which delivers energy with a constant current to the load, irrespective of the terminal voltage across the load.

Ideal d.c. voltage sources and ideal d.c. current sources are represented symbolically as shown in Figs 1.1 (a) and (b) respectively.

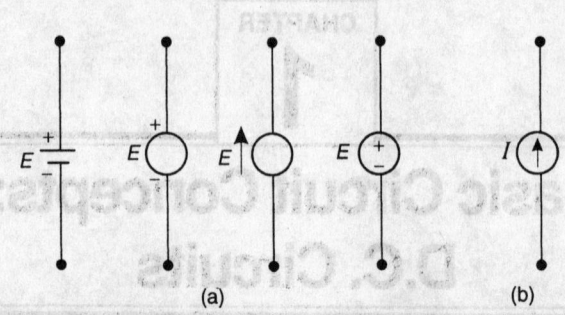

Fig. 1.1 (a) Various methods of representing ideal d.c. voltage sources. (b) Method of representing ideal d.c. current source

An ideal source does not exactly represent any physical device. A practical source always possesses a very small value of internal resistance r. The symbolic representations of a practical voltage source, and a practical current source are shown in Figs 1.2 (a) and (b) respectively.

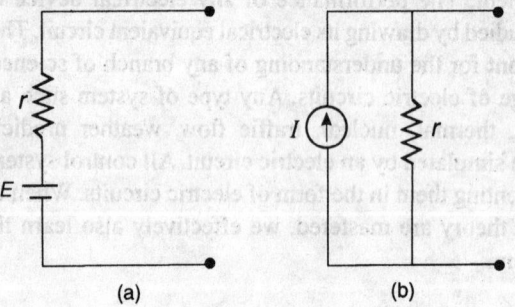

Fig. 1.2 (a) Practical voltage source
(b) Practical current source

The internal resistance of a voltage source is always connected in series with it and for a current source, it is always connected in parallel with it.

As the value of the internal resistance of a practical voltage source is very small, its terminal voltage is assumed to be almost constant within a certain limit of current flowing through the load.

A practical current source is also assumed to deliver a constant current, irrespective of the terminal voltage across the load connected to it.

An ideal a.c. voltage source and a practical voltage source are represented as shown in Fig. 1.3 (a), where z is the internal impedance of the practical voltage source. E represents the r.m.s. value of the a.c. voltage.

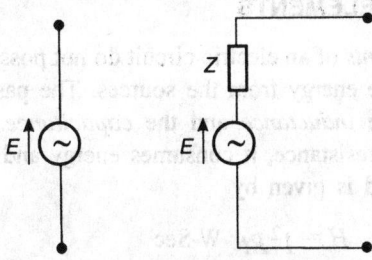

Fig. 1.3 (a) Ideal and practical a.c. voltage sources

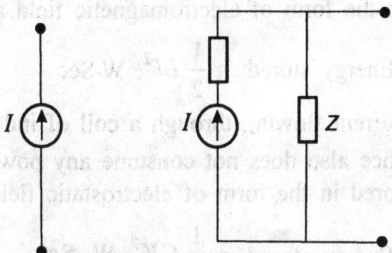

Fig. 1.3 (b) Ideal and practical a.c. current sources

Figure 1.3 (b) represents ideal and practical a.c. current *sources* where, I is the r.m.s. value of the current, and z is the internal impedance.

1.3 INDEPENDENT AND DEPENDENT SOURCES

All the sources described in section 1.2 are independent sources as the voltage of the voltage source is completely independent of current and the current of the current source is completely independent of the voltage. But, there are special kinds of sources in which the source voltage or current depends upon a current or voltage elsewhere in the circuit. Such sources are called *dependent sources* or *controlled sources*. Diamond symbol is used to represent dependent sources and circles are used to represent independent sources. Figs 1.4 (a) and (b) represent dependent voltage source and current source respectively.

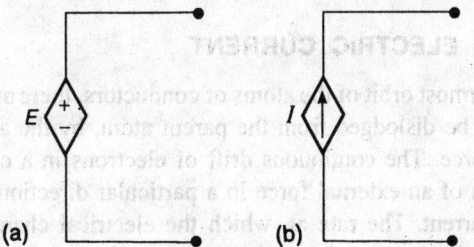

Fig. 1.4 (a) Dependent voltage source (b) Dependent current source

1.4 PASSIVE ELEMENTS

The *passive elements* of an electric circuit do not possess energy of their own. They receive energy from the sources. The passive elements are the *resistance*, the *inductance* and the *capacitance*. When current is passed through a resistance, it consumes energy and heat is produced. The heat produced is given by

$$H = I^2 Rt \quad \text{W-Sec} \tag{1.1}$$

Where, I = current flowing through a resistance R for time t sec.

A pure inductance does not consume any power and the energy given to it is stored in the form of electromagnetic field and is given by

$$\text{Energy stored} = \frac{1}{2} LI^2 \quad \text{W-Sec} \tag{1.2}$$

Where, I = current flowing through a coil of inductance L henries. A pure capacitance also does not consume any power and the energy given to it is stored in the form of electrostatic field and is given by

$$\text{Energy stored} = \frac{1}{2} CV^2 \quad \text{W-Sec} \tag{1.3}$$

Where, V = voltage applied across a capacitor of capacitance C farads.

1.5 BILATERAL AND UNILATERAL ELEMENTS

An element is said to be *bilateral*, when the same relation exists between voltage and current for the current flowing in either direction. An element is said to be *unilateral*, when the same relation does not exist between the voltage and current in either direction. There will be two entirely different laws governing the voltage-current relation in the two possible directions. The voltage source, the current source, the resistance, inductance, and the capacitance are all bilateral elements. The circuits containing them are called bilateral circuits. Vacuum diodes, silicon diodes, selenium rectifiers etc., are examples for unilateral elements. The circuits containing them are called unilateral circuits.

1.6 THE ELECTRIC CURRENT

In the outermost orbit of the atoms of conductors, there are free electrons, which can be dislodged from the parent atom, by the application of an external force. The continuous drift of electrons in a conductor on the application of an external force in a particular direction, constitutes the flow of current. The rate at, which the electrical charge is transferred across a point in a conductor is known as the current flowing through the conductor.

$$I = \frac{dq}{dt} = \frac{q}{t} \qquad (1.4)$$

The unit of current is *ampere*.

1.7 THE AMPERE

One ampere of current is defined as that current, which, when flowing through a resistance of one ohm, causes a potential difference of one volt across it.

From Eq. (1.4), one ampere of current may also be defined as the current flowing through a conductor, when a charge of one coulomb crosses across a point in the conductor in one second.

One coulomb of charge is equal to the charge of 6.242×10^{18} electrons. Hence, one ampere of current is said to be flowing through a conductor, when 6.242×10^{18} electrons cross across a point in the conductor in one second.

One ampere of current is also defined as the current; which, when flowing through two straight parallel conductors of infinite length and of negligible cross-section, placed in free space, one metre apart, producing between them, a force of 2×10^{-7} Newton per metre length of conductors.

1.8 THE ELECTRIC POTENTIAL

The electric potential always refers to a point in a charged conductor. The electric potential at any point in a charged conductor is defined as the work done to bring a unit positive charge from infinity to that point. The unit of electric potential is *volt*.

1.9 POTENTIAL DIFFERENCE (V)

The potential difference between any two points of a charged conductor is the amount of work that has to be done to bring a unit positive charge from the point of lower potential to the point of higher potential. The unit of potential difference is *volt*. The potential difference is also referred as the voltage between the two points of a conductor.

1.10 VOLT (V)

One volt is defined as the potential difference across a resistance of one ohm, through which, a current of one ampere is flowing.

1.11 E.M.F. OF A SOURCE (*E*)

The e.m.f. of a source is the voltage available across its terminals. The voltage available across the terminals of a voltage source is slightly less than its internal voltage E_i, due to the small voltage drop across its internal resistance, as shown in Fig. 1.5.

$$E = E_i - Ir \tag{1.5}$$

The unit of e.m.f. is also volts.

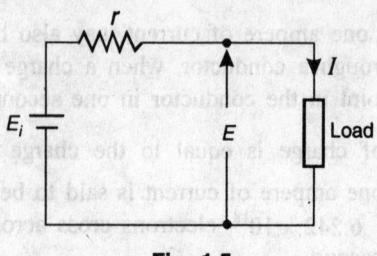

Fig. 1.5

1.12 OHM'S LAW

Statement: The temperature remaining constant, the current flowing through any conductor is directly proportional to the potential difference between the two ends of the conductor.

$I \propto V$, when temperature is constant

$$I = \frac{1}{R} V = \frac{V}{R} \tag{1.6}$$

Where, R is a constant, known as the *resistance* of the conductor. Ohm's law can be applied both for a.c. and d.c. circuits.

Limitations of Ohm's Law

(i) Ohm's law does not hold good for nonmetallic conductors such as silicon carbide. The law governing the *V-I* relation for them is given by

$$V = K I^m \tag{1.7}$$

Where K and m are constants.

(ii) Ohm's law also does not hold good for nonlinear devices such as zener diodes, voltage regulators, etc.

1.13 POWER (*P*)

Power is defined as the rate at which, the work is done. Its unit is watt.

$$P = E I = \frac{E^2}{R} = I^2 R \tag{1.8}$$

1.14 ENERGY (*W*)

Energy is the capacity to do work. It is equal to the total work done in a particular time. The unit is Watt-Sec.

$$W = EIt = \frac{E^2}{R} t = I^2 Rt \qquad (1.9)$$

Watt-sec. is a very small unit of energy. The practical unit of energy is kilo watt hour (kWH) whose trade name is *unit*.

1.15 RESISTANCE (*R*)

Resistance is the property of a conductor by virtue of which, it opposes or limits the flow of current through it. The unit of resistance is *ohm*, named after George Simon Ohm (1787-1854) a German physicist, who investigated the relation between the applied voltage across a resistance and the current flowing through it. The symbol of ohm is Ω.

The resistance of a conductor is directly proportional to its length and inversely proportional to its area of cross section.

$$R \propto \frac{l}{a} \text{ i.e. } R = \rho \frac{l}{a} \qquad (1.10)$$

Where ρ is a constant, known as the *specific resistance* or *resistivity* of the material of the conductor.

In equation (1.10), $\rho = R$, if $l = 1$ and $a = 1$. Hence, the resistivity of a conductor, may be defined as its resistance, when its length is unity and area of cross section is also unity. The unit of resistivity is ohm-metre (Ω-m). Typical values of the resistivities of various materials are listed in Table 1.1.

Table 1.1: Resistivities of materials

S. No.	Conductor	ρ in Ω-m at 20°C
1.	Silver	1.629×10^{-8}
2.	Copper	1.724×10^{-8}
3.	Gold	2.44×10^{-8}
4.	Aluminium	2.688×10^{-8}
5.	Tungsten	5.5×10^{-8}
6.	Nickel	7.8×10^{-8}
7.	Iron	9.8×10^{-8}
8.	Tantalum	15.5×10^{-8}
9.	Nichrome	100×10^{-8}
10.	Tin oxide	250×10^{-8}
11.	Carbon	3500×10^{-8}

1.16   RESISTANCES IN SERIES

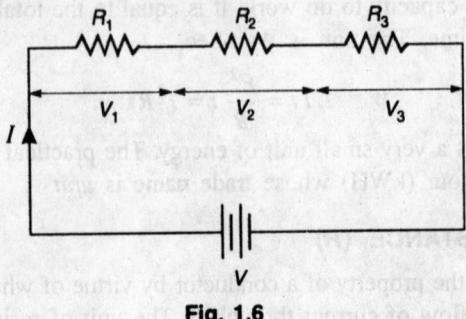

Fig. 1.6

In Fig. 1.6, three resistances R_1, R_2 and R_3 are connected in series. V is the total voltage applied across the combination and I is the current flowing through them. V_1, V_2 and V_3 are the voltage drops across R_1, R_2 and R_3 respectively.

Then $V = V_1 + V_2 + V_3$

i.e. $IR = I R_1 + I R_2 + I R_3$

or $R = R_1 + R_2 + R_3$ = Total resistance.   (1.11)

If there are n resistances connected in series, then the total resistance is given by

$$R = R_1 + R_2 + R_3 + \ldots\ldots\ldots\ldots + R_n \qquad (1.12)$$

1.17   RESISTANCES IN PARALLEL

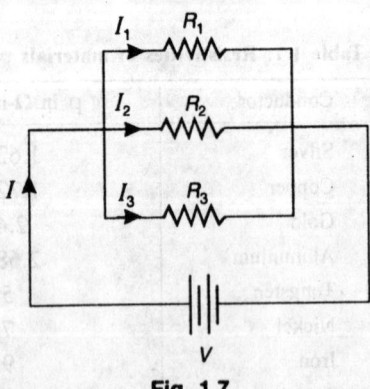

Fig. 1.7

In Fig. 1.7, three resistances R_1, R_2, and R_3 are connected in parallel across a voltage of V volts. I is the total current. I_1, I_2, and I_3 are the currents flowing through R_1, R_2, and R_3 respectively.

$$I = I_1 + I_2 + I_3$$

i.e. $$\frac{V}{R} = \frac{V}{R_1} + \frac{V}{R_2} + \frac{V}{R_3}$$

∴ $$\frac{1}{R} = \frac{1}{R_1} + \frac{1}{R_2} + \frac{1}{R_3} \qquad (1.13)$$

Where, R = Total resistance

If there are n resistances connected in parallel, the total resistance is given by

$$\frac{1}{R} = \frac{1}{R_1} + \frac{1}{R_2} + \frac{1}{R_3} + \cdots\cdots\cdots + \frac{1}{R_n} \qquad (1.14)$$

If there are only two resistances connected in parallel, then the total resistance is given by

$$\frac{1}{R} = \frac{1}{R_1} + \frac{1}{R_2} = \frac{R_1 + R_2}{R_1 R_2} \qquad \therefore R = \frac{R_1 R_2}{R_1 + R_2} \qquad (1.15)$$

Equation (1.15) is more convenient to use, when only two resistances are connected in parallel.

1.18 CURRENT IN A PARALLEL BRANCH

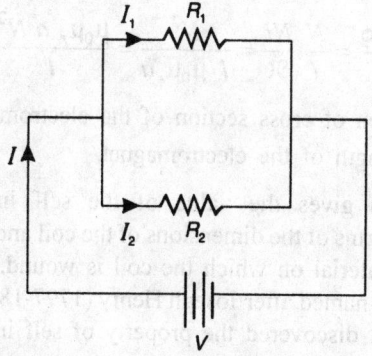

Fig. 1.8

From the circuit in Fig. 1.8.

$$V = I_1 R_1 = I_2 R_2$$

i.e. $$\frac{I_1}{I_2} = \frac{R_2}{R_1}$$

i.e.
$$\frac{I_1 + I_2}{I_2} = \frac{R_2 + R_1}{R_1}$$

i.e.
$$\frac{I}{I_2} = \frac{R_1 + R_2}{R_1}$$

or
$$I_2 = I \frac{R_1}{R_1 + R_2}$$

Similarly
$$I_1 = I \frac{R_2}{R_1 + R_2}$$

Hence, the branch current is given by

$$\text{Branch current} = \text{Total current} \times \frac{\text{The other resistance}}{\text{Sum of the two resistances}} \quad (1.16)$$

1.19 SELF INDUCTANCE (L)

The self inductance of a coil is its *property by virtue of which it always opposes any change in the value of the current flowing through it.*

The self inductance of a coil may also be defined as its *property by virtue of which an e.m.f. is induced in it, whenever an alternating current flows through it.*

The self inductance of a coil is also defined as the number of weber turns produced per ampere in the coil as per the Eq. 1.17.

$$L = \frac{N\phi}{I} = \frac{N}{I} \frac{NI}{\Re} = \frac{N^2}{l/\mu_0\mu_r a} = \frac{\mu_0\mu_r \, a \, N^2}{l} \quad (1.17)$$

Where a = area of cross section of the electromagnet.

l = length of the electromagnet.

Equation (1.17) gives the value of the self inductance of an electromagnet, in terms of the dimensions of the coil and the permeability of the magnetic material on which the coil is wound. The unit of self inductance is *henry* named after Joseph Henry (1797-1878), an American physicist, who first discovered the property of self inductance.

1.20 ENERGY STORED IN AN INDUCTOR

An inductor is an inductive coil which, possesses both inductance and a small resistance. If the resistance is neglected, it is called as an ideal inductor. We are considering such an ideal inductive coil.

Consider a coil of inductance L henrys as shown in Fig. 1.9, through which, an alternating current i is flowing. This alternating current produces an alternating flux, which links the coil, and hence an e.m.f. e is induced in the coil, which is given by

$$e = -L\frac{di}{dt} = -v$$

Where, v is the applied voltage.

$$\therefore \qquad v = L\frac{di}{dt} \quad \text{or} \quad i = \frac{1}{L}\int v.dt \qquad (1.18)$$

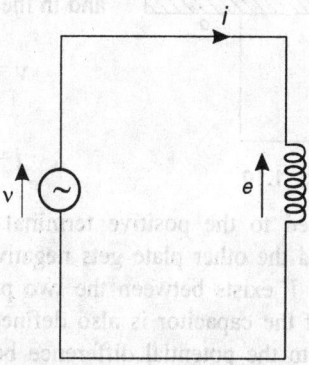

Fig. 1.9

A pure inductance does not consume any energy and the energy supplied to the coil is stored in the form of an electromagnetic field.

The induced e.m.f. opposes any change in the value of the current flowing through the coil. Hence, in order to establish a steady current of I amperes in t seconds, work has to be done to overcome the opposition due to the induced e.m.f. The work done in dt seconds is given by

$$dw = v\, i\, dt = L\frac{di}{dt}\, i\, dt = L\, i\, di$$

The work done in t seconds in given by

$$W = \int_0^t dw = \int_0^t L\, i\, di = \frac{1}{2}LI^2 \qquad (1.19)$$

This work done is stored in the coil in the form of an electromagnetic field. Hence, the energy stored in a coil of inductance L henrys in the form of an electromagnetic field is given by $1/2\, LI^2$.

1.21 CAPACITANCE (C)

It is defined as the capacity of a capacitor to store electrical energy in the form of electrostatic field or static charges.

A capacitor consists of two conducting plates separated by an insulating medium called *dielectric* medium. When the plates of the capacitor are connected to a battery of voltage V volts as shown in Fig. 1.10, the plates get charged to a charge of q coulombs.

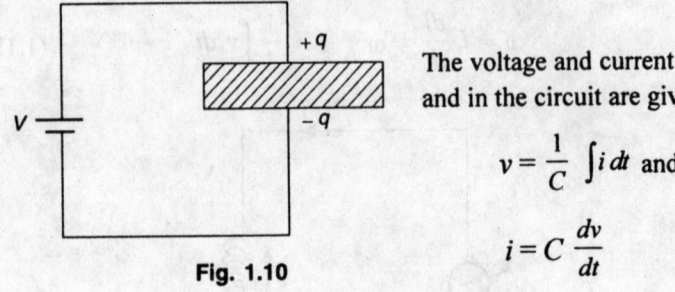

The voltage and current across and in the circuit are given by:

$$v = \frac{1}{C} \int i \, dt \text{ and}$$

$$i = C \frac{dv}{dt}$$

Fig. 1.10

The plate connected to the positive terminal of the battery gets positively charged and the other plate gets negatively charged. Hence, a potential difference V exists between the two plates.

The capacitance of the capacitor is also defined as the ratio of the charge on the plates to the potential difference between the plates.

$$C = \frac{q}{V} \tag{1.20}$$

If $V = 1$ volt, then $C = q$. Hence, the capacitance of a capacitor may also be defined as the charge that is required to create unit potential difference across its plates. The capacitance is symbolically represented as shown in Fig. 1.11.

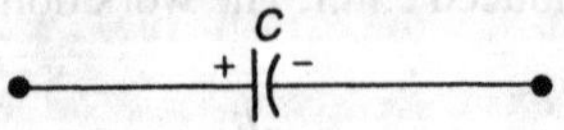

Fig. 1.11

The unit of capacitance is coulomb per volt or *Farad*. It may be defined as the capacitance of a capacitor between the plates of which, there appears a potential difference of 1 volt, when its plates are charged to 1 coulomb.

Farad is a very big unit. Hence, the capacitance of a capacitor is usually expressed in microfarads, nanofarads and picofarads.

$$1 \mu F = 10^{-6} F, \qquad 1 nF = 10^{-9} F, \qquad 1 pF = 10^{-12} F$$

1.22 ENERGY STORED IN A CAPACITOR

Whenever a capacitor is connected to a d.c. voltage, it gets charged and energy supplied during charging is stored in the dielectric medium in the form of electrostatic field.

Let at any stage of charging, the potential difference across the plates be equal to v volts. By the definition of the potential difference, this is the work done to move a unit positive charge from one plate to the other. The work that has to be done to move a charge dq from one plate to the other is given by

$$dw = v.dq = vd(Cv) = Cv.dv$$

The total work done to transfer a charge of q coulombs from one plate to the other given by

$$W = \int_0^V Cv \, dv = \frac{1}{2} CV^2$$

Where, V is the potential difference across the plates of the capacitor, corresponding to a charge of q coulombs on the plates. Equation (1.21) represents the energy stored in the capacitor in the form of electrostatic field.

$$\therefore \quad \text{Energy stored} = \frac{1}{2} C V^2 \tag{1.21}$$

1.23 KIRCHHOFF'S LAWS

Gustav Robert Kirchhoff (1824-1887), a German physicist, enunciated two laws which enable us to find the currents flowing in an electric circuit and voltages across the various elements of the circuit. These laws form the basis for the study of electrical circuits. The two laws are (i) current law and (ii) voltage law.

(i) Current law

Statement: The algebraic sum of all the currents meeting at any junction of an electrical circuit is zero.

i.e. $$\sum \bar{I} = 0 \tag{1.22}$$

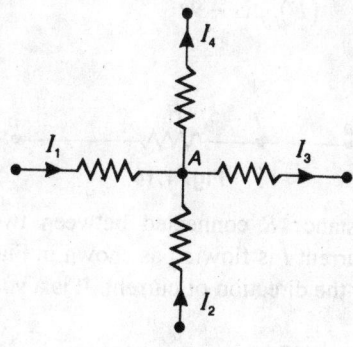

Fig. 1.12

Figure 1.12 shows the junction A of an electric circuit at, which four currents I_1, I_2, I_3, and I_4 meet. All the currents flowing towards the junction are taken as +ve and all the currents flowing away from the junction are taken as –ve. Then, according to Kirchhoff's current law,

$$I_1 + I_2 - I_3 - I_4 = 0$$

or
$$I_1 + I_2 = I_3 + I_4 \qquad (1.23)$$

From Eq. (1.23), Kirchhoff's current law can also be stated as *At any junction of an electric circuit, the sum of all the currents entering the junction is equal to the sum of all the currents leaving the junction.*

(ii) Voltage law

Statement: In any closed electrical circuit, the algebraic sum of all the e.m.f.s and the resistive drops is equal to zero.

i.e. $\qquad \sum \vec{E} + \sum I \vec{R} = 0 \qquad (1.24)$

All the voltage rises are taken as + ve and all the voltage drops are taken as –ve.

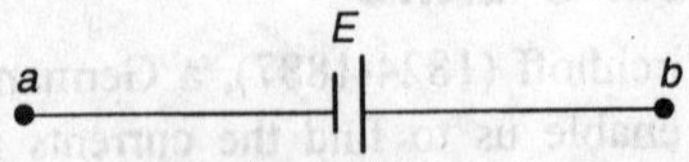

Fig. 1.13

Figure 1.13 represents a battery of e.m.f. E volts connected between two points a and b, which can be traced either from a to b or from b to a. When it is traced from a to b, the battery is traced from –ve terminal to + ve terminal. It is a voltage rise. Hence, the e.m.f. is +ve.

$\therefore \qquad (E)_{ab}$ is + ve

When the battery is traced from b to a, it is traced from +ve terminal to –ve terminal. It is a voltage fall and hence, the e.m.f. is –ve.

$\therefore \qquad (E)_{ba}$ is – ve

Fig. 1.14

Consider a resistance R connected between two points a and b, through which, a current I is flowing as shown in Fig. 1.14. The voltage drop $(IR)_{ab}$ is along the direction of current. It is a voltage fall and hence – ve.

$\therefore \qquad (IR)_{ab}$ is – ve

On the other hand, the voltage drop $(IR)_{ba}$ is against the direction of current. It is a voltage rise and hence, it is +ve.

$\therefore \qquad\qquad (IR)_{ba}$ is +ve

Consider a circuit as shown in Fig. 1.15.

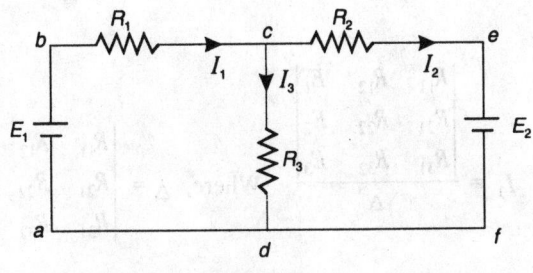

Fig. 1.15

The directions of currents I_1, I_2 and I_3 flowing in the various branches of the circuit are arbitrarily assumed. For the closed loops in the circuit, using Kirchhoff's voltage law, the equations are

$\qquad$ For loop *abcda*: $\qquad E_1 - I_1 R_1 - I_2 R_2 = 0$
$\qquad$ For loop *dcefd*: $\qquad I_2 R_2 - I_3 R_3 - E_2 = 0$

1.24 CRAMER'S RULE

This rule is very useful in solving simultaneous equations. Let the simultaneous equations having two unknowns I_1 and I_2 be written as

$\qquad R_{11} I_1 + R_{12} I_2 = E_1 \qquad$ and $\qquad R_{21} I_1 + R_{22} I_2 = E_2$

According to Cramer's rule

$$I_1 = \frac{\begin{vmatrix} E_1 & R_{12} \\ E_2 & R_{22} \end{vmatrix}}{\Delta} \quad \text{and} \quad I_2 = \frac{\begin{vmatrix} R_{11} & E_1 \\ R_{21} & E_2 \end{vmatrix}}{\Delta}$$

Where $\qquad \Delta = \begin{vmatrix} R_{11} & R_{12} \\ R_{21} & R_{22} \end{vmatrix}$

When there are three unknowns, let the simultaneous equations be:

$$R_{11} I_1 + R_{12} I_2 + R_{13} I_3 = E_1$$
$$R_{21} I_1 + R_{22} I_2 + R_{23} I_3 = E_2$$
$$R_{31} I_1 + R_{32} I_2 + R_{33} I_3 = E_3$$

According to Cramer's Rule, the unknowns I_1, I_2, and I_3 are given by

$$I_1 = \frac{\begin{vmatrix} E_1 & R_{12} & R_{13} \\ E_2 & R_{22} & R_{23} \\ E_3 & R_{32} & R_{33} \end{vmatrix}}{\Delta}, \qquad I_2 = \frac{\begin{vmatrix} R_{11} & E_1 & R_{13} \\ R_{21} & E_2 & R_{23} \\ R_{31} & E_3 & R_{33} \end{vmatrix}}{\Delta}$$

and $\qquad I_3 = \dfrac{\begin{vmatrix} R_{11} & R_{12} & E_1 \\ R_{21} & R_{22} & E_2 \\ R_{31} & R_{32} & E_3 \end{vmatrix}}{\Delta}$ Where, $\Delta = \begin{vmatrix} R_{11} & R_{12} & R_{13} \\ R_{21} & R_{22} & R_{23} \\ R_{31} & R_{32} & R_{33} \end{vmatrix}$

1.25 PROCEDURE FOR SOLVING AN ELECTRIC CIRCUIT USING KIRCHHOFF'S LAWS

Solving an electric circuit normally means that the currents in the various branches are to be found, when the e.m.f.s and the resistances of the circuit are known. The following guidelines help in solving the circuit using Kirchhoff's laws.

1. All the junctions of the circuit are named such as *a, b, c, d*, etc.

2. The currents in all the branches are arbitrarily assumed in any direction satisfying Kirchhoff's current law. While assuming currents, care is taken to see that their number is minimum.

3. Closed loops in the circuit are selected and equations are written using Kirchhoff's voltage law.

4. The number of equations written must be equal to the number of unknown currents.

5. While selecting loops, care is taken to see that all the elements of the circuit are covered in one loop or the other.

6. The simultaneous equations are solved using Cramer's rule.

7. Positive answers obtained for the currents indicate that, their assumed directions are correct. Negative answers indicate that, their assumed directions are wrong and their directions have to be reversed.

1.26 LINEAR NETWORKS

The connection of various circuit elements together is called an *electrical network*. If the network contains at least one closed path, it is also an *electric circuit*. Hence, every circuit is a network, but all networks are not circuits.

A linear element is a passive element which has a linear voltage-current relationship.

A linear dependent source is a dependent current or voltage source, whose output current or voltage is proportional only to the first power of some current or voltage variable in the circuit or to the sum of such quantities.

A linear network is a circuit consisting of only the independent sources, linear dependent sources and linear elements.

The principle of *superposition* and *homogeneity* are the two important properties of a linear network.

In any linear bilateral network containing several sources, the current through or voltage across any element is equal to the algebraic sum of all the individual currents flowing or voltages across the element, caused by separate independent sources acting alone, with all other independent voltage sources replaced by short circuits and all other independent current sources replaced by open circuits. This property is called the property of *superposition* of linear network.

If all the sources are multiplied by a constant, then the response across or through any other element of the network also gets multiplied by the same constant. This property is called the property of homogeneity of linear networks.

1.27 SOURCE TRANSFORMATION

Two sources are said to be identical, when they produce identical terminal voltage V_L and load current I_L.

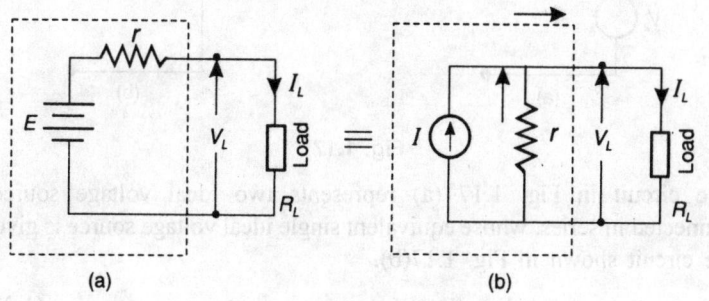

Fig. 1.16

The circuits in Fig. 1.16 (a) and (b) represent a practical voltage source and a practical current source respectively, with load connected to both the sources. The terminal voltage V_L and load current I_L across their terminals are same. Hence, the practical voltage source shown in dotted box of Fig. 1.16 (a) is equivalent to the practical current source shown in dotted box of Fig. 1.16 (b). The two equivalent sources should also provide the same open circuit voltage and short circuit current.

From Figs 1.16 (a), and (b)

$$I_L = \frac{E}{r + R_L} \tag{1.25a}$$

$$I_L = I \frac{r}{r + R_L} \tag{1.25b}$$

From Eq. (1.25a), and (1.25b), it is evident that

$$E = I r$$

or $$I = \frac{E}{r} \tag{1.26}$$

Hence, a voltage source E in series with its internal resistance r can be converted into a current source $I = E/r$, with its internal resistance r connected in parallel with it. Similarly, a current source I in parallel with its internal resistance r can be converted into a voltage source $E = Ir$, in series with its internal resistance r.

1.28 IDEAL VOLTAGE SOURCES CONNECTED IN SERIES

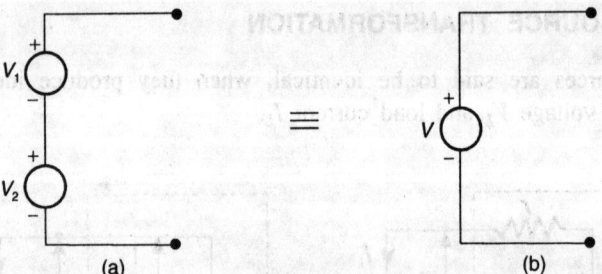

Fig. 1.17

The circuit in Fig. 1.17 (a) represents two ideal voltage sources connected in series, whose equivalent single ideal voltage source is given the circuit shown in Fig. 1.17(b).

$$\therefore \qquad V = V_1 + V_2 \tag{1.27}$$

Any number of ideal voltage sources connected in series can be represented by a single ideal voltage source taking into account the polarities connected together into consideration.

1.29 PRACTICAL VOLTAGE SOURCES CONNECTED IN SERIES

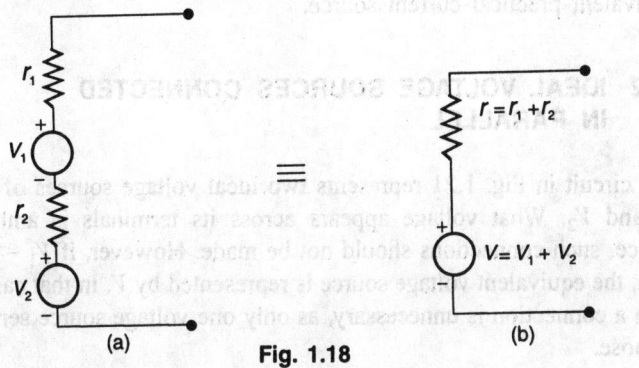

Fig. 1.18

The circuit in Fig. 1.18 (a) represents two practical voltage sources connected in series and the circuit in Fig. 1.18 (b) represents equivalent single practical voltage source.

1.30 IDEAL CURRENT SOURCES CONNECTED IN PARALLEL

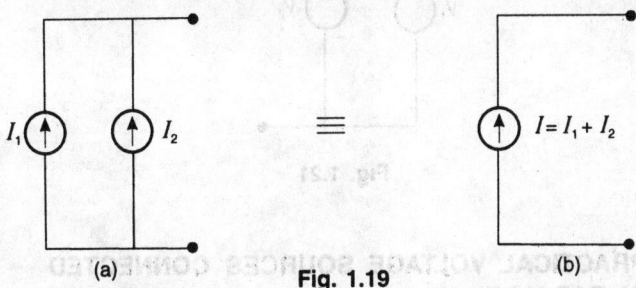

Fig. 1.19

Two ideal current sources connected in parallel can be replaced by a single equivalent ideal current source as shown in Figs 1.19 (a), (b).

1.31 PRACTICAL CURRENT SOURCES CONNECTED IN PARALLEL

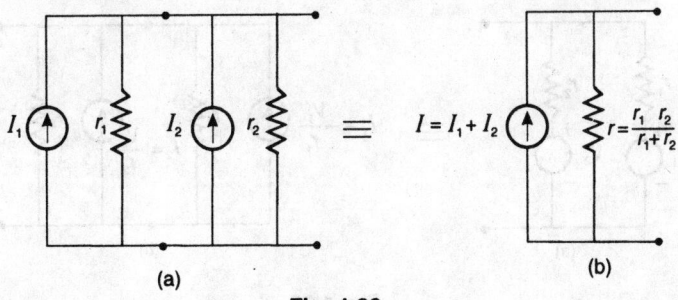

Fig. 1.20

The circuits in Figs 1.20 (a) and (b) clearly show, how two practical current sources connected in parallel may be represented by a single equivalent practical current source.

1.32 IDEAL VOLTAGE SOURCES CONNECTED IN PARALLEL

The circuit in Fig. 1.21 represents two ideal voltage sources of e.m.f.s V_1 and V_2. What voltage appears across its terminals is ambiguous Hence, such connections should not be made. However, if $V_1 = V_2 = V$, then, the equivalent voltage source is represented by V. In that case also, such a connection is unnecessary, as only one voltage source serves the purpose.

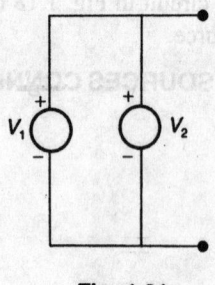

Fig. 1.21

1.33 PRACTICAL VOLTAGE SOURCES CONNECTED IN PARALLEL

The circuit in Fig. 1.22 (a) represents two practical voltage sources connected in parallel. The circuit in Fig. 1.22 (b) represents its equivalent circuit with voltage sources converted into current sources.

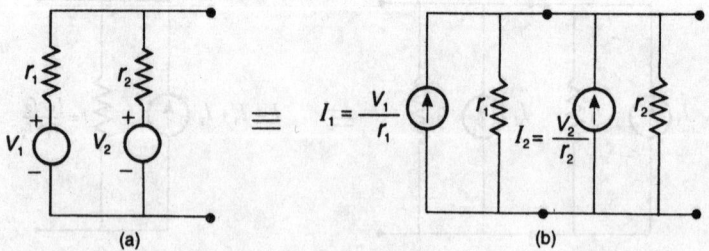

(a)   (b)

Fig. 1.22

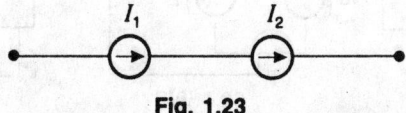

Fig. 1.22

The circuit in Fig. 1.22 (c) is the equivalent circuit of the circuit in Fig. 1.22 (b) and ultimately the circuit in Fig. 1.22 (d) represents the single equivalent voltage source.

1.34 TWO IDEAL CURRENTSOURCES CONNECTED IN SERIES

When two ideal current sources are connected in series as shown in the circuit of Fig. 1.23, what current flows through the line is ambiguous. Hence, such a connection is not permissible. However, if $I_1 = I_2 = I$, then the current in the line is I. But, such a connection is not necessary as only one current source serves the purpose.

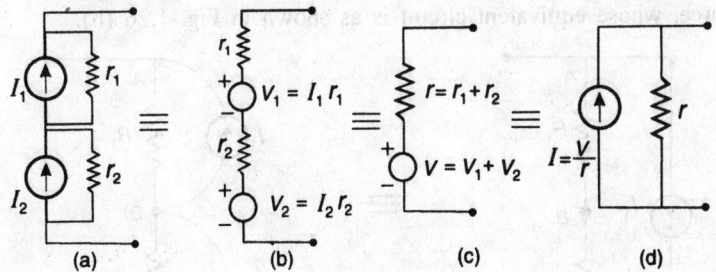

Fig. 1.23

1.35 TWO PRACTICAL CURRENT SOURCES CONNECTED IN SERIES

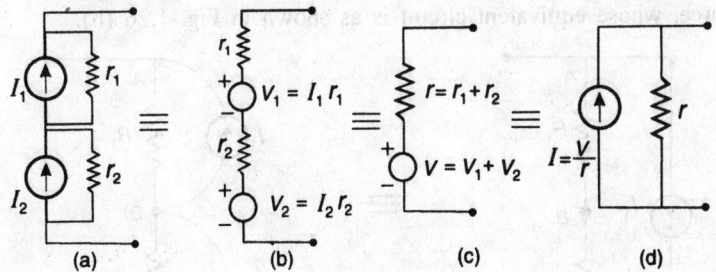

Fig. 1.24

Figures 1.24 (a), (b), (c) and (d) clearly explain the steps to be followed to convert two practical current sources connected in series into a single equivalent practical current source.

1.36 SOURCE SHIFTING

In the analysis of networks, we often encounter voltage sources without a series passive element or a current source without a parallel passive element. If it is desired to transform one kind of source to the other, it is first necessary to shift the source within the network.

Shifting of voltage source

Consider a network as shown in Fig. 1.25 (a). The single voltage source may be considered equivalent to two identical voltage sources in parallel as shown in Fig. 1.25 (b). The network in Fig. 1.25 (b) is equivalent to the network in Fig. 1.25 (c). It is observed that the source in Fig. 1.25 (a) is pushed through the node in obtaining an equivalent network as shown in Fig. 1.25 (c), in which the current through the various elements of the network remains unchanged after the transformation.

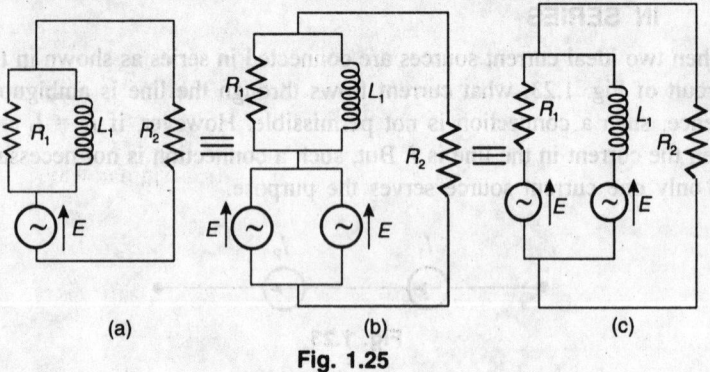

(a)   (b)   (c)

Fig. 1.25

Shifting of current source

Consider a network as shown in Fig. 1.26 containing a single current source, whose equivalent circuit is as shown in Fig. 1.26 (h).

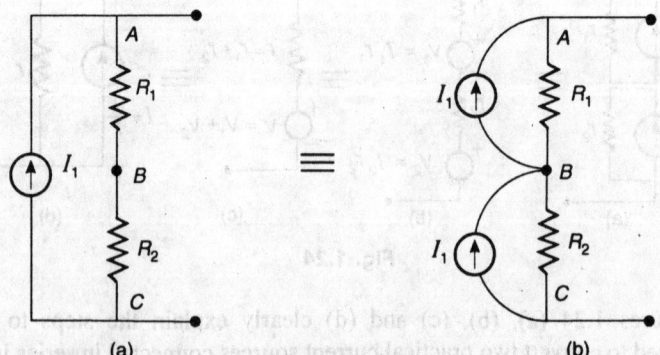

(a)   (b)

Fig. 1.26

In Fig. 1.26 (b), the current I_1 enters and leaves node B, while the currents at nodes A and C remain the same, as in Fig. 1.26 (a).

The general principle of current source shifting is to maintain the same currents at all nodes of the network after shifting.

One more example is given in Figs 1.27 (a) and (b), in which the current source is shifted from one part of the network to the other parts.

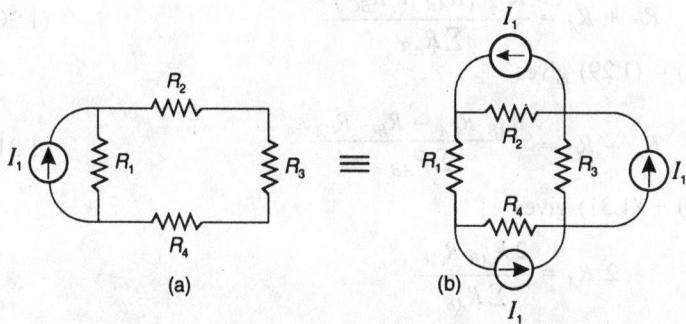

Fig. 1.27

1.37 STAR-DELTA TRANSFORMATION

When an electrical circuit consists of a large number of elements, the solution of such a network using Kirchhoff's laws is difficult, as a large number of simultaneous equations have to be solved. In such cases, the network has to be reduced to a simple network using star-delta transformation. Delta to star conversion is very frequently used.

1.38 DELTA TO STAR TRANSFORMATION (Δ TO Y)

Let R_{AB}, R_{BC} and R_{CA} be the three resistances connected in delta as shown in Fig. 1.28 (a). R_A, R_B and R_C are the three equivalent resistances connected in star as shown in Fig. 1.28 (b). The principle of conversion is that the equivalent resistance between the corresponding points in both the connections must be the same.

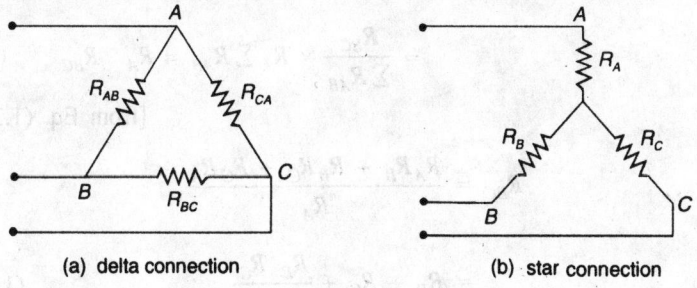

(a) delta connection   (b) star connection

Fig. 1.28

$$\therefore \qquad R_A + R_B = \frac{R_{AB}(R_{BC} + R_{CA})}{R_{AB} + R_{BC} + R_{CA}} = \frac{R_{AB}(R_{BC} + R_{CA})}{\sum R_{AB}} \qquad (1.28)$$

$$R_B + R_C = \frac{R_{BC}(R_{CA} + R_{AB})}{\sum R_{AB}} \qquad (1.29)$$

$$R_C + R_A = \frac{R_{CA}(R_{AB} + R_{BC})}{\sum R_{AB}} \qquad (1.30)$$

(1.28) – (1.29) gives

$$R_A - R_C = \frac{R_{AB} R_{CA} - R_{BC} R_{CA}}{\sum R_{AB}} \qquad (1.31)$$

(1.30) + (1.31) gives

$$2 R_A = \frac{2 R_{AB} R_{CA}}{\sum R_{AB}}$$

$$\therefore \qquad R_A = \frac{R_{AB} R_{CA}}{\sum R_{AB}} \qquad (1.32)$$

In a similar way

$$R_B = \frac{R_{BC} R_{AB}}{\sum R_{AB}} \qquad (1.33)$$

and

$$R_C = \frac{R_{BC} R_{CA}}{\sum R_{AB}} \qquad (1.34)$$

1.39 STAR TO DELTA TRANSFORMATION (Y TO Δ)

From Eqs (1.32), (1.33) and (1.34) we can write

$$R_A R_B + R_B R_C + R_C R_A = \frac{R_{AB} R_{BC} R_{CA}(R_{AB} + R_{BC} + R_{CA})}{(\sum R_{AB})^2}$$

$$= \frac{R_{BC}}{\sum R_{AB}} \times R_A \sum R_{AB} = R_A \cdot R_{BC}$$

[from Eq. (1.32)]

$$\therefore \qquad R_{BC} = \frac{R_A R_B + R_B R_C + R_C R_A}{R_A}$$

$$= R_B + R_C + \frac{R_B R_C}{R_A} \qquad (1.35)$$

In a similar way, we can write equations for R_{AB} and R_{CA}.

$$R_{AB} = R_A + R_B + \frac{R_A R_B}{R_C} \tag{1.36}$$

$$R_{CA} = R_C + R_A + \frac{R_A R_C}{R_B} \tag{1.37}$$

WORKED EXAMPLES

1.1 Two resistances 20 Ω and 40 Ω are connected in parallel. A resistance of 10 Ω is connected in series with the combination. A voltage of 200 V is applied across the circuit. Find the current in each resistance and the voltage across 10 Ω. Find also the power consumed in all the resistances.

Solution: The total resistance of the circuit is given by

$$R = 10 + \frac{20 \times 40}{20 + 40} = 23.33 \ \Omega$$

$$I = \frac{200}{23.33} = 8.57 \text{ A}$$

$$I_1 = 8.57 \times \frac{40}{40 + 20} = 5.71 \text{ A}$$

$$I_2 = 8.57 - 5.71 = 2.86 \text{ A}$$

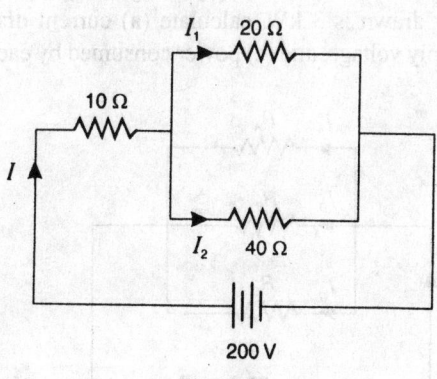

Fig. 1.29

$$V_{10 \ \Omega} = 8.57 \times 10 = 85.7 \text{ volts}$$

$$P_{10 \ \Omega} = 8.57^2 \times 10 = 734.45 \text{ watts}$$

$$P_{20\,\Omega} = 5.71^2 \times 20 = 652.08 \text{ watts}$$
$$P_{40\,\Omega} = 2.86^2 \times 40 = 327.18 \text{ watts}$$

1.2 A d.c. arc has a voltage current relation given by $V = 20 + 40/I$. It is connected in series with a resistor. The total voltage applied is 120 V. If the voltage across the arc is half the voltage across the resistor, find the value of the resistor.

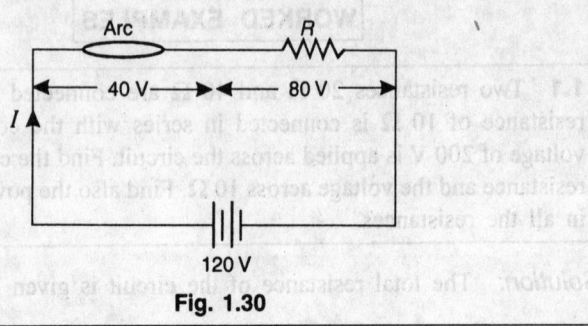

Fig. 1.30

Solution: For the arc

$$40 = 20 + \frac{40}{I}, \qquad \therefore \quad I = 2 \text{ A}$$

For the resistor

$$I R = 80, \quad \text{i.e.} \quad 2R = 80, \quad \text{i.e.} \quad \therefore R = 40\,\Omega$$

1.3 Three resistors A, B and C are connected in parallel taking a total current of 12 A from the supply. If $I_B = 2I_A$, $I_C = 3.5\,I_B$, and the total power drawn is 3 kW, calculate (a) current drawn by each resistor, (b) supply voltage, and (c) power consumed by each resistance.

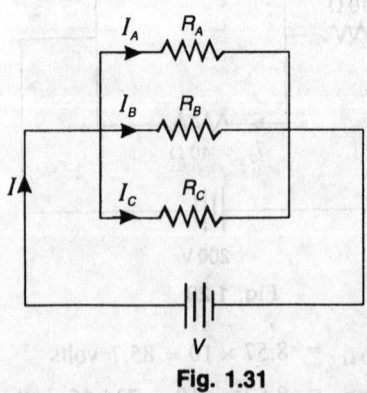

Fig. 1.31

Solution

(a)
$$I_A + I_B + I_C = I = 12 \text{ A}$$

$$I_A + 2I_A + 3.5I_B = 12$$

$$3I_A + 3.5\left(2I_A\right) = 12$$

$$\therefore I_A = 1.2 \text{ A}, \ I_B = 2.4 \text{ A} \text{ and } I_C = 8.4 \text{ A}$$

(b)
$$P = 3000 \text{ W} = V I$$

$$\therefore \quad V = \frac{3000}{12} = 250 \text{ V}$$

(c)
$$P_A = 250 \times 1.2 = 300 \text{ watts}$$

$$P_B = 250 \times 2.4 = 600 \text{ watts}$$

$$P_C = 250 \times 8.4 = 2100 \text{ watts}$$

1.4 In the series parallel circuit shown in Fig. 1.32, find (i) the voltage drop across 4 Ω resistor and (ii) the supply voltage.

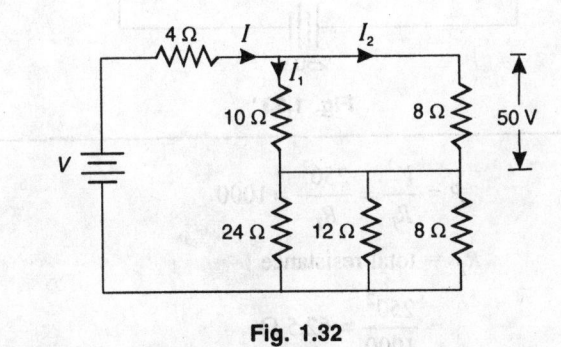

Fig. 1.32

Solution: Let the currents I, I_1, and I_2 flow in the branches as shown in Fig. 1.32.

$$I_1 = \frac{10}{50} = 5\text{A}, \qquad I_2 = \frac{50}{8} = 6.25 \text{A}$$

$$\therefore \quad I = I_1 + I_2 = 5 + 6.25 = 11.25 \text{ A}$$

(i) $V_{4\Omega} = 11.25 \times 4 = 45 \text{ V}$

(ii) $\qquad 24 \,\Omega \,\|\, 12 \,\Omega = \dfrac{24 \times 12}{24 + 12} = 8 \,\Omega$

$8\Omega \,\|\, 8 \,\Omega = \dfrac{8 \times 8}{8 + 8} = 4\Omega, \qquad 10\Omega \,\|\, 8\Omega = \dfrac{10 \times 8}{10 + 8} = 4.44\Omega$

Total resistance $= R = 4 + 4.44 + 4 = 12.44 \,\Omega$

Supply voltage $= V = 12.44 \times I = 12.44 \times 11.25 = 140$ volts

---

1.5 A resistance R is connected in series with a parallel circuit comprising 20 Ω and 48 Ω. The total power dissipated in the circuit is 1000 Ω and the applied voltage is 250 V. Calculate R.

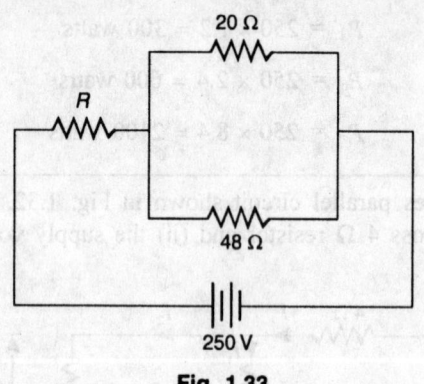

Fig. 1.33

Solution: $\qquad P = \dfrac{V^2}{R_T} = \dfrac{250^2}{R_T} = 1000$

$\therefore \qquad\qquad R_T = $ total resistance

$\qquad\qquad\qquad = \dfrac{250^2}{1000} = 62.5 \,\Omega$

$\qquad\qquad R_T = R + \dfrac{20 \times 48}{20 + 48}$

$\qquad\quad 62.5 = R + 14.12$

$\therefore \qquad\qquad R = 48.38 \,\Omega$

1.6 In the given circuit shown in Fig. 1.34, calculate (a) the total current, (b) current in 5 Ω and (c) the power dissipated in 6 Ω and 7 Ω.

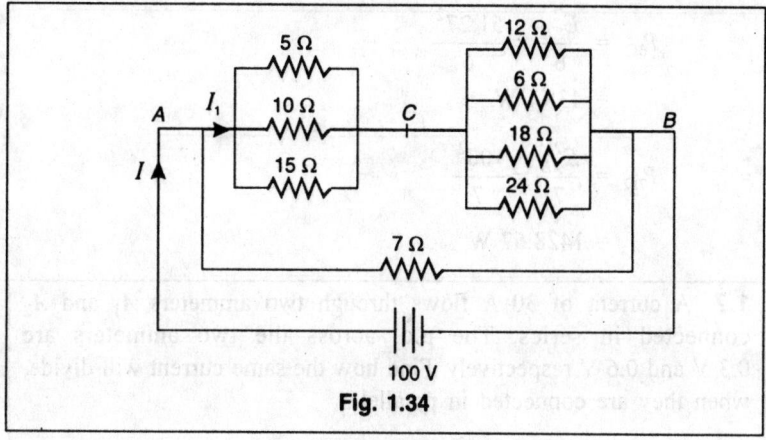

Fig. 1.34

Solution

(a) $\dfrac{1}{R_{AC}} = \dfrac{1}{5} + \dfrac{1}{10} + \dfrac{1}{15} = \dfrac{11}{30}\ \Omega, \quad \therefore R_{AC} = 2.73\ \Omega$

$\dfrac{1}{R_{CB}} = \dfrac{1}{12} + \dfrac{1}{6} + \dfrac{1}{18} + \dfrac{1}{24} = \dfrac{25}{72}\ \Omega, \quad \therefore R_{CB} = 2.88\ \Omega$

$R_{AB} = R_{AC} + R_{CB}$

$\qquad = 2.73 + 2.88 = 5.61\ \Omega$

The total resistance $= R_T = \dfrac{5.61 \times 7}{5.61 + 7} = 3.11\ \Omega$

$$I = \dfrac{100}{3.11} = 32.15\ \text{A}$$

(b) $\quad I_1 = 32.15 \times \dfrac{7}{7 + 5.61} = 17.85\ \text{A}$

$E_{AC} = 17.85 \times R_{AC} = 17.85 \times 2.73$

$\qquad = 48.73\ \text{V}$

$I_{5\Omega} \doteq \dfrac{48.73}{5} = 9.75\ \text{A}$

(c) $\quad E_{CB} = 100 - E_{AC} = 100 - 48.73$

$\qquad = 51.27\ \text{V}$

$$P_{6\Omega} = \frac{E_{CB}^2}{6} = \frac{51.27^2}{6}$$

$$= 438.1 \text{ W}$$

$$P_{7\Omega} = \frac{E_{AB}^2}{7} = \frac{100^2}{7}$$

$$= 1428.57 \text{ W}$$

1.7 A current of 30 A flows through two ammeters A_1 and A_2 connected in series. The p.d. across the two ammeters are 0.3 V and 0.6 V respectively. Find how the same current will divide, when they are connected in parallel.

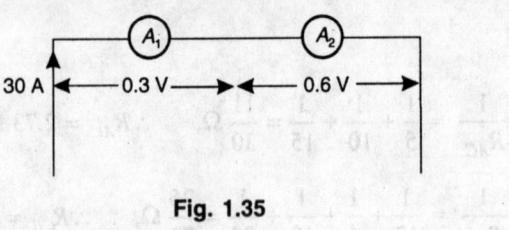

Fig. 1.35

Solution: Let R_1 and R_2 be the resistances of the ammeters A_1 and A_2 respectively.

$$30 R_1 = 0.3 \qquad \therefore \quad R_1 = 0.01 \,\Omega$$

$$30 R_2 = 0.6 \qquad \therefore \quad R_2 = 0.02 \,\Omega$$

The two ammeters are now connected in parallel as shown in Fig. 1.36.

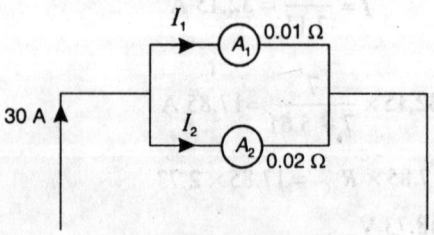

Fig. 1.36

$$I_1 = 30 \times \frac{0.02}{0.02 + 0.01} = 20 \text{ A}$$

$$I_2 = 30 - 20 = 10 \text{ A}$$

1.8 A voltage of 200 V is applied to a tapped resistor of $500\,\Omega$. Find the resistance between the tapping points connected to a load, needing 0.1 A at 25 V. Also calculate the total power consumed.

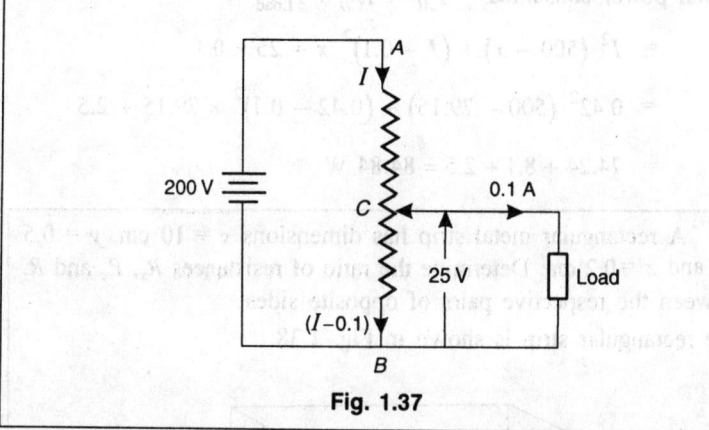

Fig. 1.37

Solution: Let the resistor be tapped between points B and C and x be the resistance between these points.

$$\therefore \quad R_{BC} = x\,\Omega$$

Then $(I - 0.1)\,x = 25$

i.e. $I\,x - 0.1\,x = 25$ (1.38)

$$R_{AC} = (500 - x)\,\Omega$$

Also $I\,(500 - x) = 200 - 25 = 175$

$\therefore \quad 500\,I - I\,x = 175$ (1.39)

Adding Eqs (1.38) and (1.39), we get

$$500\,I - 0.1\,x = 200$$

or $$I = \frac{200 + 0.1\,x}{500}$$ (1.40)

Substituting I given by Eq. (1.40) in Eq. (1.38), we get

$$\frac{200 + 0.1\,x}{500}\,x - 0.1\,x = 25$$

on simplification, we get $0.1x^2 + 150x - 12{,}500 = 0$

$$\therefore \quad x = \frac{-150 \pm \sqrt{150^2 + 4 \times 0.1 \times 12{,}500}}{2 \times 0.1}$$

$= 79.15\,\Omega$ (considering only the positive value)

Substituting this value in Eq. (1.38), we get

$$I = 0.42 \text{ A}$$

Total power consumed $= P_{AC} + P_{CB} + P_{\text{Load}}$

$$= I^2 (500 - x) + (I - 0.1)^2 \, x + 25 \times 0.1$$

$$= 0.42^2 (500 - 79.15) + (0.42 - 0.1)^2 \times 79.15 + 2.5$$

$$= 74.24 + 8.1 + 2.5 = 84.84 \text{ W}$$

1.9 A rectangular metal strip has dimensions $x = 10$ cm, $y = 0.5$ cm and $z = 0.2$ cm. Determine the ratio of resistances R_x, R_y and R_z between the respective pairs of opposite sides.

The rectangular strip is shown in Fig. 1.38.

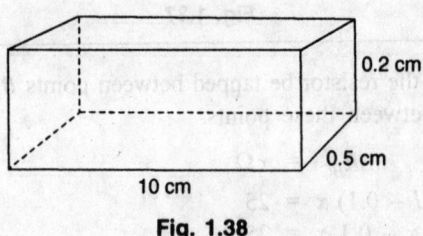

0.2 cm

0.5 cm

10 cm

Fig. 1.38

Solution:
$$R_x = \rho \, \frac{10 \times 10^{-2}}{0.5 \times 0.2 \times 10^{-4}} = 10,000 \, \rho$$

$$R_y = \rho \, \frac{0.5 \times 10^{-2}}{10 \times 0.2 \times 10^{-4}} = 25 \, \rho$$

$$R_z = \rho \, \frac{0.2 \times 10^{-2}}{10 \times 0.5 \times 10^{-4}} = 4 \, \rho$$

$$\therefore \quad R_x : R_y : R_z = 10,000 \, \rho : 25 \, \rho : 4 \, \rho$$

$$= 2500 : 6.25 : 1$$

1.10 A piece of silver wire has a resistance of 1Ω. What will be the resistance of a manganin wire half the length and half the diameter, if the specific resistance of manganin is 30 times that of silver.

Solution: Let R_1 be the resistance of silver wire.

$$R_1 = \rho_1 \frac{l_1}{a_1} = \rho_1 \frac{l_1}{\pi d_1^2 / 4} \tag{1.41}$$

Let R_2 be the resistance of manganin wire.

$$R_2 = \rho_2 \frac{l_2}{a_2} = \rho_2 \frac{l_2}{\pi d_2^2 / 4}. \tag{1.42}$$

From Eq. (1.41) and (1.42), we get

$$\frac{R_2}{R_1} = \frac{\rho_2}{\rho_1} \times \frac{l_2}{l_1} \left(\frac{d_1}{d_2}\right)^2$$

$$\frac{R_2}{1} = 30 \times \frac{1}{2} \times (2)^2 \qquad \because d_2 = \frac{1}{2} d_1$$

$$\therefore \qquad R_2 = 60 \ \Omega$$

1.11 When a certain battery is loaded by a 60 Ω resistor, its terminal voltage is 98.4 V. When it is loaded by a 90 Ω resistor, its terminal voltage is 98.9 V. What load resistance would give a terminal voltage of 90 volts?

Solution: Let r be the internal resistance of the battery.

When $\qquad R_L = 60 \ \Omega$ and $V = 98.4$ V

$$E_i = 98.4 + I_1 r \tag{1}$$

When $\qquad R_L = 90 \ \Omega$ and $V = 98.9$ V

$$E_i = 98.9 + I_2 r \tag{2}$$

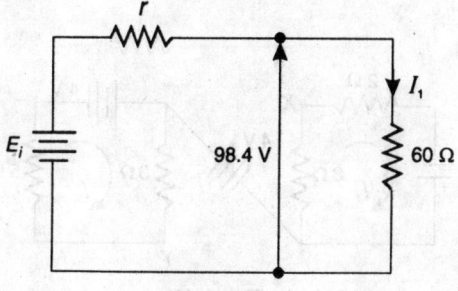

Fig. 1.39

I_1 and I_2 are given by

$$I_1 = \frac{98.4}{60} = 1.64 \text{ A}, \qquad I_2 = \frac{98.9}{90} = 1.099 \text{ A}$$

Substituting these values in (1) and (2), we get

$$E_i = 98.4 + 1.64\,r \tag{3}$$
$$E_i = 98.9 + 1.099\,r \tag{4}$$

From (3) and (4), we get

$$98.4 + 1.64\,r = 98.9 + 1.099\,r$$
$$\therefore \qquad r = 0.924 \ \Omega$$

Substituting this value in (3), we get $E_i = 99.92$ V

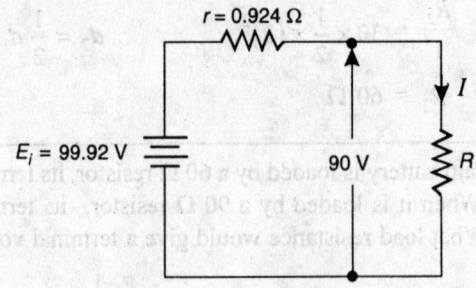

Fig. 1.40

$$I = \frac{E_i - 90}{0.924} = \frac{99.92 - 90}{0.924} = 10.736 \text{ A}$$

$$\therefore \qquad R = \frac{90}{10.736} = 8.383 \ \Omega$$

1.12 Find the p.d. between the points X and Y in the network shown in Fig. 1.41.

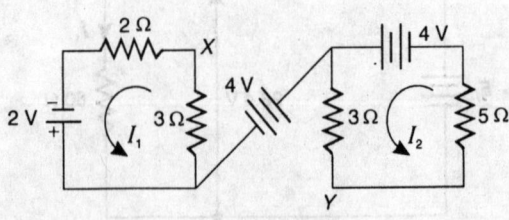

Fig. 1.41

Solution: Let I_1 and I_2 be the loop currents.

$$I_1 = \frac{2}{2+3} = 0.4 \text{ A}$$

and

$$I_2 = \frac{4}{3+5} = 0.5 \text{ A}$$

∴

$$V_{XY} = 3I_2 - 4 - 3I_1$$

$$= 3 \times 0.5 - 4 - 3 \times 0.4 = -3.7 \text{ V}.$$

Y is at a higher potential than X.

1.13 In the network shown in Fig. 1.42, determine the direction and magnitude of current flow in the milli-ammeter A, having a resistance of 10 Ω.

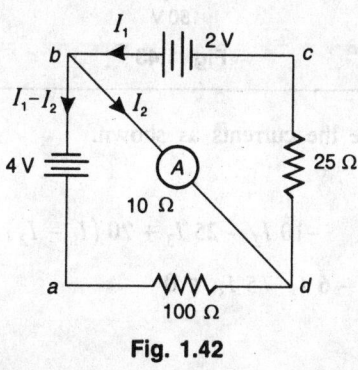

Fig. 1.42

Solution: Let the current distribution be as shown in Fig. 1.42

For *abda*

$$-4 - 10 I_2 + 100 (I_1 - I_2) = 0, \quad \text{i.e.} \quad 100I_1 - 110I_2 = 4$$

For *bcdb*

$$-2 + 25 I_1 + 10 I_2 = 0, \quad \text{i.e.} \quad 25 I_1 + 10 I_2 = 2$$

∴

$$I_2 = \frac{\begin{vmatrix} 100 & 4 \\ 25 & 2 \end{vmatrix}}{\begin{vmatrix} 100 & -110 \\ 25 & 10 \end{vmatrix}} = 0.0267 \text{A} = 26.7 \text{ mA}$$

1.14 In the network shown in Fig. 1.43, find the currents through 25 Ω in each branch using Kirchhoff's Laws.

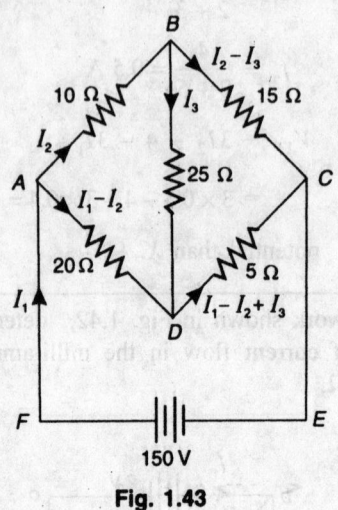

Fig. 1.43

Solution: Assume the currents as shown.

For *ABDA*

$$-10\,I_2 - 25\,I_3 + 20\,(I_1 - I_2) = 0$$

i.e.

$$4\,I_1 - 6\,I_2 - 5\,I_3 = 0 \qquad (1)$$

For *BCDB*

$$-15\,(I_2 - I_3) + 5\,(I_1 - I_2 + I_3) + 25\,I_3 = 0$$

i.e.

$$I_1 - 4\,I_2 + 9\,I_3 = 0 \qquad (2)$$

For *ADCEFA*

$$-20\,(I_1 - I_2) - 5\,(I_1 - I_2 + I_3) + 150 = 0$$

i.e.

$$5\,I_1 - 5\,I_2 + I_3 = 30 \qquad (3)$$

Solving Eqs (1), (2) and (3) using Cramer's rule, we get $I_3 = 1.72$ A.

1.15 In the circuit shown in Fig. 1.44, find the voltage across *cd* if (i) switch *S* is open, and (ii) *S* is closed.

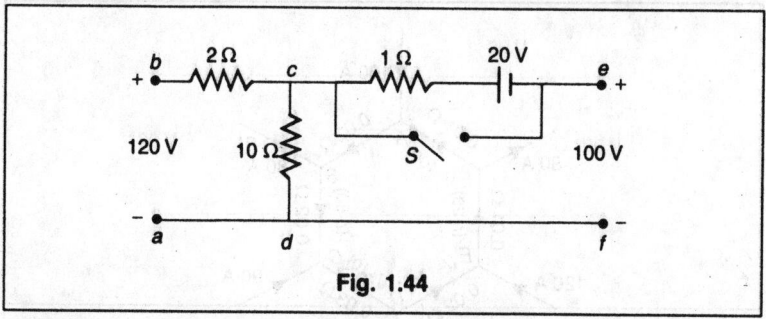

Fig. 1.44

Solution: (i) When S is open, the circuit is rewritten as below.

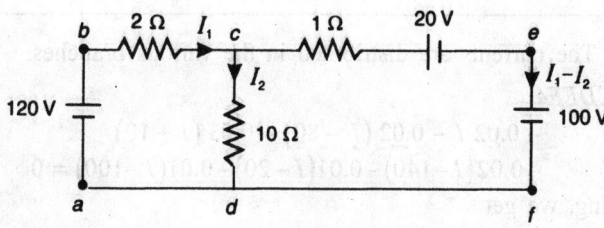

Fig. 1.45

The current distribution is as shown in Fig. 1.45
For *abcda*

$$120 - 2 I_1 - 10 I_2 = 0, \quad \text{i.e.} \quad I_1 + 5 I_2 = 60 \tag{1}$$

For *dcefd* $\qquad 10 I_2 - 1(I_1 - I_2) - 20 - 100 = 0$

i.e. $\qquad\qquad -I_1 + 11 I_2 = 120 \tag{2}$

Solving Eqs (1) and (2), we get

$$I_1 = 3.75 \text{ A} \quad \text{and} \quad I_2 = 11.25 \text{ A}$$

Voltage across $\qquad cd = V_{cd} = 10 \times I_2 = 10 \times 11.25 = 112.5$ V

(ii) When the switch S is closed, no current flows through
1 Ω and 20 V.

∴ $\qquad\qquad\qquad V_{cd} = 100$ V

1.16 Find the currents in the various branches of the given network
shown in Fig. 1.46

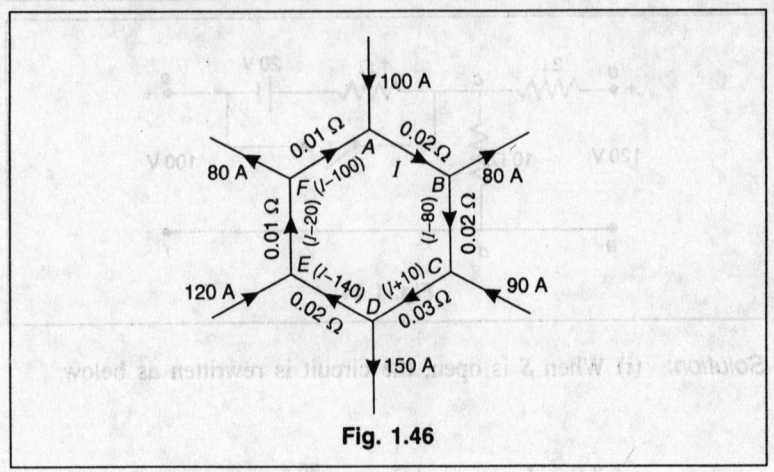

Fig. 1.46

Solution: The currents are distributed in the various branches.

For *ABCDEFA*

$$-0.02\,I - 0.02\,(I - 80) - 0.03\,(I + 10)$$
$$-0.02(I - 140) - 0.01(I - 20) - 0.01(I - 100) = 0$$

On solving, we get

$$I = 57.27\ \text{A}$$

$$\therefore \qquad I_{AB} = I = 57.27\ \text{A}$$

$$I_{BC} = (I - 80) = -22.73\ \text{A or } I_{CB} = 22.73\text{A}$$

$$I_{CD} = (I + 10) = 67.27\ \text{A}$$

$$I_{DE} = (I - 140) = -82.73\ \text{A or } I_{ED} = 82.73\ \text{A}$$

$$I_{EF} = (I - 20) = 37.27\text{A}$$

$$I_{FA} = (I - 100) = -42.73\ \text{A or } I_{AF} = 42.73\ \text{A}$$

1.17 Find the currents I_1, I_2 and I_3 and the voltages V_a, V_b in the network shown in Fig. 1.47.

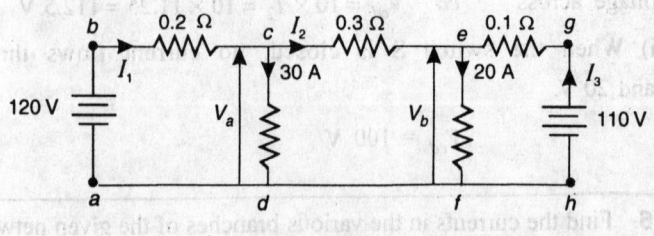

Fig. 1.47

Solution: From the circuit, we know

$$I_2 = I_1 - 30, \qquad I_3 + I_1 - 30 = 20$$

$$\text{i.e.} \quad I_3 = 50 - I_1$$

For *abceghfda*

$$120 - 0.2 I_1 - 0.3 I_2 + 0.1 I_3 - 110 = 0$$

Substituting for I_2 and I_3, we get

$$-0.6 I_1 = -24 \text{ or } I_1 = 40 \text{ A}$$

$$\therefore \qquad I_2 = 40 - 30 = 10 \text{ A},$$

$$I_3 = 50 - 40 = 10 \text{ A},$$

$$V_a = 120 - 0.20 \times 40 = 120 \text{ V}$$

and $$V_b = 110 - 0.1 \times 10 = 109 \text{ V}$$

1.18 Find the current in the 10 Ω resistor in the given network shown in Fig. 1.48, using star-delta transformation.

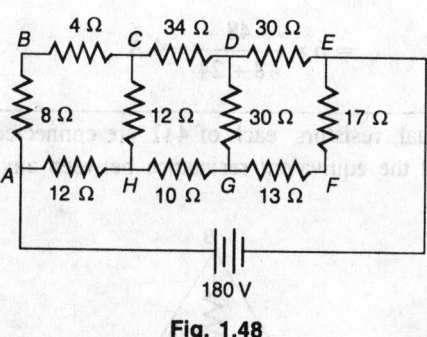

Fig. 1.48

Solution: 8 Ω and 4 Ω are in series whose total resistance is 12 Ω. Now, three 12 Ω resistances are connected in delta between *A*, *C* and *H*. They can be converted into star. Each star resistance is 4 Ω.

Similarly, 13 Ω and 17 Ω are in series, whose total resistance is 30 Ω. Three 30 Ω resistances are connected in delta between *G*, *D* and *E*. They can be converted into star. Each star resistance is 10 Ω.

After these two conversions, the circuit in Fig. 1.48 can be written as shown in Fig. 1.49.

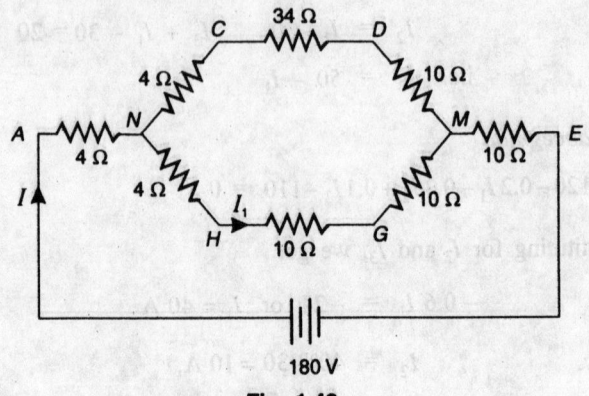

Fig. 1.49

The total resistance of the circuit is given by

$$R = 4 + \frac{48 \times 24}{48 + 24} + 10 = 30 \ \Omega$$

$$\therefore \quad I = \frac{180}{30} = 6 \text{ A}$$

$$I_1 = \text{Current through } 10 \ \Omega$$

$$= 6 \times \frac{48}{48 + 24} = 4 \text{ A}$$

1.19 Six equal resistors each of $4 \ \Omega$ are connected as shown in Fig. 1.50. Find the equivalent resistance between any two corners.

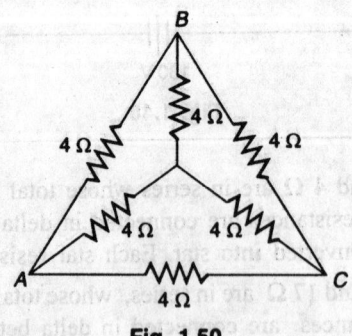

Fig. 1.50

Solution: The star network is converted into delta network. Each delta resistance is $12 \ \Omega$.

After conversion, the network in Fig. 1.50, may be written as shown in Fig. 1.51.

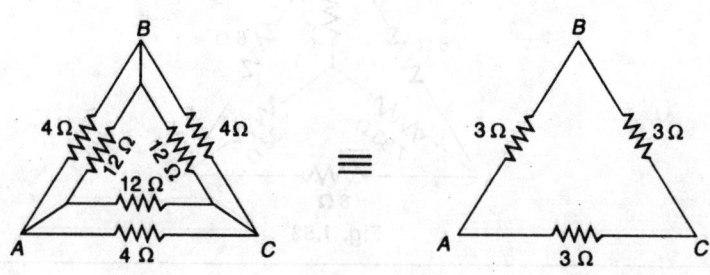

Fig. 1.51

The resistance between any two corners $= \dfrac{3 \times 6}{3 + 6} = 2\,\Omega$.

1.20 In the network shown in Fig. 1.52, the numbers represent the value of resistances in Ω. Find the resistance between the points A and B.

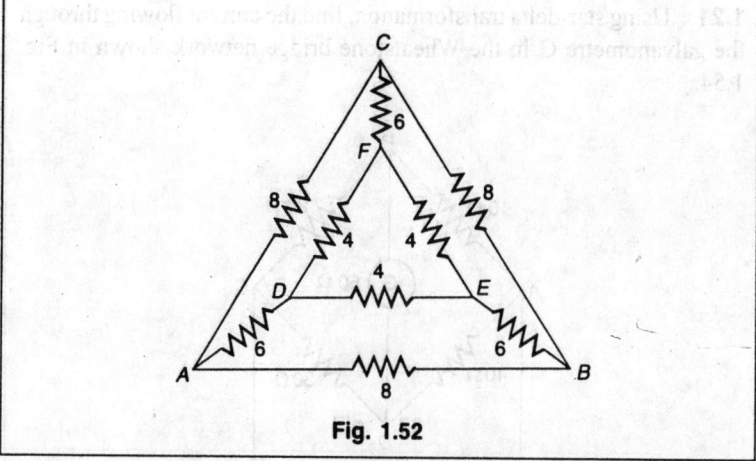

Fig. 1.52

Solution: The three, $4\,\Omega$ resistances connected in delta between D, E and F are converted into star. Each star resistance is $1.33\,\Omega$.

After conversion, $1.33\,\Omega$ is in series with $6\,\Omega$ in the resulting star network. Hence the network of Fig. 1.52, may be written by an equivalent network as shown in Fig. 1.53.

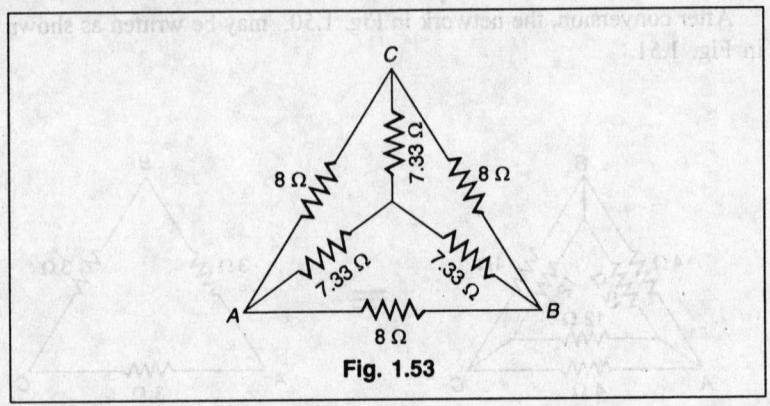

Fig. 1.53

The star network inside may be converted into delta and each delta resistance is 21.99 Ω.

Now across each branch, 8 Ω is in parallel with 21.99 Ω. Now resistance of each branch is 5.87 Ω.

Hence, the equivalent resistance between any to points is given by

$$R_{AB} = \frac{(5.87 + 5.87)\,5.87}{5.87 + 5.87 + 5.87} = 3.913 \ \Omega$$

1.21 Using star-delta transformation, find the current flowing through the galvanometre G in the Wheatstone bridge network shown in Fig. 1.54.

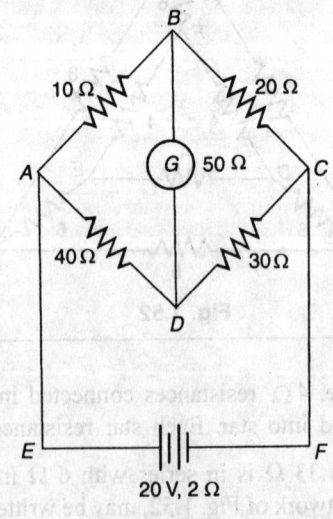

Fig. 1.54

Solution: The delta resistances connected to A, B and D are converted into star resistances.

After this conversion, the network in Fig. 1.54 can be written as shown in Fig. 1.55.

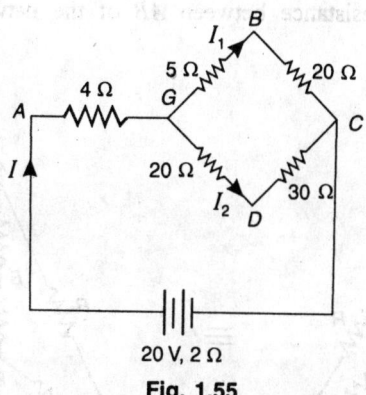

Fig. 1.55

The total resistance is given by

$$R = 2 + 4 + \frac{25 \times 50}{25 + 50} = 22.67 \; \Omega$$

The total current is given by

$$I = \frac{20}{22.67} = 0.882 \; A$$

$$I_1 = 0.882 \times \frac{50}{50 + 25} = 0.588 \; A$$

$$I_2 = 0.882 - 0.588 = 0.294 \; A$$

In order to find the current flowing through G, the p.d. between the points B and D has to be found.

Applying Kirchhoff's voltage law to the loop $GBDG$ in Fig. 1.55, we get

$$- 5 \, I_1 + V_{BD} + 20 \, I_2 = 0$$

$$V_{BD} = 5 I_1 - 20 \, I_2 = 5 \times 0.588 - 20 \times 0.2940$$

$$= - 2.94 \; volts$$

i.e. D is at a higher potential then B.

$$I_G = \frac{-2.94}{50} = -0.0588 \text{ A}$$

∴ The current flows from D to B.

1.22 Find the resistance between AB of the network shown in Fig. 1.56 (a).

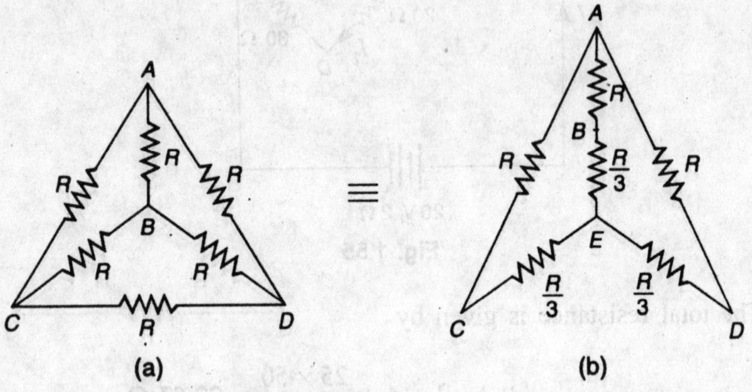

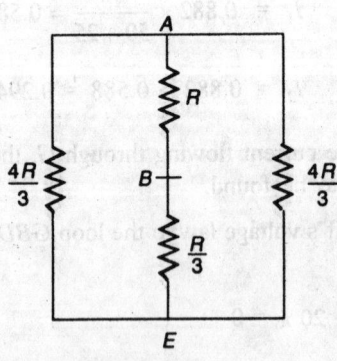

(a)   (b)

Fig. 1.56

Solution: Converting the delta resistance connected between CBD, the circuit in Fig. 1.56 (a) can be written as shown in Fig. 1.56 (b).

The circuit in Fig. 1.56 (b) may be further simplified as in Fig. 1.57.

Fig. 1.57

The circuit shown in the Fig. 1.57 can further be simplified as shown in Fig. 1.58 (a), (b) and (c).

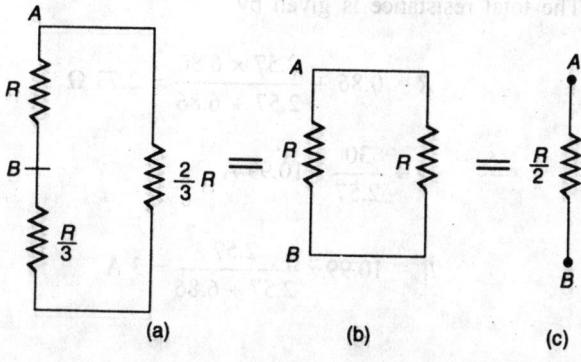

Fig. 1.58

$$\therefore \qquad R_{AB} = \frac{R}{2}\ \Omega$$

1.23 In the network shown in Fig. 1.59, find I and I_1 using Y–Δ transformation.

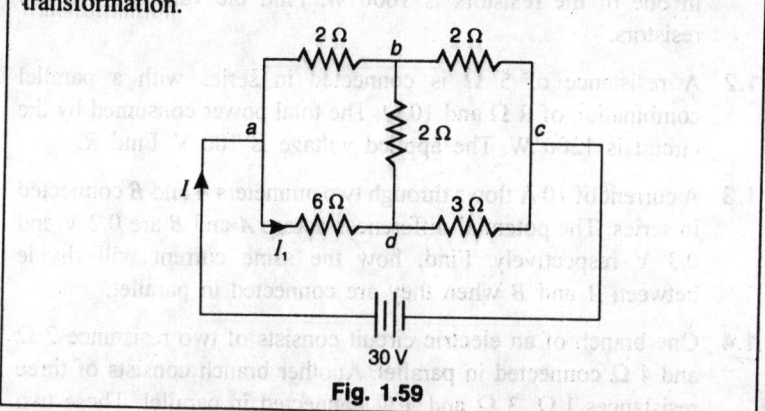

Fig. 1.59

Solution: Converting the delta resistances connected between b, c and d into star, the circuit in Fig. 1.59 may be written as shown in Fig. 1.60.

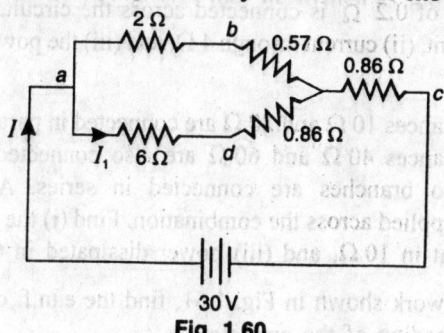

Fig. 1.60

The total resistance is given by

$$R = 0.86 + \frac{2.57 \times 6.86}{2.57 + 6.86} = 2.73 \; \Omega$$

$$I = \frac{30}{2.57} = 10.99 \; \text{A}$$

$$I_1 = 10.99 \times \frac{2.57}{2.57 + 6.86} = 3 \; \text{A}$$

NUMERICAL PROBLEMS

1.1 Two resistors are connected in parallel. The voltage applied is 220 V. The current taken is 30 A. The power dissipated in one of the resistors is 1600 W. Find the value of the two resistors.

1.2 A resistance of 5 Ω is connected in series with a parallel combination of R Ω and 10 Ω. The total power consumed by the circuit is 1200 W. The applied voltage is 100 V. Find R.

1.3 A current of 10 A flows through two ammeters A and B connected in series. The potential differences across A and B are 0.2 V and 0.3 V respectively. Find, how the same current will divide between A and B when they are connected in parallel.

1.4 One branch of an electric circuit consists of two resistance 2 Ω and 4 Ω connected in parallel. Another branch consists of three resistances 1 Ω, 3 Ω and 5 Ω connected in parallel. These two branches are connected in series. A resistance of 7 Ω is connected in parallel with this combination. A 10 V battery with an internal resistance of 0.2 Ω is connected across the circuit. Find (i) the total current, (ii) current through 4 Ω, and (iii) the power consumed in 5 Ω.

1.5 Two resistances 10 Ω and 20 Ω are connected in parallel. Another two resistances 40 Ω and 60 Ω are also connected in parallel. These two branches are connected in series. A voltage of 200 V is applied across the combination. Find (i) the total current, (ii) current in 10 Ω, and (iii) power dissipated in 60 Ω.

1.6 In the network shown in Fig. 1.61, find the e.m.f. of the battery and the reading of the ammeter.

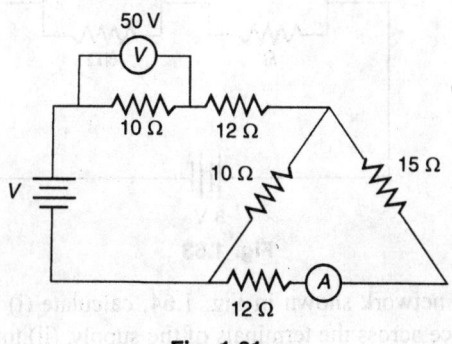

Fig. 1.61

Note: Assume that the current flowing through the voltmeter is Zero and the resistance of the ammeter is so small that it can be neglected.

1.7 Two voltmeters of resistances 15 kΩ and 20 kΩ are connected in series across 200 V supply. Both metre are 0–300 V range. Find the readings of the voltmenters.

1.8 Find the resistance across *RQ* of the network shown in Fig. 1.62.

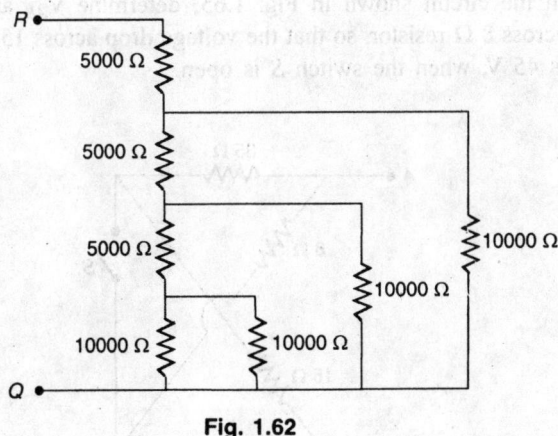

Fig. 1.62

1.9 The total power consumed by the network shown in Fig. 1.63 is 16 W. Find the value of *R* and the total current.

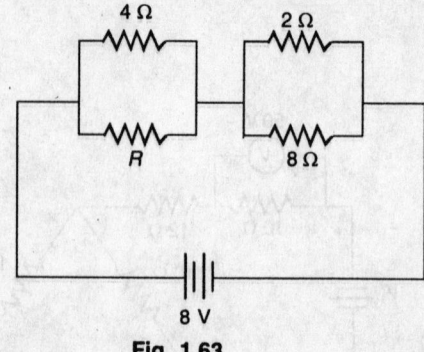

Fig. 1.63

1.10 For the network shown in Fig. 1.64, calculate (i) the equivalent resistance across the terminals of the supply, (ii) total current and (iii) power delivered to 16 Ω resistor.

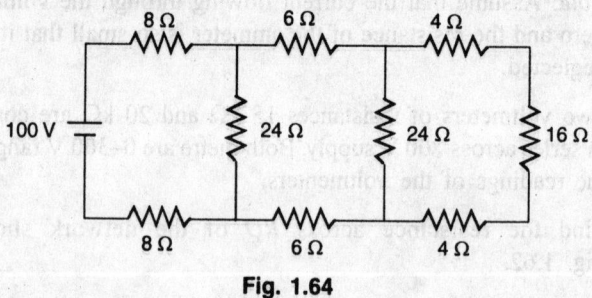

Fig. 1.64

1.11 In the circuit shown in Fig. 1.65, determine V_{AB} and voltage across 8 Ω resistor, so that the voltage drop across 15 Ω resistor is 45 V, when the switch S is open.

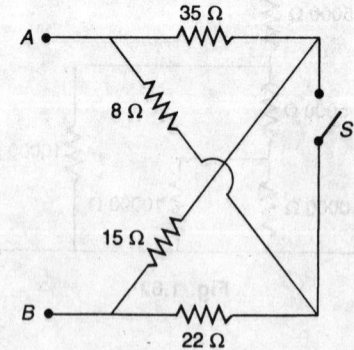

Fig. 1.65

1.12 In the circuit shown in Fig. 1.66, determine (i) the current supplied by the source and (ii) voltage across 6 Ω resistor.

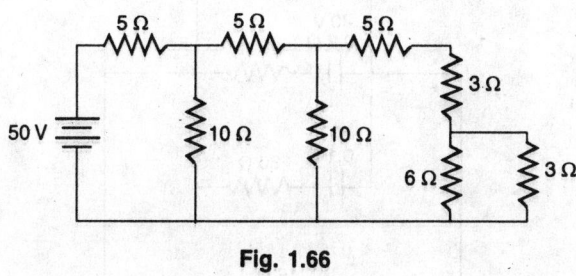

Fig. 1.66

1.13 Find the p.d. between the points A and B in the network shown in Fig. 1.67.

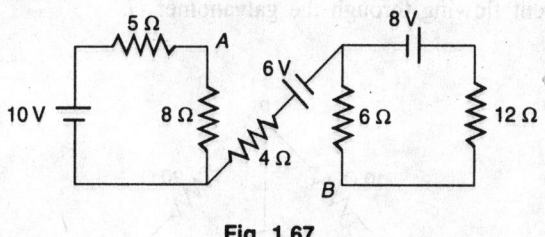

Fig. 1.67

1.14 In the given circuit shown in Fig. 1.68, find the currents flowing in the various branches of the circuit.

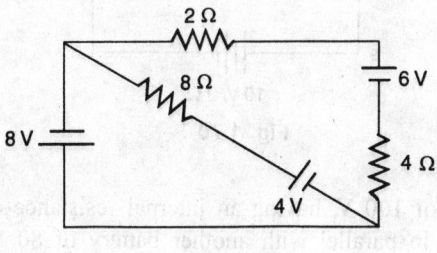

Fig. 1.68

1.15 In the network shown in Fig. 1.69, determine the current in each battery and in the 16 Ω resistor.

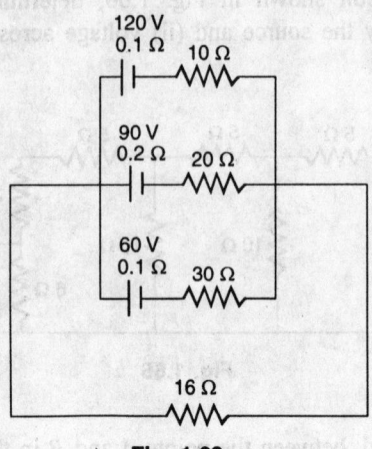

Fig. 1.69

1.16 In the Wheatstone bridge circuit shown in Fig. 1.70, find the current flowing through the galvanomer *G*.

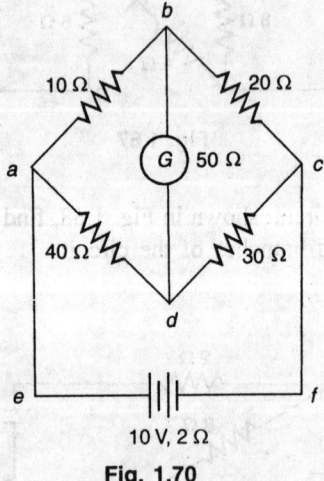

Fig. 1.70

1.17 A battery of 100 V, having an internal resistance of 0.25 Ω is connected in parallel with another battery of 80 V, having an internal resistance of 0.2 Ω. These two batteries are connected across a 220 V d.c. supply, with a regulator resistance of 10 Ω connected in series with it. Calculate the currents flowing in each battery and the total current taken from the supply.

1.18 In the given circuit shown in Fig. 1.71, find the current flowing in 3 Ω and 5 Ω.

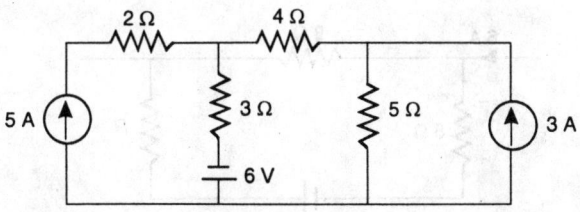

Fig. 1.71

1.19 In the given circuit shown in Fig. 1.72, find the current flowing through the galvanometer.

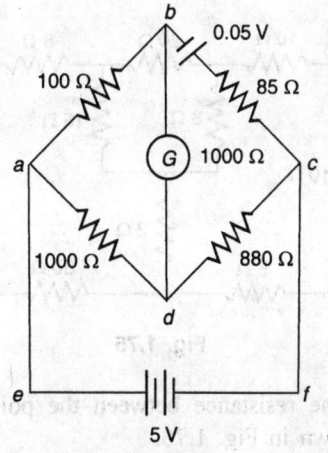

Fig. 1.72

1.20 Determine the currents I_1, I_2, and I_3 in the circuit shown in Fig. 1.73.

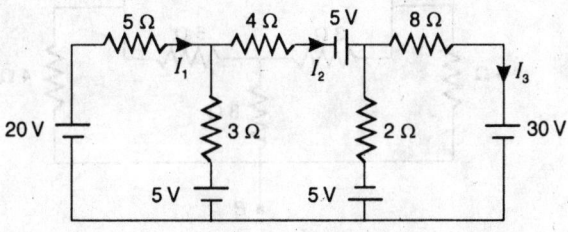

Fig. 1.73

1.21 In the circuit shown in Fig. 1.74, the voltage drop across 8 Ω resistor is 12 V having polarities as shown. Find the value of *R*.

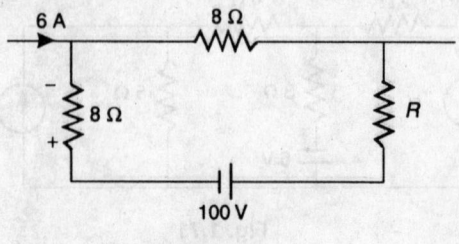

Fig. 1.74

1.22 Determine the resistance between the terminals *M* and *N* of the network shown in Fig. 1.75.

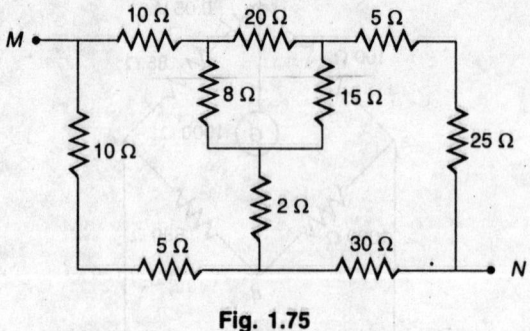

Fig. 1.75

1.23 Determine the resistance between the points *A* and *B* in the network shown in Fig. 1.76.

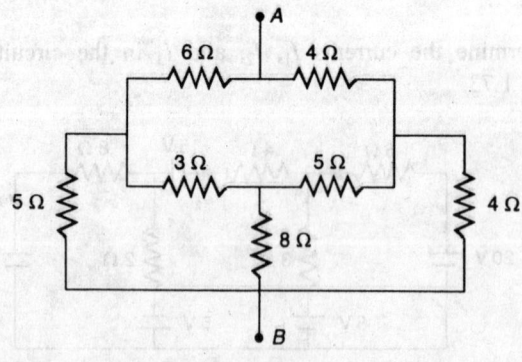

Fig. 1.76

1.24 Calculate the current in the 40 Ω resistance of the network shown in Fig. 1.77, using $Y - \Delta$ transformation.

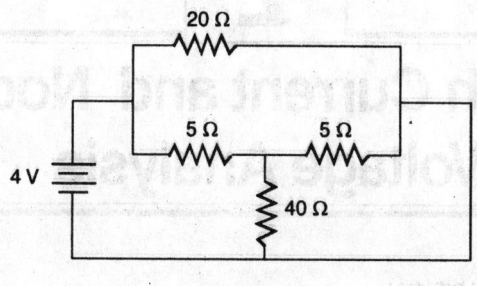

Fig. 1.77

1.25 Find the current supplied by the source of the network shown in Fig. 1.78, using $Y - \Delta$ transformation.

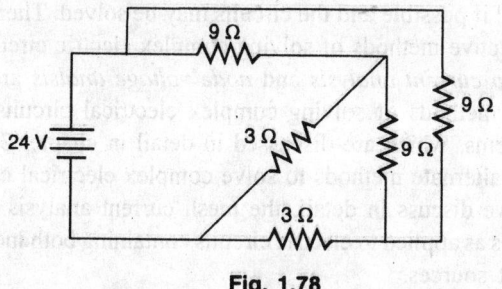

Fig. 1.78

1.26 In the network shown in Fig. 1.79, calculate the equivalent resistance between (a) A and B, (b) A and N.

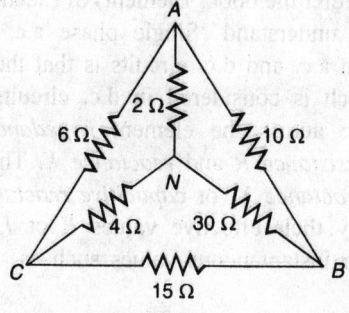

Fig. 1.79

2

Mesh Current and Node Voltage Analysis

2.1 INTRODUCTION

In the previous chapter, we learnt the analysis of electrical circuits using Ohm's law and Kirchhoff's laws, as applied to simple series and parallel circuits. If the circuits are complex, containing several sources and a large number of elements, they may be simplified using star-delta transformation, if possible and the circuits may be solved. There are also other very effective methods of solving complex electric circuits. *Mesh current* or *loop current analysis* and *node voltage analsis* are the two very effective methods of solving complex electrical circuits. Various network theorems, which are discussed in detail in chapter 3, are also very effective alternate methods to solve complex electrical circuits. In this chapter, we discuss in detail, the mesh current analysis and node voltage analysis as applied to electric circuits containing both independent and dependent sources.

2.2 A.C. FUNDAMENTALS

At this stage, it is assumed that the students have sufficient knowledge about the a.c. fundamentals. Any preliminary book on electrical science, gives a comprehensive discussion on single phase a.c. circuits. The students may also refer the book 'Elements of Electrical Science' written by the author to understand 'Single phase a.c. circuits'. The only difference between a.c. and d.c. circuits is that the resistance R is the only element which is considered in d.c. circuits, where as, in a.c. circuits, we come across the element *impedance Z*, which is the combination of *resistance R* and *reactance X*. The reactance may be either *inductive reactance X_L* or *capacitive reactance X_C*. The sources are represented by their effective values E or I, or in the form of equations for their instantaneous values such as

$$e = E_m \sin(\omega t \pm \phi) \quad \text{and} \quad i = I_m \sin(\omega t \pm \alpha)$$

Where, ϕ and α are the phase angles of voltage and current respectively.

Some problems on star-delta transformation containing impedances are also worked out in this chapter. The equations used to convert a delta network into star network are:

$$Z_A = \frac{Z_{AB}\ Z_{CA}}{\sum Z_{AB}}, \qquad Z_B = \frac{Z_{BC}\ Z_{AB}}{\sum Z_{AB}}$$

and
$$Z_C = \frac{Z_{CA}\ Z_{BC}}{\sum Z_{AB}}$$

Equations used to convert a star network into a delta network are:

$$Z_{AB} = Z_A + Z_B + \frac{Z_A\ Z_B}{Z_C}$$

$$Z_{BC} = Z_B + Z_C + \frac{Z_B\ Z_C}{Z_A}$$

$$Z_{CA} = Z_C + Z_A + \frac{Z_C\ Z_A}{Z_B}$$

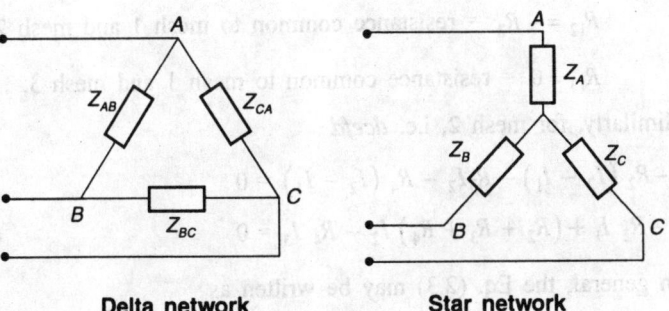

Delta network   **Star network**

2.3 MESH CURRENT ANALYSIS

A *loop* in an electric circuit is any closed path in it. A *mesh* is a loop in an electric circuit which does not contain any other loops within it. therefore, all meshes are loops but all loops are not meshes.

Consider a general electrical circuit as shown in Fig. 2.1.

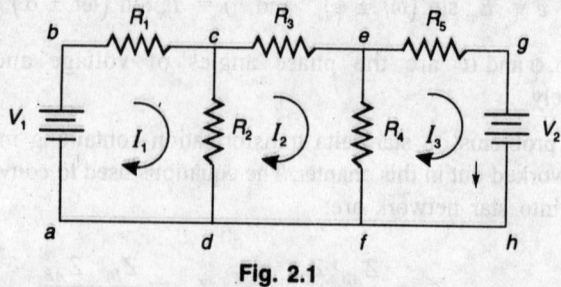

Fig. 2.1

V_1 and V_2 are the e.m.f.s of the two sources. Assume loop currents or mesh currents I_1, I_2, and I_3 as shown in Fig. 2.1, either in the clockwise direction or anti-clockwise direction. Usually, the mesh currents are assumed in clockwise direction.

For mesh 1, i.e. *abcda*

$$V_1 - I_1 R_1 - (I_1 - I_2) R_2 = 0$$

or $$(R_1 + R_2) I_1 - R_2 I_2 = V_1 \qquad (2.1)$$

In general, the Eq. (2.1) may be written as

$$R_{11} I_1 + R_{12} I_2 + R_{13} I_3 = E_1 \qquad (2.2)$$

Where, $R_{11} = (R_1 + R_2)$ = sum of all the resistances in mesh 1 = self resistance of mesh 1.

$R_{12} = -R_2$ = resistance common to mesh 1 and mesh 2.

$R_{13} = 0$ = resistance common to mesh 1 and mesh 3.

Similarly, for mesh 2, i.e. *dcefd*

$$-R_2 (I_2 - I_1) - R_3 I_2 - R_4 (I_2 - I_3) = 0$$

i.e. $$-R_2 I_1 + (R_2 + R_3 + R_4) I_2 - R_4 I_3 = 0 \qquad (2.3)$$

In general, the Eq. (2.3) may be written as

$$R_{21} I_1 + R_{22} I_2 + R_{23} I_3 = E_2 \qquad (2.4)$$

Where, $R_{21} = -R_2$ = resistance common to mesh 1 and mesh 2

$= R_{12}$, as the network is bilateral.

$R_{22} = (R_3 + R_4 + R_5)$ = self resistance of mesh 2 = sum of all resistances in mesh 2.

$R_{23} = -R_4$ = resistance common to mesh 2 and mesh 3.

For mesh 3, i.e. *feghf*

$$-R_4 \left(I_3 - I_2\right) - I_3 R_5 - V_2 = 0$$

i.e. $\quad 0\, I_1 - R_4\, I_2 + \left(R_4 + R_5\right) I_3 = -V_2 \quad\quad\quad (2.5)$

Equation (2.5) may be written as

$$R_{31}\, I_1 + R_{32}\, I_2 + R_{33}\, I_3 = E_3 \quad\quad\quad (2.6)$$

Where, $R_{31} = 0$ = resistance common to mesh 3 and mesh 1.

$R_{32} = -R_4$ = resistance common to mesh 3 and mesh 2.

$R_{33} = \left(R_4 + R_5\right)$ = self resistance of mesh 3 = sum of all resistances in mesh 3.

The self resistances R_{11}, R_{22} and R_{33} are always +ve. The mutual resistances R_{12}, R_{23}, R_{31}, etc. are +ve, if the two mesh currents through them are flowing in the same direction, otherwise, they are –ve.

$E_1, E_2,$ and E_3 are the sum of all the e.m.f.s in mesh 1, mesh 2, and mesh 3 respecitvely. They are +ve, if along the direction of their mesh currents, they are voltage rises, otherwise, they are –ve.

Equations (2.2), (2.4), and (2.6) are general mesh current equations from, which the voltage equations for the meshes considered may be written directly. In general, if there are *n* meshes, the mesh equations are:

For d.c. circuits:

$$R_{11}\, I_1 + R_{12}\, I_2 + \ldots\ldots\ldots + R_{1n}\, I_n = E_1$$

$$R_{21}\, I_1 + R_{22}\, I_2 + \ldots\ldots\ldots + R_{2n}\, I_n = E_2$$

$$\vdots$$

$$R_{n1}\, I_1 + R_{n2}\, I_2 + \ldots\ldots\ldots + R_{nn}\, I_n = E_n$$

For a.c. circuits:

$$Z_{11}\, I_1 + Z_{12}\, I_2 + \ldots\ldots\ldots + Z_{1n}\, I_n = E_1$$

$$Z_{21}\, I_1 + Z_{22}\, I_2 + \ldots\ldots\ldots + Z_{2n}\, I_n = E_2$$

$$\vdots$$

$$Z_{n1}\, I_1 + Z_{n2}\, I_2 + \ldots\ldots\ldots + Z_{nn}\, I_n = E_n \quad\quad\quad (2.7)$$

2.4 PROCEDURE FOR SOLVING AN ELECTRICAL CIRCUIT USING MESH CURRENT ANALYSIS

1. Make sure that the network given is a planar network, because, mesh current analysis cannot be used to solve a non-planar network.

2. All the nodes of the electrical circuit are named by letters either capital or small case to identify the meshes in the circuit.

3. If there are any practical current sources in the circuit, they are converted into practical voltage sources. The source transformation methods are given in sections 1.27 to 1.35 of the 1st chapter.

4. The voltage sources and current sources which are connected in series or parallel are represented by their equivalent voltage sources or current sources.

5. Any resistance or impedance in parallel with an ideal voltage source or in series with an ideal current source can be neglected.

6. The mesh currents are assumed arbitrarily in clockwise or anti-clockwise direction. Usually, they are assumed in clockwise direction.

7. The equations for the meshes are written using general mesh analysis Eq. as in (2.7).

8. The mesh Eqs are solved using Cramer's rule or any other method.

9. Once the mesh currents are calculated, the branch currents can also be calculated.

2.5 SUPER MESH ANALYSIS

If any branch of a given circuit contains an ideal current source, such as in branch de, as shown in Fig. 2.2, ordinary mesh analysis can not be applied. There are two methods of solving the circuit. One method is to assign an unknown voltage across the ideal current source, to apply Kirchhoff's voltage law around each mesh and then, relating the source current to the mesh currents and to solve the equations obtained. But, this is a difficult and lengthy approach.

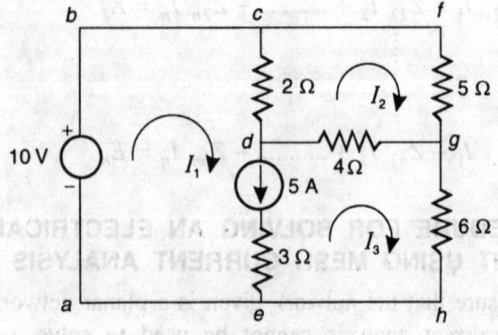

Fig. 2.2

A better technique is to create a *super mesh* from the two meshes that have the ideal current source as a common element. Thus, in Fig. 2.2, *abcdghea* forms the super mesh. Thus, the total number of meshes are reduced by 1, for each ideal current source present. If the ideal current source lies in the perimeter of the circuit, then the single mesh in which it is found is ignored because, such a current source is a common element with the mesh that encloses the outside of the entire circuit. Kirchhoff's voltage law is applied only to those meshes or super meshes in the modified network.

Applying Kirchhoff's voltage law to the supermesh *abcdghea*, we get

$$10 - 2(I_1 - I_2) - 4(I_3 - I_2) - 6 I_3 = 0$$

i.e. $$I_1 - 3I_2 + 5I_3 = 5 \tag{i}$$

For mesh *dcfgd*, we get

$$-2(I_2 - I_1) - 5I_2 - 4(I_2 - I_3) = 0$$

i.e. $$2I_1 - 11I_2 + 4I_3 = 0 \tag{ii}$$

and $$I_1 - I_3 = 5 \tag{iii}$$

Solving (i), (ii), and (iii), we get

$$I_1 = 5.625\,\text{A}, \quad I_2 = 1.25\,\text{A} \quad \text{and} \quad I_3 = 0.625\,\text{A}$$

2.6 NETWORK CONTAINING A DEPENDENT SOURCE

Consider a network as shown in Fig. 2.3, which contains a dependent current source $\dfrac{1}{10} v_x$.

10 A ideal current source is located on the perimeter of the circuit. Hence, we may eliminate mesh 1 form consideration.

$$I_1 = 10\,\text{A} \tag{1}$$

Once 10 A source is eliminated, the dependent current source $\dfrac{1}{10} v_x$ lies on the perimeter of the modified circuit and hence, mesh 3 also may be eliminated.

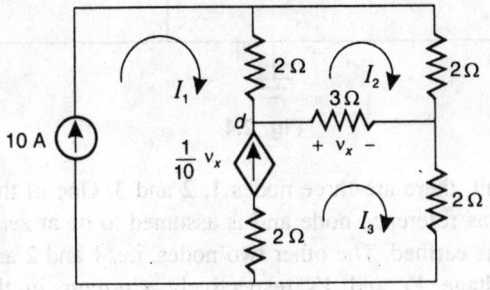

Fig. 2.3

$$I_3 - I_1 = \frac{1}{10} v_x \quad \text{but} \quad v_x = 3(I_3 - I_2)$$

$$= \frac{1}{10} \times 3(I_3 - I_2) \quad \text{i.e} \quad 10\,I_1 - 3I_2 - 7I_3 = 0 \tag{2}$$

Applying Kirchhoff's voltage law to the mesh 2, we get

$$2(I_2 - I_1) + 2I_2 + 3(I_2 - I_3) = 0$$

i.e.
$$-2I_1 + 7I_2 - 3I_3 = 0 \tag{3}$$

Solving (1), (2), and (3), we get

$$I_1 = 10\,\text{A}, \quad I_2 = 7.58\,\text{A} \quad \text{and} \quad I_3 = 11.03\,\text{A}$$

2.7 NODE VOLTAGE ANALYSIS

A principal node is a point in a network, where three or more elements meet. Node voltage analysis is one of the most effective methods of solving an electrical network. The number of equations to be solved by this method is one less than the number of equations required in mesh current analysis. Out of the total number of principal nodes existing in the network, one is taken as reference node at zero potential. Hence, an n node network will have $(n-1)$ unknown voltages and $(n-1)$ equations have to be solved. A three node network will have two unknown voltages and hence, two equations have to be solved, as the voltages of the third node is assumed to be at zero potential.

Consider a general electrical circuit as shown in Fig. 2.4, E_1 and E_2 are the e.m.f.s of the sources.

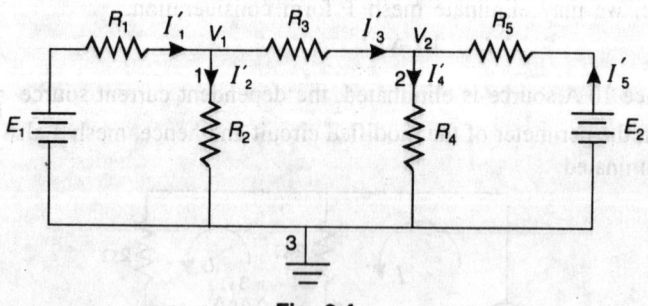

Fig. 2.4

In this circuit, there are three nodes 1, 2 and 3. One of these nodes, say 3 is taken as reference node and is assumed to be at zero potential and is shown as earthed. The other two nodes, i.e. 1 and 2 are assigned with node voltage V_1 and V_2 respectively. Currents in the various branches are assumed arbitrarily in any direction. Let these currents be

I_1', I_2', I_3', I_4' and I_5'. Then by applying Kirchhoff's current law to nodes 1 and 2, we get

For node 1

$$I_1' = I_2' + I_3'$$

i.e.
$$\frac{E_1 - V_1}{R_1} = \frac{V_1}{R_2} + \frac{V_1 - V_2}{R_3}$$

i.e.
$$\left(\frac{1}{R_1} + \frac{1}{R_2} + \frac{1}{R_3} \right) V_1 - \frac{1}{R_3} V_2 = \frac{E_1}{R_1}$$

i.e.
$$G_{11} V_1 + G_{12} V_2 = I_1 \tag{2.8}$$

Where, $G_{11} = \dfrac{1}{R_1} + \dfrac{1}{R_2} + \dfrac{1}{R_3}$ = sum of all the conductances connected to node 1, which is always + ve.

$G_{12} = -\dfrac{1}{R_3}$ = Mutual conductance between node 1 and node 2, which is always –ve.

$I_1 = \dfrac{E_1}{R_1}$ = Sum of all source currents to node 1. The source current is +ve, if it is flowing towards the node and –ve, if it is flowing away from the node.

For node 2

$$I_3' + I_5' = I_4'$$

i.e.
$$\frac{V_1 - V_2}{R_3} + \frac{E_2 - V_2}{R_5} = \frac{V_2}{R_4}$$

i.e.
$$-\frac{1}{R_3} V_1 + \left(\frac{1}{R_3} + \frac{1}{R_4} + \frac{1}{R_5} \right) V_2 = \frac{E_2}{R_5}$$

i.e.
$$G_{21} V_1 + G_{22} V_2 = I_2 \tag{2.9}$$

Where, $G_{21} = -\dfrac{1}{R_3}$ = Mutual conductance between node 2 and node 1. It is always –ve.

$G_{22} = \dfrac{1}{R_3} + \dfrac{1}{R_4} + \dfrac{1}{R_5}$ = sum of all the conductances connected to node 2. It is always +ve.

$$I_2 = \frac{E_2}{R_5} = \text{Sum of all source currents to node 2.}$$

Equations (2.8) and (2.9) are general node voltage equations, from which, the equations for nodes 1 and 2 can be directly written, without using Kirchhoff's current law. In general, if a network has nodes, $(n - 1)$ nodal equations can be written, as the nth node is taken as reference node at zero potential. The general node voltage equations are:

For d.c. circuits

$$G_{11} V_1 + G_{12} V_2 + \dots\dots + G_{1n} V_n = I_1$$

$$G_{21} V_1 + G_{22} V_2 + \dots\dots + G_{2n} V_n = I_2$$

$$\vdots$$

$$G_{n1} V_1 + G_{n2} V_2 + \dots\dots + G_{nn} V_n = I_1$$

For a.c. circuits

$$Y_{11} V_1 + Y_{12} V_2 + \dots\dots + Y_{1n} V_n = I_1$$

$$Y_{21} V_1 + Y_{22} V_2 + \dots\dots + Y_{2n} V_n = I_2$$

$$\vdots$$

$$Y_{n1} V_1 + Y_{n2} V_2 + \dots\dots + Y_{nn} V_n = I_n \qquad (2.10)$$

2.8 PROCEDURE FOR SOLVING AN ELECTRICAL CIRCUIT USING NODE VOLTAGE ANALYSIS

1. Node voltage analysis can be applied to non-planar networks also, unlike mesh current analysis.

2. All the principal nodes of the network are identified and one of them is taken as reference node at zero potential. Usually, the node to which maximum number of branches are connected is taken as reference node.

3. The remaining nodes are assigned with node voltages V_1, V_2, V_3 ..., etc.

4. The node voltage equations are written using the general node equations as given in Eq. (2.10).

5. The node voltage equations are solved using Cramer's rule or otherwise, to get V_1, V_2, V_3, etc.

6. Once the node voltages are known, the currents in all the branches of the network can be found.

2.9 SUPER NODAL ANALYSIS

The node voltage analysis described in the previous sections can not be used to solve a network which contains an ideal voltage source as one of its branches, because, the current to the node due to this source cannot be found.

Consider the network as shown in Fig. 2.5.

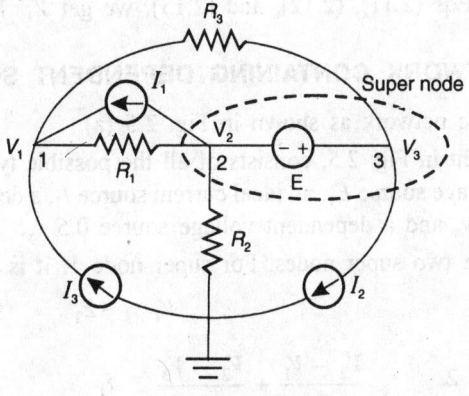

Fig. 2.5

In the circuit shown in Fig. 2.5, we find that there is an ideal voltage source between nodes 2 and 3. We do not know the current flowing from this source either to node 2 or node 3. Hence, the ordinary node voltage analysis can't be used to solve this network.

There are two methods of solving this network. An unknown current is assigned to the branch which contains the ideal voltage source. Kirchhoff's voltage law is applied to the three nodes and three equations are obtained. Applying Kirchhoff's voltage law between nodes 2 and 3, we get one more equation. Then, we have 4 equations and 4 unknowns, whose solution is tedious and lengthy.

The other method is easier and results in only 3 equations, which can be easily solved. In this method, the current in the branch containing the ideal voltage source is completely avoided. The nodes 2 and 3, along with the ideal voltage source is considered as a "super node" and Kirchhoff's current law is applied to both the nodes at the same time. The super is enclosed in the dotted envelope as shown:

First equation is written for node 1.

i.e.
$$\frac{V_1 - V_3}{R_3} + \frac{V_1 - V_2}{R_1} = I_1 + I_3 \qquad (2.11)$$

The second equation is written for the super node 2-3.

$$\frac{V_1 - V_3}{R_3} + \frac{V_1 - V_2}{R_1} = I_1 + I_2 + \frac{V_2}{R_2} \qquad (2.12)$$

The third equation is obvious

$$V_3 - V_2 = E \qquad (2.13)$$

Solving Eqs (2.11), (2.12), and (2.13), we get V_1, V_2, and V_3.

2.10 NETWORK CONTAINING DEPENDENT SOURCES

Consider the network as shown in Fig. 2.5 (a).

The circuit in Fig. 2.5, consists of all the possible types of sources, an ideal voltage source E_1, an ideal current source I_1, a dependent current source $0.2\ v_x$ and a dependent voltage source $0.5\ v_y$.

There are two super nodes. For super node 1, it is obvious that

$$V_1 = -E_1$$

For node 2, $\qquad \dfrac{V_2 - V_1}{R_1} + \dfrac{V_2 - V_3}{R_2} = I_1 \qquad (2.14)$

For super node 3-4

$$\frac{V_3 - V_2}{R_2} + \frac{V_4}{R_3} + \frac{V_4 - V_1}{R_4} = 0.2\ v_x = 0.2\ (V_2 - V_1) \qquad (2.15)$$

Then we have, $V_3 - V_4 = 0.5 v_y = 0.5 \left(V_4 - V_1\right) \qquad (2.16)$

Solving the above equations, V_1, V_2, V_3 and V_4 can be found.

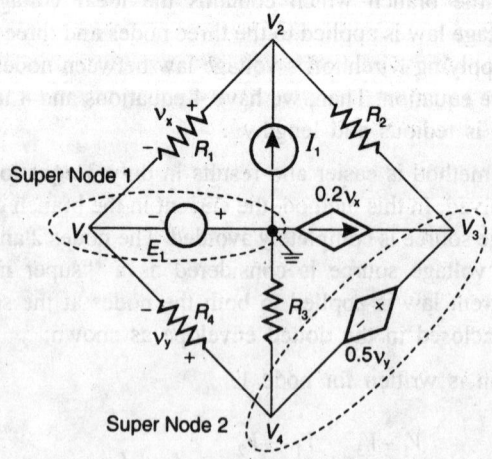

Fig. 2.5 (a)

WORKED EXAMPLES

2.1 Obtain the star connected equivalent network for the given delta connected network shown in Fig. 2.6 (a).

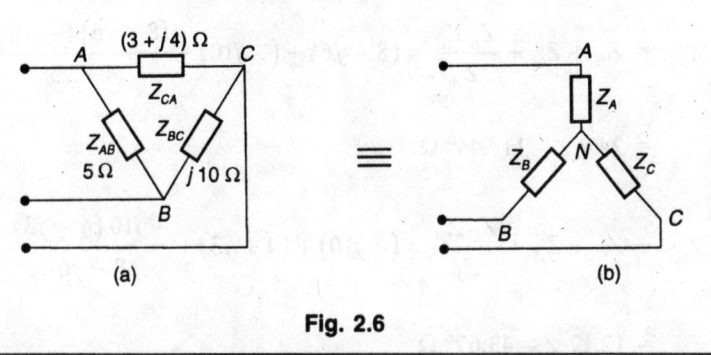

Fig. 2.6

Solution: The star equivalent network is shown in Fig. 2.6 (b), whose impedances are given by

$$Z_A = \frac{Z_{AB} \, Z_{AC}}{Z_{AB} + Z_{BC} + Z_{CA}} = \frac{5 \, (3 + j \, 4)}{5 + j \, 10 + (3 + j4)} = 1.55 \angle -7.13° \ \Omega$$

$$Z_B = \frac{Z_{BC} \, Z_{AB}}{Z_{AB} + Z_{BC} + Z_{CA}} = \frac{j10 \, (5)}{5 + j \, 10 + (3 + j4)} = 3.1 \angle 29.74° \ \Omega$$

$$Z_C = \frac{Z_{CA} \, Z_{BC}}{Z_{AB} + Z_{BC} + Z_{CA}} = \frac{(3 + j4) \, j \, 10}{5 + j \, 10 + (3 + j4)} = 3.1 \angle 82.87° \ \Omega$$

2.2 Obtain the delta connected equivalent network for the given star connected network shown in Fig. 2.7 (a).

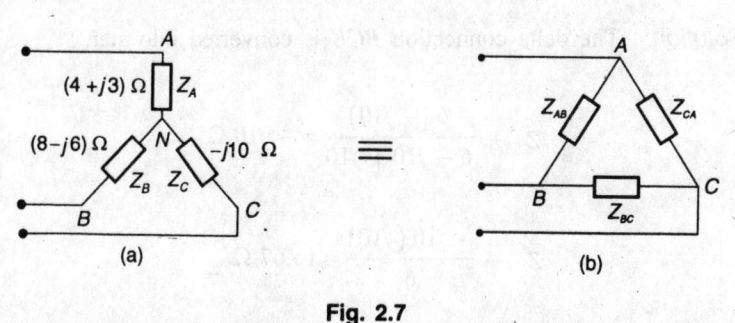

Fig. 2.7

Solution: The delta connected equivalent network is shown in Fig. 2.7 (b), whose impedances are given by

$$Z_{AB} = Z_A + Z_B + \frac{Z_A Z_B}{Z_C} = (4 + j3) + (8 - j6) + \frac{(4 + j3)(8 - j6)}{-j10}$$

$$= 12.17 \angle 9.46° \ \Omega$$

$$Z_{BC} = Z_B + Z_C + \frac{Z_B Z_C}{Z_A} = (8 - j6) + (-j10) + \frac{(8 - j6)(-j10)}{4 + j3}$$

$$= 24.3 \angle -117.44° \ \Omega$$

$$Z_{CA} = Z_C + Z_A + \frac{Z_C Z_A}{Z_B} = (-j10) + (4 + j3) + \frac{-j10(4 + j3)}{8 - j6}$$

$$= 12.17 \angle -43.67° \ \Omega$$

2.3 Obtain the delta connected equivalent network for the network shown in Fig. 2.8.

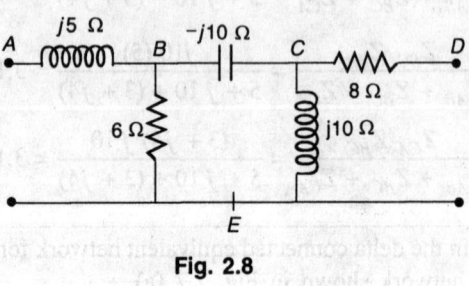

Fig. 2.8

Solution: The delta connection *BCE* is converted into star.

$$Z_B = \frac{6(-j10)}{6 - j10 + j10} = -j10 \ \Omega$$

$$Z_C = \frac{-j10(j10)}{6} = 16.67 \ \Omega$$

$$Z_E = \frac{6(j10)}{6} = j10 \ \Omega$$

After this conversion, the equivalent circuit for the circuit shown in Fig. 2.8, is as shown in Fig. 2.9.

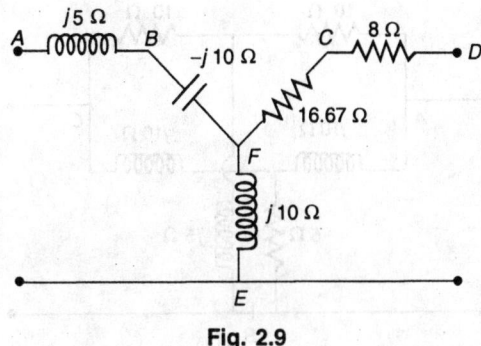

Fig. 2.9

The circuit in Fig. 2.9 is further simplified as in Fig. 2.10 (a). The delta equivalent network is as shown in Fig. 2.10 (b).

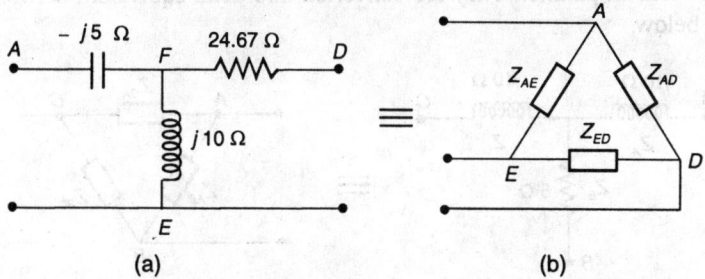

(a)   (b)

Fig. 2.10

$$Z_{AE} = Z_A + Z_E + \frac{Z_A\, Z_E}{Z_D} = -j5 + j10 + \frac{-j5\,(j10)}{24.67}$$

$$= 5.4\ \angle 67.93°\ \Omega$$

$$Z_{AD} = Z_A + Z_D + \frac{Z_A\, Z_D}{Z_E} = -j5 + 26.47 + \frac{-j5\,(26.47)}{j10}$$

$$= 14.15\ \angle -20.69°\ \Omega$$

$$Z_{ED} = Z_E + Z_D + \frac{Z_E\, Z_D}{Z_A} = j10 + 26.47 + \frac{j10\,(26.47)}{-j5}$$

$$= 28.3\ \angle 159.3°\ \Omega$$

2.4 Obtain the delta connected equivalent network for the network shown in Fig. 2.11.

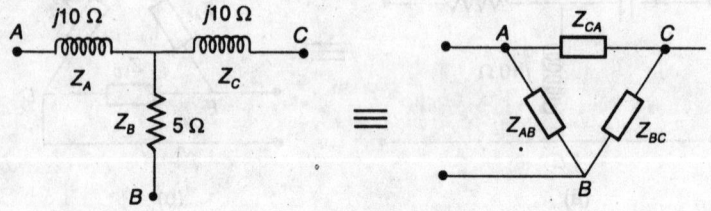

10 Ω   10 Ω

A $j10\ \Omega$ $j10\ \Omega$ C

5 Ω $j5\ \Omega$

B

Fig. 2.11

Solution: The network shown in Fig. 2.11, consists of two star networks in parallel. They are converted into delta equivalent networks as below.

$$Z_{AB} = Z_A + Z_B + \frac{Z_A Z_B}{Z_C} = j10 + 5 + \frac{j10(5)}{j10}$$

$$= (10 + j10)\ \Omega$$

$$Z_{BC} = Z_B + Z_C + \frac{Z_B Z_C}{Z_A} = 5 + j10 + \frac{5(j10)}{j10}$$

$$= (10 + j10)\ \Omega$$

$$Z_{AC} = Z_C + Z_A + \frac{Z_C Z_A}{Z_B} = j10 + j10 + \frac{j10(j10)}{5}$$

$$= (-20 + j20)\ \Omega$$

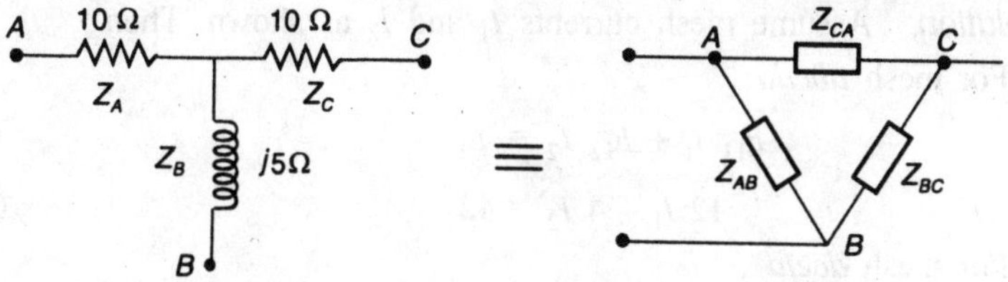

Fig. 2.11 (b)

$$Z_{AB} = 10 + j5 + \frac{10\,(j5)}{10} = (10 + j10)\,\Omega$$

$$Z_{BC} = j5 + 10 + \frac{10\,(j5)}{10} = (10 + j10)\,\Omega$$

$$Z_{CA} = 10 + 10 + \frac{10 \times 10}{j5} = (20 - j20)\,\Omega$$

The two converted delta networks are in parallel. The impedances of the equivalent delta network are:

$$Z_{AB} = \frac{(10 + j10)\,(10 + j10)}{10 + j10 + 10 + j10} = (5 + j5)\,\Omega$$

$$Z_{BC} = \frac{(10 + j10)\,(10 + j10)}{10 + j10 + 10 + j10} = (5 + j5)\,\Omega$$

$$Z_{CA} = \frac{(-20 + j20)\,(20 - j20)}{-20 + j20 + 20 - j\,20} = \infty$$

2.5 Using mesh current analysis, find the currents in the various branches of the circuit shown in Fig. 2.12.

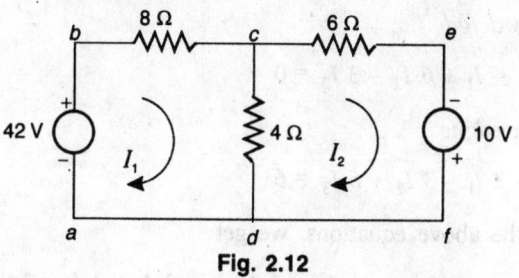

Fig. 2.12

Solution: Assume mesh currents I_1 and I_2 as shown. Then

For mesh *abcda*

$$R_{11} I_1 + R_{12} I_2 = E_1$$

i.e. $$12 I_1 - 4 I_2 = 42 \qquad (1)$$

For mesh *dcefd*

$$R_{21} I_1 + R_{22} I_2 = E_2$$

i.e. $$- 4 I_1 + 10 I_2 = 10 \qquad (2)$$

Solving the above equations, we get

$$I_1 = 4.42 \text{ A and } I_2 = 2.77 \text{ A}$$

The branch currents are

$$I_{ab} = I_{bc} = I_1 = 4.42\,\text{A}, \ I_{cd} = I_1 - I_2 = 1.65 \text{ A}, \ I_{cefd} = I_2 = 2.77\,\text{A}$$

2.6 In the circuit shown in Fig. 2.13, find the mesh current I_1, I_2, and I_3.

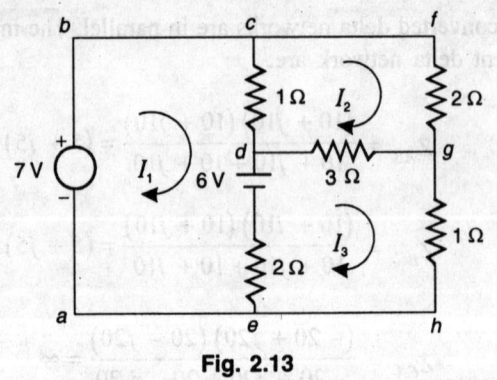

Fig. 2.13

Solution: For mesh *abcdea*

$$3 I_1 - I_2 - 2 I_3 = 1 \qquad (1)$$

For mesh *dcfgd*

$$- I_1 + 6 I_2 - 3 I_3 = 0 \qquad (2)$$

For mesh *edghe*

$$- 2 I_1 - 3 I_2 + 6 I_3 = 6 \qquad (3)$$

Solving the above equations, we get

$$I_1 = 3 \text{ A}, \ I_2 = 2 \text{ A and } I_3 = 3\text{A}$$

2.7 Use mesh analysis to find I_x in the circuit shown in Fig. 2.14.

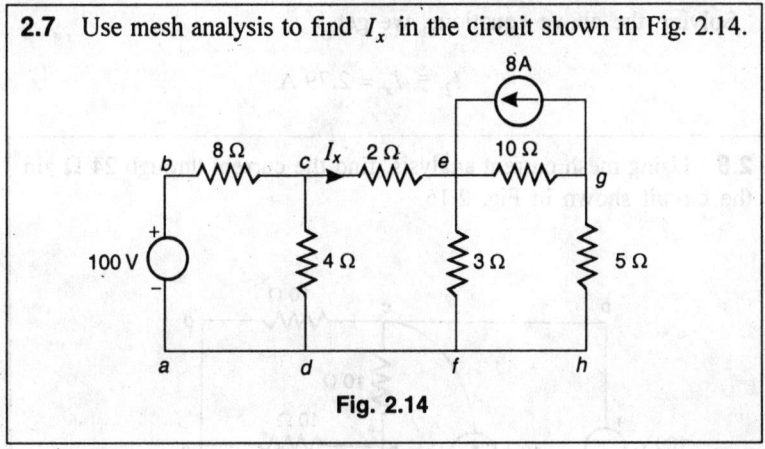

Fig. 2.14

Solution: There is a practical current source of 8 A connected between e and g. This is converted into a practical voltage source, and the circuit in Fig. 2.14 is rewritten as below.

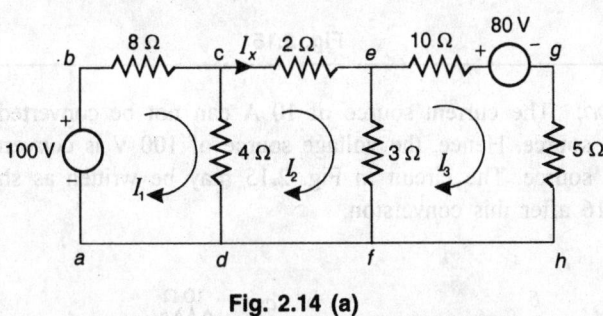

Fig. 2.14 (a)

The mesh currents I_1, I_2, and I_3 are assumed as shown.

For mesh *abcda*

$$12 I_1 - 4 I_2 + 0 I_3 = 100 \tag{1}$$

For mesh *dcefd*

$$- 4 I_1 + 9 I_2 - 3 I_3 = 0 \tag{2}$$

For mesh *feghf*

$$0 I_1 - 3 I_2 + 18 I_3 = - 80 \tag{3}$$

Solving the above equations, we get

$$I_2 = I_x = 2.79 \text{ A}$$

2.8 Using mesh current analysis, find the current through 24 Ω zin the circuit shown in Fig. 2.15.

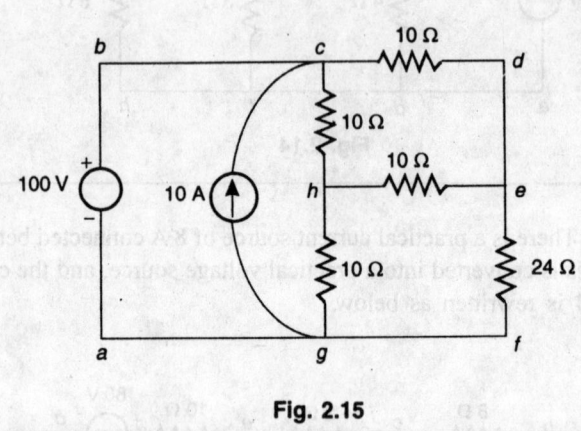

Fig. 2.15

Solution: The current source of 10 A can not be converted into a voltage source. Hence, the voltage source of 100 V is converted into current source. The circuit in Fig. 2.15 may be written as shown in Fig. 2.16 after this conversion.

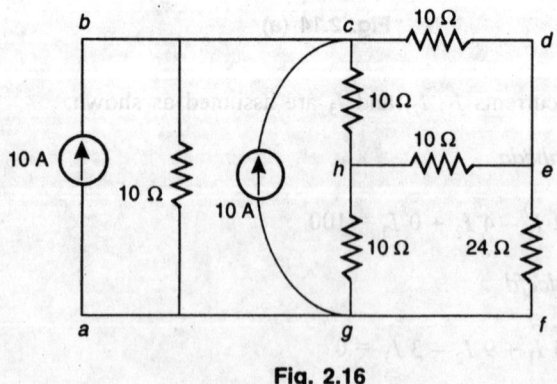

Fig. 2.16

The two current sources, which are in parallel are converted into a single current source as shown in Fig. 2.17.

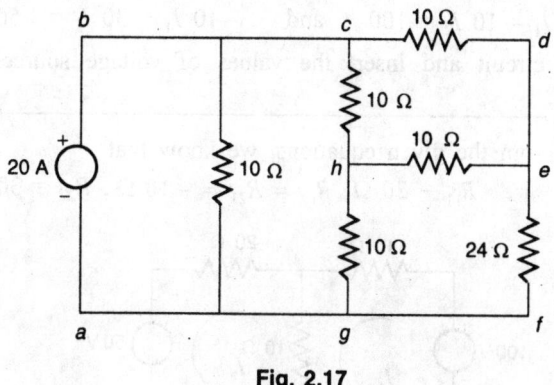

Fig. 2.17

Now, the current source is replaced by a voltage source as shown in Fig. 2.18 and mesh currents I_1, I_2 and I_3 are assumed as shown.

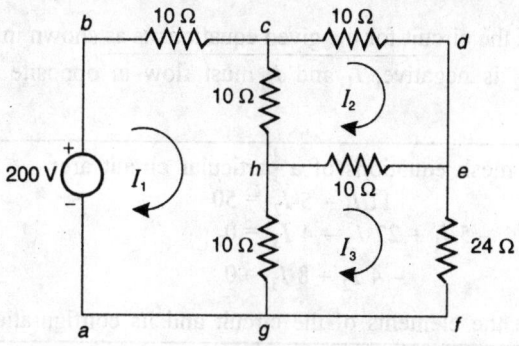

Fig. 2.18

For mesh *abcga*

$$30 I_1 - 10 I_2 - 10I_3 = 200 \qquad (1)$$

For mesh *hcdeh*

$$-10 I_1 + 30 I_2 - 10 I_3 = 0 \qquad (2)$$

For mesh *ghefg*

$$-10 I_1 - 10 I_2 + 44 I_3 = 0 \qquad (3)$$

Solving the above equations, we get

$$I_{24\Omega} = I_3 = 2.94 \text{ A}$$

2.9 The mesh equations of a particular circuit are:

$$20 I_1 - 10 I_2 = 100 \quad \text{and} \quad -10 I_1 + 30 I_2 = -50$$

Draw the circuit and insert the values of voltage sources and resistances.

Solution: From the given equations, we know that

$$R_{11} = 20 \ \Omega, \ R_{12} = R_{21} = -10 \ \Omega, \ R_{22} = 30 \ \Omega$$

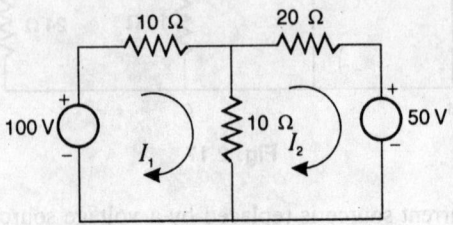

Fig. 2.19

Therefore, the circuit for the given equations is as shown in Fig. 2.19. *Note:* As R_{12} is negative, I_1 and I_2 must flow in opposite directions through it.

2.10 The mesh equations of a particular circuit are:
$$11 I_1 - 5 I_2 = 50$$
$$-5 I_1 + 27 I_2 - 4 I_3 = 0$$
$$-4 I_2 + 8 I_3 = 0$$

Determine the elements of the circuit and its configuration.

Solution: From the equations, we find that

$R_{11} = 11 \ \Omega,$	$R_{12} = -5 \ \Omega,$	$R_{13} = 0$	and $E_1 = 50 \ \text{V}$
$R_{21} = -5 \ \Omega,$	$R_{22} = 27 \ \Omega,$	$R_{23} = -4 \ \Omega$	and $E_2 = 0$
$R_{31} = 0 \ \Omega,$	$R_{32} = -4 \ \Omega,$	$R_{33} = 8 \ \Omega$	and $E_3 = 0$

Hence, the circuit is as shown in Fig. 2.20.

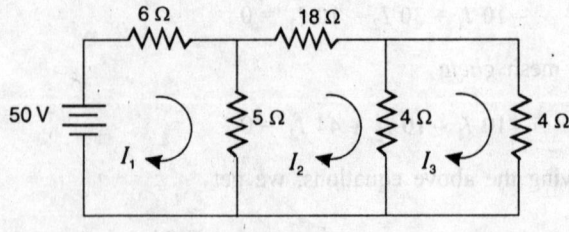

Fig. 2.20

2.11 Find the output power from the voltage source using mesh current analysis, for the circuit shown in Fig. 2.21.

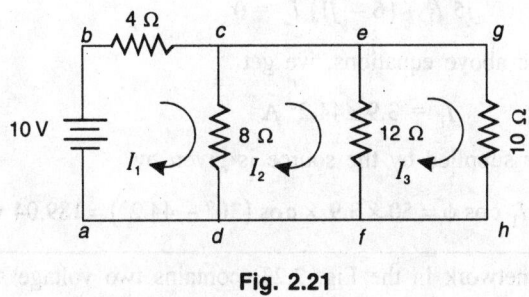

Fig. 2.21

Solution: For the mesh *abcda*

$$12 I_1 - 8 I_2 + 0 I_3 = 10 \tag{1}$$

For the mesh *dcefd*

$$-8 I_1 + 20 I_2 - 12 I_3 = 0 \tag{2}$$

For the mesh *feghf*

$$0 I_1 - 12 I_2 + 22 I_3 = 0 \tag{3}$$

Solving (1), (2) and (3), using Cramers rule, we get $I_1 = 1.38$ A, $I_2 = 0.82$ A and $I_3 = 0.447$ A.

Power delivered by the source = sum of the powers consumed by all resistances $= 1.38^2 \times 4 + (1.38 - 0.82)^2 \times 8$

$$+ (0.82 - 0.47)^2 \times 12 + 0.447^2 \times 10 = 13.8 \text{ watts}$$

2.12 Using mesh current analysis, find the power supplied by the source, for the circuit shown in Fig. 2.22.

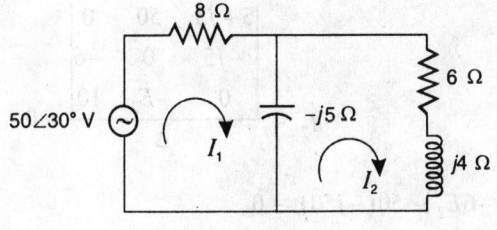

Fig. 2.22

Solution: The mesh equations are:

$$(8 - j5) I_1 + j5 I_2 = 50 \angle 30°$$

$$j5 I_1 + (6 - j1) I_2 = 0$$

Solving the above equations, we get

$$I_1 = 3.9 \angle 44.2° \text{ A}$$

The power supplied by the source is given by

$$P = E I_1 \cos \phi = 50 \times 3.9 \times \cos(30° - 44.2°) = 189.04 \text{ watts}$$

2.13 The network in the Fig. 2.23, contains two voltage sources E_1 and E_2. If $E_1 = 50 \angle 0°$ V, find E_2 such that the current in $(4 + j3)$ Ω is zero. Use mesh current analysis.

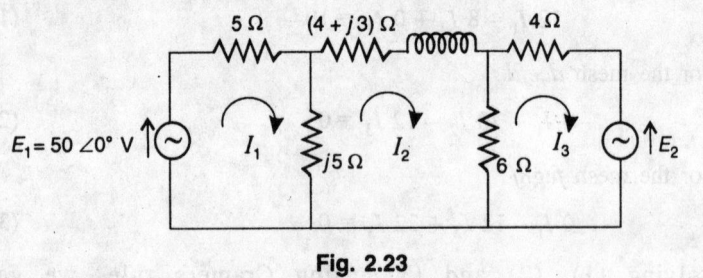

Fig. 2.23

Solution: The three mesh equations are:

$$(5 + j5) I_1 - j5 I_2 + 0 I_3 = 50$$

$$- j5 I_1 + (10 + j8) I_2 - 6 I_3 = 0$$

$$0 I_1 - 6 I_2 + 10 I_3 = -E_2$$

From the above equations:

$$I_2 = \frac{\begin{vmatrix} 5+j5 & 50 & 0 \\ -j5 & 0 & -6 \\ 0 & -E_2 & 10 \end{vmatrix}}{\Delta} = 0$$

$$\therefore \quad (5+j5)(-6E_2) - 50(-j50) = 0$$

$$\therefore \quad E_2 = 58.925 \angle 45° \text{ volts}$$

Alternate Solution

As there is no current through $(4 + j\,3)$ Ω, it amounts to o.c.

∴ Voltage drop across $j\,5$ Ω = voltage drop across 6 Ω

∴ $$-j\,5\,I_1 = 6\,I_3 \tag{1}$$

But $$I_1 = \frac{50\,\angle 0°}{5 + j5} = 7.071\,\angle -45°\,\text{A}$$

and $$I_3 = -\frac{E_2}{10}$$

Substituting these values in Eq. (1), we get

$$E_2 = 58.925\,\angle 45°\,\text{volts}$$

2.14 In the circuit shown in Fig. 2.24, find the mesh current I_3.

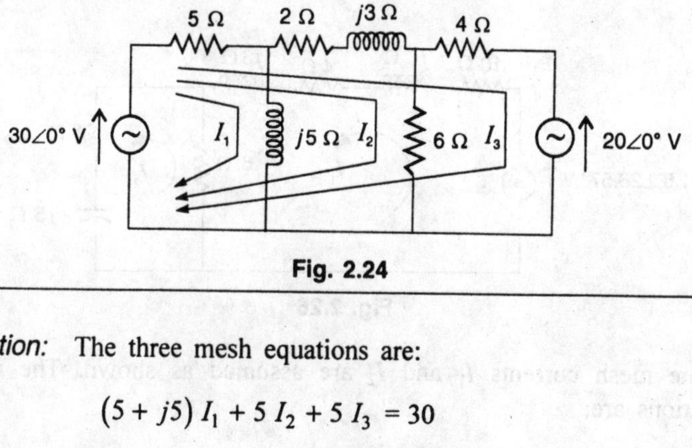

Fig. 2.24

Solution: The three mesh equations are:

$$(5 + j5)\,I_1 + 5\,I_2 + 5\,I_3 = 30$$

$$5\,I_1 + (13 + j3)\,I_2 + (7 + j3)\,I_3 = 30$$

$$5\,I_1 + (7 + j3)\,I_2 + (11 + j3)\,I_3 = 10$$

∴ $$I_3 = \frac{\begin{vmatrix} 5+j5 & 5 & 30 \\ 5 & 13+j3 & 30 \\ 5 & 7+j3 & 10 \end{vmatrix}}{\begin{vmatrix} 5+j5 & 5 & 5 \\ 5 & 13+j3 & 7+j3 \\ 5 & 7+j3 & 11+j3 \end{vmatrix}} = 1.38\,\angle -209.15°\,\text{A}$$

2.15 Find the current flowing through $(4 + j\,3)\,\Omega$ using mesh current analysis, in the circuit shown in Fig. 2.25.

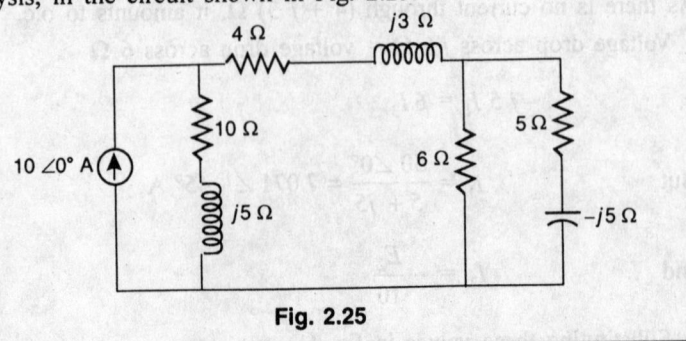

Fig. 2.25

Solution: The current source is first converted into the voltage source.

$$E = 10 \angle 0° (10 + j5) = 111.8 \angle 26.57° \text{ V}$$

The circuit is re-written as shown in Fig. 2.26.

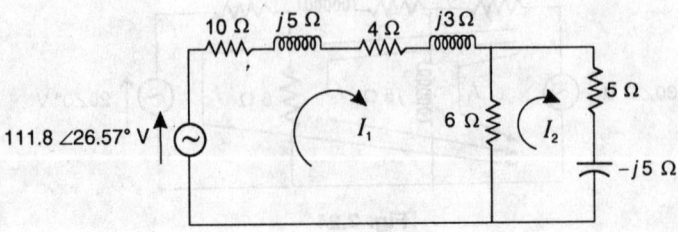

Fig. 2.26

The mesh currents I_1 and I_2 are assumed as shown. The mesh equations are:

$$(20 + j8)\, I_1 - 6 I_2 = 111.8 \angle 26.57°$$

$$-6 I_1 + (11 - j5)\, I_2 = 0$$

Solving the above equation, the current I_1 through $(4 + j\,3)\,\Omega$ works out to be $6.022 \angle 5.19°$ A.

2.16 A balanced delta connected load consists of an impedance of $20 \angle 45°\ \Omega$ in each of its branch. A three phase balanced supply of 400 V is applied to the load. Using mesh current analysis, find the line currents.

Solution: The balanced delta network is shown in Fig. 2.27.

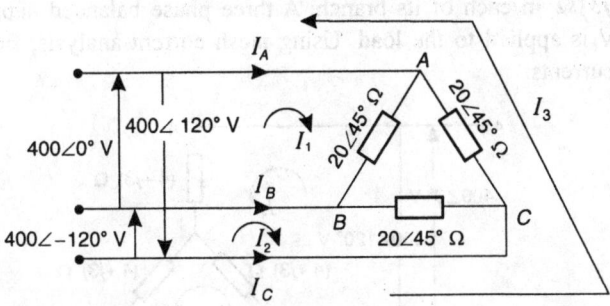

Fig. 2.27

The loop currents I_1, I_2 and I_3 are assumed as shown.

The line voltages are:

$$V_{AB} = 400 \angle 0° \text{ V}$$

$$V_{BC} = 400 \angle -120° \text{ V}$$

and $\qquad V_{CA} = 400 \angle 120° \text{ V}$

The mesh equations are

$$20 \angle 45° I_1 + 0 I_2 + 0 I_3 = 400 \angle 0° \text{ V}$$

$$0 I_1 + 20 \angle 45° I_2 + 0 I_3 = 400 \angle -120° \text{ V}$$

$$0 I_1 + 0 I_2 + 20 \angle 45° I_3 = 400 \angle 120° \text{ V}$$

The currents are: $\quad I_1 = \dfrac{400 \angle 0°}{20 \angle 45°} = 20 \angle -45° \text{ A}$

$$I_2 = \dfrac{400 \angle -120°}{20 \angle 45°} = 20 \angle -165° \text{ A}$$

and $\qquad I_3 = \dfrac{400 \angle 120°}{20 \angle 45°} = 20 \angle 75° \text{ A}$

The line currents are:

$$I_A = I_1 - I_3 = 20 \angle -45° - 20 \angle 75° = 34.64 \angle -75° \text{ A}$$

$$I_B = I_2 - I_1 = 20 \angle -165° - 20 \angle -45° = 34.64 \angle 165° \text{ A}$$

$$I_C = I_3 - I_2 = 20 \angle 75° - 20 \angle -165° = 34.64 \angle 45° \text{ A}$$

2.17 A balanced star connected load consists of an impedance $(4+j3)\,\Omega$ in each of its branch. A three phase balanced supply of 440 V is applied to the load. Using mesh current analysis, find the line currents.

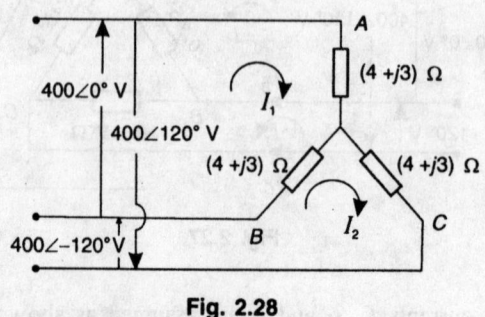

Fig. 2.28

Solution: The voltages are:

$$V_{AB} = 440\angle 0°\ \text{V}$$

$$V_{BC} = 440\angle -120°\ \text{V}$$

and

$$V_{CA} = 440\angle 120°\ \text{V}$$

The two mesh currents I_1 and I_2 are assumed as shown in Fig. 2.28. The two mesh equations are:

$$(8+j6)\,I_1 - (4+j3)\,I_2 = 440 \tag{1}$$

$$-(4+j3)\,I_1 + (8+j6)\,I_2 = 440\angle -120° \tag{2}$$

$$\therefore\ I_1 = \dfrac{\begin{vmatrix} 440 & -(4+j3) \\ 440\angle -120° & 8+j6 \end{vmatrix}}{\begin{vmatrix} 8+j6 & -(4+j3) \\ -(4+j3) & 8+j6 \end{vmatrix}} = 50.81\angle -66.87°\ \text{A} = I_A$$

Substituting the value of I_1 in Eq. (1), we get

$$I_2 = \dfrac{(8+j6) \times 50.81\angle -66.87° - 440}{4+j3} = 50.81\angle -126.87°\ \text{A}$$

The line currents are:

$$I_A = I_1 = 50.81\angle -66.87°\ \text{A},$$

$$I_B = I_2 - I_1 = 50.81 \angle 173.14° \text{ A}$$

and $$I_C = -I_2 = 50.81 \angle -306.87° \text{ A}$$

2.18 Use mesh current analysis to find the various currents flowing in the network shown in Fig. 2.29.

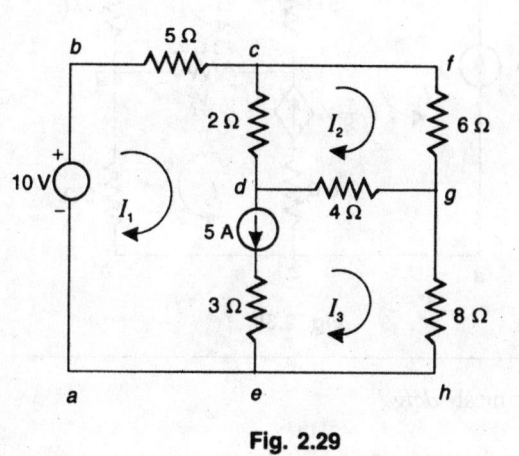

Fig. 2.29

Solution: Assume loop currents I_1, I_2 and I_3 as shown. As the branch de contains an ideal current source, *abcdghea* forms a super loop. The equation for this loop is:

$$10 - 5I_1 - 2(I_1 - I_2) - 4(I_3 - I_2) - 8I_3 = 0$$

i.e. $$7I_1 - 6I_2 + 12I_3 = 10 \qquad (1)$$

For loop *dcfgd*

$$-2(I_2 - I_1) - 6I_2 - 4(I_2 - I_3) = 0$$

i.e. $$2I_1 - 12I_2 + 4I_3 = 0 \qquad (2)$$

and $$I_1 - I_3 = 5 \qquad (3)$$

Solving Eqs (1), (2) and (3), we get

$$I_1 = 3.75 \text{ A}, \quad I_3 = -1.25 \text{ A}$$

and $$I_2 = 3\text{A}$$

2.19 Find the loop currents I_1, I_2 and I_3 in the circuit shown in Fig. 2.30.

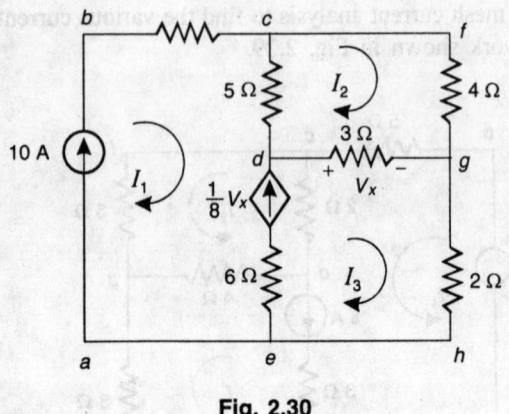

Fig. 2.30

Solution: For mesh *dcfgd*

$$-5\left(I_2 - I_1\right) - 4\,I_2 - 3\left(I_2 - I_3\right) = 0$$

i.e. $$5\,I_1 - 12\,I_2 + 3\,I_3 = 0$$

But $$I_1 = 10 \text{ A}$$

$\therefore$ $$12\,I_2 - 3\,I_3 = 50 \tag{1}$$

$$\frac{1}{8}v_x = I_3 - I_1 = I_3 - 10 = \frac{1}{8}\,3\left(I_3 - I_2\right)$$

i.e. $$3\,I_2 + 5\,I_3 = 80 \tag{2}$$

Solving Eqs (1) and (2), we get

$$I_2 = 7.1 \text{ A}$$

and $$I_3 = 11.73 \text{ A}$$

2.20 Use mesh current analysis to find V_3 in the circuit shown in Fig. 2.30 (a), if the element *A* is (a) a short circuit, (b) a 20 V independent voltage source, +ve reference on the right and

(c) dependent voltage source, +ve reference on the right, labeled $15i_1$.

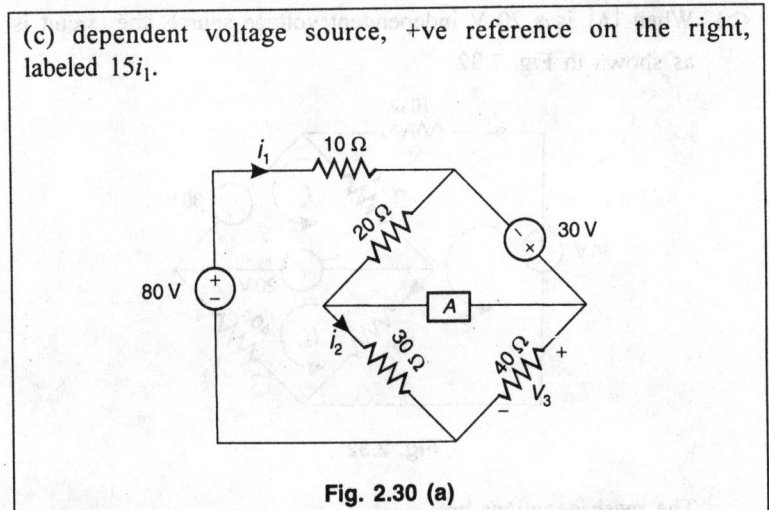

Fig. 2.30 (a)

Solution: (a) Where $\boxed{A}$ is a short circuit, the circuit can be re-written as shown in Fig. 2.31.

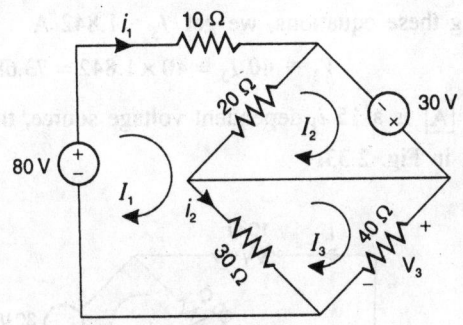

Fig. 2.31

The mesh currents I_1, I_2 and I_3 are assumed as shown.

The mesh equations are:

$$60\,I_1 - 20\,I_2 - 30\,I_3 = 80$$

$$-\,20\,I_1 + 20\,I_2 + 0\,I_3 = 30$$

$$-\,30\,I_1 + 0\,I_2 + 70\,I_3 = 0$$

Solving these equations, we get $I_3 = 1.737$ A

$$V_3 = 40\,I_3 = 40 \times 1.737 = 69.48 \text{ V}$$

(b) When $\boxed{A}$ is a 20 V independent voltage source, the circuit is as shown in Fig. 2.32.

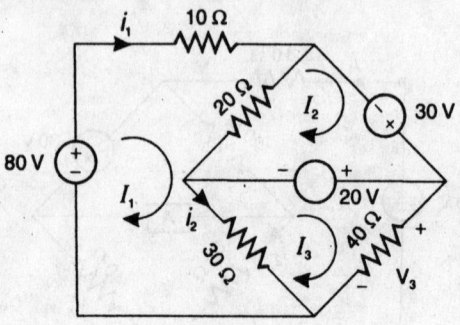

Fig. 2.32

The mesh equations are:

$$60\,I_1 - 20\,I_2 - 30\,I_3 = 80$$
$$-20\,I_1 + 20\,I_2 + 0\,I_3 = 10$$
$$-30\,I_1 + 0\,I_2 + 70\,I_3 = 20$$

Solving these equations, we get $I_3 = 1.842$ A

$$V_3 = 40\,I_3 = 40 \times 1.842 = 73.68 \text{ V}$$

(c) When $\boxed{A}$ is a 15 i_1 dependent voltage source, the circuit is as shown in Fig. 2.33.

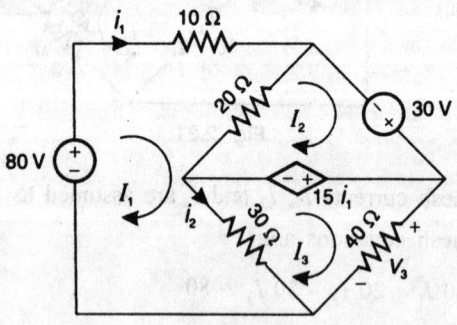

Fig. 2.33

The mesh equations are:

$$60\,I_1 - 20\,I_2 - 30\,I_3 = 80 \tag{1}$$
$$-20\,I_1 + 20\,I_2 + 0\,I_3 = 30 - 15i_1$$

but
$$i_1 = I_1$$

∴ $$-5\,I_1 + 20\,I_2 + 0\,I_3 = 30 \qquad (2)$$

$$-30\,I_1 + 0\,I_2 + 70\,I_3 = 15\,i_1 = 15\,I_1$$

∴ $$-45\,I_1 + 0\,I_2 + 70\,I_3 = 0 \qquad (3)$$

Solving the Eqs (1), (2) and (3), we get
$$I_3 = 1.98 \text{ A}$$

∴ $$V_3 = 40\,I_3 = 40 \times 1.98 = 79.2 \text{ V}$$

2.21 The mesh equations of a given network are:
$$(8 - j2)\,I_1 - 3\,I_2 = 10\angle 30°$$
$$-3\,I_1 + (8 + j5)\,I_2 - 5\,I_3 = 0$$
$$-5\,I_2 + (7 - j2)\,I_3 = 0$$

Draw the network satisfying the above equations.

Solution: Here $Z_{11} = (8 - j2)\,\Omega$, $Z_{12} = -3\,\Omega$, $Z_{13} = 0\,\Omega$

$$Z_{22} = (8 - j5)\,\Omega, \ Z_{23} = -5\,\Omega, \ Z_{33} = (7 - j2)\,\Omega$$

Hence, the resulting network is as shown in Fig. 2.34.

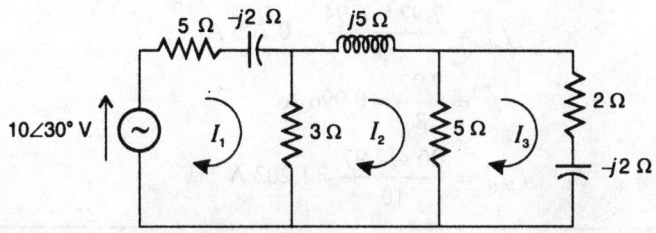

Fig. 2.34

2.22 Using node voltage analysis, find the currents in each branch of the network shown in Fig. 2.35.

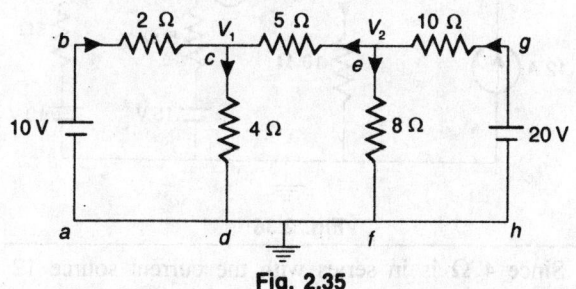

Fig. 2.35

Solution: One node is taken as reference node and the other nodes are assigned with potentials V_1 and V_2 respectively.

The node voltage equations are:

$$\left(\frac{1}{2} + \frac{1}{4} + \frac{1}{5}\right) V_1 - \frac{1}{5} V_2 = \frac{10}{2}$$

i.e. $\qquad\qquad 0.95\ V_1 - 0.2\ V_2 = 5 \qquad\qquad\qquad\qquad (1)$

$$-\frac{1}{5} V_1 + \left(\frac{1}{5} + \frac{1}{8} + \frac{1}{10}\right) V_2 = \frac{20}{10}$$

i.e. $\qquad\qquad -0.2\ V_1 + 0.425\ V_2 = 2 \qquad\qquad\qquad\qquad (2)$

Solving Eqs (1) and (2), we get

$$V_1 = 6.94\ \text{V} \qquad \text{and} \qquad V_2 = 7.97\ \text{V}$$

The currents in the various branches are:

$$I_{abc} = \frac{10 - 6.94}{2} = 1.53\ \text{A}$$

$$I_{cd} = \frac{6.94}{4} = 1.735\ \text{A}$$

$$I_{ec} = \frac{7.97 - 6.94}{5} = 0.206\ \text{A}$$

$$I_{ef} = \frac{7.97}{8} = 0.996\ \text{A}$$

$$I_{fhge} = \frac{20 - 7.97}{10} = 1.203\ \text{A}$$

2.23 Using node voltage analysis, find the voltages V_1 and V_2 of the circuit shown in Fig. 2.36.

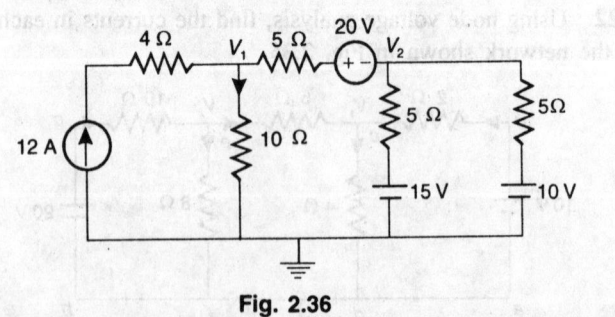

Fig. 2.36

Solution: Since 4 Ω is in series with the current source 12 A, it is neglected.

The nodal equations are:

$$\left(\frac{1}{10}+\frac{1}{5}\right)V_t -\frac{1}{5}V_2 = 12 + \frac{20}{5}$$

i.e. $\qquad 0.3\,V_1 - 0.2\,V_2 = 16$ $\hfill (1)$

$$-\frac{1}{5}V_1 + \left(\frac{1}{5}+\frac{1}{5}+\frac{1}{5}\right)V_2 = -\frac{20}{5} + \frac{15}{5} - \frac{10}{5}$$

i.e. $\qquad -0.2\,V_1 + 0.6\,V_2 = -3$ $\hfill (2)$

Solving Eqs (1) and (2), we get

$$V_1 = 64.3 \text{ V}$$

and $\qquad\qquad V_2 = 16.43 \text{ V}$

2.24   Find the current flowing through 4 Ω in the circuit given in Fig. 2.37.

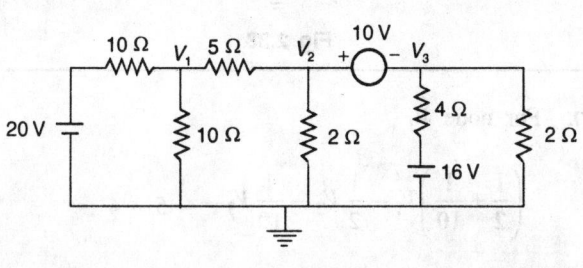

Fig. 2.37

Solution

For node $V_1, \left(\dfrac{1}{10}+\dfrac{1}{10}+\dfrac{1}{5}\right)V_1 - \dfrac{1}{5}V_2 = \dfrac{20}{10}$

i.e. $\qquad\qquad 0.4\,V_1 - 0.2\,V_2 = 2$ $\hfill (1)$

For the super node $V_2 - V_3$

$$\frac{V_1 - V_2}{5} - \frac{V_2}{2} - \frac{V_3}{2} + \frac{16 - V_3}{4} = 0$$

i.e. $0.2\,V_1 - 0.7\,V_2 - 0.75\,V_3 = -4$ $\hfill (2)$

Also, $\qquad\qquad V_2 - V_3 = 10$ $\hfill (3)$

Solving Eqs (1), (2) and (3), we get

$$V_3 = -0.74 \text{ V}$$

$$\therefore \quad I_{4\Omega} = \frac{16 - V_3}{4} = 4.185 \text{ A towards node 3}$$

2.25   Find the node voltages V_1, V_2 and V_3 in the circuit shown in Fig. 2.38.

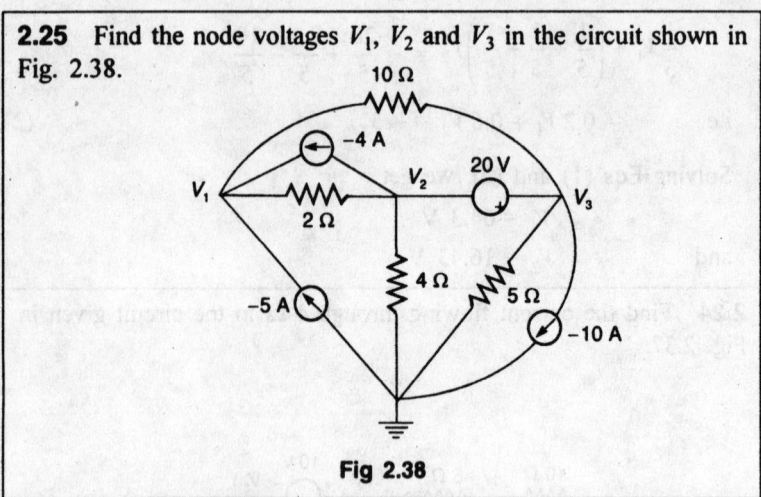

Fig 2.38

Solution: For node V_1

$$\left(\frac{1}{2} + \frac{1}{10} \right) V_1 - \frac{1}{2} V_2 - \frac{1}{10} V_3 = -5 - 4$$

i.e. $$0.6 V_1 - 0.5 V_2 - 0.1 V_3 = -9 \qquad (1)$$

For super node $(V_2 - V_3)$

$$\frac{V_2 - V_1}{2} - 4 + \frac{V_2}{4} + \frac{V_3 - V_1}{10} + \frac{V_3}{5} - 10 = 0$$

i.e. $$-0.6 V_1 + 0.75 V_2 + 0.3 V_3 = 14 \qquad (2)$$

and $$V_3 - V_2 = 20 \qquad (3)$$

Solving Eqs (1), (2) and (3), we get

$$V_1 = -9.44 \text{ V}, \qquad V_2 = 2.226 \text{ V}$$

and $$V_3 = 22.226 \text{ V}$$

2.26 Find the voltages V_1, V_2, V_3, and V_4 in the circuit shown in Fig. 2.39.

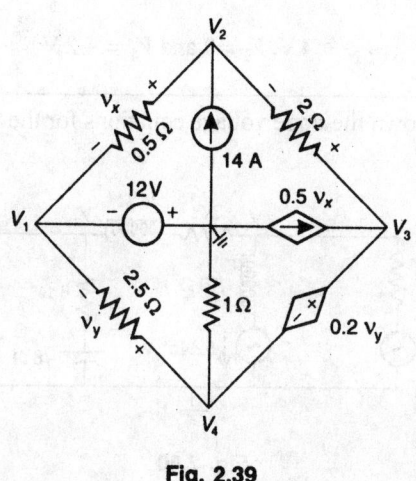

Fig. 2.39

Solution: It is obvious that, $V_1 = -12$ V (1)

For super node $V_3 - V_4$

$$\frac{V_3 - V_2}{2} - 0.5\,v_x + \frac{V_4}{1} + \frac{V_4 - V_1}{2.5} = 0$$

but, $v_x = V_2 - V_1$

$\therefore$
$$\frac{V_3 - V_2}{2} - 0.5\,(V_2 - V_1) + \frac{V_4}{1} + \frac{V_4 - V_1}{2.5} = 0$$

but, $V_1 = -12$ V

$\therefore$
$$\frac{V_3 - V_2}{2} - 0.5\,(V_2 + 12) + V_4 + \frac{V_4 + 12}{2.5} = 0$$

i.e. $-V_2 + 0.5\,V_3 + 1.4\,V_4 = 1:2$ (2)

Also $V_3 - V_4 = 0.2\,v_y = 0.2\,(V_4 - V_1) = 0.2\,(V_4 + 12)$

i.e. $V_3 - 1.2\,V_4 = 2.4$ (3)

For node V_2, $14 = \dfrac{V_2 - V_1}{0.5} + \dfrac{V_2 - V_3}{2} = \dfrac{4V_2 - 4V_1 + V_2 - V_3}{2}$

i.e. $28 = 4V_2 + 48 + V_2 - V_3 = 5V_2 - V_3 + 48$

i.e. $\qquad V_3 - 5V_2 = 20 \qquad\qquad\qquad$ (4)

Solving Eqs (2), (3), and (4), we get

$$V_2 = 4\,V,\ V_3 = 0 \text{ and } V_4 = -2\,V$$

2.27 Write down the node voltage equations for the network shown in Fig. 2.40.

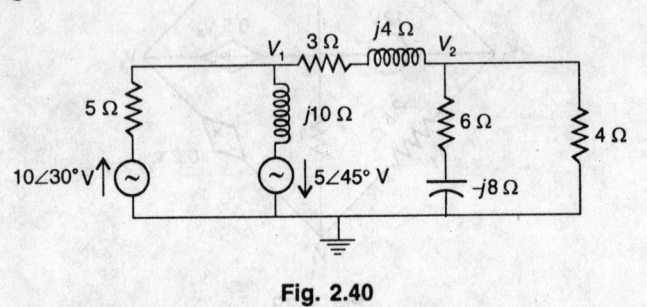

Fig. 2.40

Solution: The node voltage equations are:

$$\left(\frac{1}{5} + \frac{1}{j10} + \frac{1}{3+j4}\right)V_1 - \left(\frac{1}{3+j4}\right)V_2 = \frac{10\angle 30°}{5} - \frac{5\angle 45°}{j10} \quad \text{and}$$

$$-\left(\frac{1}{3+j4}\right)V_1 + \left(\frac{1}{3+j4} + \frac{1}{6-j8} + \frac{1}{4}\right)V_2 = 0$$

2.28 In the network shown in Fig. 2.41, find V_{AB}.

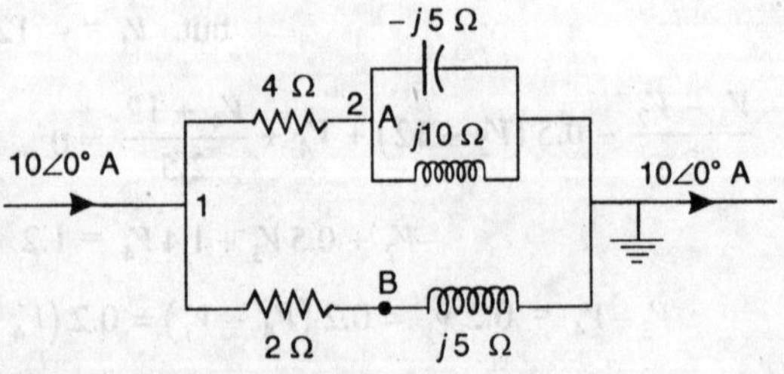

Fig. 2.41

Solution

For node 1, $\qquad \left(\frac{1}{4} + \frac{1}{2+j5}\right)V_1 - \frac{1}{4}V_2 = 10$

i.e. $\qquad 0.362\angle-28.33°V_1-0.25V_2 = 10$ $\qquad\qquad$ (1)

For node 2, $\qquad -\dfrac{1}{4}V_1 +\left(\dfrac{1}{-j5}+\dfrac{1}{4}+\dfrac{1}{j10}\right)V_2 = 0$

i.e. $\qquad -0.25V_1 + 0.391\angle-50.19°V_2 = 0$ $\qquad\qquad$ (2)

Solving Eqs (1) and (2), we get

$$V_1 = 27.38\angle53.75°\ V$$

$$V_2 = 17.52\angle103.86°\ V$$

$$V_B = \dfrac{V_1}{2+j5}\times j5 = \dfrac{27.38\angle53.75°}{2+j5}\times j5$$

$$= 25.42\angle75.55°\,V$$

$\therefore \qquad V_{AB} = V_A - V_B = 17.52\angle103.86° - 25.42\angle75.55°$

$$= 13\angle-144.17°\,V$$

2.29 In the circuit shown in Fig. 2.42, find the currents I_1, I_2 and I_3.

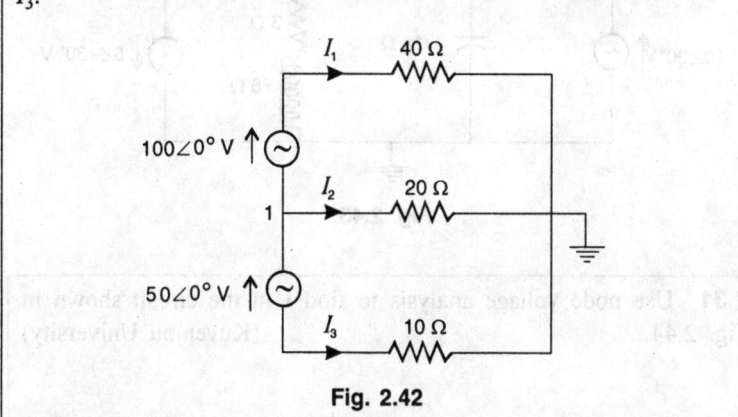

100∠0° V

1

50∠0° V

I_1 40 Ω

I_2 20 Ω

I_3 10 Ω

Fig. 2.42

Solution

For node 1, $\qquad \left(\dfrac{1}{40}+\dfrac{1}{20}+\dfrac{1}{10}\right)V_1 = \dfrac{50}{10}-\dfrac{100}{40}$

$$\therefore \qquad V_1 = 14.286\ V$$

$$I_1 = \dfrac{V_1+100}{40} = 2.857\ A$$

$$I_2 = \dfrac{V_1}{20} = 0.714\ A$$

$$I_3 = \frac{V_1 - 50}{10} = -3.571\text{A}$$

Note that $I_1 + I_2 + I_3 = 0$

2.30 For the given node voltage equations, write the configuration of the circuit and insert the elements.

$$\left(\frac{1}{6} + \frac{1}{j5} + \frac{1}{-j10}\right)V_1 - \frac{1}{j5}V_2 = \frac{10\angle 30°}{6}$$

$$-\frac{1}{j5}V_1 + \left(\frac{1}{j5} + \frac{1}{3+j8} + \frac{1}{4}\right)V_2 = -\frac{5\angle -30°}{4}$$

Solution: The circuit is as shown in Fig. 2.43.

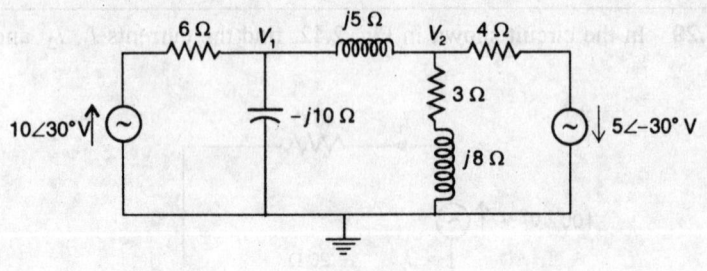

Fig. 2.43

2.31 Use node voltage analysis to find I in the circuit shown in Fig. 2.44. (Kuvempu University)

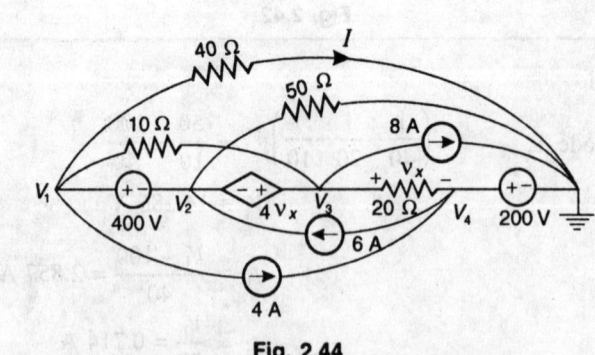

Fig. 2.44

Solution: Take node 5 as reference and assign potentials V_1, V_2, V_3 and V_4 for the other nodes as shown in Fig. 2.44.

From the circuit, it is evident that

$$V_4 = 200 \text{ V}, \quad V_1 - V_2 = 400 \text{ V}, \quad V_3 - V_4 = v_x$$

$$V_2 - V_3 = 4v_x = 4(V_3 - V_4)$$

$\therefore \qquad V_2 - 5V_3 = -4V_4 = -800 \text{ V}$

but $\qquad V_2 = V_1 - 400$

$(V_1 - 400) - 5V_3 = -800$

i.e. $\qquad V_1 - 5V_3 = -400 \qquad\qquad\qquad\qquad (1)$

There is an ideal voltage source between nodes V_1 and V_2. Also there is a dependent voltage source between nodes V_2 and V_3. Therefore, V_1-V_2-V_3 is a super node.

For super node V_1-V_2-V_3

$$4 + \frac{V_1 - V_3}{10} + \frac{V_1}{40} + \frac{V_2}{50} - 6 + \frac{V_3 - V_4}{20} + \frac{V_3 - V_1}{10} + 8 = 0$$

i.e. $\qquad \dfrac{V_1}{40} + \dfrac{V_2}{50} + \dfrac{V_3}{20} - \dfrac{V_4}{20} = -6$

$$\text{but, } V_4 = 200 \text{ V}$$

$\therefore \qquad 0.025 V_1 + 0.02 V_2 + 0.05 V_3 = 4$

$\therefore \; 0.025 V_1 + 0.02 (V_1 - 400) + 0.05 V_3 = 4$

On simplification, we get

$\qquad 4.5 V_1 + 5 V_3 = 1200 \qquad\qquad\qquad\qquad (2)$

Solving (1) and (2), we get

$$V_1 = 145.45 \text{ V}$$

$$I = \frac{V_1}{40} = \frac{145.45}{40} = 3.63 \text{ A}$$

2.32 Reduce the circuit given in Fig. 2.45 (a) into a single voltage source in series with a resistance between the terminals AB.

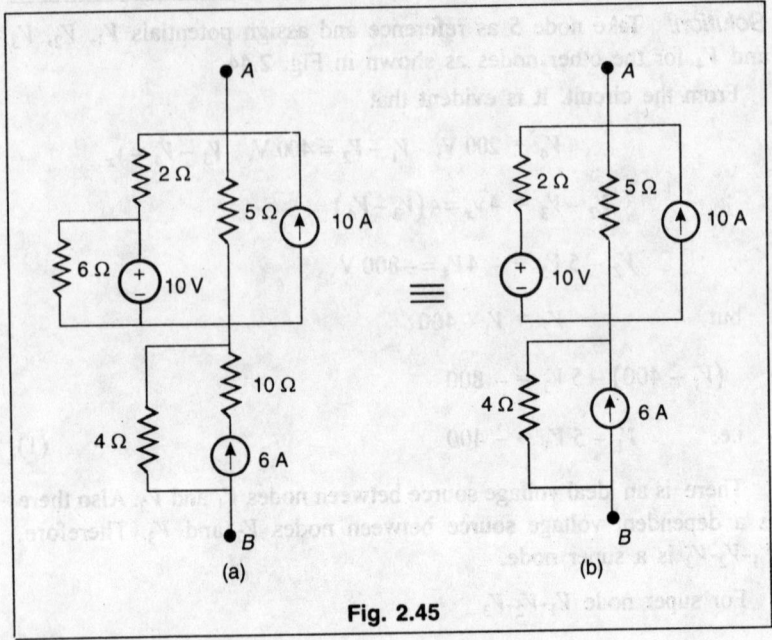

Fig. 2.45

Solution: 6 Ω in parallel with 10 volts is neglected and 10 Ω in series with 6 A is neglected. The circuit in Fig. 2.45 (a) is re-written as shown in Fig. 2.45 (b). The 10 volts voltage source is converted into a current source and 6 A is converted into a voltage source. The circuit in Fig. 2.45 (b), is re-written as shown in Fig. 2.45 (c).

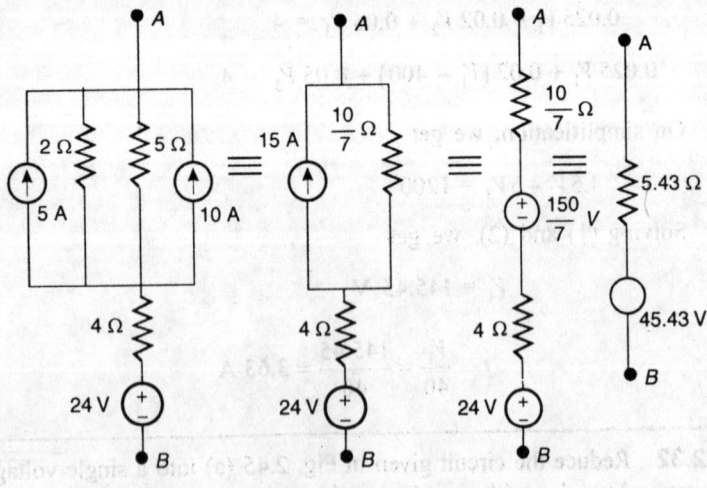

Fig. 2.45 (c)

2.33 Reduce the circuit given in Fig. 2.46 into a single equivalent current source.

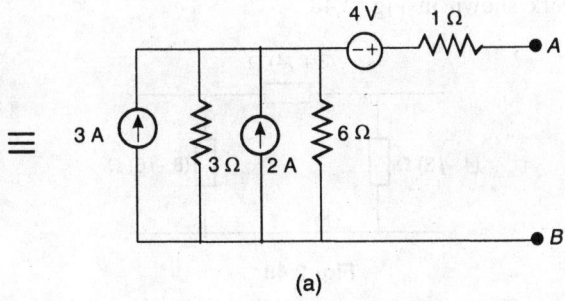

Fig. 2.46

Solution:

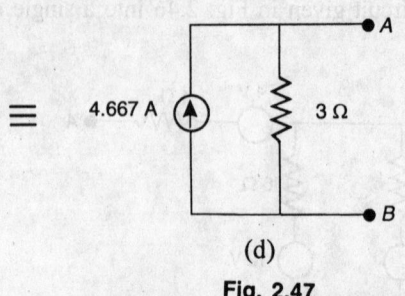

(d)

Fig. 2.47

NUMERICAL PROBLEMS

2.1 Obtain the star connected equivalent of the delta connected network shown in Fig. 2.48.

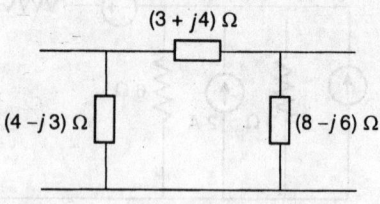

Fig. 2.48

2.2 The network shown in Fig. 2.49 consists of two star connected networks in parallel. Obtain the single delta connected equivalent circuit.

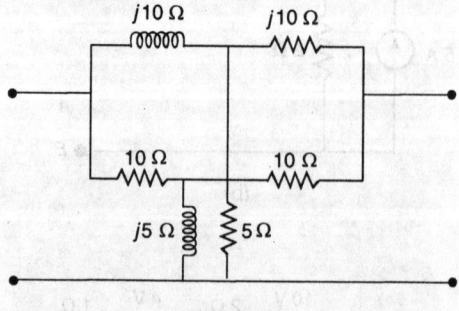

Fig. 2.49

2.3 A balanced delta connected network with an impedance of $5 \angle 45° \, \Omega$ is in parallel with a balanced star connected network with an inpedance of $(4 - j3)\Omega$. Obtain the star connected equivalent circuit.

2.4 For the network shown in Fig. 2.50, find the voltage at node A relative to ground by (a) loop current analysis and (b) node voltage analysis. (Bangalore University)

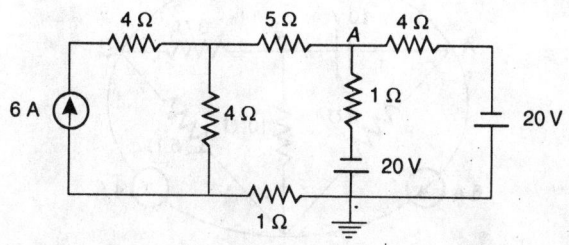

Fig. 2.50

2.5 Use node voltage method to find the current through the load resistance R_L of the circuit shown in Fig. 2.51.

(Kuvempu University)

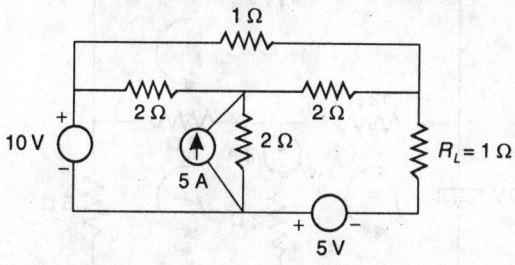

Fig. 2.51

2.6 Using node voltage analysis, find v_x and power delivered to 50 Ω resistor in the circuit shown in Fig. 2.52. (Mysore University)

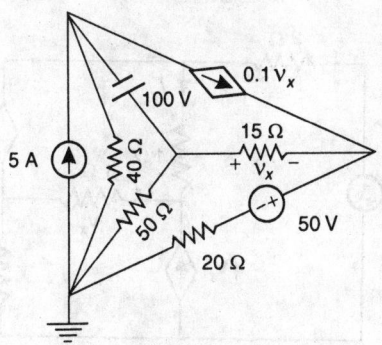

Fig. 2.52

2.7 In the network shown in Fig. 2.53, find all the node potentials.

(Bangalore University)

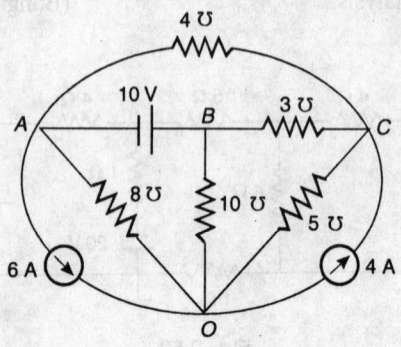

Fig. 2.53

2.8 Find the currents I_1, I_2 and I_3 in the circuit shown in Fig. 2.54.

(Mangalore University)

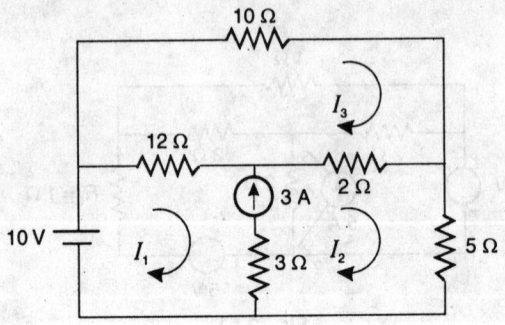

Fig. 2.54

2.9 Using mesh analysis, find the power absorbed by 2 Ω resistor in the circuit shown in Fig. 2.55. (Gulbarga University)

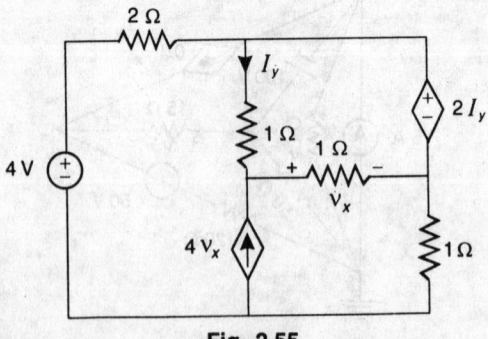

Fig. 2.55

2.10 For the circuit shown in Fig. 2.56 (a), find the potentials. All resistances are in ℧. (Kuvempu University)

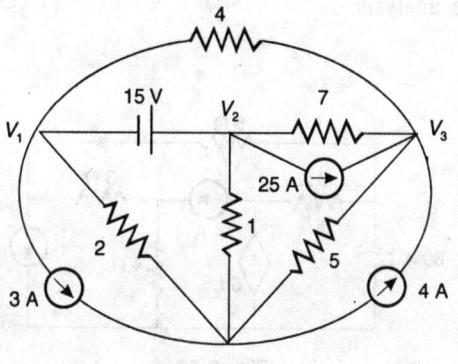

Fig. 2.56 (a)

2.11 For the circuit shown in Fig. 2.56 (b), find the power supplied by 5 V source and the output of the dependent voltage source. Take all the values in mho. (Kuvempu University)

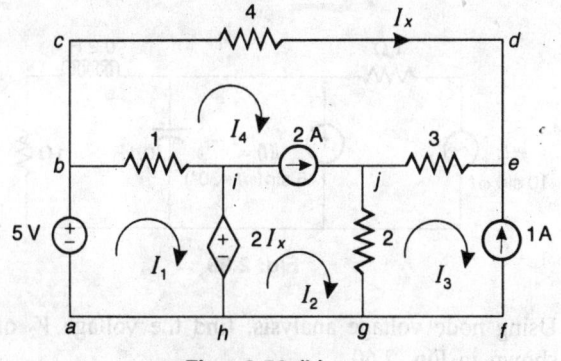

Fig. 2.56 (b)

2.12 Obtain the voltage V_x in the network shown in Fig. 2.57, using mesh current analysis. (Karnataka University)

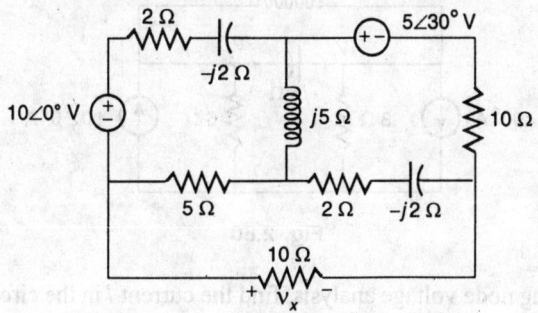

Fig. 2.57

2.13 Calculate the power dissipated in the 2 Ω resistor of the circuit shown in Fig. 2.58, using (a) mesh current analysis and (b) node voltage analysis. (Mysore University)

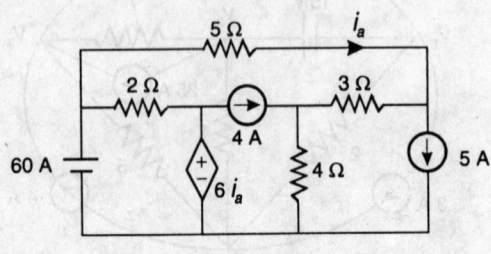

Fig. 2.58

2.14 Using node voltage analysis, find v (t) in the circuit shown in Fig. 2.59, ω = 314 rad/sec. (Kuvempu University)

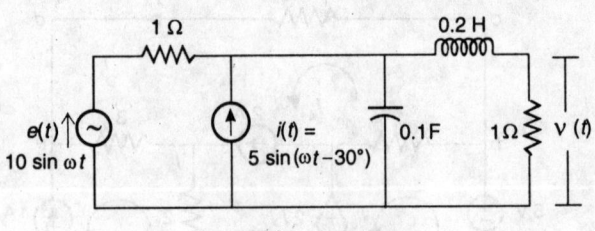

Fig. 2.59

2.15 Using node voltage analysis, find the voltage V_2 of the circuit shown in Fig. 2.60.

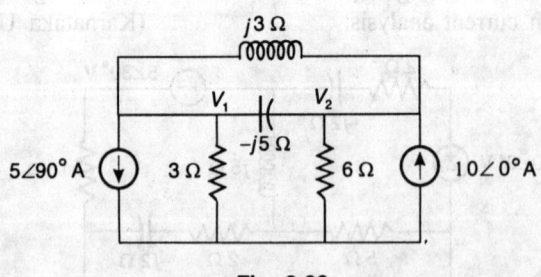

Fig. 2.60

2.16 Using node voltage analysis, find the current I in the circuit shown in Fig. 2.61. (Mysore University)

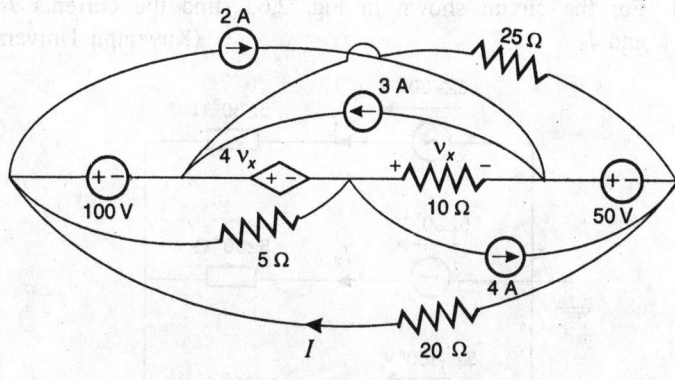

Fig. 2.61

2.17 Determine the line currents flowing, when a 400 V, 3 phase supply is applied to a balanced star connected load of impedance $(3 - j4)\,\Omega$, using mesh current analysis. (Mysore University)

2.18 Using mesh current analysis, find the line currents, when a 400 V, 3 phase supply is applied to a balanced delta connected load of impedance $10 \angle 30°\,\Omega$. (Kuvempu University)

2.19 The mesh equations of a circuit are

$$(5 + j5)\,I_1 - j\,5I_2 = 30\angle0°\text{ V},$$

$$- j5\,I_1 + (8 + j8)\,I_2 - 6I_3 = 0$$

and $\qquad -6I_2 + 10\,I_3 = -20\angle0°\text{ V}$

Derive the circuit and insert the values of voltage sources and other elements. (Karnataka University)

2.20 Using node voltage analysis, find the magnitude of V_1 in the circuit shown in Fig. 2.62, when $V_2 = 20$ volts.

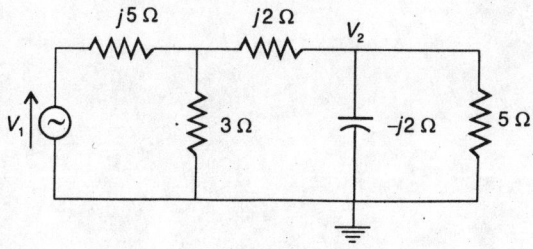

Fig. 2.62

2.21 For the circuit shown in Fig. 2.63, find the currents I_1, I_2 and I_3. (Kuvempu University)

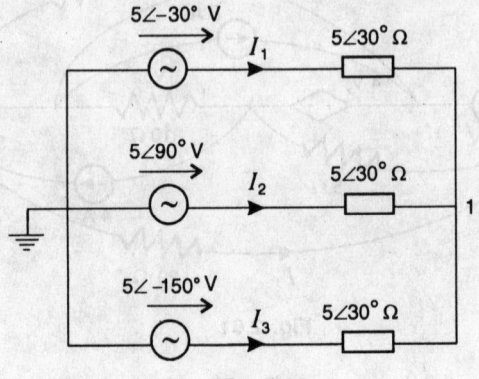

Fig. 2.63

2.22 The node voltage equations of a network are:

$$\left(\frac{1}{5} + \frac{1}{j2} + \frac{1}{4}\right) V_1 - \frac{1}{4} V_2 = \frac{50 \angle 0°}{5}$$

and $$-\frac{1}{4} V_1 + \left(\frac{1}{4} + \frac{1}{-j2} + \frac{1}{2}\right) V_2 = \frac{50 \angle 90°}{2}$$

Derive the network. (Kuvempu University)

Network Theorems

3.1 INTRODUCTION

Kirchhoff's laws and Ohm's law are the basic laws which are used to solve electrical networks. Maxwell's loop current analysis and node voltage analysis are the other two methods which are more easier methods to solve eletrical networks. There are also several theorems evolved, which in certain networks, when used, provide simple and more easier methods to solve electrical networks. In this chapter all the theorems are discussed in detail.

3.2 SUPERPOSITION THEOREM

Statement: Any linear, bilateral network containing more than one independent source, the response in any element is equal to the algebraic sum of all the responses due to each independent source acting independently, setting all the other independent sources to zero.

If it is a voltage source, it is replaced by its internal impedance or a short circuit. If it is a current source, it is replaced by an open circuit.
Explanation: Consider an electrical network as shown in Fig. 3.1.

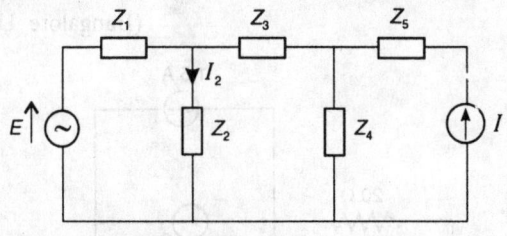

Fig. 3.1

Let I_2 be the current flowing through Z_2, when both the sources E and I are present in the circuit.

Now consider the source E only, replacing the current source by an open circuit. The resulting network is shown in Fig. 3.2.

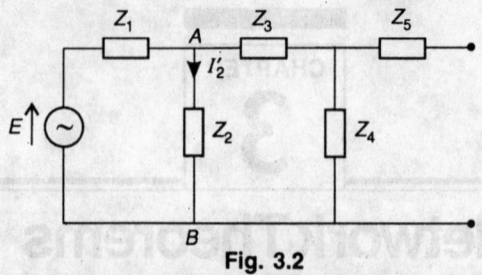

Fig. 3.2

Let I_2' be the current flowing through Z_2.

Next, consider the source I only, replacing the voltage source by a short-circuit as shown in Fig. 3.3.

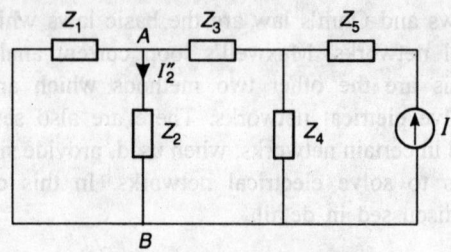

Fig. 3.3

Let I_2'' be the current flowing through Z_2. Then, according to superposition theorem

$$I_2 = I_2' + I_2'' \qquad (3.1)$$

WORKED EXAMPLES

3.1 Find the components of V_x caused by each source acting alone in the circuit shown Fig. 3.4. What is V_x, when all the sources are active?　　　　　　　　　　　　　　　　　　(Bangalore University)

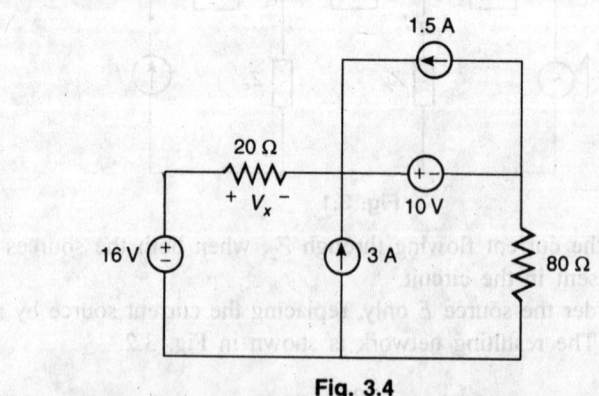

Fig. 3.4

Solution: When only 16 V is acting, the resulting network is as shown in Fig. 3.5.

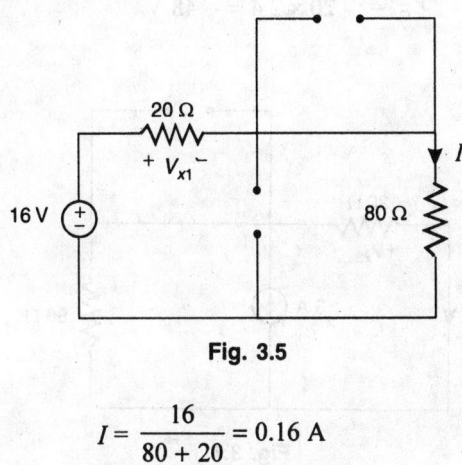

Fig. 3.5

$$I = \frac{16}{80 + 20} = 0.16 \text{ A}$$

$$V_{x1} = 20\ I = 20 \times 0.16 = 3.2 \text{ V}$$

When only 10 V is acting, the resulting network is as shown in Fig. 3.6.

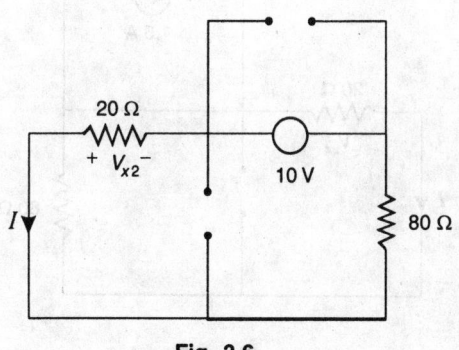

Fig. 3.6

$$I = \frac{10}{20+80} = 0.1 \text{A}$$

$$V_{x2} = -20 \times 0.1 = -2 \text{ V}$$

When only 3 A is acting, the resulting network is as shown in Fig. 3.7.

$$I = 3 \times \frac{80}{20 + 80} = 2.4 \text{ A}$$

$$\therefore \qquad V_{x_3} = -20 \times 2.4 = -48 \text{ V}$$

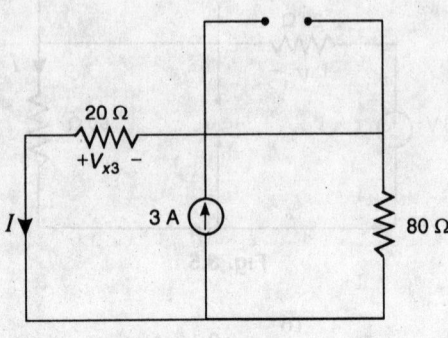

Fig. 3.7

When only 1.5 A is acting, the resulting network is as shown in Fig. 3.8.

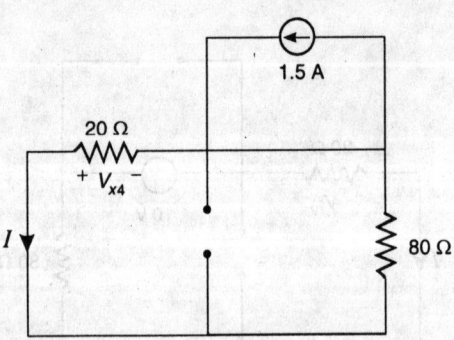

Fig. 3.8

$$I = 0$$

$$\therefore \qquad V_{x_4} = 0$$

When all the sources are active

$$V_x = V_{x_1} + V_{x_2} + V_{x_3} + V_{x_4}$$

$$= 3.2 - 2 - 48 + 0 = -46.8 \text{ V}$$

3.2 Using superposition theorem, find the current in 8 Ω, for the circuit shown in Fig. 3.9. (Kuvempu University)

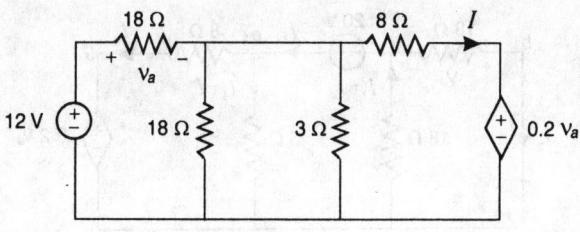

Fig. 3.9

Solution: When only 12 V is present, the resulting network is as shown in Fig. 3.10.

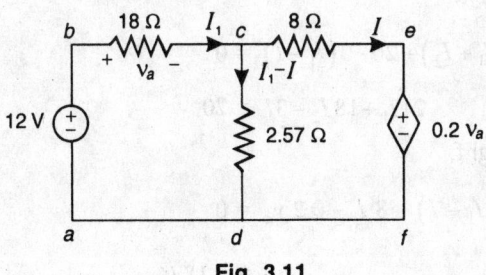

Fig. 3.10

$$18 \, \Omega \parallel 3 \, \Omega = \frac{18 \times 3}{18 + 3} = 2.57 \, \Omega$$

The circuit in Fig. 3.10 can be re-written as in Fig. 3.11.

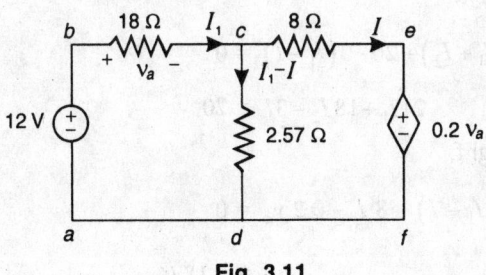

Fig. 3.11

For loop *abcda*

$$12 - 18 I_1 - 2.57 \left(I_1 - I \right) = 0$$

i.e.
$$20.57\,I_1 - 2.57\,I = 12 \tag{1}$$

For loop *dcefd*
$$(I_1 - I)2.57 - 8I - 0.2v_a = 0$$

but
$$v_a = 18\,I_1$$

$$2.57\,I_1 - 2.57\,I - 8I - 0.2(18I_1) = 0$$

i.e.
$$1.03\,I_1 + 10.57\,I = 0 \tag{2}$$

Solving Eqs (1) and (2), we get
$$I = -0.056 \text{ A}$$

When only 20 V is considered, the circuit in Fig. 3.9 may be written as in Fig. 3.12.

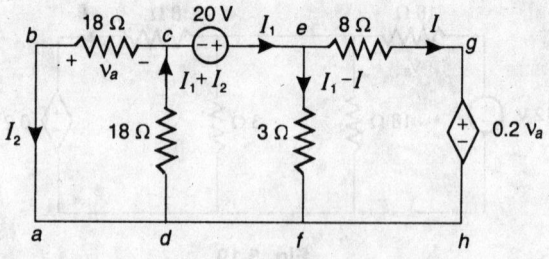

Fig. 3.12

For loop *abcda*
$$18\,I_2 + 18(I_1 + I_2) = 0$$

i.e.
$$18\,I_1 + 36\,I_2 + 0\,I = 0 \tag{1}$$

For loop *dcefd*
$$-18(I_1 + I_2) + 20 - 3(I_1 - I) = 0$$

i.e.
$$21\,I_1 + 18\,I_2 - 3I = 20 \tag{2}$$

For loop *feghf*,
$$3(I_1 - I) - 8I - 0.2\,v_a = 0$$

but
$$v_a = -18\,I_2$$

$$3(I_1 - I) - 8I - 0.2(-18I_2) = 0$$

i.e.
$$3I_1 + 3.6\,I_2 - 11I = 0 \tag{3}$$

Solving Eqs (1), (2), and (3), we get

$$I = -0.221 \text{ A}$$

Hence, the current flowing through $8\,\Omega$, when both independent sources are present $= 0.056 + 0.221 = 0.277$ A, from right to left.

3.3 Using superposition theorem, find the current flowing through load resistance $R_L = 10\,\Omega$ in the circuit shown in Fig. 3.13.

(Kuvempu University)

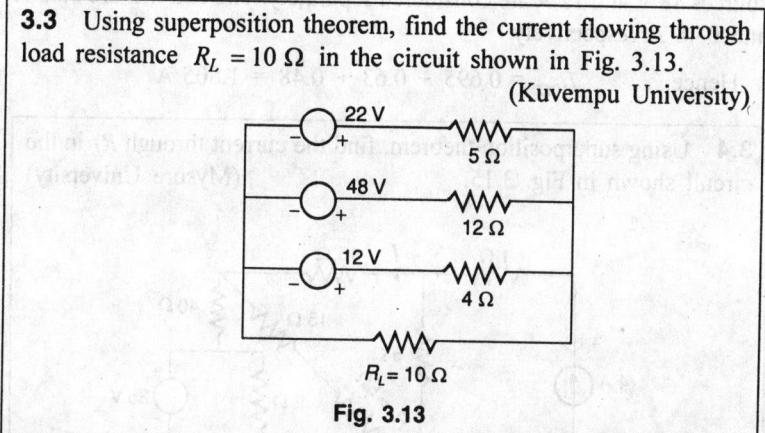

Fig. 3.13

Solution: Considering only 22 V, the circuit in Fig. 3.13 may be written as in Fig. 3.14.

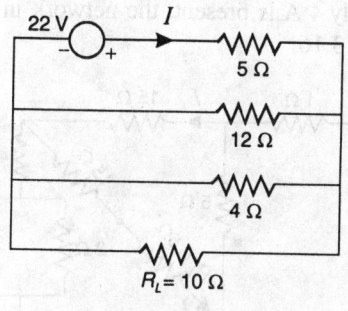

Fig. 3.14

$$R_T = \text{Total resistance} = 5 + \cfrac{1}{\cfrac{1}{12} + \cfrac{1}{4} + \cfrac{1}{10}}$$

$$= 7.31\,\Omega$$

$$\therefore \qquad I = \frac{22}{7.31} = 3.01 \text{ A}$$

$$\therefore \qquad I_{10\Omega} = \frac{3.01 \times \dfrac{12 \times 4}{12 + 4}}{10 + \dfrac{12 \times 4}{12 + 4}} = 0.695 \text{ A}$$

Similar procedure can be used to find currents through R_L, when sources 48 V and 12 V are considered separately. The currents are 0.63 A and 0.48 A respectively.

Hence, $\qquad I_{10\Omega} = 0.695 + 0.63 + 0.48 = 1.805$ A

3.4 Using superposition theorem, find the current through R_L in the circuit shown in Fig. 3.15. (Mysore University)

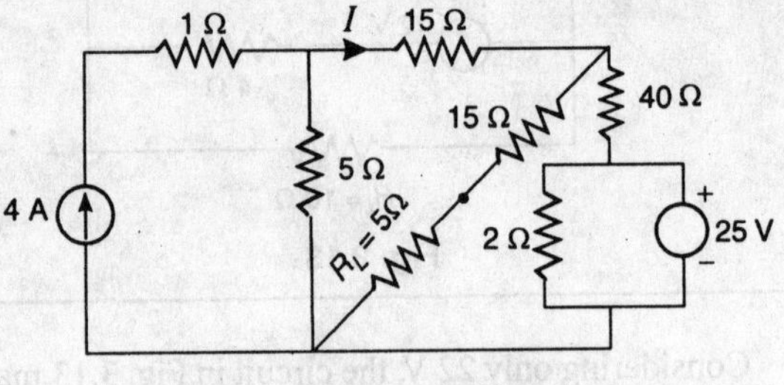

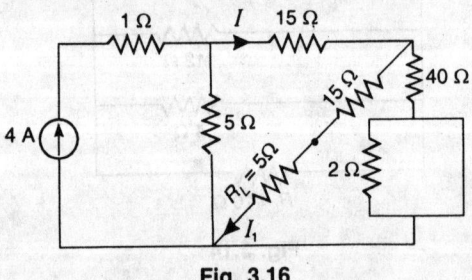

Fig. 3.15

Solution: When only 4 A is present, the network in Fig. 3.15 may be re-written as in Fig. 3.16.

Fig. 3.16

2 Ω is neglected, as it is across a short-circuit.

40 Ω ∥ 20 Ω = 13.33 Ω

13.33 Ω in series with 15 Ω = 28.33 Ω

$$I = 4 \times \frac{5}{5 + 28.33} = 0.6 \text{ A}$$

$$\therefore \qquad I_1 = 0.6 \times \frac{40}{40 + 20} = 0.4 \text{ A}$$

When only 25 V is considered, the network in Fig. 3.15 may be written as in Fig. 3.17.

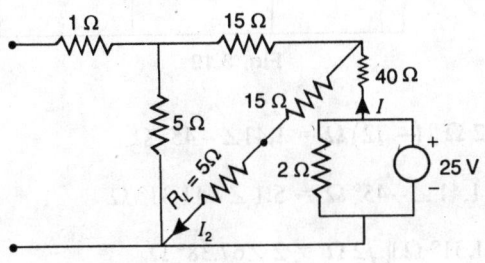

Fig. 3.17

2 Ω across 25 V can be neglected.

$$R_T = 40 + \frac{20 \times 20}{20 + 20} = 50 \text{ Ω}$$

$$\therefore \qquad I = \frac{25}{50} = 0.5 \text{ A}$$

$$\therefore \qquad I_2 = 0.5 \times \frac{20}{20 + 20} = 0.25 \text{ A}$$

Hence, $\qquad I_L = I_1 + I_2 = 0.4 + 0.25 = 0.65 \text{ A}$

3.5 In the circuit shown in Fig. 3.18, find I_x using superposition theorem. (Kuvempu University)

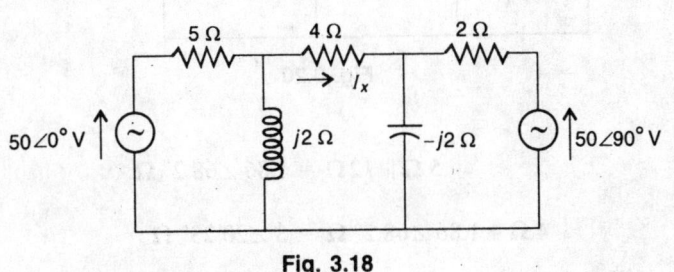

Fig. 3.18

Solution: Considering only $50 \angle 0° \text{ V}$, the circuit in Fig. 3.18 may be re-written as shown in Fig. 3.19.

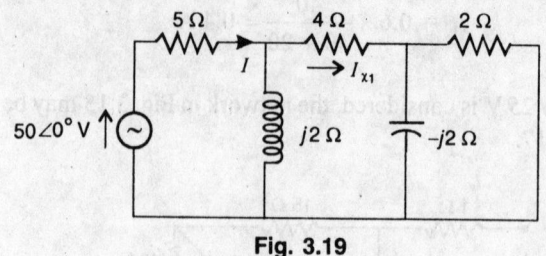

Fig. 3.19

$$2\,\Omega \,\|\,(-\,j2)\,\Omega = 1.41\angle-45°\,\Omega$$

$$4\,\Omega + 1.41\angle-45°\,\Omega = 5.1\angle-11.31°\,\Omega$$

$$5.1\angle-11.31°\,\Omega\,\|\,j2\,\Omega = 2\angle67.38°\,\Omega$$

$$Z_T = 5\,\Omega + 2\angle67.38°\,\Omega = 6.06\angle17.76°\,\Omega$$

$$\therefore \quad I = \frac{50}{6.06\,\angle17.78°\,\Omega} = 8.25\angle-17.78°\,\text{A}$$

$$\therefore \quad I_{x1} = 8.25\angle-17.78°\times\frac{j2}{j2 + 5.1\angle-11.31°} = 3.24\angle60.91°\,\text{A}$$

Considering only $50\angle90°$ V, the circuit in Fig. 3.18 may be re-written as shown in Fig. 3.20.

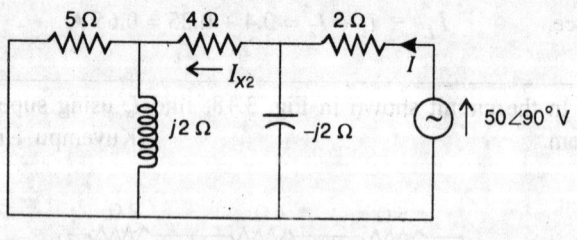

Fig. 3.20

$$5\,\Omega\,\|\,j2\Omega = 1.86\angle68.2°\,\Omega$$

$$4\,\Omega + 1.86\angle68.2°\,\Omega = 5\angle20.25°\,\Omega$$

$$-j\,2\,\Omega\,\|\,5\angle20.25°\,\Omega = 2.34\angle-66.46°\,\Omega$$

$$Z_T = 2\,\Omega + 2.34\angle-66.46°\,\Omega = 3.64\angle-36.18°\,\Omega$$

$$I = \frac{50\angle 90°}{3.64\angle -36.18°} = 13.74\angle 126.18° \text{ A}$$

$$I_{x2} = 13.74\angle 126.18° \times \frac{-j2}{-j2 + 5\angle 20.25°}$$

$$= 5.85\angle 39.47° \text{ A}$$

$$\therefore \qquad I_x = I_{x1} - I_{x2} = 3.24\angle 60.91° - 5.85\angle 39.47°$$

$$= 3.07\angle -163.12° \text{ A}$$

3.6 Using superposition theorem, find the voltage across $(4 + j3)$ Ω in the circuit shown in Fig. 3.21. (Mangalore University)

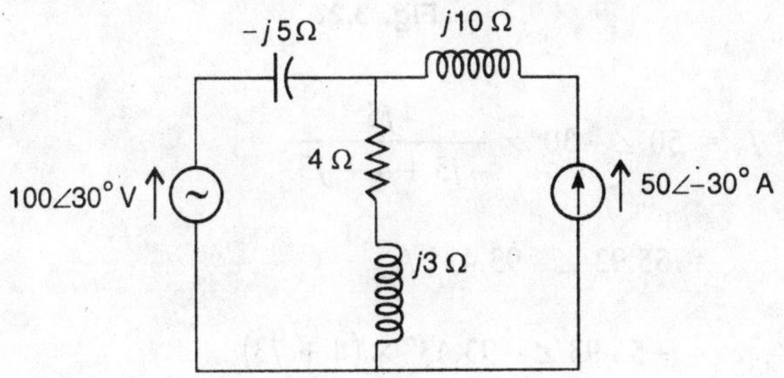

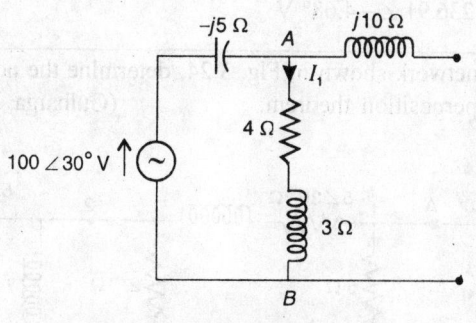

Fig. 3.21

Solution: Considering only $100\angle 30°$ V, the circuit in Fig. 3.21 may be re-written as in Fig. 3.22.

Fig. 3.22

$$I_1 = \frac{100\angle 30°}{4 - j2} = 22.37\angle 56.57° \text{ A}$$

$$V_{AB1} = +22.37\angle 56.57° \times (4 + j3)$$

$$= 111.85\angle -93.44° \text{ V}$$

Considering only $50 \angle -30°$ A, the circuit in Fig. 3.21 may be rewritten as in Fig. 3.23.

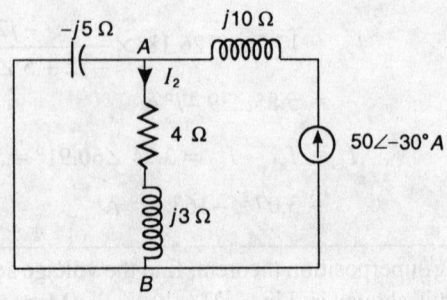

Fig. 3.23

$$I_2 = 50 \angle -30° \times \frac{-j5}{-j5 + 4 + j3}$$

$$= 55.93 \angle -93.43° \text{ A}$$

$$V_{AB2} = +55.93 \angle -93.43° \times (4 + j3)$$

$$= 279.65 \angle -29.56° \text{ V}$$

$$\therefore \quad V_{AB} = V_{AB1} + V_{AB2} = 111.85 \angle -93.44° + 279.65 \angle -29.56°$$

$$= 236.91 \angle -4.68° \text{ V}$$

3.7 In the network shown in Fig. 3.24, determine the node voltage V_2, using superposition theorem.　　　(Gulbarga University)

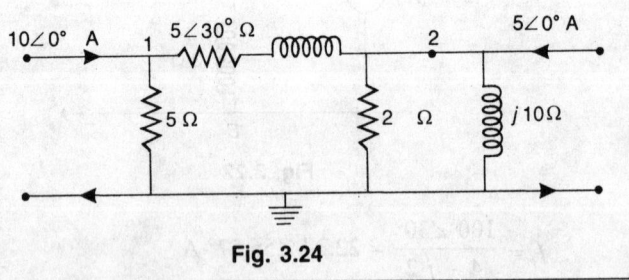

Fig. 3.24

Solution: Considering only $10 \angle 0°$ A, the network in Fig. 3.24 may be re-written as in Fig. 3.25.

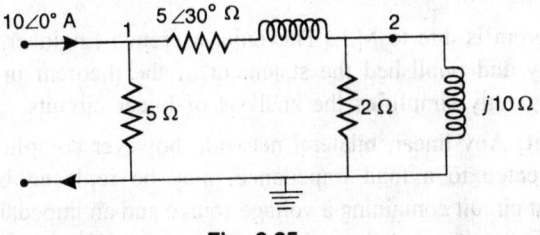

Fig. 3.25

The node voltage equations are:

$$\left(\frac{1}{5} + \frac{1}{5 \angle 30°}\right) V_1 - \frac{1}{5 \angle 30°} V_2 = 10 \angle 0° \tag{1}$$

$$-\frac{1}{5 \angle 30°} V_1 + \left(\frac{1}{2} + \frac{1}{j10} + \frac{1}{5 \angle 30°}\right) V_2 = 0 \tag{2}$$

Solving Eqs (1) and (2), we get

$$V_2 = 8.51 \angle -3.35° \text{ V}$$

Considering only $5 \angle 0°$ A, the network in Fig. 3.24 may be re-written as shown in Fig. 3.26.

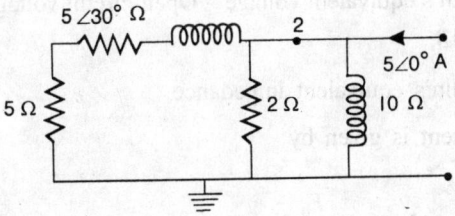

Fig. 3.26

The node voltage equation is:

$$\left(\frac{1}{5 + 5 \angle 30°} + \frac{1}{2} + \frac{1}{j10}\right) V_2 = 5$$

$$V_2 = 8.15 \angle 11.93° \text{ V}$$

∴

∴ When both sources are present

$$V_2 = 8.51 \angle -3.35° + 8.15 \angle 11.93° = 16.52 \angle 4.1° \text{ V}$$

3.3   THEVENIN'S THEOREM

This theorem is due to M.L. Thevenin, a French Engineer, working in telegraphy and published the statement of the theorem in 1883. This theorem greatly simplifies the analysis of linear circuits.

Statement: Any linear, bilateral network, however complicated it may be, connected to a load impedance, may be replaced by a simple equivalent circuit containing a voltage source and an impedance in series with it. The voltage of the voltage source is equal to the open-circuit voltage across the load terminals and the value of the impedance is equal to the equivalent impedance of the network as viewed from the load terminals into the network, replacing all the voltage sources by short-circuits or internal impedances and all current sources by open-circuits.

Figure 3.27 (a) represents a complicated network connected to a load impedance Z_L and Fig. 3.27, (b) represents Thevenin's equivalent circuit.

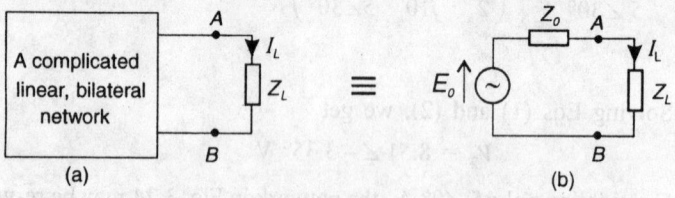

Fig. 3.27

E_0 = Thevenin's equivalent voltage = Open-circuit voltage across the load terminals *AB*.

Z_0 = Thevenin's equivalent impedance.

The load current is given by

$$I_L = \frac{E_0}{Z_0 + Z_L} \tag{3.2}$$

3.4   NORTON'S THEOREM

This theorem is due to E.L. Norton of Bell Telephone laboratories.

Statement: Any linear, bilateral network, however complicated it may be, connected to a load impedance, may be replaced by a simple equivalent circuit containing a current source and an impedance in parallel with it. The current of the current source is equal to the short-circuit current across the load terminals and the value of the impedance is equal to the equivalent impedance of the network as viewed from the load terminals into the network, replacing all the voltage sources by short-circuits or internal impedances and all current sources by open circuits.

Figure 3.28 (a) represents a complicated network connected to a load impedance Z_L and Fig. 3.28 (b) represents Norton's equivalent circuit.

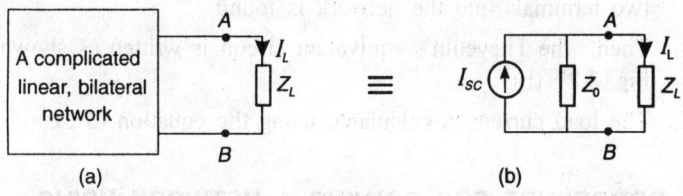

Fig. 3.28

The load current is given by

$$I_L = I_{SC} \frac{Z_0}{Z_0 + Z_L} \tag{3.3}$$

Equations (3.2) and (3.3), are the equations for the load current I_L and hence, are equal.

$$\therefore \qquad E_0 = I_{SC} \, Z_0 \tag{3.4}$$

or $\qquad I_{SC} = E_0 / Z_0 \tag{3.5}$

As for as the load impedance Z_L is concerned, the networks in 3.27 (b) and 3.28 (b) are similar.

Fig. 3.28

Thus, Thevenin's equivalent circuit can be converted into Norton's equivalent circuit by using (3.5) and Norton's equivalent circuit can be converted into Thevenin's equivalent circuit by using equation (3.4).

3.5 PROCEDURE FOR SOLVING A NETWORK USING THEVENIN'S THEOREM

1. The load impedance through which the current is to be found is removed and an open-circuit is created across its terminals.

2. The open-circuit voltage E_0 across these terminals is found.

3. All the voltage sources are replaced by short-circuits or their internal impedances and all the current sources are replaced by open-circuits. The equivalent impedance Z_0 as looked from these two terminals into the network is found.

4. Then, the Thevenin's equivalent circuit is written as shown in Fig. 3.27 (b).

5. The load current is calculated using the equation (3.2).

3.6 PROCEDURE FOR SOLVING A NETWORK USING NORTON'S THEOREM

1. The load impedance through which the current is to be found is removed and a short-circuit is created across its terminals.

2. The current I_{SC} through the short-circuit is found.

3. Z_0 is found as explained in section 3.5 (3).

4. The Norton's equivalent circuit is written as shown in Fig. 3.28 (b).

5. The load current is calculated using Eq. (3.3).

WORKED EXAMPLES

3.8 For the circuit shown in Fig. 3.29 (a), find (i) Thevenin's equivalent circuit, (ii) Norton's equivalent circuit and (iii) Power dissipated in a 5 Ω resistor connected between the terminals A and B. (Bangalore University)

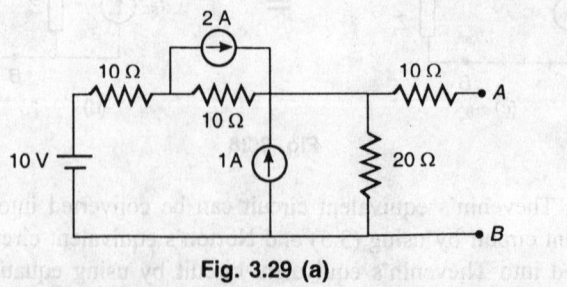

Fig. 3.29 (a)

Solution: No current flows through 10 Ω, connected to point A and hence, it can be neglected. The circuit in Fig. 3.29 (a) is re-written as in Fig. 3.29 (b).

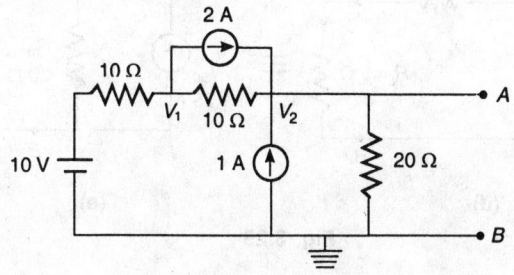

Fig. 3.29 (b)

Node voltages V_1 and V_2 are assumed as shown. V_2 is the open-circuit voltage across AB.

The node voltage equations are:

$$\left(\frac{1}{10} + \frac{1}{10}\right) V_1 - \frac{1}{10} V_2 = \frac{10}{10} - 2 = -1 \tag{1}$$

$$-\frac{1}{10} V_1 + \left(\frac{1}{10} + \frac{1}{20}\right) V_2 = 2 + 1 = 3 \tag{2}$$

Solving Eqs (1) and (2), we get

$$V_2 = E_0 = 25 \text{ V}$$

For finding R_0, the circuit in Fig. 3.29 (a) is re-written as in Fig. 3.29 (c).

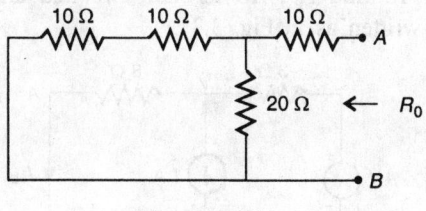

Fig. 3.29 (c)

$$R_0 = \frac{20 \times 20}{20 + 20} + 10 = 20 \ \Omega$$

The Thevenin's and Norton's equivalent circuits are written as in Fig. 3.29 (d) and (e) respectively.

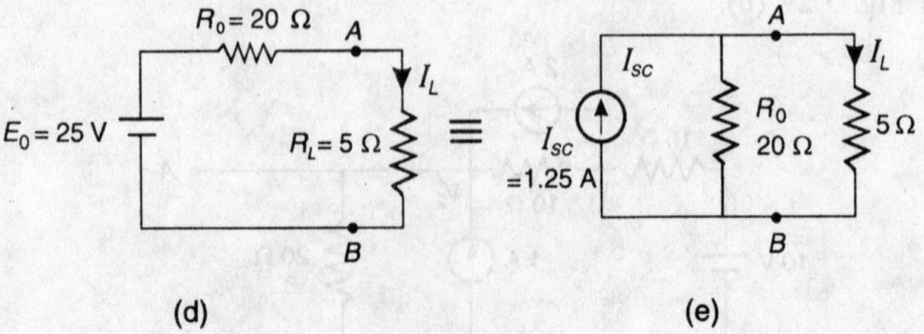

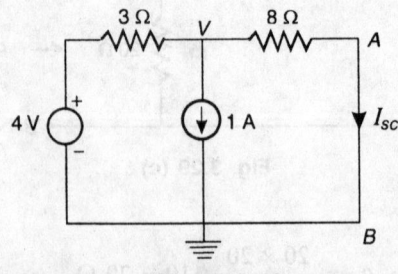

(d)   (e)

Fig. 3.29

$$I_{SC} = \frac{E_0}{R_0} = \frac{25}{20} = 1.25 \text{ A}, \qquad I_L = \frac{25}{20 + 5} = 1 \text{ A}$$

$$P_{5\Omega} = I^2 \times 5 = 5 \text{ watts}$$

3.9   Obtain Norton's equivalent circuit for the network shown in Fig. 3.30.   (Kuvempu University)

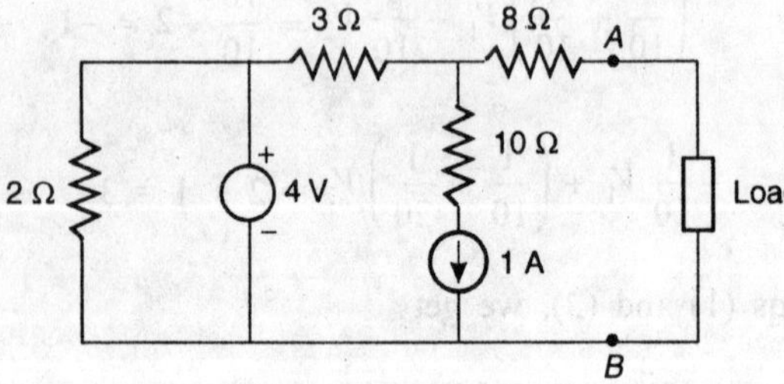

Fig. 3.30

Solution: 2 Ω connected across 4 V has no significance and hence, can be neglected. 10 Ω resistance in series with 1 A current source is also neglected. To find I_{SC}, AB is short-circuited and the circuit in Fig. 3.30 is re-written as in Fig. 3.31.

Fig. 3.31

The equation for V is

$$\left(\frac{1}{3} + \frac{1}{8}\right) V = \frac{4}{3} - 1 = \frac{1}{3}$$

From which $\quad V = 0.727$ V

$$\therefore \qquad I_{SC} = \frac{0.727}{8} = 0.0908 \text{ A}$$

$$R_0 = 8 + 3 = 11 \ \Omega$$

Hence, the Norton's equivalent circuit is as shown in Fig. 3.2.

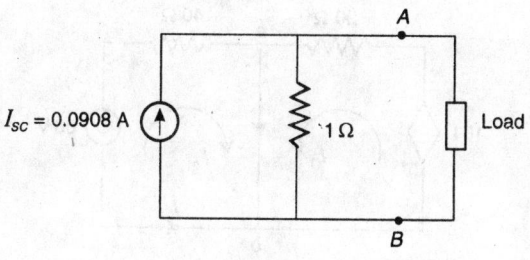

Fig. 3.32

3.10 Find Thevenin's and Norton's equivalent circuit across the terminals a and b for the circuit shown in Fig. 3.33.

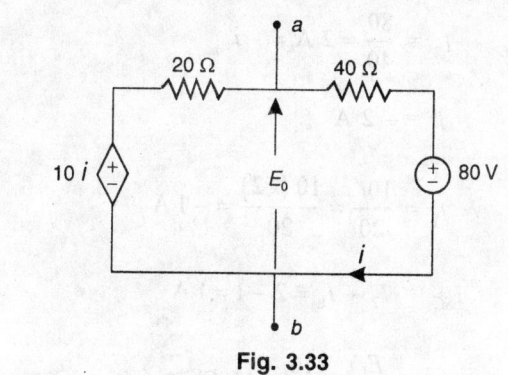

Fig. 3.33

Solution: Note that $10i$ is a dependent voltage source.

Applying Kirchhoff's voltage law to the circuit

$$80 + 60\,i - 10\,i = 0$$

$$\therefore \qquad i = -\frac{80}{50} = -1.6 \text{ A}$$

Then $\qquad 80 + 40\,i - E_0 = 0$

$$\therefore \qquad E_0 = 80 + 40\,(-1.6) = 16 \text{ V}$$

As there is dependent voltage source, R_0 cannot be found directly. I_{SC} is calculated.

When there is a short circuit between a and b, the circuit in Fig. 3.33 may be written as in Fig. 3.34 (a).

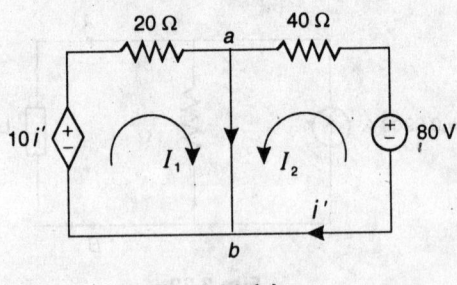

Fig. 3.34 (a)

Note that the current i changes to i'. Assume loop currents I_1 and I_2 as shown.

$$I_2 = \frac{80}{40} = 2 \text{ A} = -i'$$

$$\therefore \qquad i' = -2 \text{ A}$$

$$I_1 = \frac{10i'}{20} = \frac{10\,(-2)}{20} = -1 \text{ A}$$

$$\therefore \qquad I_{SC} = I_2 + I_1 = 2 - 1 = 1 \text{ A}$$

$$\therefore \qquad R_0 = \frac{E_0}{I_{SC}} = \frac{16}{1} = 16 \ \Omega$$

Thevenin's and Norton's equivalent circuits are as shown in Figs. 3.34 (b) and (c) respectively.

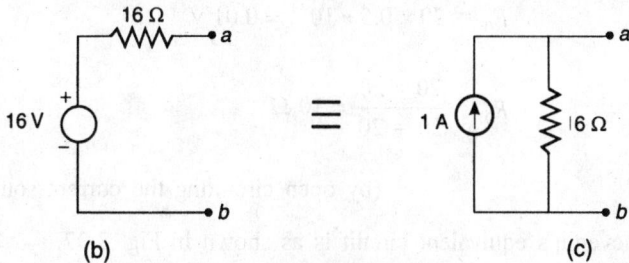

Fig. 3.34

3.11 The network shown in Fig. 3.35 is that of a light meter. P is a photocell and G is a galvanometer of conductance 10 siemens. If a known light intensity causes the photocell to produce a current of 1 mA, determine the current in the galvanometer using Thevenin's theorem. (Bangalore University)

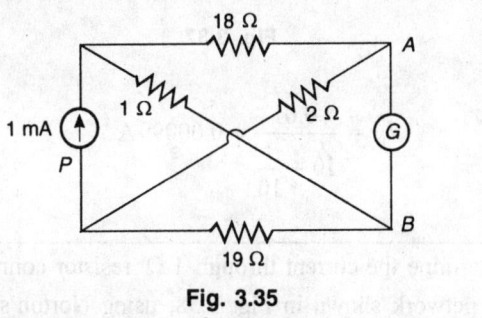

Fig. 3.35

Solution: To find E_0, ☐G☐ is removed and an open circuit is created. The circuit in Fig. 3.35 is written as in Fig. 3.36.

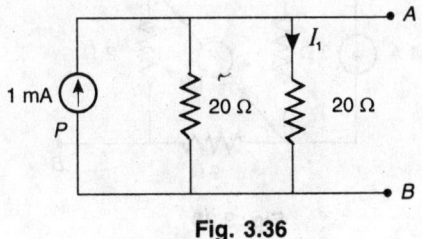

Fig. 3.36

$$I_1 = 1 \times \frac{20}{20 + 20} = 0.5 \text{ mA}$$

$$E_0 = 20 \times 0.5 \times 10^{-3} = 0.01 \text{ V}$$

$$R_0 = \frac{20 \times 20}{20 + 20} = 10 \ \Omega$$

(by open circuiting the current source)

∴ Thevenin's equivalent circuit is as shown in Fig. 3.37.

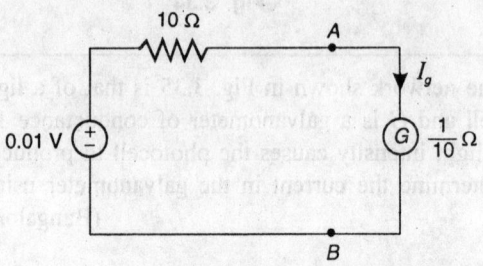

Fig. 3.37

$$I_g = \frac{0.01}{10 + \dfrac{1}{10}} = 0.00099 \text{ A}$$

3.12 Determine the current through $1 \ \Omega$ resistor connected across *AB* in the network shown in Fig. 3.38, using Norton's theorem.

(Bangalore Unviersity)

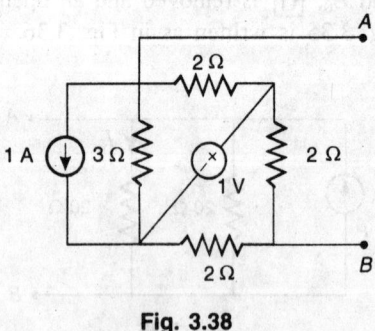

Fig. 3.38

Solution: The circuit in Fig. 3.38 is re-written after short circuiting *AB* in Fig. 3.39.

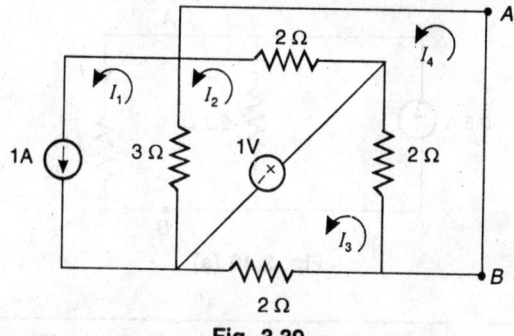

Fig. 3.39

The loop currents I_1, I_2, I_3 and I_4 are assumed as shown.
The loop equations are

$$-3I_1 + 5I_2 - 2I_4 = 1,$$

but $\qquad\qquad I_1 = 1 \text{ A}$

$\therefore \qquad\qquad 5I_2 - 2I_4 = 4 \qquad\qquad\qquad\qquad (1)$

$$4I_3 - 2I_4 = -1 \qquad\qquad\qquad\qquad (2)$$

$$-2I_2 - 2I_3 + 4I_4 = 0 \qquad\qquad\qquad\qquad (3)$$

Solving the above equations, we get

$$I_4 = I_{SC} = 0.5 \text{ A}$$

To find R_0, the circuit is as in Fig. 3.40.

$$R_0 = \frac{3 \times 2}{3 + 2} + \frac{2 \times 2}{2 + 2} = 2.2 \ \Omega$$

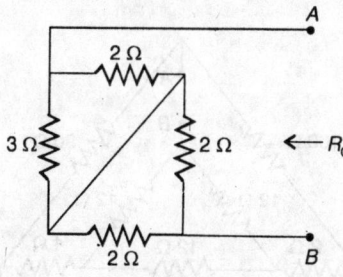

Fig. 3.40

Norton's equivalent circuit is shown in Fig. 3.40 (a),

$$I_L = 0.5 \times \frac{2.2}{2.2 + 1} = 0.344 \text{ A}$$

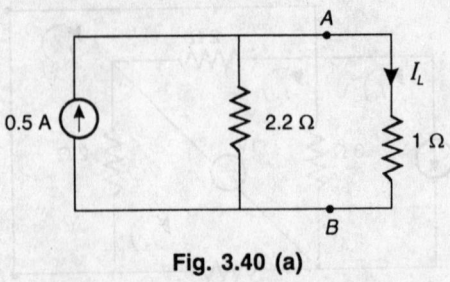

Fig. 3.40 (a)

3.13 Find the current through the galvanometer G of resistacne 50 Ω in the circuit shown in Fig. 3.41, using Thevenin's theorem.

(Bangalore University)

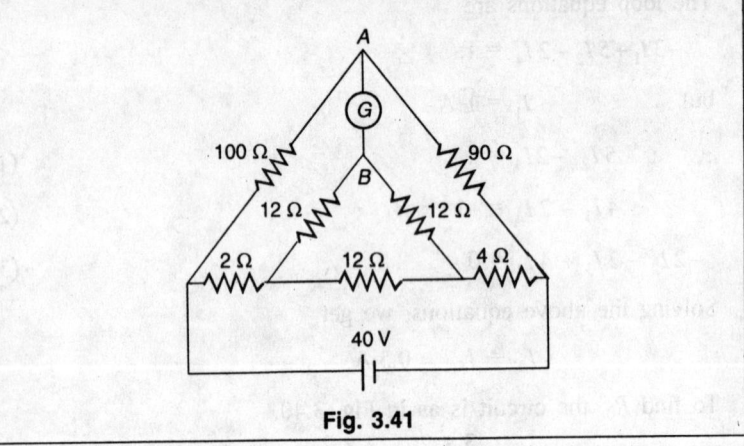

Fig. 3.41

Solution: The galvanometer is removed and an open circuit is created.

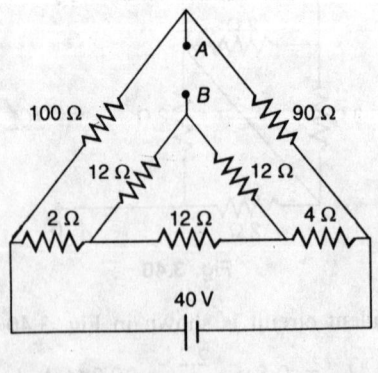

Fig. 3.42

The delta connection consisting of three 12 Ω resistances is connected into star network and the circuit is written as in Fig. 3.43.

No current flows through 4 Ω connected to B and hence, is neglected.

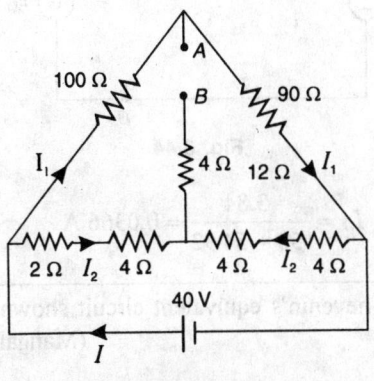

Fig. 3.43

The total resistance of the circuit is given by

$$R_T = \frac{190 \times 14}{190 + 14} = 13.04 \ \Omega,$$

$$I = \frac{40}{13.04} = 3.07 \ A$$

$$I_1 = 3.07 \times \frac{14}{190 + 14} = 0.21 \ A$$

$$I_2 = 3.07 - 0.21 = 2.86 \ A$$

$$-E_0 + 100 I_1 - 6I_2 = 0$$

i.e.

$$E_0 = 100 I_1 - 6I_2$$

$$= 100 \times 0.21 - 6 \times 2.86 = 3.84 \ V$$

$$R_0 = 4 + \frac{106 \times 98}{106 + 98} = 54.92 \ \Omega$$

The Thevenin's equivalent circuit is as in Fig. 3.44.

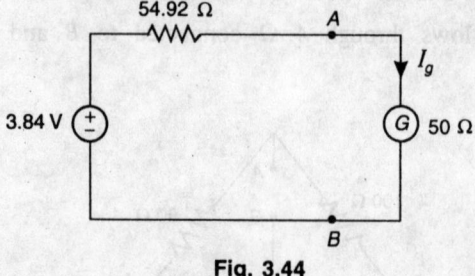

Fig. 3.44

$$\therefore \qquad I_g = \frac{3.84}{50 + 54.92} = 0.0366 \text{ A}$$

3.14 Find the Thevenin's equivalent circuit shown in Fig. 3.45 at the terminals *AB*.   (Mangalore University)

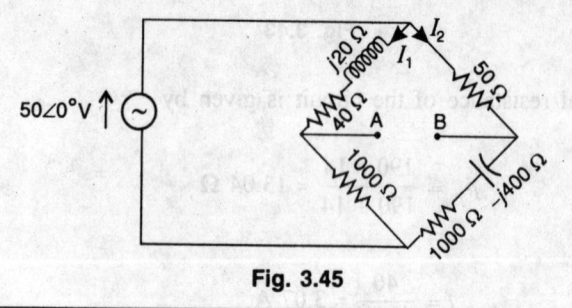

Fig. 3.45

Solution

$$I_1 = \frac{50 \angle 0°}{1040 + j20} = 0.0481 \angle -1.1° \text{ A}$$

$$I_2 = \frac{50 \angle 0°}{1050 - j400} = 0.0445 \angle 20.85° \text{ A}$$

$$E_0 = V_{AB} = 50 \, I_2 - (40 + j20) \, I_1$$

$$= 50 \times 0.0445 \angle 20.85° - 44.72 \angle 26.57° \times 0.0481 \angle -1.1°$$

$$= 0.191 \angle -44.15° \text{ V}$$

$$Z_0 = \frac{1000 \times (40 + j20)}{1040 + j20} + \frac{50 \, (1000 - j400)}{1050 - j400}$$

$$= 88.517 \angle 11.53° \, \Omega$$

Thevenin's equivalent circuit is as in Fig. 3.46.

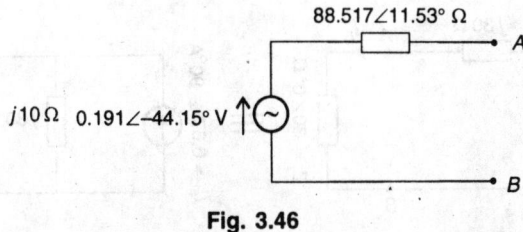

$j10\,\Omega$ $0.191\angle-44.15°\,V$

88.517∠11.53° Ω

Fig. 3.46

3.15 Obtain Norton's and Thevenin's equivalent for the circuit shown in Fig. 3.47 at the terminals AB and hence, find the current through the load impedance of $Z_L = 30 \angle 0° \, \Omega$.

(Kuvempu University)

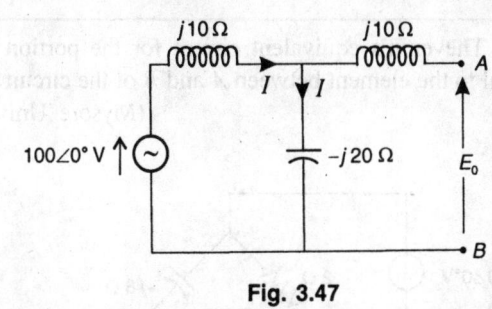

$j10\,\Omega$ $j10\,\Omega$

$100\angle 0°\,V$ $-j20\,\Omega$ E_0

Fig. 3.47

Solution

$$I = \frac{100 \angle 0°}{j10 - j20} = \frac{100}{-j10} = 10 \angle 90° \, A$$

$\therefore$

$$E_{AB} = E_0 = I \left(- j20\right)$$

$$= 10 \angle 90° \times - j20 = 200 \, V$$

$$Z_0 = j10 + \frac{j10 \left(- j20\right)}{j10 - j20} = j30 \, \Omega,$$

$$I_{SC} = \frac{E_0}{Z_0} = \frac{200}{j30} = 6.67 \angle - 90° \, A$$

The Thevenin's and Norton's equivalent circuits are as shown in Fig. 3.48 (a) and (b) respectively.

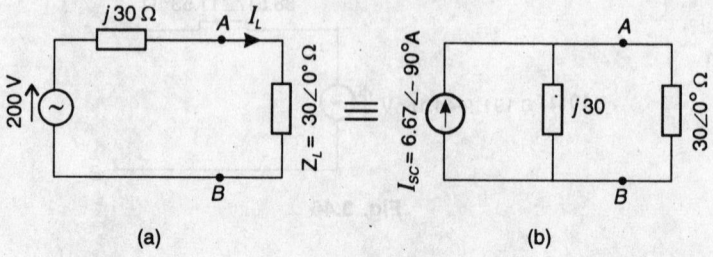

(a) (b)

Fig. 3.48

$$I_L = \frac{E_0}{Z_0 + Z_L} = \frac{200}{j\,30 + 30} = 4.71 \angle -45° \text{ A}$$

3.16 Find the Thevenin's equivalent circuit for the portion of the network external to the element between A and B of the circuit shown in Fig. 3.49. (Mysore University)

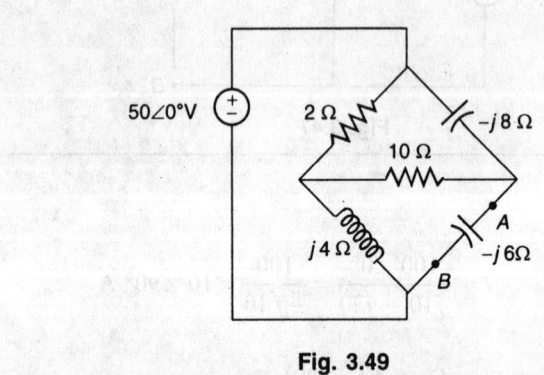

Fig. 3.49

Solution: To find E_0, $-j6\,\Omega$ is removed and O.C. is created between AB and the resulting circuit is as shown in Fig. 3.50.

Loop currents I_1 and I_2 are assumed as shown. The loop equations are:

$$(2 + j4)\,I_1 - 2\,I_2 = 50 \tag{1}$$

$$-2I_1 + (12 - j8)\,I_2 = 0 \tag{2}$$

Solving Eqs (1) and (2), we get

$$I_2 = 0.6045 \angle 29.74° \text{ A}$$

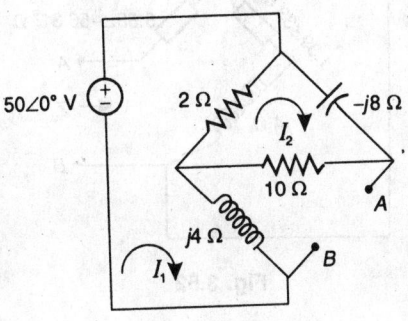

Fig. 3.50

Then $50 - (-j8) I_2 - E_0 = 0$

or $\qquad E_0 = 50 + j8 (0.6045 \angle 29.74°)$

$$= 30.60 \angle 38.83° \text{ V}$$

For finding Z_0, $50 \angle 0°$ V is short circuited and the circuit is as shown in Fig. 3.51.

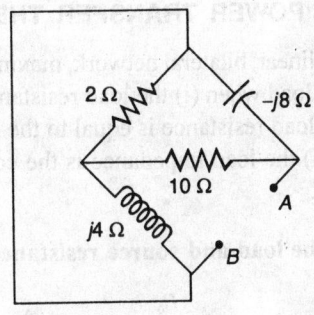

Fig. 3.51

Converting the delta network into star, the resulting network is as shown in Fig. 3.52.

$$Z_0 = 5.55 \angle - 56.31° + \frac{1.11 \angle - 56.31° \times (1.39 \angle 33.69° + j4)}{1.11 \angle - 56.31° + (1.39 \angle 33.69° + j4)}$$

$$= 6.83 \angle - 54.17° \; \Omega$$

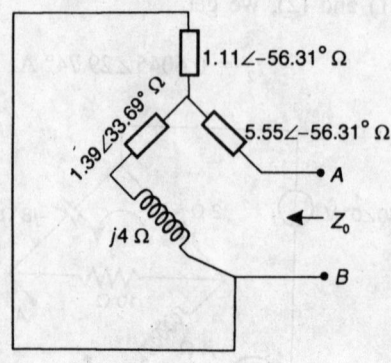

Fig. 3.52

Thevenin's equivalent circuit is as shown in Fig. 3.53.

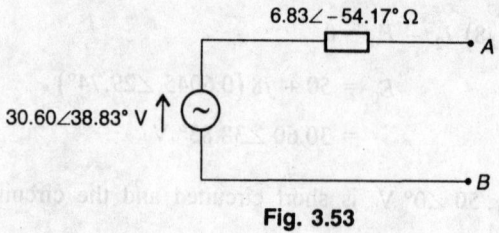

Fig. 3.53

3.7 MAXIMUM POWER TRANSFER THEOREM

Statement: In any linear, bilateral network, maximum power is transferred from source to the load when (i) the load resistance is equal to the source resistance, (ii) the load resistance is equal to the magnitude of the source impedance and (iii) the load impedance is the complex conjugate of the source impedance.

Proof: (i) When the load and source resistances are purely resistive.

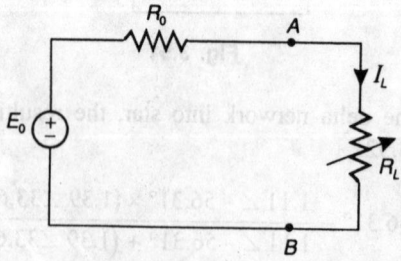

Fig. 3.54

From the circuit in Fig. 3.54.

$$I_L = \frac{E_0}{R_0 + R_L}$$

The power transferred from source to the load is given by

$$P = I_L^2 \, R_L = \frac{E_0^2}{\left(R_0 + R_L\right)^2} \, R_L$$

The power transferred is maximum, when

$$\frac{dp}{dR_L} = 0$$

$$\frac{dp}{dR_L} = \frac{\left(R_0 + R_L\right)^2 E_0^2 - E_0^2 \, R_L \times 2 \left(R_0 + R_L\right)}{\left(R_0 + R_L\right)^4} = 0$$

$$\therefore \qquad \left(R_0 + R_L\right)^2 - 2 R_L \left(R_0 + R_L\right) = 0$$

i.e. $\qquad R_L = R_0$

Hence, maximum power is transferred from source to the load, when $R_L = R_0$.

Under this condition $\qquad\qquad I_L = \dfrac{E_0}{R_0 + R_L} = \dfrac{E_0}{2 \, R_L}$

and maximum power transferred is given by

$$P_m = I_L^2 \, R_L = \frac{E_0^2}{4 \, R_L^2} \times R_L = \frac{E_0^2}{4 \, R_L} \tag{3.6}$$

(ii) When load is purely resistive and the source has impedance

Let $Z_0 = R_0 + jX_0$ be the internal impedance of the source E_0, as shown in the circuit of Fig. 3.55.

$$I_L = \frac{E_0}{R_L + R_0 + jX_0} = \frac{E_0}{\sqrt{\left(R_0 + R_L\right)^2 + X_0^2}}$$

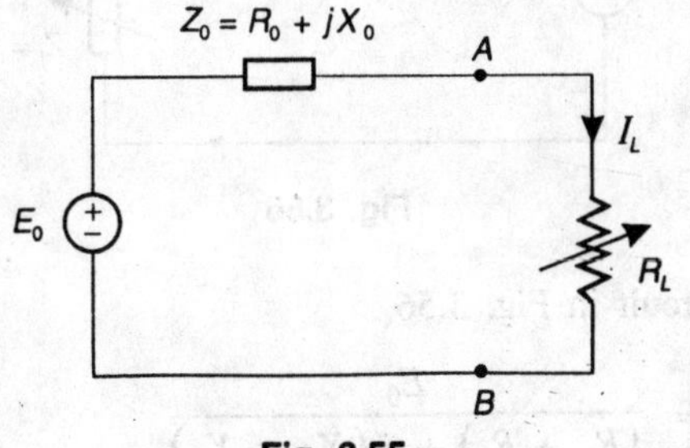

Fig. 3.55

The power consumed by the load is given by

$$P = I_L^2 \, R_L = \frac{E_0^2}{\left(R_0 + R_L\right)^2 + X_0^2} \, R_L$$

The power transferred is maximum, when

$$\frac{dP}{dR_L} = 0$$

$$\frac{dP}{dR_L} = \frac{\{\left(R_0 + R_L\right)^2 + X_0^2\} \, E_0^2 - E_0^2 \, R_L \, \{2\left(R_0 + R_L\right)\}}{\{\left(R_0 + R_L\right)^2 + X_0^2\}^2} = 0$$

$$\therefore \left(R_0 + R_L\right)^2 + X_0^2 = 2 R_L \left(R_0 + R_L\right), \text{ from which, we get}$$

$$R_L^2 = R_0^2 + X_0^2$$

or

$$R_L = \sqrt{R_0^2 + X_0^2} = |Z_0|$$

Thus, the power transferred from the source to the load is maximum, when

$$R_L = |Z_0|$$

$$I_L = \frac{E_0}{R_0 + j X_0 + R_L} \tag{3.7}$$

and

$$P_m = I_L^2 \, R_L \tag{3.8}$$

(iii) When both the load and the source have impedances

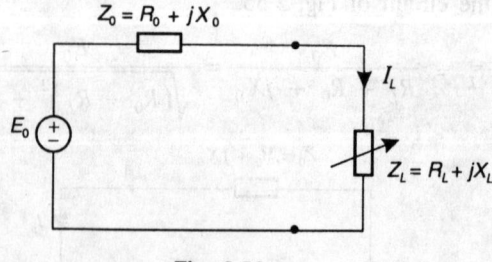

Fig. 3.56

From the circuit in Fig. 3.56.

$$I_L = \frac{E_0}{\left(R_L + R_0\right) + j\left(X_L + X_0\right)}$$

$$= \frac{E_0}{\sqrt{(R_0 + R_L)^2 + (X_L + X_0)^2}}$$

$$P = I_L^2 \, R_L = \frac{E_0^2}{(R_L + R_0)^2 + (X_L + X_0)^2} \, R_L$$

The power transferred is maximum, when

$$X_L = -X_0 \qquad\qquad\qquad (1)$$

$$\therefore \qquad P_1 = \frac{E_0^2}{(R_0 + R_L)^2} \, R_L$$

The power transferred is further maximum, when

$$\frac{dP_1}{dR_L} = 0$$

i.e. when, $R_L = R_0$ (2) as proved earlier in case (i)

From conditions (1) and (2), it is evident that the power transferred is maximum, when

$$R_L + jX_L = R_0 - jX_0$$

i.e. when, $\qquad Z_L = \overset{*}{Z}_0$

The maximum power transferred is given by,

$$P_m = \frac{E_0^2}{4 \, R_L} \qquad \left(\because I_L = \frac{E_0}{2R_L} \right)$$

<hr>

WORKED EXAMPLES

<hr>

3.17 Find the value of R for which the power transferred across AB of the circuit shown in Fig. 3.57 is maximum and the maximum power transferred.   (Kuvempu University)

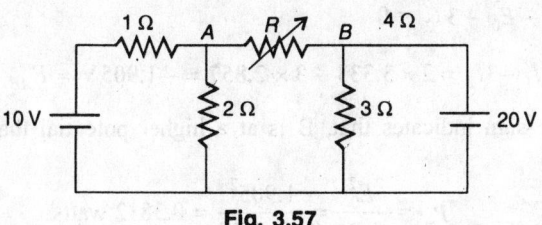

Fig. 3.57

Solution: To find R_0, an O.C. is created across *AB*, and voltage sources are short circuited as in Fig. 3.58.

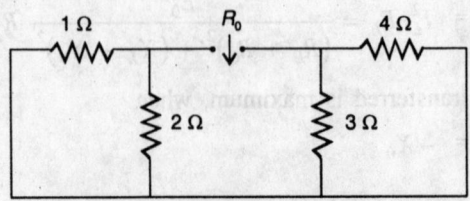

Fig. 3.58

$$R_0 = \frac{1 \times 2}{1 + 2} + \frac{3 \times 4}{3 + 4} = 2.38 \ \Omega$$

Hence, power transferred is maximum, when

$$R_L = R_0 = 2.38 \ \Omega$$

To find E_0, the circuit is as shown in Fig. 3.59.

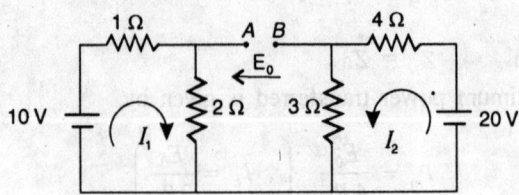

Fig. 3.59

$$I_1 = \frac{10}{3} = 3.333 \ \text{A}$$

$$I_2 = \frac{20}{7} = 2.857 \ \text{A}$$

Then, $2I_1 - E_0 - 3I_2 = 0$

$$\therefore \ E_0 = 2I_1 - 3I_2 = 2 \times 3.333 - 3 \times 2.857 = -1.905 \ \text{V} = E_{AB}$$

The –ve sign indicates that, B is at a higher potential than *A*.

$$P_m = \frac{E_0^2}{4R_L} = \frac{1.905^2}{4 \times 2.38} = 0.3812 \ \text{watts}$$

3.18 In the circuit shown in Fig. 3.60, find the maximum power received by 12 Ω resistance. (Gulbarga University)

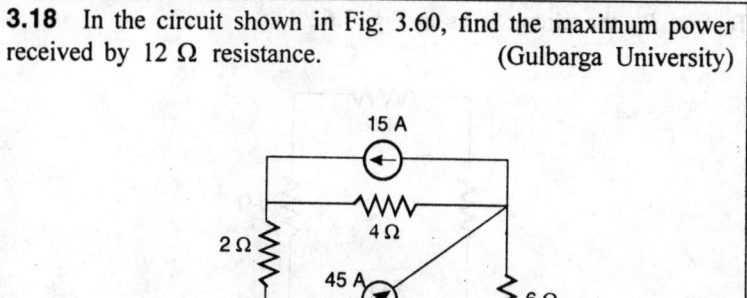

Fig. 3.60

Solution: To find E_0, an O.C. is created between AB and the circuit is as in Fig. 3.61.

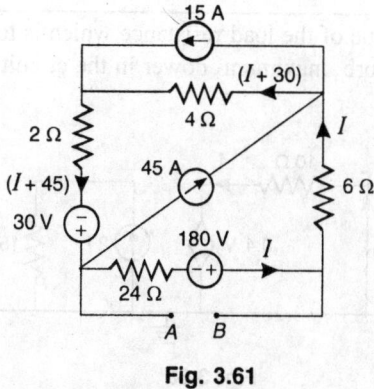

Fig. 3.61

The current distribution in the network is as shown. Then

$$-30\,I + 180 - 4\,(I + 30) - 2\,(I + 45) + 30 = 0$$

∴ $$I = 0$$

∴ $$E_0 = 180 \text{ V (B is at a higher potential than } A)$$

To find R_0, the circuit is as in Fig. 3.62.

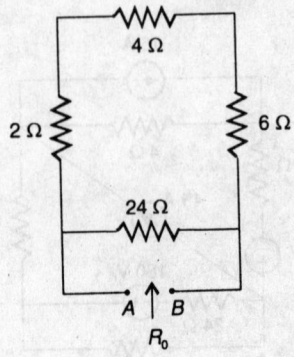

Fig. 3.62

$$R_0 = \frac{12 \times 24}{12 + 24} = 8\ \Omega$$

$$P_m = \frac{E_0^2}{4R_L} = \frac{180^2}{4 \times 8} = 1012.5\ \text{watts}$$

3.19 Find the value of the load resistance which is to be connected across PQ to absorb maximum power in the circuit shown in Fig. 3.63.

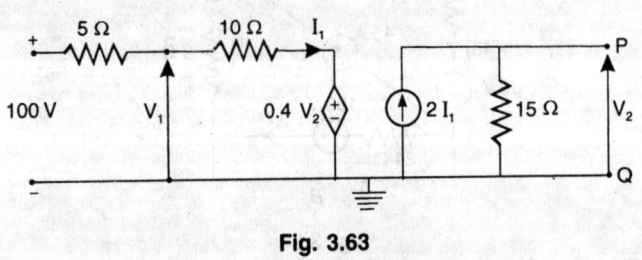

Fig. 3.63

Solution: From the first part of the circuit, we can write

$$100 - 15I_1 - 0.4\,V_2 = 0 \qquad \text{but} \qquad V_2 = 15 \times 2I_1$$

$$\therefore \qquad 100 - 15I_1 - 0.4 \times 30I_1 = 0$$

$$\therefore \qquad I_1 = \frac{100}{27} = 3.7\ \text{A}$$

$$\therefore \qquad V_2 = E_0 = 15 \times 2I_1 = 30 \times 3.7 = 111\ \text{volts}$$

When *PQ* is short-circuited, all the current 2 I_1 flows through the short-circuit.

$\therefore$ $\qquad I_{SC} = 2I_1 = 2 \times 3.7 = 7.4$ A

$\therefore$ $\qquad R_0 = \dfrac{E_0}{I_{SC}} = \dfrac{111}{7.4} = 15 \ \Omega = R_L$

3.20 Find the maximum power transferred to the load impedance Z_L shown in the circuit of Fig. 3.64. (Mysore University)

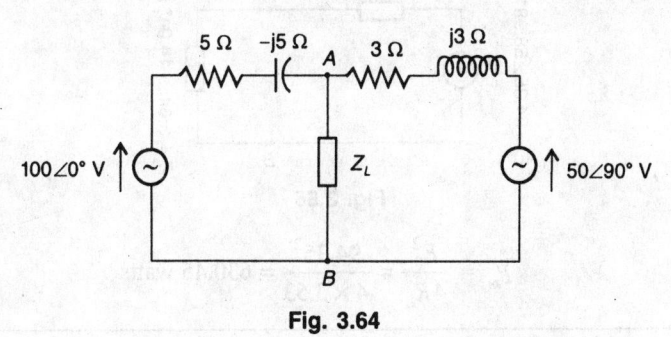

Fig. 3.64

Solution: Z_L is removed and O.C. is created as in Fig. 3.65.

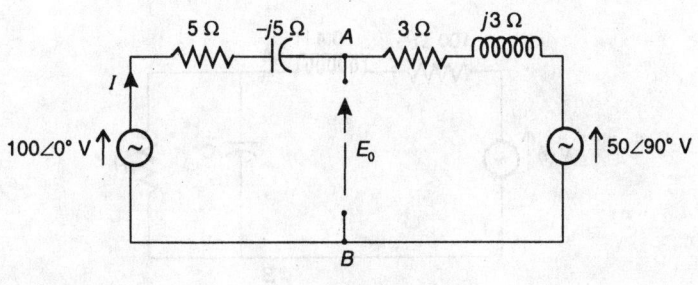

Fig. 3.65

$$I = \frac{100 \ \angle 0° - 50 \ \angle 90°}{8 - j2} = 13.55 \ \angle -12.57° \text{ A}$$

Then, $100 - (5 - j5) \ I_1 - E_0 = 0$

$\therefore$ $\qquad E_0 = 100 - (5 - j5) \times 13.55 \ \angle -12.57°$

$\qquad\qquad = 94.35 \ \angle 58.98° \text{ V}$

$$Z_0 = \frac{(5 - j5)(3 + j3)}{8 - j2} = 3.64 \angle 14.04° \ \Omega$$

For maximum power transfer

$$Z_L = \overset{*}{Z_0} = 3.64 \angle -14.04° \ \Omega$$

$$= (3.53 - j\,0.88) \ \Omega$$

Thevenin's equivalent circuit is as in Fig. 3.66.

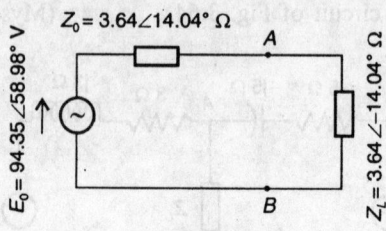

Fig. 3.66

$$P_m = \frac{E_0^2}{4R_L} = \frac{94.35^2}{4 \times 3.53} = 630.45 \text{ watts}$$

3.21 Find the values of R and C of the circuit shown in Fig. 3.67, such that the power delivered to R is maximum, and calculate this power. $v(t) = 250 \sin 500t$ volts. (Mangalore University)

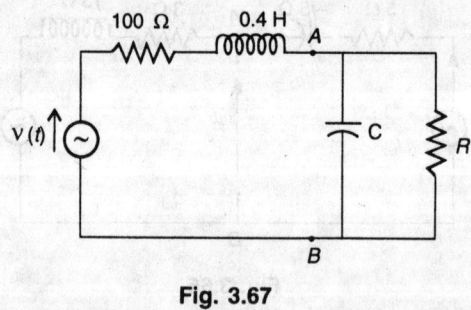

Fig. 3.67

Solution: Consider the load impedance Z_L to be the parallel combination of R and C.

$$\therefore \quad Z_L = \frac{R(-j\,X_C)}{R - j\,X_C} = \frac{-jR\,X_C}{(R - j\,X_C)} \times \frac{R + j\,X_C}{R + j\,X_C}$$

$$= \frac{R\,X_C^2}{R^2 + X_C^2} - j\,\frac{R^2\,X_C}{R^2 + X_C^2} \tag{1}$$

$$\therefore \qquad X_0 = \omega L = 500 \times 0.4 = 200\,\Omega$$

$$Z_0 = (100 + j\,200)\,\Omega$$

For maximum power transfer

$$\therefore \quad Z_L = \overset{*}{Z}_0 = (100 - j\,200)\,\Omega \tag{2}$$

Equating the real and imaginary parts of equations (1) and (2)

$$\frac{R\,X_C^2}{R^2 + X_C^2} = 100 \tag{3}$$

and

$$\frac{R^2\,X_C}{R^2 + X_C^2} = 200 \tag{4}$$

$$\therefore \qquad 2R\,X_C^2 = R^2\,X_C$$

or

$$2X_C = R$$

Substituting this in (3), we get

$$\frac{2X_C^3}{5X_C^2} = 100$$

or

$$X_C = 250\,\Omega$$

$$\therefore \qquad R = 500\,\Omega$$

$$C = \frac{1}{500 \times X_C} = \frac{1}{500 \times 250} = 8\,\mu F$$

3.22 Find the value of the resistance R_L, which receives maximum power of the circuit shown in Fig. 3.68, and find this power.

(Karnataka University)

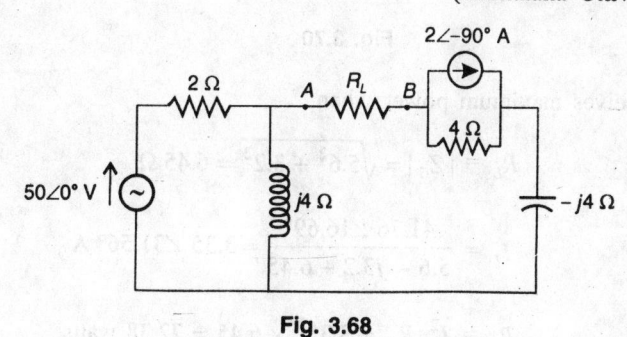

Fig. 3.68

Solution: The resistance R_L is removed and an o.c. is created between A and B. the current source is converted into voltage source and the circuit is re-written as in Fig. 3.69.

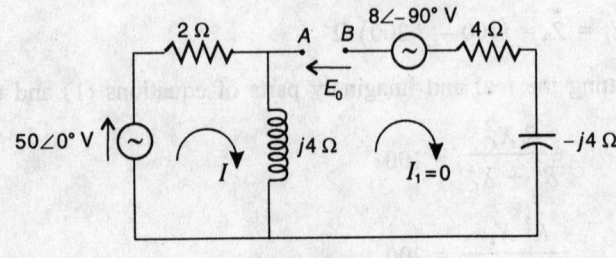

Fig. 3.69

$$I = \frac{50 \angle 0°}{2 + j4} = 1.19 \angle -63.43° \text{ A}$$

Then, $\qquad j\,4 \times I_1 - E_0 + 8 \angle -90° = 0$

$\therefore E_0 = j4 \times 1.19 \angle -63.43° + 8 \angle -90° = 41.76 \angle 16.69° \text{ V}$

$\therefore \qquad Z_0 = (4 - j4) + \dfrac{2 \times j4}{2 + j4} = (5.6 - j3.2) \,\Omega$

Thevenin's equivalent circuit is as in Fig. 3.70.

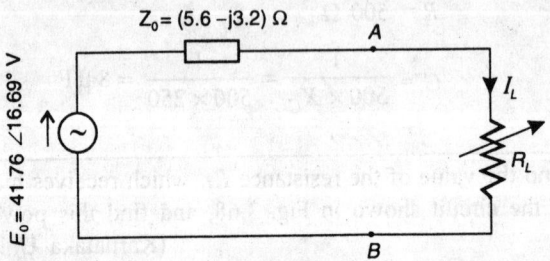

Fig. 3.70

R_L receives maximum power when

$$R_L = |Z_0| = \sqrt{5.6^2 + 3.2^2} = 6.45 \,\Omega$$

$$I_L = \frac{41.76 \angle 16.69°}{5.6 - j3.2 + 6.45} = 3.35 \angle 31.56° \text{ A}$$

$\therefore \qquad P_m = I_L^2\, R_L = 3.35^2 \times 6.45 = 72.38 \text{ watts}$

3.23 Determine the maximum power, which can be absorbed by a pure resistor, when placed across the terminals *AB* of the network shown in Fig. 3.71. (Bangalore University)

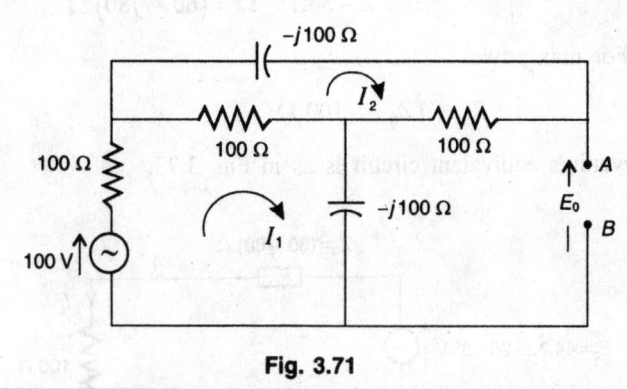

Fig. 3.71

Solution: Assume loop currents I_1 and I_2 as shown. The loop equations are

$$(200 - j100)\, I_1 - 100\, I_2 = 100 \tag{1}$$

$$-100\, I_1 + (200 - j100)\, I_2 = 0 \tag{2}$$

Solving Eqs (1) and (2), we get

$$I_1 = 0.5 \angle 36.86° \text{ A}$$

and $$I_2 = 0.2236 \angle 63.43° \text{ A}$$

Then, $-j\,100\, I_1 + 100\, I_2 - E_0 = 0$

∴ $$E_0 = 100 \times 0.2236 \angle 63.43° - j\,100 \times 0.5 \angle 36.86°$$

$$= 44.7 \angle -26.56° \text{ V}$$

To find Z_0, the voltage source is short-circuited and the star network consisting of 100 Ω, 100 Ω, and $-j\,100$ Ω is converted into delta. The resulting network is as in Fig. 3.72.

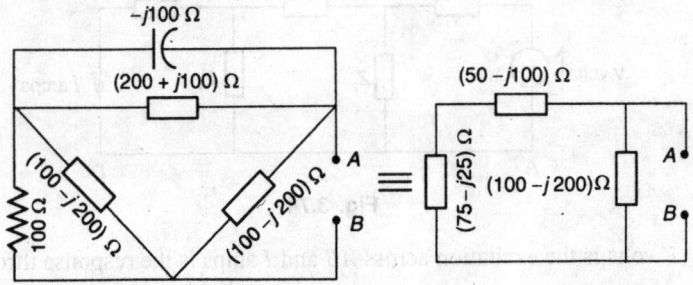

Fig. 3.72

$$Z_0 = \frac{(125 - j\,125)\,(100 - j200)}{125 - j\,125 + 100 - j200}$$

$$= 100 \angle -53.15° \; \Omega = (60 - j80) \; \Omega$$

∴ For max power

$$R_L = |Z_0| = 100 \; \Omega$$

Thevenin's equivalent circuit is as in Fig. 3.73.

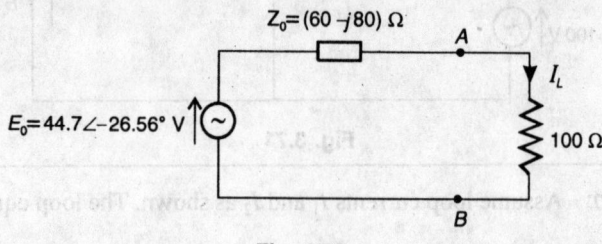

Fig. 3.73

$$I_L = \frac{44.7 \angle -26.56°}{(60 - j80) + 100} = 0.25 \; A$$

$$P_m = I_L^2 \; R_L = 0.25^2 \times 100 = 6.25 \; \text{watts}$$

3.8 RECIPROCITY THEOREM

Statement: In any linear, bilateral network containing only one independent source, the ratio of excitation to response remains constant, when their positions are interchanged.

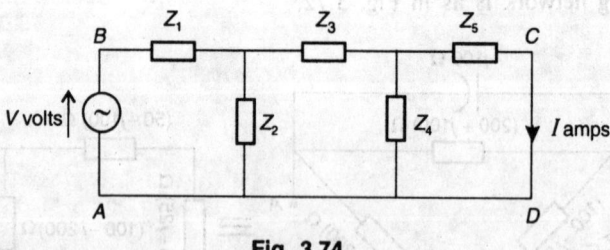

Fig. 3.74

V volts is the excitation across AB and I amps is the response through CD. The ratio of excitation to response is $\dfrac{V}{I}$.

Now, interchange the position of excitation and response as shown in Fig. 3.75.

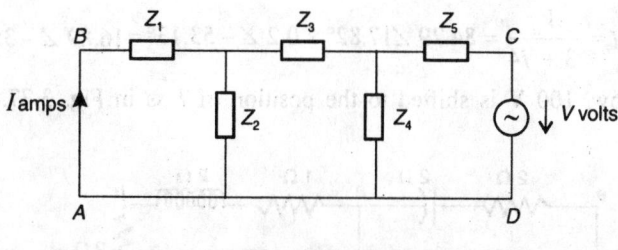

Fig. 3.75

Then, according to reciprocity theorem, the ratio of excitation to response, i.e. $\dfrac{V}{I}$ remains the same. In otherwords, if a voltage of V volts across AB produces a current of I amperes through CD, then, if the voltage of V volts is placed across CD, it produces the same current I through AB.

WORKED EXAMPLES

3.24 Verify Reciprocity theorem by finding the current through the branch bc in the circuit shown in Fig. 3.76.

(Mysore University)

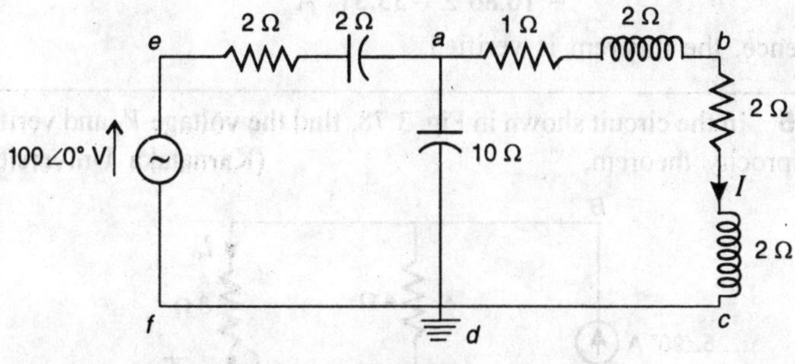

Fig. 3.76

Solution: Taking d as reference node, the nodal equation at a is

$$\left(\frac{1}{2 - j2} + \frac{1}{-j10} + \frac{1}{3 + j4}\right) V_a = \frac{100}{2 - j2}$$

from which

$$V_a = 84.29 \angle 17.82° \text{ V}$$

$$I = \frac{V_a}{3 + j4} = 84.29 \angle 17.82° \times 0.2 \angle -53.13° = 16.89 \angle -35.31° \text{ A}$$

Now, 100 V is shifted to the position of I as in Fig. 3.77.

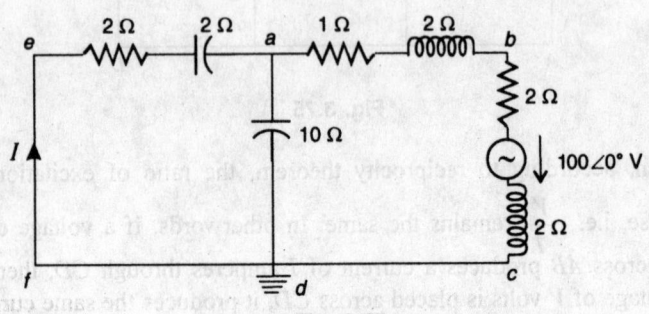

Fig. 3.77

The nodal equation at a is

$$\left(\frac{1}{2 - j2} + \frac{1}{-j10} + \frac{1}{3 + j4} \right) V_a = -\frac{100}{3 + j4} = -20.2 \angle -53.13°$$

$$\therefore \qquad V_a = -47.62 \angle -80.31° \text{ V}$$

$$I = -\frac{V_a}{2 - j2} = \frac{47.62 \angle -80.31°}{2 - j2}$$

$$= 16.86 \angle -35.31° \text{ A}$$

Hence, the theorem is verified.

3.25 In the circuit shown in Fig. 3.78, find the voltage V_x and verifiy reciprocity theorem. (Karnataka University)

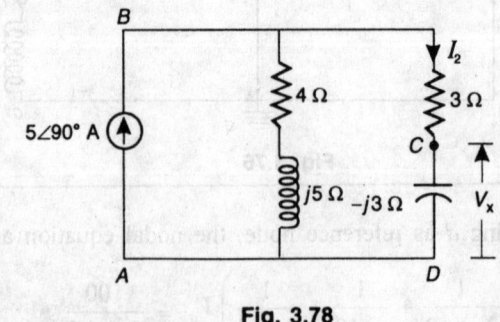

Fig. 3.78

Solution:
$$I_2 = 5 \angle 90° \times \frac{4 + j5}{7 + j2} = 4.4 \angle 125.39° \text{ A}$$

$$V_x = I_2 \, (-j3) = 13.2 \angle 35.39° \text{ V}$$

Now, interchange the positions of $5 \angle 90°$ A and V_x and the circuit is as in Fig. 3.79.

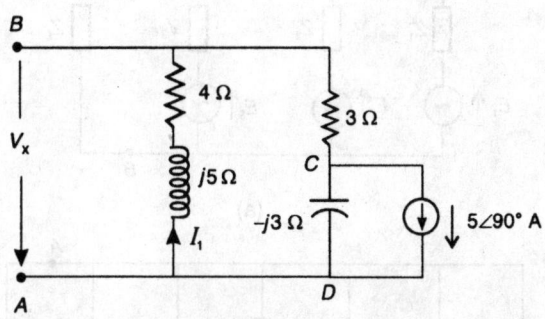

Fig. 3.79

$$I_1 = 5 \angle 90° \times \frac{-j3}{7 + j2} = 2.06 \angle -15.95° \text{ A}$$

$$V_x = (4 + j5) \, I_1 = 13.2 \angle 35.39° \text{ V}$$

3.9 MILLMAN'S THEOREM

Statement: Several voltage sources $E_1, E_2, E_3 \ldots E_n$ with their internal impedances $Z_1, Z_2, Z_2 \ldots Z_m$, connected in parallel may be replaced by a single voltage source E with the internal impedance Z, where

$$E = \frac{\dfrac{E_1}{Z_1} + \dfrac{E_2}{Z_2} + \dfrac{E_3}{Z_3} + \ldots + \dfrac{E_n}{Z_n}}{\dfrac{1}{Z_1} + \dfrac{1}{Z_2} + \dfrac{1}{Z_3} + \ldots + \dfrac{1}{Z_n}}$$

$$= \frac{E_1 Y_1 + E_2 Y_2 + E_3 Y_3 + \ldots + E_n Y_n}{Y_1 + Y_2 + Y_3 + \ldots + Y_n} = \frac{\Sigma EY}{\Sigma Y}$$

and
$$\frac{1}{Z} = \frac{1}{Z_1} + \frac{1}{Z_2} + \frac{1}{Z_3} + \ldots + \frac{1}{Z_n}$$

or
$$Y = Y_1 + Y_2 + Y_3 + \ldots + Y_n = \Sigma Y$$

Proof: Consider the circuit as shown in Fig. 3.80 (a), where three voltage sources E_1, E_2 and E_3 along with their internal impedances Z_1, Z_2, and Z_3 are connected in parallel across the load impedance Z_L. The voltage sources are converted into current sources and the circuit is rewritten as in Fig. 3.80 (b).

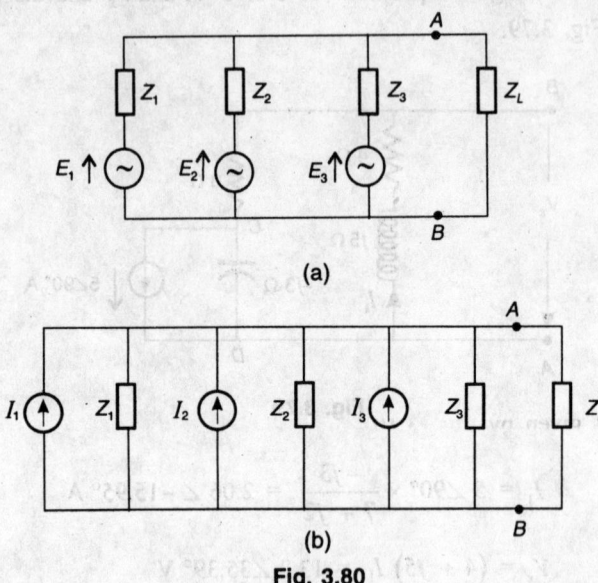

(a)

(b)

Fig. 3.80

Where, $I = I_1 + I_2 + I_3$ and $\dfrac{1}{Z} = \dfrac{1}{Z_1} + \dfrac{1}{Z_2} + \dfrac{1}{Z_3}$

Figure 3.81 (a) is equivalent to Fig. 3.81 (b)

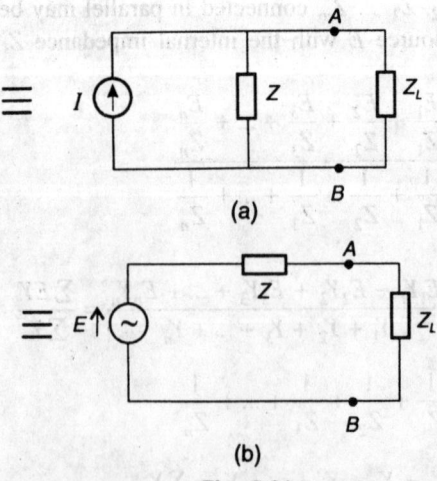

(a)

(b)

Fig. 3.81

Where $E = IZ = \dfrac{I_1 + I_2 + I_3}{\dfrac{1}{Z}}$

$$= \dfrac{\dfrac{E_1}{Z_1} + \dfrac{E_2}{Z_2} + \dfrac{E_3}{Z_3}}{\dfrac{1}{Z_1} + \dfrac{1}{Z_2} + \dfrac{1}{Z_3}} = \dfrac{E_1 Y_1 + E_2 Y_2 + E_3 Y_3}{Y_1 + Y_2 + Y_3} \qquad (3.9)$$

and $\qquad \dfrac{1}{Z} = \dfrac{1}{Z_1} + \dfrac{1}{Z_2} + \dfrac{1}{Z_3} \qquad\qquad (3.10)$

i.e. $\qquad Y = Y_1 + Y_2 + Y_3 \qquad\qquad\qquad\qquad (3.11)$

Thus, three voltage sources can be replaced by a single voltage source, whose voltage is given by Eq. (3.9) and whose internal impedance or admittance is given by Eqs (3.10) and (3.11) respectively.

WORKED EXAMPLES

3.26 Find the output voltage E_0 of the circuit shown in Fig. 3.82 (a), using Millman's theorem. (Bangalore University)

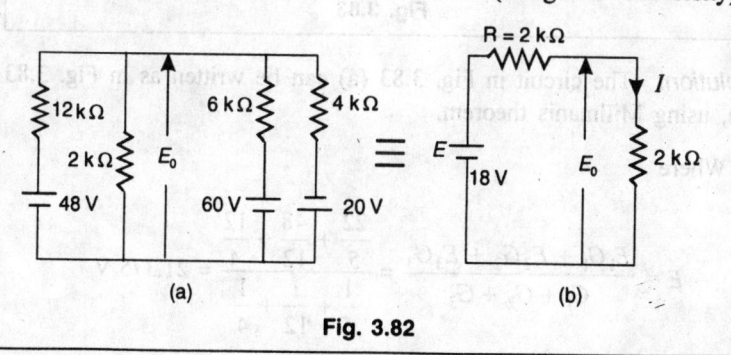

Fig. 3.82

Solution:

$$E = \dfrac{E_1 Y_1 + E_2 Y_2 + E_3 Y_3}{Y_1 + Y_2 + Y_3} = \dfrac{\dfrac{48}{12 \times 10^3} + \dfrac{60}{6 \times 10^3} - \dfrac{20}{4 \times 10^3}}{\dfrac{1}{12 \times 10^3} + \dfrac{1}{6 \times 10^3} + \dfrac{1}{4 \times 10^3}} = 18\ \text{V}$$

$$\frac{1}{R} = \frac{1}{12} + \frac{1}{6} + \frac{1}{4} = \frac{1}{2} \, k\Omega$$

or
$$R = 2 \, k\Omega$$

The equivalent circuit is as shown in Fig. 3.82 (b)

$$I = \frac{18}{4 \times 10^3} = 4.5 \times 10^{-3} \, A$$

∴
$$E_0 = 18 - 2 \times 10^3 \times 4.5 \times 10^{-3} = 9 \, V$$

3.27 Using Millman's theorem, find the current through the load resistance $R_L = 100 \, \Omega$ in the circuit of Fig. 3.83 (a).

(Mysore University)

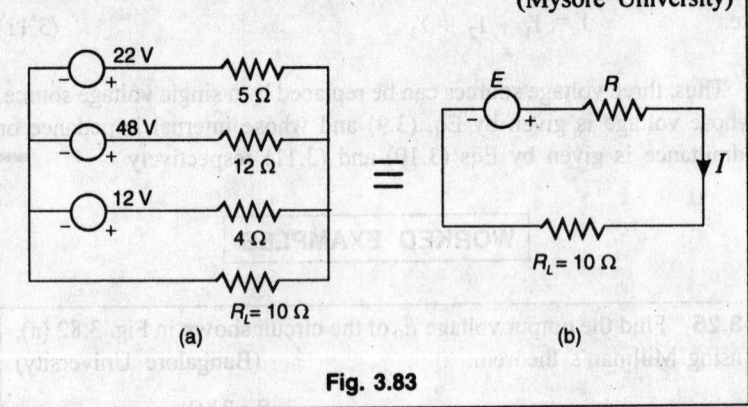

Fig. 3.83

Solution: The circuit in Fig. 3.83 (a) can be written as in Fig. 3.83 (b), using Millmanis theorem.

Where

$$E = \frac{E_1 G_1 + E_2 G_2 + E_3 G_3}{G_1 + G_2 + G_3} = \frac{\dfrac{22}{5} + \dfrac{48}{12} + \dfrac{12}{4}}{\dfrac{1}{5} + \dfrac{1}{12} + \dfrac{1}{4}} = 21.375 \, V$$

$$G = G_1 + G_2 + G_3 = \frac{1}{5} + \frac{1}{12} + \frac{1}{4} = 0.53333 \, \mho$$

∴
$$R = 1.875 \, \Omega$$

∴
$$I = \frac{E}{R + R_L} = \frac{21.375}{1.875 + 10} = 1.8 \, A$$

3.28 Using Millman's theorem, find the current flowing through $(4 + j\,3)\,\Omega$ of the circuit shown in Fig. 3.84 (a).

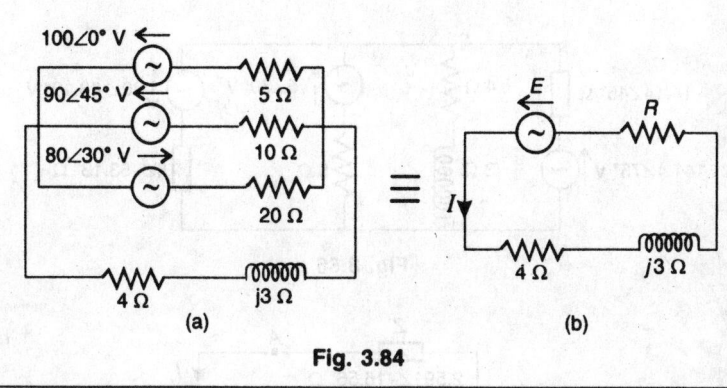

Fig. 3.84

Solution: The equivalent circuit is shown in Fig. 3.84 (b).

$$E = \frac{E_1 G_1 + E_2 G_2 + E_3 G_3}{G_1 + G_2 + G_3} = \frac{\dfrac{100\angle 0°}{5} + \dfrac{90\angle 45°}{10} - \dfrac{80\angle 30°}{20}}{\dfrac{1}{5} + \dfrac{1}{10} + \dfrac{1}{20}}$$

$$= 66.6\angle 10.78° \text{ V}$$

$$R = \frac{1}{G} = \frac{1}{G_1 + G_2 + G_3} = \frac{1}{\dfrac{1}{5} + \dfrac{1}{10} + \dfrac{1}{20}} = 2.86\ \Omega$$

$$\therefore \quad I = \frac{E}{R + (4 + j3)} = \frac{66.6\angle 10.78°}{6.86 + j3} = 8.89\ \angle -12.84°\text{ A}$$

3.29 Using Millman's theorem, find the current flowing through $(4 + j3)\,\Omega$ in the circuit of Fig. 3.85.

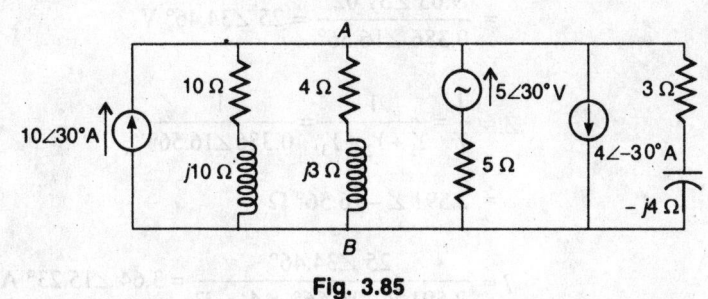

Fig. 3.85

Solution: The two current sources are converted into voltage sources and the circuit is re-written as in Fig. 3.86.

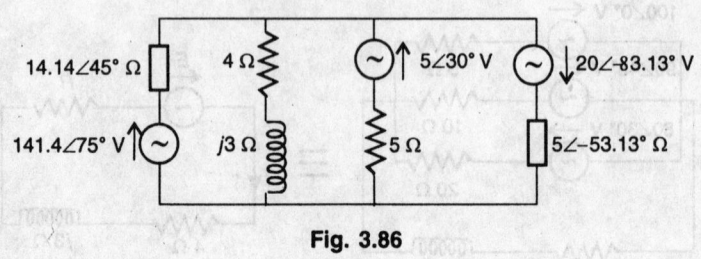

Fig. 3.86

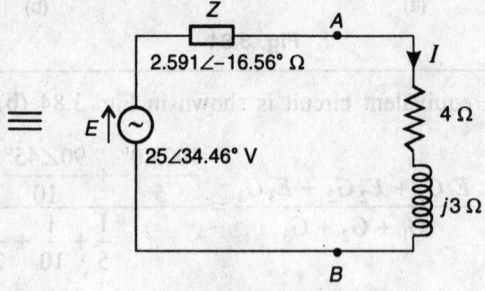

Fig. 3.87

Where,
$$E = \frac{E_1 Y_1 + E_2 Y_2 - E_3 Y_3}{Y_1 + Y_2 + Y_3}$$

$$= \frac{\dfrac{141.4\angle 75°}{14.14\angle 45°} + \dfrac{5\angle 30°}{5} - \dfrac{20\angle -83.13°}{5\angle -53.13°}}{\dfrac{1}{14.14\angle 45°} + \dfrac{1}{5} + \dfrac{1}{5\angle -53.13°}}$$

$$= \frac{9.65\angle 51.02°}{0.386\angle 16.56°} = 25\angle 34.46° \text{ V}$$

$$Z = \frac{1}{Y} = \frac{1}{Y_1 + Y_2 + Y_3} = \frac{1}{0.386\angle 16.56°}$$

$$= 2.591\angle -16.56° \ \Omega$$

$$\therefore \quad I = \frac{25\angle 34.46°}{2.591\angle -16.56° + 4 + j3} = 3.64\angle 15.23° \text{ A}$$

3.10 COMPENSATION THEOREM (SUBSTITUTION THEOREM)

Statement: In any linear, bilateral network, the voltage drop across any impedance may be replaced by a compensation e.m.f., whose magnitude and phase is equal to the voltage drop across the impedance.

Consider an impedance Z as in Fig. 3.88 (a), through which a current I is flowing and according to compensation theorem, the voltage drop IZ can be replaced by a compensation e.m.f. V_C as shown in Fig. 3.88 (b).

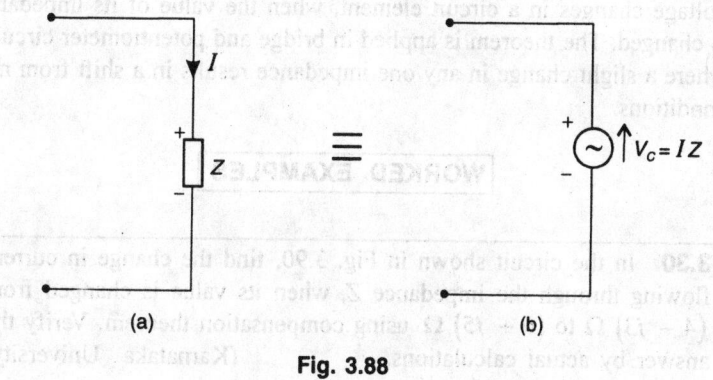

(a)   (b)

Fig. 3.88

Care must be taken to see that the polarities of IZ and $\mathbf{V}_C$ are the same. V_C is a dependent voltage source.

If any change in the impedance Z occurs, which in turn changes the current I, then the compensation source also must be changed accordingly.

Consider an impedance Z connected across V volts, and let I be the current flowing as shown in Fig. 3.89 (a).

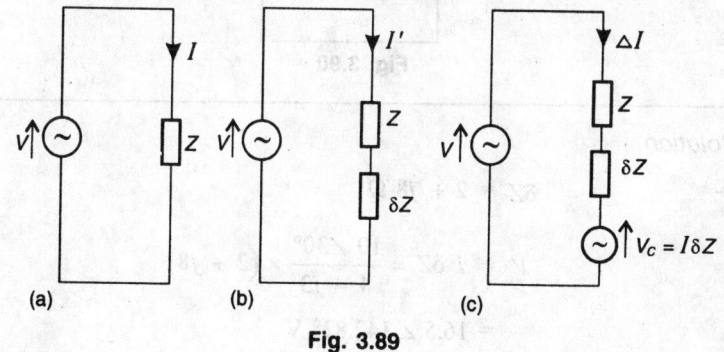

(a)   (b)   (c)

Fig. 3.89

Now, let the impedance be changed to $Z + \delta Z$ and the current flowing is $I' = \dfrac{V}{Z + \delta Z}$ as shown in Fig. 3.89 (b).

The change in current is $\Delta I = I' - I$. The changes that take place in the circuit due to change in the impedance is shown in Fig. 3.2 (c), setting the original voltage source to zero, by inserting a compensation e.m.f.

$$V_C = I \ \delta Z. \quad \text{Then,} \quad \Delta I = I' - I = -\frac{V_c}{Z + \delta Z}$$

The compensation theorem is useful in determining the current and voltage changes in a circuit element, when the value of its impedance is changed. The theorem is applied in bridge and potentiometer circuits, where a slight change in any one impedance results in a shift from null conditions.

WORKED EXAMPLES

3.30 In the circuit shown in Fig. 3.90, find the change in current flowing through the impedance Z, when its value is changed from $(4 - j3)$ Ω to $(6 + j5)$ Ω using compensation theorem. Verify the answer by actual calculations. (Karnataka University)

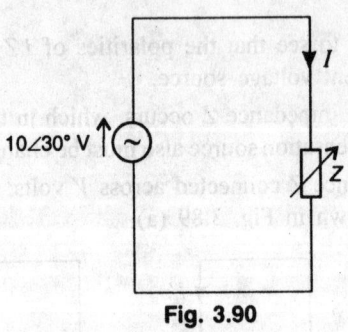

Fig. 3.90

Solution

$$\delta Z = 2 + j8 \ \Omega$$

$$V_c = I \ \delta Z = \frac{10 \ \angle 30°}{4 - j3} \times (2 + j8)$$

$$= 16.5 \ \angle 142.83° \ V$$

As per Fig. 3.89 (c)

$$\Delta I = -\frac{V_c}{Z + \delta Z} = -\frac{16.5 \ \angle 142.83°}{6 + j5} = -2.11 \ \angle 102.93° \ A$$

By actual calculation

$$I' = \frac{10 \angle 30°}{6 + j5} = 1.28 \angle -9.81° \text{ A}$$

$$I = \frac{10 \angle 30°}{4 - j5} = 2 \angle 66.87° \text{ A}$$

$$\Delta I = I' - I = 1.28 \angle -9.81° - 2 \angle 66.87° = -2.11 \angle 102.93° \text{ A}$$

3.11 TELLEGAN'S THEOREM

Statement: The algebraic sum of all the powers in the branches of a circuit is zero, or the sum of the powers delivered by the active elements is equal to the sum of the powers absorbed by the passive elements.

$$\sum_{k=1}^{b} V_K I_K = 0$$

where, b = number of branches.

Consider the circuit shown in Fig. 3.91.

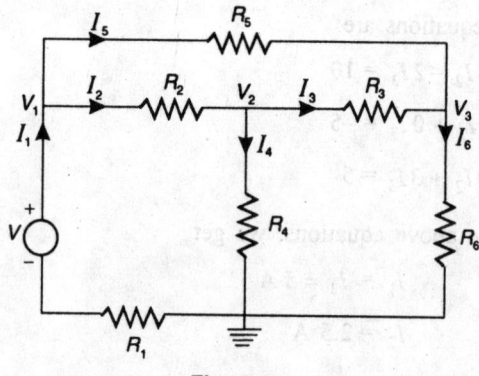

Fig. 3.91

Power delivered by the source = VI_1

Power consumed by all the branches

$$= (V - V_1) I_1 + (V_1 - V_2) I_2 + (V_2 - V_3) I_3 + V_2 I_4$$
$$+ (V_1 - V_3) I_5 + V_3 6$$

$$= VI_1 + V_1 (I_2 + I_5 - I_1) + V_2 (I_3 + I_4 - I_2) + V_3 (I_6 - I_5 - I_3)$$

$$= VI_1 + V_1 (0) + V_2 (0) + V_3 (0) = VI_1$$

∴ Power delivered by the source = power consumed by all the branches and hence, the proof.

WORKED EXAMPLE

3.31 Illustrate Tellegan's theorem for the circuit shown in Fig. 3.92.
(Kuvempu University)

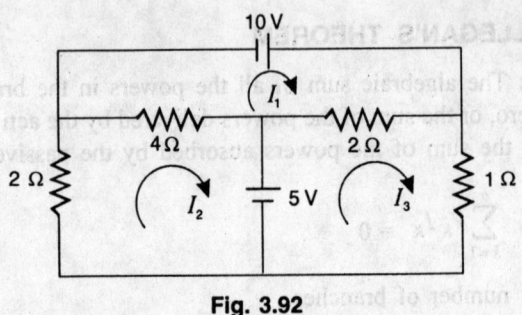

Fig. 3.92

Solution: Assume the loop currents as shown

The loop equations are

$$6I_1 - 4I_2 - 2I_3 = 10 \tag{1}$$

$$-4I_1 + 6I_2 + 0I_3 = -5 \tag{2}$$

$$-2I_1 + 0I_2 + 3I_3 = 5 \tag{3}$$

Solving the above equations, we get

$$I_1 = I_3 = 5 \text{ A}$$

and $$I_2 = 2.5 \text{ A}$$

Power delivered by the sources

$$= 10 I_1 + 5 (I_3 - I_2)$$

$$= 10 \times 5 + 5 \times 2.5 = 62.5 \text{ watts}$$

Power received by the resistors

$$= 2 I_2^2 + 4 (I_1 - I_2)^2 + 2 (I_1 - I_2)^2 + 1 I_3^2$$

$$= 2 \times 2.5^2 + 4 \times (5 - 2.5)^2 + 2 (5 - 5)^2 + 1 \times 5^2$$

$$= 62.5 \text{ watts}.$$

NUMERICAL PROBLEMS

3.1 Using superposition theorem, find the current flowing through 6 Ω resistance, in the circuit shown in Fig. 3.93.

(Karnataka University)

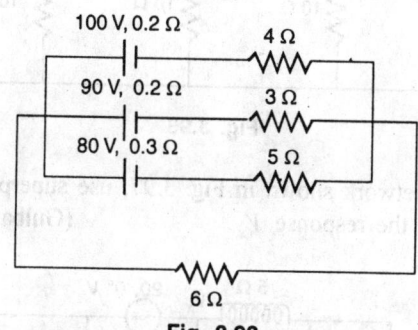

Fig. 3.93

3.2 Using superposition theorem, find the voltage drop across 2 Ω of the circuit shown in Fig. 3.94. (Kuvempu University)

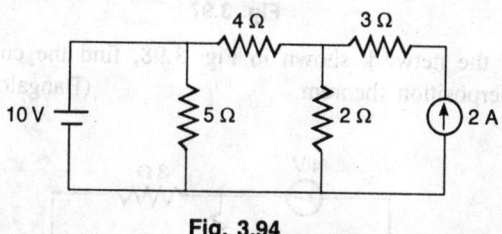

Fig. 3.94

3.3 Find the current through 10 Ω resistor using superposition theorem, in the circuit shown in Fig. 3.95.

(Mysore University)

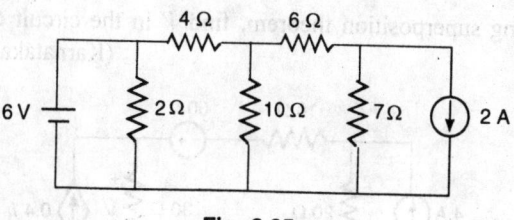

Fig. 3.95

3.4 In the network shown in Fig. 3.96, determine V_b using Superposition theorem. (Mangalore University)

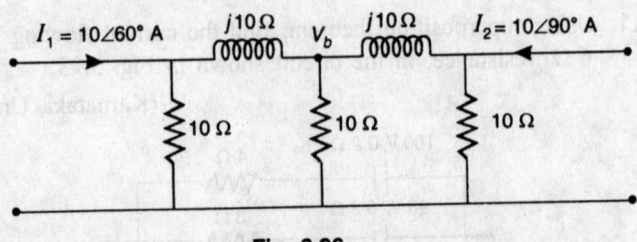

Fig. 3.96

3.5 For the network shown in Fig. 3.97, use superposition theorem to obtain the response V_x. (Gulbarga University).

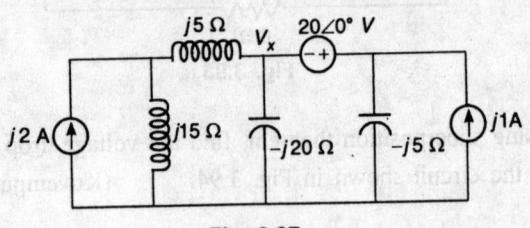

Fig. 3.97

3.6 For the network shown in Fig. 3.98, find the current I, using superposition theorem. (Bangalore Univesity)

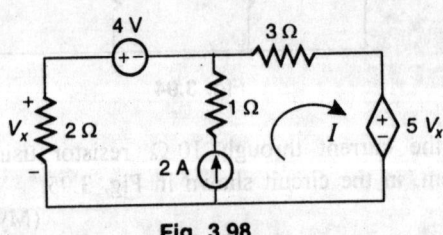

Fig. 3.98

3.7 Using superposition theorem, find V in the circuit of Fig. 3.99. (Karnataka University)

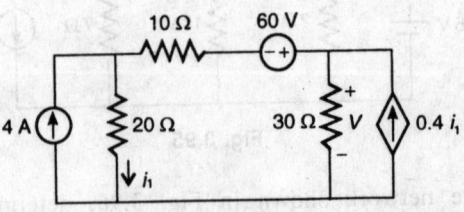

Fig. 3.99

3.8 In the circuit shown in Fig. 3.100, find the current through 10 Ω, using superposition principle. (Bangalore University)

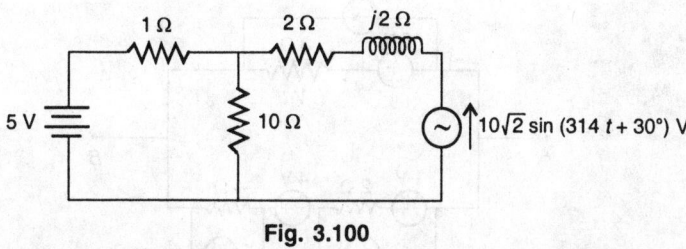

Fig. 3.100

3.9 Obtain the Thevenin's equivalent circuit between the terminals *AB* of the circuit shown in Fig. 3.101 and find the current through 5 Ω connected to *AB*. (Kuvempu University)

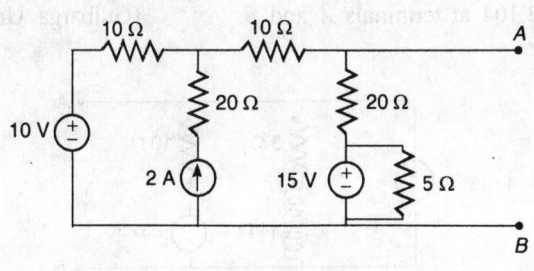

Fig. 3.101

3.10 Find the current I_1 using Norton's theorem in the circuit shown in Fig. 3.102. (Mysore University)

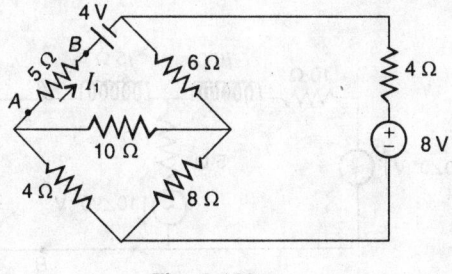

Fig. 3.102

3.11 Obtain Thevenin's equivalent circuit at the terminals *AB* of the network shown in Fig. 3.103. (Karnataka University)

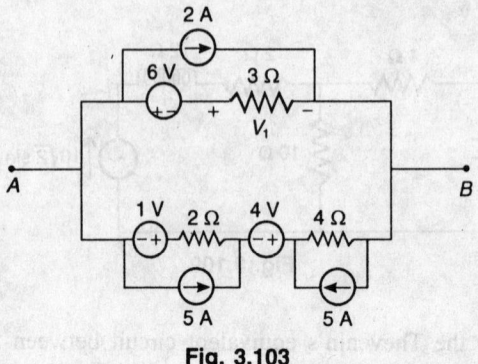

Fig. 3.103

3.12 Obtain Norton's equivalent circuit for the network shown in Fig. 3.104 at terminals A and B. (Gulbarga University)

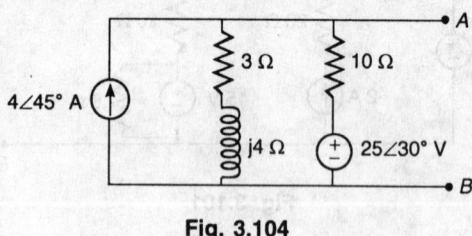

Fig. 3.104

3.13 Find the current flowing through $(7.5 - j7.5)$ Ω in the circuit of Fig. 3.105, using Thevenin's theorem. (Kuvempu University)

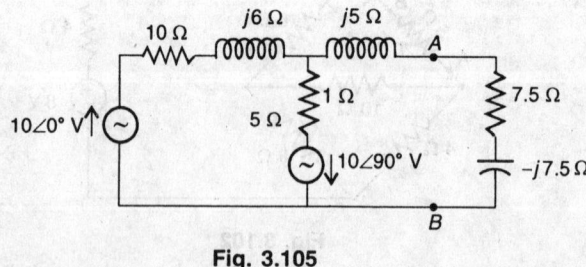

Fig. 3.105

3.14 Obtain Norton's equivalent circuit at the terminals AB of the circuit shown in Fig. 3.106. (Mysore University)

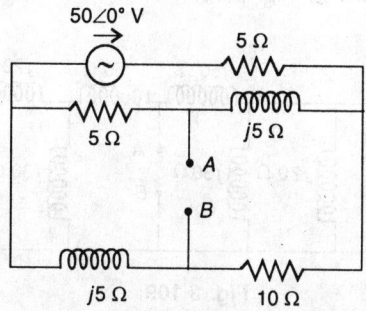

Fig. 3.106

3.15 Determine the Thevenin's equivalent circuit of the network shown
in Fig. 3.107.   (Gulbarga University)

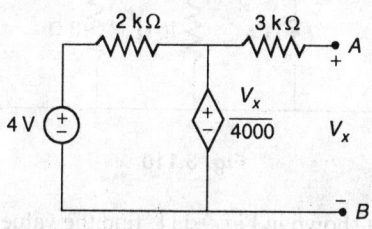

Fig. 3.107

3.16 Obtain Norton's equivalent circuit of the network shown in
Fig. 3.108 at terminals A and B.   (Mangalore University)

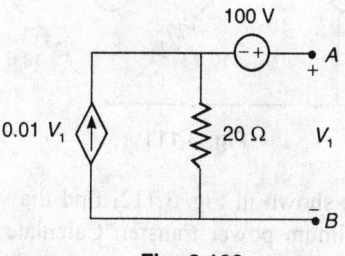

Fig. 3.108

3.17 Find Thevenin's equivalent circuit at the terminals *AB* of the network shown in Fig. 3.109. (Mangalore University)

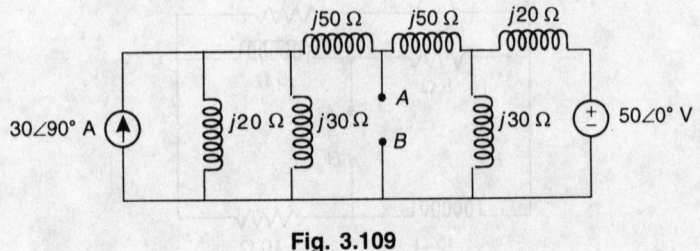

Fig. 3.109

3.18 In the network shown in Fig. 3.110, obtain the value of R_L for which maximum power is transferred and find this power.
(Kuvempu University)

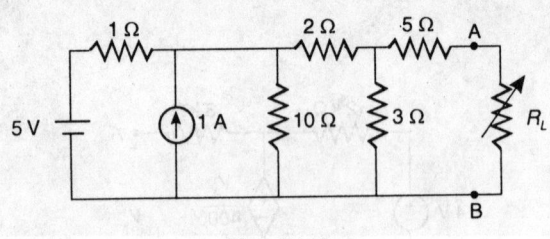

Fig. 3.110

3.19 In the circuit shown in Fig. 3.111, find the value of R that receives maximum power and find this power. (Bangalore University)

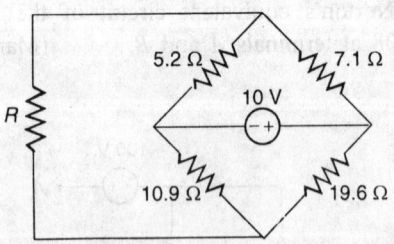

Fig. 3.111

3.20 In the network shown in Fig. 3.112, find the value of R_L which results in maximum power transfer. Calculate the value of this power. (Mysore University)

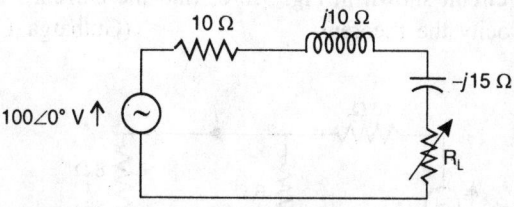

Fig. 3.112

3.21 Determine the maximum power in a load resistor connected across the terminals *a b* of the network shown in Fig. 3.113.

(Bangalore University)

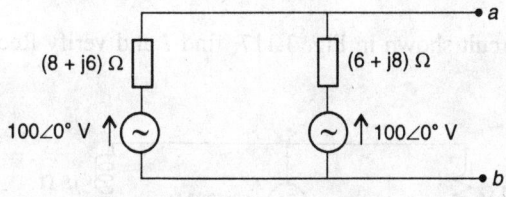

Fig. 3.113

3.22 Determine the value of Z_L for which it receives Maximum power and calculate this power in the circuit shown in Fig. 3.114.

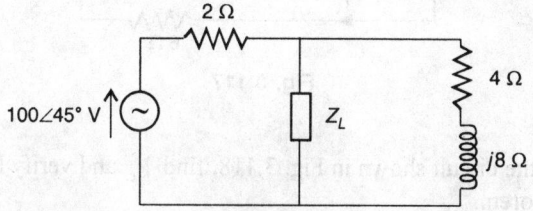

Fig. 3.114

3.23 For the circuit shown in Fig. 3.115, find the value of Z_L for receiving maximum power and find this power.

(Bangalore University)

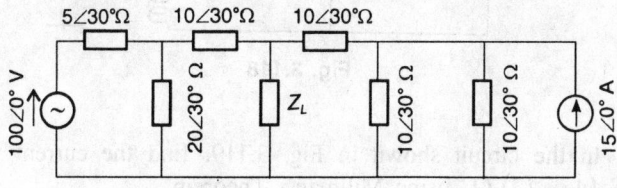

Fig. 3.115

3.24 In the circuit shown in Fig. 3.116, find the current I and verify Reciprocity the theorem. (Gulbarga University)

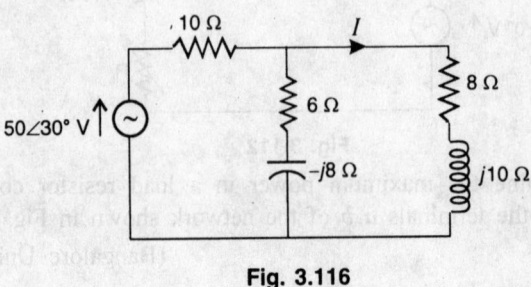

Fig. 3.116

3.25 In the circuit shown in Fig. 3.117, find I and verify Reciprocity theorem.

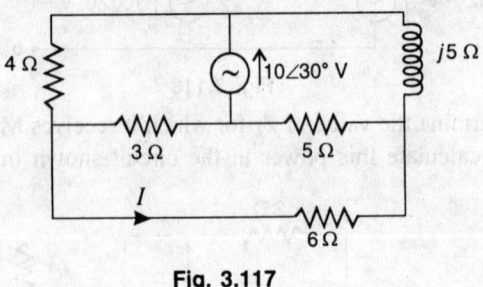

Fig. 3.117

3.26 In the circuit shown in Fig. 3.118, find V_x and verify Reciprocity theorem.

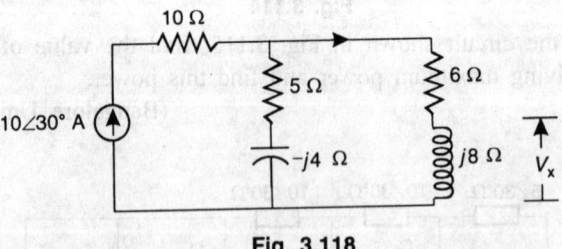

Fig. 3.118

3.27 In the circuit shown in Fig. 3.119, find the current through $(4 - j3)\ \Omega$, using Millman's Theorem.

(Kuvempu University)

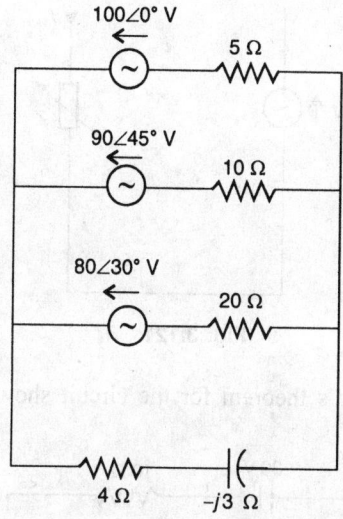

Fig. 3.119

3.28 Using Millman's theorem, find the current flowing through $(1\cdot0 + j10)\,\Omega$ in the circuit of Fig. 3.120.

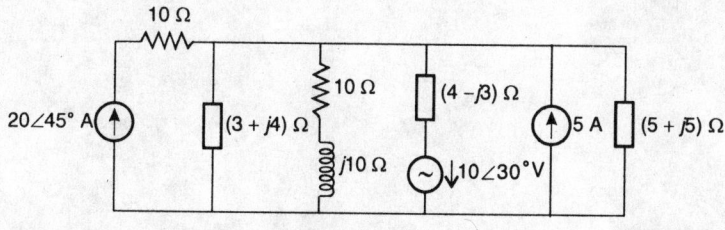

Fig. 3.120

3.29 In the circuit shown in Fig. 3.121, find the change in the current flowing through the impedance Z, when its value is changed from $(4 + j3)\,\Omega$ to $(4 + j8)\,\Omega$ using compensation theorem. Verify the answer by actual calculations.

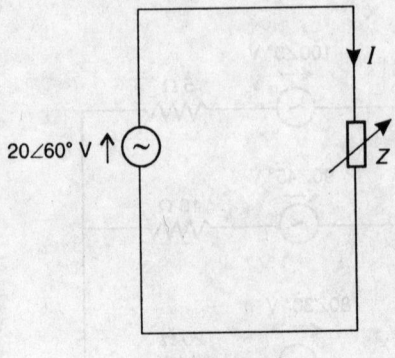

Fig. 3.121

3.30 Verify Tellegan's theorem for the circuit shown in Fig. 3.122.

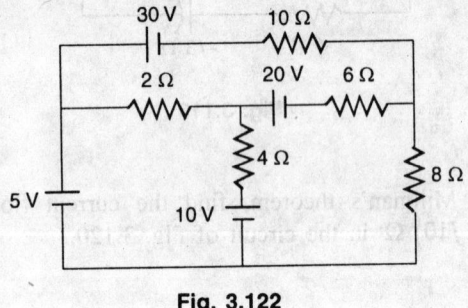

Fig. 3.122

Resonance

4.1 INTRODUCTION

Resonance is a phenomenon which occurs in a.c. circuits containing all the three elements R, L and C. When the circuit is under series resonance, the current is maximum for the frequency f_r, known as *resonant frequency*, decreasing to the left and right of this frequency as shown in Fig. 4.1.

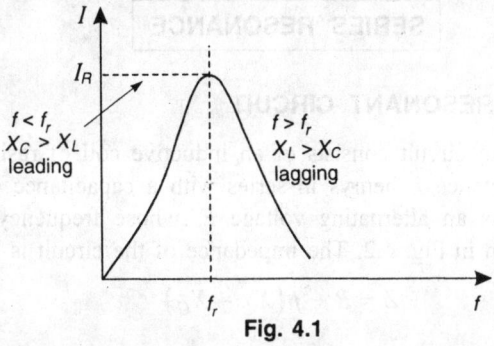

Fig. 4.1

The radio or television receiver has a response curve for each broadcasting station as shown in Fig. 4.1. The receiver is tuned to this frequency to obtain signals from that particular station.

The concept of resonance is not limited to electrical or electronic systems. When a mechanical impulse is applied to a mechanical system at the proper frequency, the system will enter a state of resonance at which, the sustained vibrations of very large magnitude will develop. The frequency at which this occurs is known as the *natural frequency* of the system. In 1940, Tacoma Narrows Bridge was built over the river Puget Sound in Washington State. It had a suspended span of 2800 feet. Four months after the bridge was built, a 42 mph gale set the bridge into oscillations at its natural frequency. The amplitude of the oscillations increased to the point where, the main span broke up and fell into the water below. A new bridge was built in 1950.

A resonant circuit must have an inductance and a capacitance. The resistance will be always present either due to lack of ideal elements or due to the presence of the resistive element itself. When resonance occurs, at any instant, the energy absorbed by one reactive element is exactly equal to the energy released by another reactive element, within the system. The total apparent power is simply the average power dissipated by the resistive elements. The average power absorbed by the system will also be maximum at resonance.

The resonant condition in a.c. circuits may be achieved by varying the frequency of the supply, keeping the network elements constant or by varying L or C, keeping the frequency constant. There are two types of resonant circuits.

(i)   Series resonant circuit.

(ii) Parallel resonant circuit.

SERIES RESONANCE

4.2   SERIES RESONANT CIRCUIT

A series resonant circuit consists of an inductive coil of resistance R ohms and inductance L henrys in series with a capacitance C farads connected across an alternating voltage E, whose frequency can be varied, as shown in Fig. 4.2. The impedance of the circuit is given by

$$Z = R + j \left(X_L - X_C \right)$$

We know that $X_L = 2 \pi f L$ and $X_C = \dfrac{1}{2 \pi f C}$

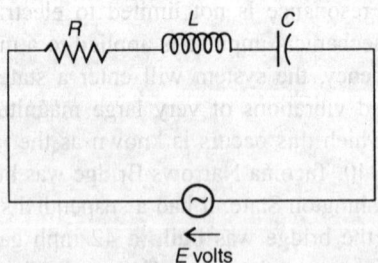

Fig. 4.2

By varying the frequency of the supply, X_L is made equal to X_C, then $Z = R$. The current is in phase with the voltage. The p.f. of the circuit is unity. Under these conditions, the circuit is said to be in *series resonance*. Let f_r be the frequency at which $X_L = X_C$.

Then

$$2\,\pi\,f_r\,L = \frac{1}{2\,\pi\,f_r C}$$

or $$f_r = \frac{1}{2\pi\sqrt{LC}} \tag{4.1}$$

f_r is known as the resonant frequency. At resonance, the current through the circuit is maximum and is given by

$$I_r = \frac{E}{R} = I_m \tag{4.2}$$

For values of frequencies less than f_r, $X_C > X_L$ and for frequencies more than f_r, $X_L > X_C$. These conditions are also indicated in Fig. 4.1.

4.3 THE QUALITY FACTOR (Q_S)

During series resonance, the voltage across the reactive elements, i.e. inductance and capacitance increase to many times more than the applied voltage itself.

At resonance $$I_r = \frac{E}{R} = I_m$$

The voltage across the inductance L is

$$V_L = I_m\,X_L = \frac{E}{R}\,X_L = \frac{E}{R}\,\omega_r L = \frac{X_{Lr}}{R}\,E = Q_s E \tag{4.3}$$

Where $$Q_s = \frac{\omega_r L}{R} = \frac{X_{Lr}}{R} \tag{4.4}$$

Q_s is known as the quality factor of the series resonant circuit or simply the quality factor of the coil.

The voltage across C is

$$V_C = I_m\,X_C = \frac{E}{R}\,\frac{1}{\omega_r C} = \frac{1}{\omega_r CR}\,E = Q_s E$$

Where, $$Q_s = \frac{1}{\omega_r CR} = \frac{X_{cr}}{R} \tag{4.5}$$

The quality factor of a coil, in view of equations (4.4) and (4.5), may be defined as the ratio of the inductive reactance or capacitive reactance at resonance to the resistance of the coil.

$$Q_s = \frac{\omega_r L}{R} = \frac{2\pi f_r L}{R} = \frac{2\pi L}{R} \times \frac{1}{2\pi\sqrt{LC}} = \frac{1}{R}\sqrt{\frac{L}{C}} \qquad (4.6)$$

From Eq. (4.6), we understand that the quality factor depends on the resistance of the resonant circuit. Higher the resistance, smaller will be the value of Q_s.

4.4 VARIATION OF REACTANCES WITH FREQUENCY

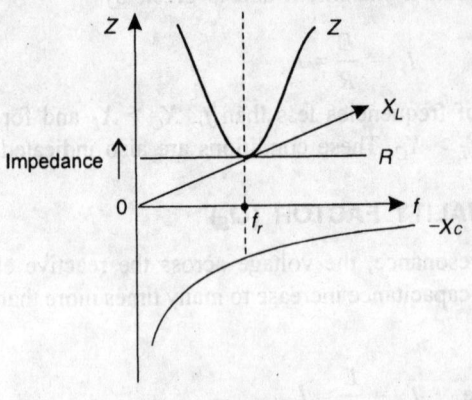

Fig. 4.3

Figure 4.3 shows the variation of various reactances with frequency. R remains constant for all frequencies. As $X_L = 2\pi f L$, it increases linearly with the increase of frequency. As $X_c = \dfrac{1}{2\pi fC}$, $-X_C$ has the hyperbolic variation. At f_r, $X_L = X_C$. The variation of Z is as shown, which tends to increase for lower values of f due to increased values of X_C and which also tends to increase for higher values of f due to increased values of X_L, both to the left and right of f_r. At f_r, $Z = R$.

4.5 SELECTIVITY AND BANDWIDTH

Figure 4.4 represents the frequency response curve of a series resonant circuit, i.e. the variation of I w.r.t. f, when E is kept constant. The frequencies f_1 and f_2, coresponding to $I = \dfrac{I_m}{\sqrt{2}} = 0.707 I_m$ are called *band frequencies* or *cut-off frequencies* or *half power frequencies*.

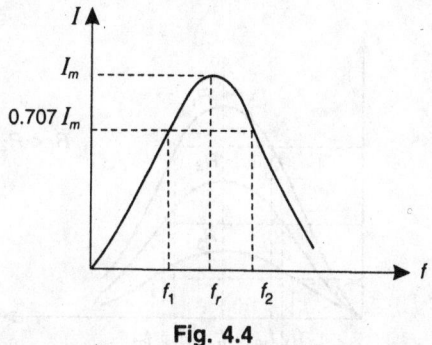

Fig. 4.4

The range of frequencies between these two cut off frequencies, i.e. $(f_2 - f_1)$ is called the *Bandwidth* (BW) of the resonant circuit. f_1 and f_2 are also called as half power frequencies because, the power delivered by the circuit at these frequencies is half of the power delivered by it at resonant frequency. This fact can be proved as follows.

Let P_m = maximum power delivered by the circuit.

= power delivered at resonant frequency = $I_m^2 R$

$$\text{Power delivered at } f_1 \text{ or } f_2 = \left(\frac{I_m}{\sqrt{2}}\right)^2 R = \frac{I_m^2 R}{2}$$

A resonant circuit is always adjusted to select a band of frequencies lying between f_1 and f_2. Hence, the frequency response curve shown in Fig. 4.4 is also known as the *selectivity curve*. The smaller the band width, higher is the selectivity.

Effect of *R* on frequency response curve

The shape of the curve shown in Fig. 4.4, depends on the elements R, L and C. If the resistance is decreased, keeping L and C constant, then the bandwidth decreases and the selectivity increases as shown in Fig. 4.5. On the other hand, if R is increased, keeping L and C constant, then bandwidth increases and hence the selectivity decreases.

f_{11} and f_{12} are the cut off frequencies for R_1.

f_{12} and f_{22} are the cut off frequencies for R_2.

f_{13} and f_{23} are the cut off frequencies for R_3.

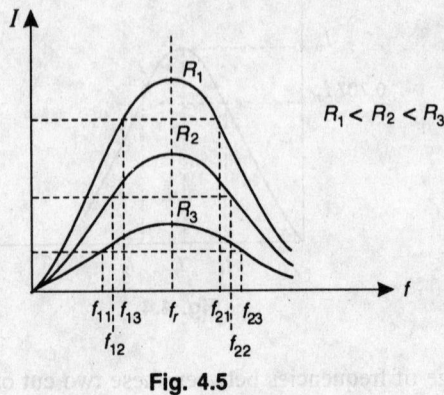

Fig. 4.5

$R_1 < R_2 < R_3$

Effect of $\dfrac{L}{C}$ on frequency response curve

If the ratio $\dfrac{L}{C}$ increases with fixed resistance, the bandwitdth again decreases and selectivity increases as shown in Fig. 4.6.

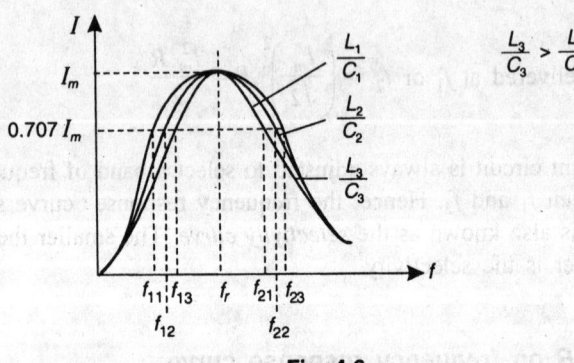

$\dfrac{L_3}{C_3} > \dfrac{L_2}{C_2} > \dfrac{L_1}{C_1}$

Fig. 4.6

f_{11} and f_{12} are the cut off frequencies for $\dfrac{L_1}{C_1}$.

f_{12} and f_{22} are the cut off frequencies for $\dfrac{L_2}{C_2}$.

f_{13} and f_{23} are the cut off frequencies for $\dfrac{L_3}{C_3}$.

4.6 EXPRESSIONS FOR f_1 AND f_2

If $Q_s \geq 10$, the resonant curve is almost symmetrical about resonant frequency. Then f_1 and f_2 are equidistant from f_r as shown in Fig. 4.7.

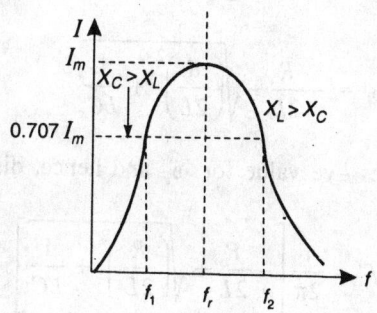

Fig. 4.7

At f_1 and f_2, the current is $\dfrac{I_m}{\sqrt{2}}$ and hence, the impedance is $\sqrt{2}$ times the value of impedance at f_r. At f_r, $Z = R$.

∴ At f_1 and f_2, $Z = \sqrt{2}\, R$

In general, the impedance of the circuit is given by

$$Z = \sqrt{R^2 + (X_L - X_C)^2}$$

At f_1 and f_2, $\sqrt{2}\, R = \sqrt{R^2 + (X_L - X_C)^2}$

$$2R^2 = R^2 + (X_L - X_C)^2$$

or $R^2 = (X_L - X_C)^2$ or $R = X_L - X_C$ (4.7)

At f_1, $X_C > X_L$. Hence, Eq. (4.7) can be written as

$$R = X_C - X_L \qquad (4.8)$$

The solution of Eq. (4.8) gives f_1

$$R = \frac{1}{\omega_1 C} - \omega_1 L = \frac{1 - \omega_1^2 LC}{\omega_1 C}$$

i.e. $R\omega_1 C - 1 + \omega_1^2 LC = 0 \div LC$

i.e. $\omega_1^2 + \dfrac{R}{L}\, \omega_1 - \dfrac{1}{LC} = 0 \qquad (4.9)$

The solution of Eq. (4.9) is given by

$$\omega_1 = \frac{-\dfrac{R}{L} \pm \sqrt{\dfrac{R^2}{L^2} + \dfrac{4}{LC}}}{2}$$

or

$$\omega_1 = -\frac{R}{2L} + \sqrt{\left(\frac{R}{2L}\right)^2 + \frac{1}{LC}}$$

The −ve sign gives −ve value for ω_1 and hence, discorded.

$$\therefore \qquad f_1 = \frac{1}{2\pi}\left[-\frac{R}{2L} + \sqrt{\left(\frac{R}{2L}\right)^2 + \frac{1}{LC}}\right] \qquad (4.10)$$

At f_2, $X_L > X_C$

$$\therefore \qquad R = X_L - X_C \text{ as per Eq. (4.7)}$$

$$R = \omega_2 L - \frac{1}{\omega_2 C}$$

or

$$R = \frac{\omega_2^2 LC - 1}{\omega_2 C} \quad \text{i.e. } \omega_2^2 LC - 1 - R\omega_2 C = 0 \div LC$$

$$\omega_2^2 - \frac{R}{L}\omega_2 - \frac{1}{LC} = 0 \qquad (4.11)$$

The solution of Eq. (4.11), gives

$$\omega_2 = \frac{\dfrac{R}{L} \pm \sqrt{\left(\dfrac{R}{L}\right)^2 + \dfrac{4}{LC}}}{2} = \frac{R}{2L} \pm \sqrt{\left(\frac{R}{2L}\right)^2 + \frac{1}{LC}}$$

The −ve sign gives negative value for ω_2 and hence, discorded.

$$\therefore \qquad f_2 = \frac{1}{2\pi}\left[\frac{R}{2L} + \sqrt{\left(\frac{R}{2L}\right)^2 + \frac{1}{LC}}\right] \qquad (4.12)$$

From Eqs (4.10) and (4.12), the bandwidth is given by

$$\text{Bandwidth} = f_2 - f_1 = \frac{R}{2\pi L} \quad \text{or} \quad \omega_2 - \omega_1 = \frac{R}{L} \qquad (4.13)$$

But at resonance, $Q_s = \dfrac{X_{Lr}}{R} = \dfrac{2\pi f_r L}{R} = 2\pi f_r \left(\dfrac{1}{R/L}\right)$

$$= 2\pi f_r \times \dfrac{1}{2\pi(f_2 - f_1)} = \dfrac{f_r}{f_2 - f_1} \qquad (4.14)$$

$\therefore$ Bandwidth $= f_2 - f_1 = \dfrac{f_r}{Q_s}$ \qquad (4.15)

Equation (4.14) may also be written as

$$\dfrac{f_2 - f_1}{f_r} = \dfrac{1}{Q_s} \qquad (4.16)$$

Sometimes, $\dfrac{f_2 - f_1}{f_r}$ is referred as *fractional bandwidth*.

4.7 RELATION BETWEEN f_r, f_1 AND f_2

The impedances of an *RLC* resonant circuit at f_1 and f_2 are given by

$$Z_1 = \sqrt{R^2 + \left(X_{C1} - X_{L1}\right)^2}$$

and $\qquad\qquad Z_2 = \sqrt{R^2 + \left(X_{L2} - X_{C2}\right)^2}$

But $\qquad\qquad Z_1 = Z_2$

$\therefore \qquad R^2 + \left(X_{C1} - X_{L1}\right)^2 = R^2 + \left(X_{L2} - X_{C2}\right)^2$

or $\qquad\qquad X_{C1} - X_{L1} = X_{L2} - X_{C2}$

i.e. $\qquad\qquad X_{C1} + X_{C2} = X_{L1} + X_{L2}$

i.e. $\qquad\qquad \dfrac{1}{\omega_1 C} + \dfrac{1}{\omega_2 C} = \omega_1 L + \omega_2 L$

i.e. $\qquad\qquad \dfrac{1}{C}\left[\dfrac{\omega_1 + \omega_2}{\omega_1 \omega_2}\right] = L\left(\omega_1 + \omega_2\right)$

i.e. $\qquad\qquad \omega_1 \omega_2 = \dfrac{1}{LC} = \omega_r^2$

$\therefore \qquad\qquad \omega_r = \sqrt{\omega_1 \omega_2} \text{ or } f_r = \sqrt{f_1 f_2} \quad (4.17)$

4.8 RESONANCE BY VARYING CIRCUIT ELEMENTS

(i) By Varying Inductance

Consider and RLC series resonant circuit, resonant condition being obtained by varying L, as shown in Fig. 4.8.

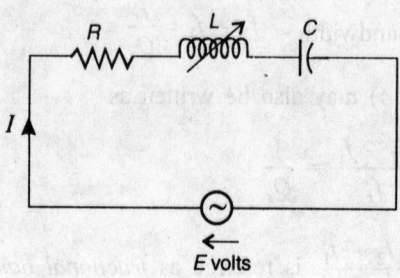

Fig. 4.8

At resonance $\qquad X_L = X_C$

i.e. $\qquad \omega L_r = \dfrac{1}{\omega C}$

or $\qquad L_r = \dfrac{1}{\omega^2 C} \qquad\qquad (4.18)$

Where, L_r = inductance at resonance.

Let, L_1 = inductance at f_1

$\qquad$ At $f_1, X_C - X_L = R$

i.e. $\qquad \dfrac{1}{\omega C} - \omega L_1 = R$

or $\qquad L_1 = \dfrac{1}{\omega^2 C} - \dfrac{R}{\omega} \qquad\qquad (4.19)$

Let L_2 = inductance at f_2

$\qquad$ At $f_2, X_L - X_C = R$

i.e. $\qquad \omega L_2 - \dfrac{1}{\omega C} = R$

or $$L_2 = \frac{1}{\omega^2 C} + \frac{R}{\omega} \qquad (4.20)$$

For frequencies less than f_r, $X_C > X_L$, and hence, $V_C > V_L$. At f_r, $V_C = V_L$. For frequencies more than f_r, $X_L > X_C$ and hence, $V_L > V_C$. The variation of voltages V_L, V_C w.r.t. L are as shown in Fig. 4.9.

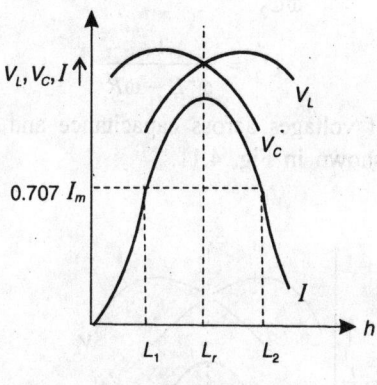

Fig. 4.9

(ii) By Varying Capacitance

Consider an *RLC* series resonant circuit, resonant condition being obtained by varying C, as shown in Fig. 4.10.

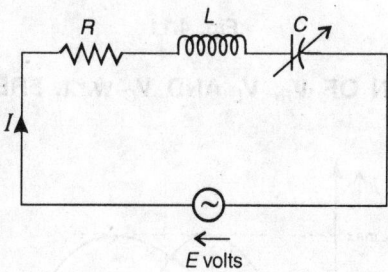

Fig. 4.10

At resonance $X_C = X_L$

i.e. $$\frac{1}{\omega C_r} = \omega L$$

or $$C_r = \frac{1}{\omega^2 L} \qquad (4.21)$$

Where, C_r = capacitance at resonance.

At $f_1, \dfrac{1}{\omega C_1} - \omega L = R$ (since $X_C > X_L$)

$$\therefore \qquad C_1 = \frac{1}{\omega^2 L + \omega R} \qquad (4.22)$$

At $f_2, \ \omega L - \dfrac{1}{\omega C_2} = R$ (since $X_L > X_C$)

$$\therefore \qquad C_2 = \frac{1}{\omega^2 L - \omega R} \qquad (4.22a)$$

The variation of voltages across capacitance and inductance w.r.t. capacitance is as shown in Fig. 4.11.

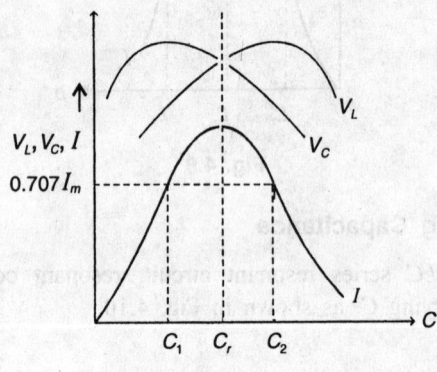

Fig. 4.11

4.9 VARIATION OF V_R, V_L AND V_C w.r.t. FREQUENCY

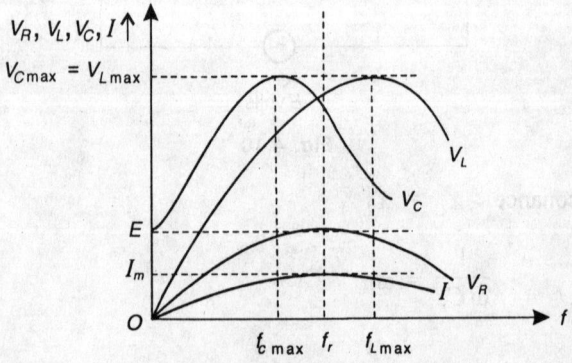

Fig. 4.12

Figure 4.12 shows the variation of V_R, V_L and V_C w.r.t. frequency. The V_R curve has the same shape as the I curve having the peak value of E volts, as $V_R = E = IR$. When the frequency is zero, X_C is infinite, current is zero and hence the capacitance acts as an open-circuit and the entire applied voltage exists across this open-circuit. Therefore, when $f = 0$, $V_C = E$.

We know that, $V_C = I \, X_C = \dfrac{1}{\omega C} = \dfrac{I}{2\pi f C}$

As the frequency is increased, V_C decreases due to $\dfrac{1}{\omega C}$, but I increases at a faster rate than $\dfrac{1}{\omega C}$ and hence, V_C increases. As resonant condition approaches, the rate of change of I decreases. $\dfrac{1}{\omega C}$ overcomes the rate of change of I and hence, V_C starts decreasing. The peak value of V_C occurs at a frequency f_{Cmax} just before resonance. After resonance, V_C drops in magnitude and approaches zero.

At $f = 0$, $X_L = 0$ and hence, $V_L = 0$

As f increases, V_L also increases and reaches its maximum value slightly after f_r. For values more than f_r, V_L decreases and approaches E as frequency increases.

For higher values of Q_s, i.e. $Q_s \geq 10$, f_{Cmax} approaches f_r and $V_{Cmax} \cong Q_s E$. Similarly f_{Lmax} also approaches f_r and $V_{Lmax} \cong Q_s E$. For any value of $Q_s \geq 10$, the variation of V_R, V_L and V_C w.r.t. frequency are as shown in Fig. 4.13.

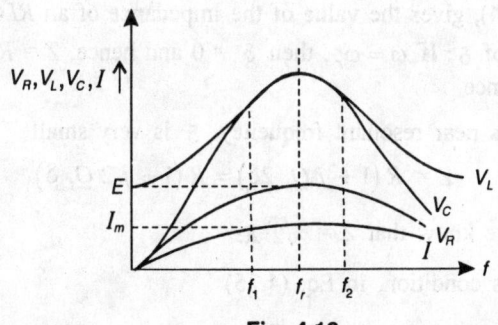

Fig. 4.13

4.10 FREQUENCY DEVIATION (δ)

The frequency deviation of an *RLC* series circuit is defined as the ratio of the difference between the operating frequency and resonant frequency to the resonant frequency.

$$\delta = \frac{\omega - \omega_r}{\omega_r} = \frac{f - f_r}{f_r} \qquad (4.23)$$

Where ω = operating frequency is rad/sec.

              ω_r = resonant frequency in rad/sec.

              $f =$ operating frequency in Hz.

and f_r = resonant frequency in Hz.

The impedance of an *RLC* series circuit is given by

$$Z = R + j X_L - j X_C = R + j \omega L - j \frac{1}{\omega C}$$

$$= R \left[1 + j \left(\frac{\omega L}{R} - \frac{1}{\omega CR} \right) \right] = R \left[1 + j \left(\frac{\omega_r L}{R} \frac{\omega}{\omega_r} - \frac{1}{\omega_r CR} \frac{\omega_r}{\omega} \right) \right]$$

$$= R \left[1 + j \left(Q_s \frac{\omega}{\omega_r} - Q_s \frac{\omega_r}{\omega} \right) \right] = R \left[1 + j Q_s \left(\frac{\omega}{\omega_r} - \frac{\omega_r}{\omega} \right) \right]$$

$$\therefore \quad = R \left[1 + j Q_s \left(1 + \delta - \frac{1}{1 + \delta} \right) \right] \quad \left(\because \delta = \frac{\omega - \omega_r}{\omega_r} \text{ or } \frac{\omega}{\omega_r} = 1 + \delta \right)$$

$$= R \left[1 + j Q_s \left(\frac{1 + \delta^2 + 2\delta - 1}{1 + \delta} \right) \right] = R \left[1 + j Q_s \delta \left(\frac{2 + \delta}{1 + \delta} \right) \right] \quad (4.24)$$

Equation (4.24), gives the value of the impedance of an *RLC* series circuit in terms of δ. If $\omega = \omega_r$, then $\delta = 0$ and hence, $Z = R$, which is true at resonance.

At frequencies near resonant frequency, δ is very small

$$\therefore \qquad\qquad Z = R \left(1 + j Q_s 2\delta \right) = R \left(1 + j 2 Q_s \delta \right) \qquad (4.25)$$

At f_1 or f_2, we know that $Z = \sqrt{2} R$.

To satisfy this condition, in Eq. (4.25)

        $Q_s \delta = 0.5$ at f_2 and $Q_s \delta = -0.5$ *at* f_1

$\therefore$ At f_2, $Q_s \delta = 0.5$ and at f_1, $Q_s \delta = -0.5$

$\therefore$ At f_1, δ is –ve. From equation (4.23), we get

$$-\delta = \frac{f_1 - f_r}{f_r} \text{ or } f_1 = f_r (1 - \delta) \qquad (4.26)$$

At f_2, δ is +ve.

$$\delta = \frac{f_2 - f_r}{f_r}, \text{ i.e. } f_2 = f_r \,(1 + \delta) \tag{4.27}$$

4.11 EXPRESSIONS FOR $f_{C\text{max}}$ AND $f_{L\text{max}}$

$f_{C\text{max}}$ is the frequency at which, $V_{C\text{max}}$ occurs. $V_{C\text{max}}$ occurs earlier to f_r, for which $X_C > X_L$.

$$V_C = I\,X_C = \frac{I}{\omega C} = \frac{E}{Z}\,\frac{1}{\omega C} = \frac{E}{\sqrt{R^2 + (X_C - X_L)^2}}\,\frac{1}{\omega C}$$

$$V_C^2 = \frac{E^2}{R^2 + \left(\dfrac{1}{\omega C} - \omega L\right)^2}\,\frac{1}{\omega^2 C^2}$$

$$= \frac{E^2}{\omega^2 C^2 R^2 + \omega^2 C^2 \left(\dfrac{1}{\omega^2 C^2} + \omega^2 L^2 - 2\dfrac{L}{C}\right)}$$

$$= \frac{E^2}{\omega^2 C^2 R^2 + \left(1 + \omega^4 L^2 C^2 - 2\omega^2 LC\right)} = \frac{E^2}{\omega^2 C^2 R^2 + \left(\omega^2 LC - 1\right)^2}$$

V_C is maximum, when $\dfrac{dV_C^2}{d\omega} = 0$

$$\frac{dV_C^2}{d\omega} = \frac{0 - E^2 \{2\,\omega C^2 R^2 + 2\left(\omega^2 LC - 1\right)2\,\omega LC\}}{\{\omega^2 C^2 R^2 + \left(\omega^2 LC - 1\right)^2\}^2} = 0$$

As $E \neq 0$, $2\omega\,C^2 R^2 + 4\,\omega^3 L^2 C^2 - 4\,\omega LC = 0$

i.e. $\qquad 2\,\omega C\left(CR^2 + 2\omega^2 L^2 C - 2L\right) = 0$

i.e. $\qquad CR^2 + 2\omega^2 L^2\,C - 2L = 0$

$\therefore \quad \omega^2 = \dfrac{1}{LC} - \dfrac{R^2}{2L^2}$

or
$$\omega = \sqrt{\frac{1}{LC} - \frac{R^2}{2L^2}}$$

$$\therefore \qquad f_{C_{max}} = \frac{1}{2\pi} \sqrt{\frac{1}{LC} - \frac{R^2}{2L^2}} \qquad (4.28)$$

f_{Lmax} is the frequency at which V_{Lmax} occurs.

V_{Lmax} occurs after f_r, for which $X_L > X_C$.

$$V_L = I\ X_L = \frac{E}{\sqrt{R^2 + \left(X_L - X_C\right)^2}} \quad \omega L = \frac{E\omega L}{\sqrt{R^2 + \left(\omega L - \frac{1}{\omega C}\right)^2}}$$

$$= \frac{E\omega^2 LC}{\sqrt{\omega^2 C^2 R^2 + \left(\omega^2 LC - 1\right)^2}}, \quad \text{i.e.}\ \ V_L^2 = \frac{E^2 \omega^4 L^2 C^2}{\omega^2 C^2 R^2 + \left(\omega^2 LC - 1\right)^2}$$

V_L is maximum, when $\dfrac{dV_L^2}{d\omega} = 0$

$$\frac{dV_L^2}{d\omega} = \frac{\{\omega^2 C^2 R^2 + (\omega^2 LC - 1)^2\}\ 4\omega^3 E^2 L^2 C^2}{\{\omega^2 C^2 R^2 + \omega^2\ LC - 1^2\}^2} = 0$$

i.e. $4\left\{\omega^2 C^2 R^2 + \left(\omega^2 LC - 1\right)^2\right\} - \omega\left\{2\omega C^2 R^2 + 4\omega^3 L^2 C^2 - 4\omega LC\right\} = 0$

i.e. $2\ \omega^2 C^2 R^2 - 4\omega^2 LC + 4 = 0$

or $4\ \omega^2 LC - 2\ \omega^2 C^2 R^2 = 4$

or $2\ \omega^2\ LC - \omega^2 C^2 R^2 = 2$

i.e.
$$\omega^2 = \frac{2}{2\,LC - C^2 R^2} = \frac{1}{LC - \dfrac{R^2 C^2}{2}}$$

$\therefore$
$$\omega = \sqrt{\frac{1}{LC - \dfrac{R^2 C^2}{2}}}$$

$\therefore$
$$f_{L_{max}} = \frac{1}{2\pi} \sqrt{\frac{1}{LC - \dfrac{R^2 C^2}{2}}} \qquad (4.29)$$

PARALLEL RESONANCE

4.12 PRACTICAL PARALLEL RESONANT CIRCUIT

A practical parallel resonant circuit as shown in Fig. 4.14, consists of an inductive coil of resistance R and inductance L, placed in parallel with a capacitance C and connected to an alternating supply of voltage E, of variable frequency f. The impedance of the coil is given by

$$Z_L = R + j\omega L$$

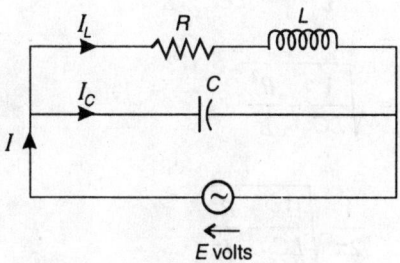

Fig. 4.14

The admittance of the coil is

$$Y_L = \frac{1}{Z_L} = \frac{1}{R + j\omega L} \times \frac{R - j\omega L}{R - j\omega L} = \frac{R - j\omega L}{R^2 + \omega^2 L^2}$$

$$Z_C = -j\,\frac{1}{\omega C}$$

or
$$Y_C = \frac{1}{Z_C} = j\omega C$$

The total admittance of the circuit is

$$Y = Y_L + Y_C = \frac{R - j\omega L}{R^2 + \omega^2 L^2} + j\omega C$$

$$= \frac{R}{R^2 + \omega^2 L^2} + j\left(\omega C - \frac{\omega L}{R^2 + \omega^2 L^2}\right)$$

For the circuit to be at resonance, the impedance of the circuit should be purely resistive or the admittance must be purely conductive. Hence, the imaginary part of the admittance must be zero.

$\therefore$ At resonance, $\omega_r C - \dfrac{\omega_r L}{R^2 + \omega_r^2 L^2} = 0$

or $\omega_r C = \dfrac{\omega_r L}{R^2 + \omega_r^2 L^2}$ or $R^2 + \omega_r^2 L^2 = \dfrac{L}{C}$

$\therefore$
$$\omega_r^2 = \frac{\dfrac{L}{C} - R^2}{L^2} = \frac{1}{LC} - \frac{R^2}{L^2}$$

$\therefore$
$$\omega_r = \sqrt{\frac{1}{LC} - \frac{R^2}{L^2}}$$

and hence,
$$f_r = \frac{1}{2\pi}\sqrt{\frac{1}{LC} - \frac{R^2}{L^2}} \tag{4.30}$$

Where, f_r = resonant frequency of a practical parallel circuit.

At resonance, the admittance of the circuit is purely conductive.

$\therefore$
$$Y_r = \frac{R}{R^2 + \omega_r^2 L^2} \quad \text{but} \quad R^2 + \omega_r^2 L^2 = \frac{L}{C}$$

$\therefore$
$$Y_r = \frac{RC}{L} \quad \text{or} \quad Z_r = \frac{L}{RC} \tag{4.31}$$

Z_r is the impedance of the practical parallel circuit at resonance and is known as the *dynamic resistance*.

The current at resonance is given by

$$I_r = EY_r = E\frac{RC}{L} \qquad (4.32)$$

4.13 PARALLEL RESONANT CIRCUIT CONSIDERING THE CAPACITANCE TO HAVE RESISTANCE

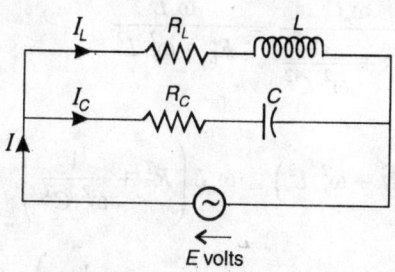

Fig. 4.15

Consider a parallel circuit as shown in Fig. 4.15.

$$Z_L = R_L + j\omega L$$

i.e.
$$Y_L = \frac{1}{R_L + j\omega L} = \frac{R_L - j\omega L}{R_L^2 + \omega^2 L^2}$$

Similarly
$$Z_C = R_C - j\frac{1}{\omega C}$$

i.e.
$$Y_C = \frac{1}{R_C - j\dfrac{1}{\omega C}} = \frac{R_C + j\dfrac{1}{\omega C}}{R_C^2 + \dfrac{1}{\omega^2 C^2}}$$

The total admittance Y is given by

$$Y = Y_L + Y_C = \frac{R_L - j\omega L}{R_L^2 + \omega^2 L^2} + \frac{R_C + j\dfrac{1}{\omega C}}{R_C^2 + \dfrac{1}{\omega^2 C^2}}$$

$$= \left(\frac{R_L}{R_L^2 + \omega^2 L^2} + \frac{R_C}{R_C^2 + \frac{1}{\omega^2 C^2}} \right) + j \left(\frac{1/\omega C}{R_C^2 + \frac{1}{\omega^2 C^2}} - \frac{\omega L}{R_L^2 + \omega^2 L^2} \right)$$

$$(4.33)$$

At resonance, the admittance is purely conductive. Hence, the imaginary part of Eq. (4.33) is zero.

$$\therefore \qquad \frac{\frac{1}{\omega_r C}}{R_C^2 + \frac{1}{\omega_r^2 C^2}} = \frac{\omega_r L}{R_L^2 + \omega_r^2 L^2}$$

i.e.
$$\frac{1}{\omega_r C} \left(R_L^2 + \omega_r^2 L^2 \right) = \omega_r L \left(R_C^2 + \frac{1}{\omega_r^2 C^2} \right)$$

i.e.
$$\frac{1}{LC} \left(R_L^2 + \omega_r^2 L^2 \right) = \omega_r^2 \left(R_C^2 + \frac{1}{\omega_r^2 C^2} \right)$$

i.e.
$$\frac{R_L^2}{LC} + \omega_r^2 \frac{L}{C} = \omega_r^2 R_C^2 + \frac{1}{C^2}$$

i.e.
$$\omega_r^2 \left(R_C^2 - \frac{L}{C} \right) = \frac{R_L^2}{LC} - \frac{1}{C^2} = \frac{1}{LC} \left(R_L^2 - \frac{L}{C} \right)$$

$$\therefore \qquad \omega_r^2 = \frac{\frac{1}{LC} \left(R_L^2 - \frac{L}{C} \right)}{R_C^2 - \frac{L}{C}}$$

$$\therefore \qquad \omega_r = \frac{1}{\sqrt{LC}} \sqrt{\frac{R_L^2 - \frac{L}{C}}{R_C^2 - \frac{L}{C}}}$$

or
$$f_r = \frac{1}{2\pi\sqrt{LC}} \sqrt{\frac{R_L^2 - \frac{L}{C}}{R_C^2 - \frac{L}{C}}}$$

$$(4.34)$$

Equation (4.34) gives the resonant frequency for the parallel circuit shown in Fig. 4.15. The admittance at resonance is purely conductive.

$$\therefore \quad Y_r = \frac{R_L}{R_L^2 + \omega_r^2 L^2} + \frac{R_C}{R_C^2 + \dfrac{1}{\omega^2 C^2}}$$

The current at resonance is given by

$$I_r = EY_r = E\left[\frac{R_L}{R_L^2 + \omega_r^2\, L^2} + \frac{R_C}{R_C^2 + \dfrac{1}{\omega^2 C^2}}\right] \tag{4.35}$$

The frequency response curve of a parallel resonant circuit is as shown in Fig. 4.16.

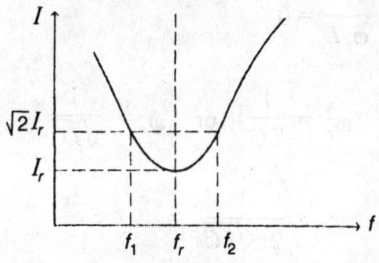

Fig. 4.16

From the Fig. 4.16, we find that the current is minimum at resonance. Hence, the impedance of the circuit is maximum at resonance. Since the current at resonance is minimum, the parallel circuit at resonance is called as a *rejector circuit*. The half power points or cut off frequencies of the rejector circuit are given by the points at which the current is $\sqrt{2}\, I_r$.

4.14 A GENERAL PARALLEL RESONANT CIRCUIT

A general parallel resonant circuit consisting of ideal elements R, L and C is shown in Fig. 4.17.

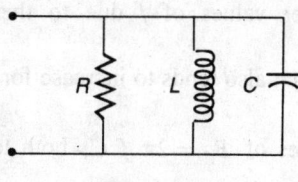

Fig. 4.17

The conductance of R is $G = \dfrac{1}{R}$.

The susceptance of L is $-jB_L = -j\dfrac{1}{X_L} = -j\dfrac{1}{\omega L}$.

The susceptance of C is $jB_C = j\dfrac{1}{X_C} = j\omega C$.

The total admittance of the circuit is given by, $Y = G + j\left(\omega C - \dfrac{1}{\omega L}\right)$.

For the circuit to be at resonance, Y should be a pure conductance. Hence, the imaginary part of Y is zero.

$$\therefore \qquad \omega_r C - \frac{1}{\omega_r L} = 0$$

or $$\omega_r^2 = \frac{1}{LC} \quad \text{or} \quad \omega_r = \frac{1}{\sqrt{LC}}$$

or $$f_r = \frac{1}{2\pi\sqrt{LC}} \tag{4.36}$$

4.15 VARIATION OF SUSCEPTANCES WITH FREQUENCY

Figure 4.18, shows the variation of the conductance G, inductive susceptance B_L, capacitive susceptance B_C and admittance Y, w.r.t. f.

G remains constant for all frequencies. As $B_L = \dfrac{1}{X_L} = \dfrac{1}{2\pi f L}$, as the frequency increases, the inductive susceptance decreases and has an hyperbolic variation. As $B_C = \omega C = 2\pi f C$, it increases linearly with frequency. The variation of Y with frequency is as shown, which tends to increase for lower values of f due to the increased values of

$B_L = \dfrac{1}{2\pi\, fL}$ and which also tends to increase for higher values of f due

to the increased values of $B_C = 2\pi f C$, both to the left and right of

f_r. At f_r, $Y = G$.

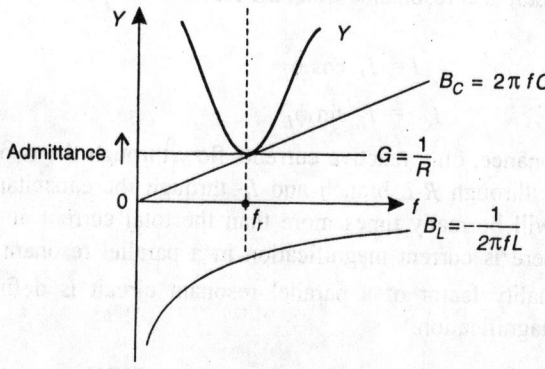

Fig. 4.18

4.16 Q FACTOR OF A PARALLEL RESONANT CIRCUIT (Q_P)

Consider a practical parallel resonant circuit as shown in Fig. 4.19.

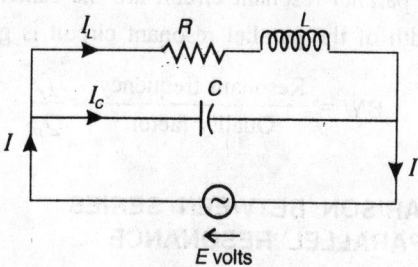

Fig. 4.19

The vector diagram for this circuit is as shown in Fig. 4.20. The current I_L lags E by an angle ϕ_L.

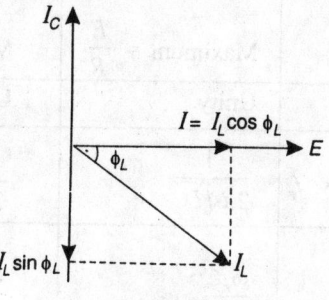

Fig. 4.20

The circuit is at resonance when the reactive component of the current I_L is zero.

i.e. $\qquad\qquad I = I_L \cos \phi_L$

and $\qquad\qquad I_C = I_L \sin \phi_L$

At resonance, only reactive currents flow through the two branches, $I_L \sin \phi_L$ through R-L branch and I_C through the capacitance. These currents will be many times more than the total current at resonance. Hence, there is current magnification in a parallel resonant circuit.

The quality factor of a parallel resonant circuit is defined as the current magnification.

$$\therefore \quad Q_p = \frac{\text{Current through capacitance at resonance}}{\text{Total current at resonance}}$$

$$= \frac{I_{Cr}}{I_r} = \frac{E\omega_r C}{E/Z_r} = Z_r \omega_r C = \frac{L}{RC} \times \omega_r C = \frac{\omega_r L}{R} = Q_s \qquad (4.37)$$

Hence, the equations for the quality factors of a series resonant circuit and a practical parallel resonant circuit are the same.

The bandwidth of the parallel resonant circuit is given by

$$\text{BW} = \frac{\text{Resonant frequency}}{\text{Quality factor}} = \frac{f_r}{Q_p} \qquad (4.38)$$

4.17 COMPARISON BETWEEN SERIES AND PARALLEL RESONANCE

Parameter	Series circuit	Parallel circuit
Impedance at resonance, Z_r	Minimum $= R$	Maximum $= \dfrac{L}{CR}$
Current at resonance, I_r	Maximum $= \dfrac{E}{R}$	Minimum $= \dfrac{ECR}{L}$
p.f. at resonance	Unity	Unity
Resonant frequency, f_r	$\dfrac{1}{2\pi\sqrt{LC}}$	$\dfrac{1}{2\pi}\sqrt{\dfrac{1}{LC} - \dfrac{R^2}{L^2}}$
Quality factor, $Q_s = Q_p$	$\dfrac{\omega_r C}{R}$	$\dfrac{\omega_r L}{R}$

WORKED EXAMPLES

4.1 A series *RLC* circuit has $R = 10\ \Omega$, $L = 0.1$ H, and $C = 100\ \mu$F and is connected across a 200 V, variable frequency source. Find (a) the resonant frequency, (b) impedance at this frequency, (c) the voltage drops across inductance and capacitance at this frequency, (d) quality factor and (e) bandwidth.

(Kuvempu University)

Solution

(a) $\quad f_r = \dfrac{1}{2\pi\sqrt{LC}} = \dfrac{1}{2 \times 3.14\ \sqrt{0.1 \times 100 \times 10^{-6}}} = 50.36$ Hz

(b) $\quad Z_r = R = 10\ \Omega, \qquad I_r = \dfrac{V}{R} = \dfrac{200}{10} = 20$ A

(c) $\quad X_{Lr} = 2 \times 3.14 \times 50.36 \times 0.1 = 31.63\ \Omega$

$\quad\quad V_{Lr} = V_{Cr} = I\,X_{Lr} = 31.63 \times 20 = 632.52$ V

(d) $\quad Q_s = \dfrac{X_L\ (\text{at resonance})}{R} = \dfrac{31.63}{10} = 3.163$

(e) $\quad$ bandwidth $= \dfrac{f_r}{Q_s} = \dfrac{50.36}{3.163} = 15.92$ Hz

4.2 For the example described in 4.1, find (a) the frequency at which the voltage across inductance is maximum and this voltage, (b) the frequency at which, the voltage across the capacitance is maximum and this voltage. (Mysore University)

Solution

(a) $\quad f_{L_{max}} = \dfrac{1}{2\pi} \sqrt{\dfrac{1}{LC - \dfrac{R^2 C^2}{2}}}$

$\quad\quad = \dfrac{1}{2 \times 3.14} \sqrt{\dfrac{1}{0.1 \times 100 \times 10^{-6} - \dfrac{10^2 \times \left(100 \times 10^{-6}\right)^2}{2}}}$

$\quad\quad = 51.66$ Hz

The impedance at this frequency is

$$Z = 10 + j2 \times 3.14 \times 51.66 \times 0.1$$

$$= (10 + j\,32.44)\ \Omega$$

$$= 33.95 \angle 72.87°\ \Omega$$

$$V_{L_{max}} = I\,X_L = \frac{E}{Z}\,X_L = \frac{200}{33.95} \times (32.44)$$

$$= 191.12\ \text{volts}$$

(b) $$f_{C_{max}} = \frac{1}{2\pi}\sqrt{\frac{1}{LC} - \frac{R^2}{2L^2}}$$

$$= \frac{1}{2 \times 3.14}\sqrt{\frac{1}{0.1 \times 100 \times 10^{-6}} - \frac{10^2}{2 \times 0.1^2}}$$

$$= 49.08\ \text{Hz}$$

The impedance at this frequency is

$$Z = 10 - j\,\frac{1}{2 \times 3.14 \times 49.08 \times 100 \times 10^{-6}}$$

$$= (10 - j\,32.44)\ \Omega = 33.95 \angle -72.87°\ \Omega$$

$$V_{C_{max}} = I\,X_C = \frac{E}{Z}X_C = \frac{200}{33.95} \times 32.44 = 191.12\ \text{volts}$$

4.3 A coil of resistance 20 Ω and inductance 0.2 H is connected in series with a capacitor across 230 V supply. Find (a) the value of the capacitance for which resonance occurs at 100 Hz, (b) the current through and voltage across the capacitor, and (c) Q factor of the coil. (Bangalore University)

Solution

(a) $$f_r = \frac{1}{2\pi\sqrt{LC}} \quad \text{i.e.}\quad 100 = \frac{1}{2 \times 3.14 \sqrt{0.2C}}$$

∴ $$C = 12.67\ \mu\text{F}$$

(b) $$I_r = \frac{E}{R} = \frac{230}{20} = 11.5\ \text{A}$$

$$V_{Cr} = I X_{Cr} = 11.5 \times \frac{1}{2 \times 3.14 \times 100 \times 12.67 \times 10^{-6}}$$

$$= 1445.3 \text{ V}$$

(c) $\quad Q_s = \dfrac{X_{Lr}}{R} = \dfrac{2\pi \, f_r L}{R} = \dfrac{2 \times 3.14 \times 100 \times 0.2}{20} = 6.28$

4.4 An *RLC* series circuit has a resistance of $10\,\Omega$, a capacitance of $100\,\mu\text{F}$ and a variable inductance. Find (a) the value of the inductance for which the voltage across the resistance is maximum, (b) Q factor, (c) voltage drops across R, L and C. The applied voltage is 230 V, 50 Hz. (Mangalore University)

Solution: (a) The voltage across the resistance is maximum at resonance:

$$\therefore \qquad f_r = \frac{1}{2\,\pi\,\sqrt{LC}}$$

i.e. $\quad 50 = \dfrac{1}{2 \times 3.14 \sqrt{L \times 100 \times 10^{-6}}}$

$\therefore \quad L_r$ = inductance at resonance

$$= 0.10142 \text{ H} = 101.42 \text{ mH}$$

(b) $\qquad Q_s = \dfrac{2\pi f L_r}{R} = \dfrac{2 \times 3.14 \times 50 \times 0.10142}{10} = 3.185$

(c) $\qquad I_r$ = current at resonance $= \dfrac{V}{R} = \dfrac{230}{10} = 23 \text{ A}$

$$V_R = 23 \times 10 = 230 \text{ V} = E$$

$$V_C = V_L = Q_s E = 3.185 \times 230 = 732.55 \text{ volts}$$

4.5 An a.c. series circuit consists of a coil connected in series with a capacitance. The circuit draws a maximum current of 10 A, when connected to 200 V, 50 Hz supply. If the voltage across the capacitor is 500 V at resonance, find the parameters of the circuit and quality factor. (Karnataka University)

Solution: Maximum current is drawn by the circuit under resonant conditions.

$$V_{Cr} = I_r \times X_C = 500,$$

$$\therefore \quad X_C = \frac{500}{10} = 50 \ \Omega$$

$$\therefore \quad C = \frac{1}{2\pi f X_C} = \frac{1}{2 \times 3.14 \times 50 \times 50} = 63.69 \ \mu F$$

At resonance, $X_L = X_C = 50 \ \Omega$

$$\therefore \quad L = \frac{X_L}{2\pi f} = \frac{50}{2 \times 3.14 \times 50} = 0.159 \ H$$

$$V_R = I_r R$$

i.e. $\quad 200 = 10 \ R$

$$\therefore \quad R = 20 \ \Omega$$

$$Q_s = \frac{V_L \text{ or } V_C}{E} = \frac{500}{200} = 2.5$$

4.6 A series *RLC* circuit has a resistance of 10 Ω, an inductance of 0.3 H and a capacitance of 100 μF. The applied voltage is 230 V. Find (a) the resonant frequency, (b) the quality factor, (c) lower and upper cut off frequencies, (d) bandwidth, (e) current at resonance, (f) currents at f_1, f_2 and (g) voltage across inductance at resonance.

(Gulbarga University)

Solution

(a) $\quad f_r = \dfrac{1}{2\pi\sqrt{LC}} = \dfrac{1}{2 \times 3.14 \ \sqrt{0.3 \times 100 \times 10^{-6}}} = 29.07$ Hz

(b) $\quad Q_s = \dfrac{X_{Lr}}{R} = \dfrac{2 \times 3.14 \times 29.07 \times 0.3}{10} = 5.48$

At f_1 $\qquad Q_s\delta = -0.5$

$\therefore \qquad \delta = \dfrac{-0.5}{5.48} = -0.0912$

But $\qquad \delta = \dfrac{f_1 - f_r}{f_r}$

$$\therefore \quad f_1 = \delta \, f_r + f_r = -0.0912 \times 29.07 + 29.07 = 26.42 \text{ Hz}$$

At $\qquad f_2, \; Q_s \; \delta = 0.5$

$$\therefore \qquad\qquad\qquad \delta = 0.0912.$$

(c) $\quad f_2 = \delta \, f_r + f_r = f_r \, (1 + \delta) = 29.07 \, (1 + 0.0912) = 31.72 \text{ Hz}$

(d) Bandwidth $= f_2 - f_1 = 31.72 - 26.42 = 5.3 \text{ Hz}$

(e) $\qquad I_r = \dfrac{E}{R} = \dfrac{230}{10} = 23 \text{ A}$

(f) Currents at f_1 and $f_2 = \dfrac{I_r}{\sqrt{2}} = 0.707 \times 23 = 16.261 \text{ A}$

(g) Voltage across inductance at resonance $= I_r \times \omega_r L$

$$= 23 \times (2 \times 3.14 \times 29.07 \times 0.3) = 1259.66 \text{ volts}$$

Note. Calculations of f_1 and f_2, using δ are approximate. The exact values of f_1 and f_2 may be calculated as follows.

$$f_1 = \frac{1}{2\pi} \left[-\frac{R}{2L} + \sqrt{\left(\frac{R}{2L}\right)^2 + \frac{1}{LC}} \right]$$

$$= \frac{1}{2 \times 3.14} \left[-\frac{10}{2 \times 0.3} + \sqrt{\left(\frac{10}{2 \times 0.3}\right)^2 + \frac{1}{0.3 \times 100 \times 10^{-6}}} \right]$$

$$= 26.54 \text{ Hz}$$

$$f_2 = \frac{1}{2\pi} \left[\frac{R}{2L} + \sqrt{\left(\frac{R}{2L}\right)^2 + \frac{1}{LC}} \right]$$

$$= \frac{1}{2 \times 3.14} \left[\frac{10}{2 \times 0.3} + \sqrt{\left(\frac{10}{2 \times 0.3}\right)^2 + \frac{1}{0.3 \times 100 \times 10^{-6}}} \right]$$

$$= 31.6 \text{ Hz}$$

4.7 A coil of resistance 20 Ω and inductance 4 H is connected in series with a capacitor across a supply of variable frequency. The resonance occurs at 60 Hz and the current at resonance is 2 A. Find the frequency, when the current is 1 A.   (Mangalore University)

Solution

$$f_r = \frac{1}{2\pi\sqrt{LC}}$$

i.e.

$$60 = \frac{1}{2 \times 3.14 \sqrt{4C}}$$

∴

$$C = 1.76 \ \mu F$$

At resonance, $E = I R = 20 \times 2 = 40$ V

When $I = 1$ A, $Z = \dfrac{E}{I} = \dfrac{40}{1} = 40 \ \Omega$

$$\omega L - \frac{1}{\omega C} = \sqrt{Z^2 - R^2} = \sqrt{40^2 - 20^2} = \pm 34.64 \ \Omega$$

Case (i) when, $\omega L - \dfrac{1}{\omega C} = 34.64$

$$\omega^2 LC - 1 = 34.64 \ \omega C = 34.64 \ \omega \times 1.76 \times 10^{-6}$$

$$= 60.97 \times 10^{-6} \ \omega$$

i.e. $\omega^2 \times 4 \times 1.76 \times 10^{-6} - 1 - 60.97 \times 10^{-6} \ \omega = 0$

i.e. $7.04 \times 10^{-6} \ \omega^2 - 60.97 \times 10^{-6} \ \omega - 1 = 0$

i.e. $7.04 \ \omega^2 - 60.97 \ \omega - 10^6 = 0$

$$\omega = \frac{60.97 \pm \sqrt{60.97^2 + 4 \times 7.04 \times 10^6}}{2 \times 7.04}$$

For +ve sign, we get $\omega = 381.24$ rad/sec and $f = 60.71$ Hz.

Case (ii) when, $\omega L - \dfrac{1}{\omega C} = -34.64$

∴

$$\omega^2 LC - 1 = -34.64 \omega C = -34.64 \times \omega \times 1.76 \times 10^{-6}$$

or $7.04 \ \omega^2 + 60.97 \ \omega - 10^6 = 0$

On solving we get, $\omega = 372.58$ rad/sec and $f = 59.33$ Hz.

4.8 A coil of resistance 20 Ω and inductance 10 mH is in series with a capacitance and is supplied with a constant voltage and variable frequency source. The maximum current is 2 A at 1000 Hz. Find the cut off frequencies. (Kuvempu University)

Solution

$$Q_s = \frac{\omega_r L}{R} = \frac{2\pi f_r L}{R} = \frac{2 \times 3.14 \times 1000 \times 10 \times 10^{-3}}{20} = 3.14$$

$$\omega_2 - \omega_1 = \frac{\omega_r}{Q_s} = \frac{2 \times 3.14 \times 1000}{3.14} = 2000 \text{ rad/sec}$$

$$f_2 - f_1 = \frac{2000}{2\pi} = 318.47 \text{ Hz} \tag{1}$$

$$f_1 f_2 = f_r^2 = 1000^2$$

$$f_2 + f_1 = \sqrt{(f_2 - f_1)^2 + 4 f_1 f_2}$$

$$= \sqrt{(318.47)^2 + 4 \times 1000^2}$$

$$= 2025.2 \text{ Hz} \tag{2}$$

Solving (1) and (2), we get

$$\therefore \qquad f_2 = 1171.84 \text{ Hz}$$

$$f_1 = 853.37 \text{ Hz}$$

4.9 A voltage of $e = 100 \sin \omega t$ is applied to an *RLC* series circuit. At resonant frequency, the voltage across the capacitor was found to be 400 V. The bandwidth is 75 Hz. The impedance at resonance is 100 Ω. Find the resonant frequency and the constants of the circuit. (Mysore University)

Solution

$$E = \frac{100}{\sqrt{2}} = 70.7 \text{ volts}$$

$$Q_s = \frac{V_C}{E} = \frac{400}{70.7} = 5.66$$

At resonance, $\quad R = Z = 100 \ \Omega$

$$\text{Bandwidth} = \frac{f_r}{Q_s}$$

i.e. $\qquad f_r = 5.66 \times 75 = 424.5$ Hz

But $\qquad f_2 - f_1 = \dfrac{R}{2\pi L}$,

$\therefore \qquad L = \dfrac{R}{2\pi(f_2 - f_1)} = \dfrac{100}{2 \times 3.14 \times 75} = 212.3$ mH

$$Q_s = \dfrac{1}{\omega_r CR}$$

i.e. $\qquad 5.66 = \dfrac{1}{2 \times 3.14 \times 424.5 \times 100\, C}$

$\therefore \qquad C = 0.663\ \mu$F

4.10 A constant voltage at a frequency of 1 MHz is applied to an inductor in series with a variable capacitor. When the capacitor is set at 500 pF, the current has its maximum value, while it is reduced to one half, when the capacitance is 600 pF. Find (i) the resistance and inductance of the coil and (ii) Q factor. (Bangalore University)

Solution: When $C = 500$ pF, current is maximum, representing resonant condition.

$\therefore \qquad f_r = \dfrac{1}{2\pi\sqrt{LC}} = \dfrac{1}{2 \times 3.14 \sqrt{L \times 500 \times 10^{-12}}} = 10^6$

$\therefore \qquad L = 0.05$ mH

and $\qquad X_L = 2 \times 3.14 \times 10^6 \times 0.05 \times 10^{-3}\ \ 314\ \Omega$

When $\qquad C = 600$ pF

$$X_C = \dfrac{1}{2 \times 3.14 \times 10^6 \times 600 \times 10^{-12}} = 265.4\ \Omega$$

$$X_L - X_C = 314 - 265.4 = 48.6\ \Omega$$

When the current is $\dfrac{I_m}{2}$, $Z = 2R$

$\therefore \qquad R^2 + (48.6)^2 = (2R)^2$

$\therefore \qquad R = \dfrac{48.6}{\sqrt{3}} = 28.06\ \Omega$

$$Q_s = \dfrac{\omega_r L}{R} = \dfrac{314}{28.06} = 11.19$$

4.11 A coil of resistance 40 Ω and inductance 0.75 H forms a part of a series circuit for which, the resonant frequency is 55 Hz. If the supply is 250 V, 50 Hz, find (a) the line current, (b) the p.f., and (c) the voltage across the coil. (Gulbarga University)

Solution: At resonance

$$\frac{1}{\omega_r C} = \omega_r L = 2 \times 3.14 \times 55 \times 0.75 = 259.05 \ \Omega$$

$$= X_C \text{ at 55 Hz}$$

$$X_L \text{ at 50 Hz} = 2 \times 3.14 \times 50 \times 0.75 = 235.5 \ \Omega$$

$$X_C \text{ at 50 Hz} = 259.05 \times \frac{55}{50} = 284.96 \ \Omega$$

$$Z \text{ at 50 Hz} = 40 + j \left(X_L - X_C \right) = 40 + j \left(235.5 - 284.96 \right)$$

$$= 40 - j \ 49.46 = 63.61 \ \angle - 51.04° \ \Omega$$

(a) Current $= \dfrac{250}{63.61} = 3.93$ A

(b) p.f. $= \cos 51.04° = 0.629$ leading

(c) The voltage across the coil $= I \ Z_{\text{coil}}$

$$= I \sqrt{R^2 + X_L^2} = 3.93 \sqrt{40^2 + 235.5^2} = 938.77 \text{ V}$$

4.12 A coil of 20 Ω resistance has an inductance of 0.2 H and is connected in parallel with a 100 µF capacitor. Calculate the frequency at which the circuit will act as a non-inductive resistance of R ohms. Find also the value of R. (Mangalore University)

Solution

$$f_r = \frac{1}{2\pi} \sqrt{\frac{1}{LC} - \frac{R^2}{L^2}} = \frac{1}{2 \times 3.14} \sqrt{\frac{1}{0.2 \times 100 \times 10^{-6}} - \frac{20^2}{0.2^2}}$$

$$= 31.85 \text{ Hz}$$

The dynamic resistance is given by

$$Z_r = \frac{L}{CR} = \frac{0.2}{100 \times 10^{-6} \times 20} = 100 \ \Omega$$

4.13 A coil of resistance 10 Ω and inductance 0.5 H is connected in series with a capacitor. On applying a sinusoidal voltage, the current is maximum, when the frequency is 50 Hz. A second capacitor is connected in parallel with the circuit. What capacitance must it have, so that the combination acts as a non-inductive resistor at 100 Hz? Calculate the total current supplied in each case, if the applied voltage is 220 V.   (Bangalore University)

Solution: At 50 Hz, current is maximum, i.e. the circuit is under resonance.

$\therefore$ $X_C = X_L = 2 \times 3.14 \times 50 \times 0.5 = 157 \ \Omega$

The inductive reactance at 100 Hz will be 2 times the inductive reactance at 50 Hz.

$\therefore$ X_L at 100 Hz $= 2 \times 157 = 314 \ \Omega$

The capacitive reactance at 100 Hz is 1/2 of the capacitive reactance at 50 Hz.

$\therefore$ X_C at 100 Hz $= \dfrac{157}{2} = 78.5 \ \Omega$

The impedance at 100 Hz is given by

$$Z = 10 + j\,(314 - 78.5) = 10 + j\,235.5$$

$$= 235.7 \ \angle 87.57° \ \Omega$$

$\therefore$ $Y = \dfrac{1}{Z} = 0.0042426 \ \angle -87.57°$

$$= (0.0001798 - j\,0.0042388) \ \mho$$

$\therefore$ The capacitive susceptance $= \omega C = 0.0042388$
$\therefore$ The required value of capacitance is given by

$$C = \frac{0.0042388}{2 \times 3.14 \times 100} = 6.75 \ \mu F$$

At 50 Hz, $I = \dfrac{E}{R} = \dfrac{220}{10} = 22$ A

At 100 Hz, $I = E \times$ Real part of Y

$$= 220 \times 0.0001798 = 0.04 \ A$$

4.14   A circuit has an inductive reactance of 20 Ω at 50 Hz in series with a resistance of 15 Ω. For an applied voltage of 200 V at 50 Hz, calculate (a) phase angle between current and voltage, (b) the current, (c) the value of shunting capacitance to bring the circuit to resonance and the current at resonance.

                                        (Karnataka University)

Solution

(a) $Z = R + j X_L = 15 + j20 = 25 \angle 53.13° \, \Omega$

$\therefore \quad \phi = 53.13° = $ phase difference

(b) $\qquad I = \dfrac{V}{Z} = \dfrac{200}{25} = 8 \text{ A}$

(c) $\qquad Y = \dfrac{1}{Z} = \dfrac{1}{25 \angle 53.13°} = 0.04 \angle -53.13°$

$\qquad\qquad = 0.024 - j0.032 = G - j \, \omega C$

For the circuit to be at resonance

$$\omega C = 0.032$$

$\therefore \qquad\qquad C = \dfrac{0.032}{2 \times 3.14 \times 50} = 101.9 \, \mu F$

Current at resonance $= E \times$ real part of $Y = 200 \times 0.024 = 4.8$ A

4.15 An inductive coil of resistance 6 Ω and inductance 1 mH is connected in parallel with another branch consisting of a resistance of 4 Ω in series with a capacitance of 20 μF. Find the resonant frequency and the corresponding current, when the applied voltage is 200 V. (Kuvempu University)

Solution

$$f_r = \frac{1}{2\pi\sqrt{LC}} \sqrt{\frac{R_L^2 - \dfrac{L}{C}}{R_C^2 - \dfrac{L}{C}}}$$

$$= \frac{1}{2 \times 3.14\sqrt{10^{-3} \times 20 \times 10^{-6}}} \sqrt{\frac{6^2 - \dfrac{10^{-3}}{20 \times 10^{-6}}}{4^2 - \dfrac{10^{-3}}{20 \times 10^{-6}}}} = 722.93 \text{ Hz}$$

$$\omega_r = 2 \times 3.14 \times 722.93 = 4540 \text{ rad/sec}$$

$$I_r = E \left[\frac{R_L}{R_L^2 + \omega_r^2 \, L^2} + \frac{R_C}{R_C^2 + \dfrac{1}{\omega_r^2 C^2}} \right]$$

$$= 200 \left[\frac{6}{6^2 + 4540^2 \times 10^{-6}} + \frac{4}{4^2 + \dfrac{1}{4540^2 \times 20^2 \times 10^{-12}}} \right]$$

$$= 27.03 \text{ A}$$

4.16 Find the values of L for which, the circuit given in Fig. 4.21 resonates at $\omega = 5000$ rad/sec. (Gulbarga University)

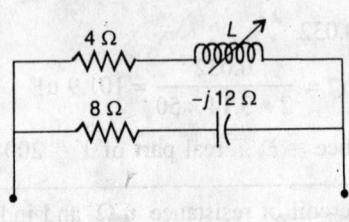

Fig. 4.21

Solution: The total admittance of the circuit is given by

$$Y = \left(\frac{1}{4 + j\,X_L} + \frac{1}{8 - j\,12} \right) \mho \qquad \text{on rationalising}$$

$$= \frac{4 - jX_L}{4^2 + X_L^2} + \frac{8 + j12}{8^2 + 12^2}$$

$$= \left(\frac{4}{4^2 + X_L^2} + \frac{8}{8^2 + 12^2} \right) + j \left(\frac{12}{8^2 + 12^2} - \frac{X_L}{4^2 + X_L^2} \right)$$

At resonance, the imaginary part of Y is zero.

$$\therefore \quad \frac{12}{8^2 + 12^2} = \frac{X_L}{4^2 + X_L^2}, \text{ on simplifying, we get } 3X_L^2 - 52\,X_L + 48 = 0$$

on solving, we get

$$X_L = \frac{52 \pm \sqrt{52^2 - (4 \times 3 \times 48)}}{2 \times 3} = 16.36 \ \Omega \text{ or } 0.978 \ \Omega$$

$$\therefore \qquad L = \frac{16.36}{5000} \text{ mH} = 3.27 \text{ mH}$$

or
$$\frac{0.978}{5000} \text{ mH} = 0.196 \text{ mH}$$

4.17 Find the values of C for which the circuit given in Fig. 4.22 resonates at 750 Hz. (Mysore University)

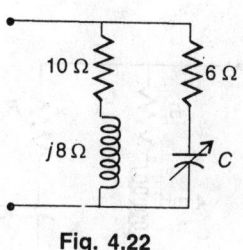

Fig. 4.22

Solution: The admittance of the circuit is given by

$$Y = Y_1 + Y_2 = \frac{1}{10 + j8} + \frac{1}{6 - j X_C}, \text{ on rationalising, we get}$$

$$Y = \frac{10 - j8}{10^2 + 8^2} + \frac{6 + j X_C}{6^2 + X_C^2} = \left(\frac{10}{164} + \frac{6}{36 + X_C^2}\right) + j\left(\frac{X_C}{36 + X_C^2} - \frac{8}{164}\right)$$

At resonance, the imaginary part of Y is zero.

$$\therefore \qquad \frac{X_C}{36 + X_C^2} - \frac{8}{164} = 0$$

i.e. $\qquad 164 X_C - 8 X_C^2 - 288 = 0$

i.e. $\qquad 2 X_C^2 - 41 X_C + 72 = 0$

$$X_C = \frac{41 \pm \sqrt{41^2 - 4 \times 2 \times 72}}{2 \times 2}$$

$$= \frac{41 \pm 33.242}{4} = 18.56 \ \Omega, 1.94 \ \Omega$$

$$\therefore \qquad X_{C1} = 18.56 \ \Omega$$

i.e. $\qquad C_1 = \dfrac{1}{2 \times 3.14 \times 750 \times 18.56} = 11.44 \ \mu F$

$$\therefore \qquad X_{C2} = 1.94 \ \Omega$$

i.e.
$$C_2 = \frac{1}{2 \times 3.14 \times 750 \times 1.94} = 109.45\ \mu F$$

4.18 Determine R_L and R_C for which the circuit shown in Fig. 4.23 resonates at all frequencies.   (Kuvempu University)

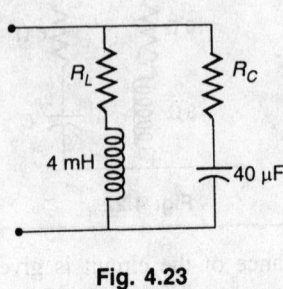

Fig. 4.23

Solution: The circuit is resonant at

$$f_r = \frac{1}{2\pi\sqrt{LC}} \sqrt{\frac{R_L^2 - L/C}{R_C^2 - L/C}}$$

The circuit can resonate at any frequency, if $R_L^2 = R_C^2 = \dfrac{L}{C}$

$\therefore$
$$R_L = R_C = \sqrt{\frac{L}{C}} = \sqrt{\frac{4 \times 10^{-3}}{40 \times 10^{-6}}} = 10\ \Omega$$

4.19 Find the value of R_L for which, the circuit shown in Fig. 4.24 is resonant.

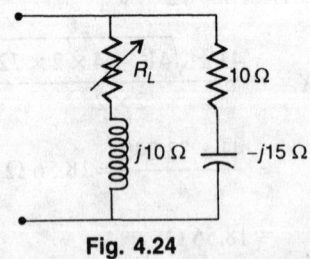

Fig. 4.24

Solution: The admittance of the circuit is given by

$$Y = \frac{1}{R_L + j10} + \frac{1}{10 - j15} = \frac{R_L - j10}{R_L^2 + 100} + \frac{10 + j15}{100 + 225}$$

$$= \left(\frac{R_L}{R_L^2 + 100} + \frac{10}{325} \right) + j \left(\frac{15}{325} - \frac{10}{R_L^2 + 100} \right)$$

For the circuit to be at resonance, the imaginary part of Y is zero.

$$\therefore \qquad \frac{15}{325} - \frac{10}{R_L^2 + 100} = 0$$

On solving, we get $\qquad\qquad R_L = 10.8\ \Omega$

4.20 Two impedances $(10 + j12)\ \Omega$ and $(20 - j15)\ \Omega$ are connected in parallel and this combination is connected in series with an impedance $(5 - jX_C)\ \Omega$. Find the value of X_C, for which resonance occurs.

(Mysore University)

Solution

$$Z = (5 - j\,X_C) + \frac{(10 + j\,12)(20 - j\,15)}{(10 + j\,12) + (20 - j\,15)}$$

$$= 17.24 + j\,(4.22 - X_C)$$

For the circuit to be at resonance, the imaginary part of Z is zero.

$$\therefore \qquad X_C = 4.22\ \Omega$$

NUMERICAL PROBLEMS

4.1 An *RLC* series circuit has $R = 5\ \Omega$, $L = 100$ mH and $C = 150\ \mu$F and is connected across a 230 V, variable frequency supply. Find (a) the resonant frequency, (b) impedance at this frequency, (c) the drops across inductance and capacitance at this frequency, (d) Q factor and (e) bandwidth.

4.2 For the problem given in 4.1, find (a) f_1, f_2, (b) BW, (c) Q factor, (d) the frequency at which maximum value of V_C occurs and this V_C, (e) the frequency at which maximum value of V_L occurs and this V_L.

4.3 A coil of resistance 10 Ω and inductance 0.1 H is connected in series with a capacitance across 230 V supply. Find (a) the value of the capacitance for which resonance occurs at 1000 rad/sec, (b) the voltage across the coil and (c) Q factor of the coil.

4.4 An *RLC* series circuit has $R = 20\ \Omega$, $C = 150\ \mu$F and an inductance of L henrys. It is connected across 200 V, 50 Hz supply. Find (a) L for which resonance occurs, (b) Q factor, (c) voltage across R, L and C, (d) bandwidth.

4.5 An RLC series circuit has an inductive coil of $R \, \Omega$ and inductance L henrys in series with a capacitance of C farads. The circuit draws a maximum current of 15 A, when connected to 230 V, 50 Hz supply. If the Q factor is 5, find the parameters of the circuit.

4.6 A series RLC circuit has a resistance of 15 Ω, and inductance of 100 mH and a capacitance of 100 μF. The appplied voltage is 200 V. Find (a) the resonant frequency, (b) quality factor, (c) f_1 and f_2 using the principle of deviation or otherwise (d) bandwidth, (e) current at resonance, (f) currents at f_1 and f_2 and (g) voltage across R, L and C at resonance.

4.7 A coil of resistance 10 Ω and inductance L henrys is connected in series with a capacitor of 2 μF across a supply of variable frequency. The resonance occurs at 100 Hz and the current at resonance is 5 A. Find the frequency, when the current is 2 A.

4.8 A coil of resistance 10 Ω is connected in series with a capacitance of 10 μF and is supplied with a constant voltage, variable frequency source. The maximum current is 5 A at 500 Hz. Find f_1 and f_2.

4.9 A voltage of $100 \sqrt{2} \sin \omega t$ is applied to an RLC series circuit. At resonant frequency, the voltage across the inductance is 500 V. The bandwidth is 50 Hz. The impedance at resonance is 75 Ω. Find the resonant frequency and the constants of the circuit.

4.10 A constant voltage of frequency 0.5 MHz is applied to an inductor in series with a variable capacitor. When the capacitor is set to 600 pF, resonance occurs. When it is set to 800 pF, the current in the circuit is 1/2 of that at resonance. Find (a) the resistance and inductance of the coil, and (b) Q factor.

4.11 A coil of resistance 20 Ω and inductance 1 H is connected in series with a capacitance C. The resonant frequency is 100 rad/sec. If the supply is 230 V, 50 Hz, find (a) the line current, (b) p.f., and (c) the voltage across the coil and capacitance.

4.12 A practical parallel resonant circuit consists of an inductor of resistance 10 Ω and reactance 314 Ω, connected in parallel with a capacitance of 100 μF. Find the resonant frequency and the dynamic resistance of the circuit. The supply is at 50 Hz.

4.13 A coil of resistance 20 Ω and inductance 0.2 H is connected in series with a capacitor. Resonance occurs at 50 Hz. An inductance L is connected across the circuit. What is the value of L, so that the resonance occurs at 25 Hz? Calculate the total current supplied in each case, if the applied voltage is 100 V.

4.14 An inductor of resistance 10 Ω and inductance 0.1 H is connected across a capacitance C. If the applied voltage is 230 V at 50 Hz, find the value of C at which resonance occurs. Also find the current at resonance.

4.15 An inductive coil of resistance 10 Ω and inductance 2 mH is connected in parallel with another branch consisting of a resistance of 20 Ω in series with a capacitance of 50 μF. Find the resonant frequency and the corresponding current, when the applied voltage is 230 V.

4.16 Find the values of L for which the circuit given in Fig. 4.25 resonates at 1000 Hz.

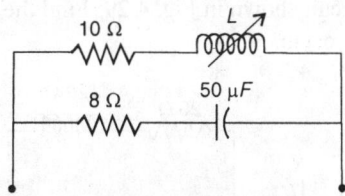

Fig. 4.25

4.17 Find the values of C for which the circuit given in Fig. 4.26 resonantes at 3000 rad/sec.

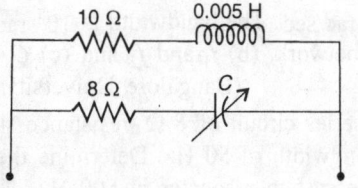

Fig. 4.26

4.18 Determine R_L and R_C for which the circuit shown in Fig. 4.27 resonates at all frequencies.

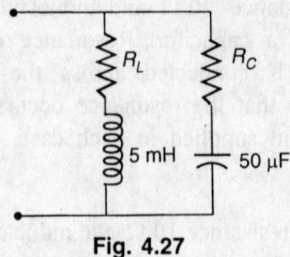

Fig. 4.27

4.19 Find the value of R_L for which the circuit shown in Fig. 4.28 is resonant. The supply frequency is 50 Hz.

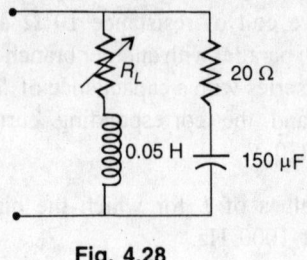

Fig. 4.28

4.20 For the circuit shown in Fig. 4.29, Find the value of X_C for which resonance occurs.

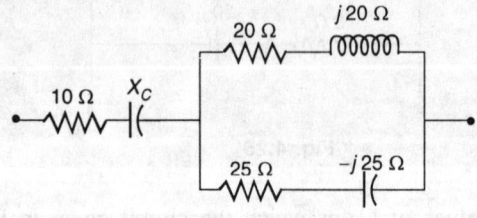

Fig. 4.29

4.21 In a series circuit, $R = 6\,\Omega$. The resonant frequency is 4.1×10^6 rad/sec. The bandwidth is 10^5 rad/sec. Find (a) L and C of the network, (b) f_1 and f_2 and (c) Q of the circuit.

(Bangalore University, April 85, Dec. 90)

4.22 An *RLC* series circuit of $8\,\Omega$ resistance should be designed to have a bandwidth of 50 Hz. Determine the values of L and C, so that the system resonates at 500 Hz.

(Bangalore University, April 91)

4.23 Determine R_L and R_C, which cause the circuit shown in Fig. 4.30 to be resonant at all frequencies.

(Bangalore University Jan. 83 and Karnataka University Dec. 94)

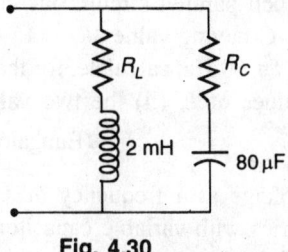

Fig. 4.30

4.24 A series circuit consisting of $R = 16\ \Omega$, $L = 0.4$ H and $C = 17.6\ \mu$F is connected to a variable frequency source, the potential of which is maintained constant at 120 V. If the frequency is varied through a range of 40 to 80 Hz, calculate (a) the resonant frequency, (b) the current and p.f. at 40 Hz, (c) the current and p.f. at 80 Hz.

4.25 For the network shown in Fig. 4.31, find the values of C, for a resonant frequency of 80 Hz. (Bangalore University March 83)

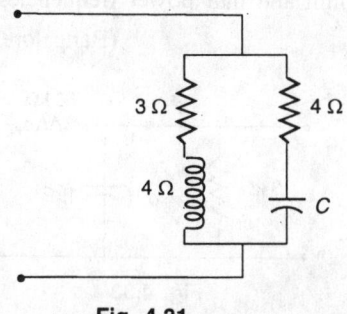

Fig. 4.31

4.26 Find the values of C, which results in resonance for the circuit shown in Fig. 4.32. The resonant frequency is 5000 rad/sec.

(Bangalore University Oct. 83)

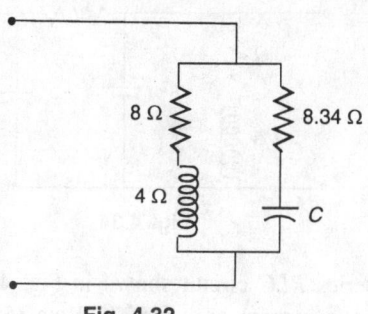

Fig. 4.32

4.27 For a two branch parallel circuit, one consisting of $R_L - L$ and the other $R_L - C$ having values $R_L = 15\ \Omega$, $R_C = 30\ \Omega$, $X_C = 30\ \Omega$, $V = 120$ V, $f = 60$ Hz, calculate, for the condition of resonance (i) the two values of L, (ii) the two values of total current.

(Bangalore University, Oct. 94)

4.28 A constant voltage at a frequency of 1 MHz is applied to an inductor in series with variable capacitor. When the capacitor is set to 500 µF, the current has a maximum value, while it is reduced to one half, when the capacitor is set to 600 µF. Find the resistance, inductance and Q of the inductor.

4.29 A coil with $R = 10\ \Omega$ and $L = 0.2$ H is in series with a capacitor of 20 µF . Determine (i) resonant frequency, (ii) Q factor and (iii) bandwidth. (Bangalore University, Aug. 94)

4.30 For the network shown in Fig. 4.33, determine (i) resonant frequency, (ii) input admittance at resonance, (iii) quality factor, (iv) bandwidth and half power frequencies.

(Bangalore University, Jan. 94)

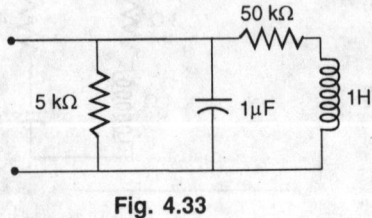

Fig. 4.33

4.31 Find the resonant frequency of the two terminal network shown in Fig. 4.34.

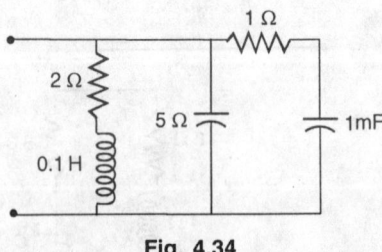

Fig. 4.34

4.32 In the series *RLC* circuit shown in Fig. 4.35, the instantaneous voltage and current are $v = 70.7 \sin (500\ t + 30°)$ volts and

$i = 2.83 \sin (500\, t + 30°)$ amps. Find R and C.

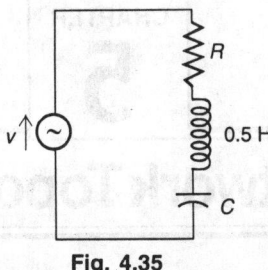

Fig. 4.35

4.33 In the circuit shown in Fig. 4.36, the impedance of the source is $(5 + j3)\,\Omega$, and the source frequency is 2000 Hz. At what value of C, will the power in 10 Ω resistor be maximum?

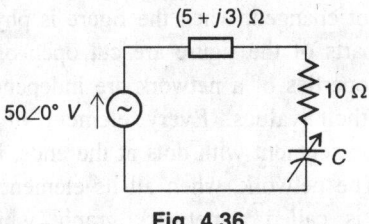

Fig. 4.36

4.34 A series *RLC* circuit with $L = 25$ mH and $C = 75$ μF, has a lagging phase angle of 25° at $\omega = 2000$ rad/sec. At what frequency will the phase angle be 25° leading. Find ω_r.

4.35 An *RLC* series circuit with $L = 0.5$ H has an instantaneous voltage $v = 70.7 \sin (500t + 30°)$ volts and an instantaneous current $i = 1.5 \sin 500t$ amps. Find the values of R and C. At what frequency ω_r, will the circuit be resonant.

4.36 A circuit with $R = 10\,\Omega$, $L = 0.2$ H and $C = 40\,\mu$F, has an applied voltage with a variable frequency. Find the frequencies at which, the current leads the voltage by 30°, is in phase, and lags the voltage by 30°, respectively.

4.37 A series *RLC* circuit with $R = 25\,\Omega$ and $L = 0.6$ H results in a leading phase angle of 60° at a frequency of 40 Hz. At what frequency will the circuit be resonant?

CHAPTER
5

Network Topology

5.1 INTRODUCTION

The various methods of solving electric circuits are, by applying Kirchhoff's laws, using mesh current analysis, node voltage analysis and several theorems. Network Topology is another method of solving electric circuits, which is a generalised approach. Topology is a branch of geometry, which is concerned with the properties of a geometrical figure, which are not changed, when the figure is physically distorted, provided that, no parts of the figure are cut open or joined together. The geometrical properties of a network are independent of the types of elements and their values. Every element of the network is represented by a line segment with dots at the ends, irrespective of its nature and value. The network, when all its elements are represented by line segments is called a network graph, which is explained in more detail in the next sections. The binary approach is used to solve the network, which is represented by a graph. Hence, the topological approach is highly suitable for solving circuits using a computer.

5.2 DEFINITIONS

(i) **Node:** A node is a point in an electrical network to which two or more elements are connected.

(ii) **Branch:** A branch is a line segment which represents a network element or a combination of elements connected between two nodes.

In general, a branch may represent a single resistance R or a capacitor C or an inductor L or a combination of these elements. The flexibility of choosing elements is illustrated in Fig. 5.1.

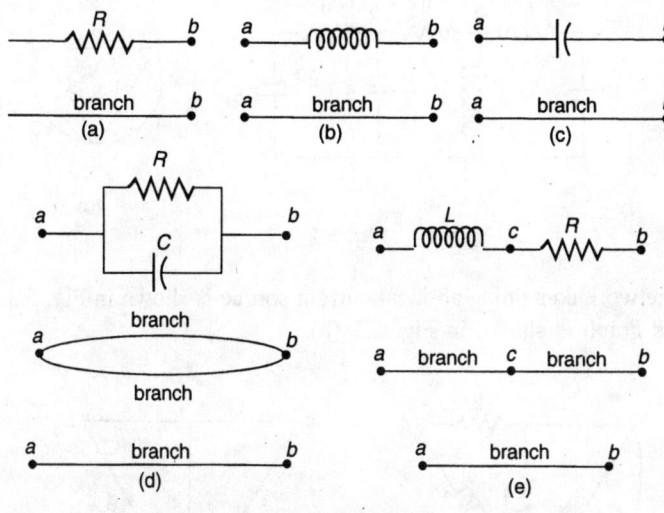

Fig. 5.1

a, b, c are the nodes. The parallel combination of R and C in Fig. 5.1 (d) can be regarded as a combination of two branches or a single branch. Similarly, the series combination of L and R as shown in Fig. 5.1 (e) may be regarded as a combination of two branches or a single branch.

(iii) **Path:** The path in a network constitutes a set of elements traversed such that no node is passed through again.

(iv) **Loop:** Any closed path in a network.

(v) **Mesh:** Any loop which does not contain any other loops in it, i.e. an independent loop is a mesh.

(vi) **Planar circuit:** A circuit which may be drawn on a plane surface such that, no branch passes over or under any other branch.

(vii) **Non-planar circuit:** Any circuit which is not planar.

5.3 GRAPH OF A NETWORK

The graph of a network is a geometric figure in which, all the passive elements are represented by line segments, all the ideal voltage sources are represented by short-circuits and all the ideal current sources are represented by open-circuits, retaining all the nodes.

Consider an electrical network as shown in Fig. 5.2 (a). The graph of this network is shown in Fig. 5.2 (b).

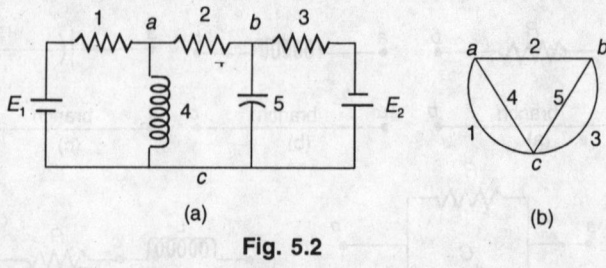

Fig. 5.2

A network containing an ideal current source is shown in Fig. 5.3 (a) and its graph is shown in Fig. 5.3 (b).

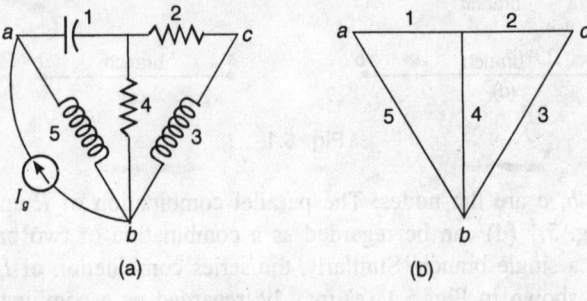

Fig. 5.3

A network containing a transformer is shown in Fig. 5.4 (a) and its graph is shown in Fig. 5.4 (b). As the nodes c_1 and c_2 are at the same potential, they are represented by a single node c in the graph.

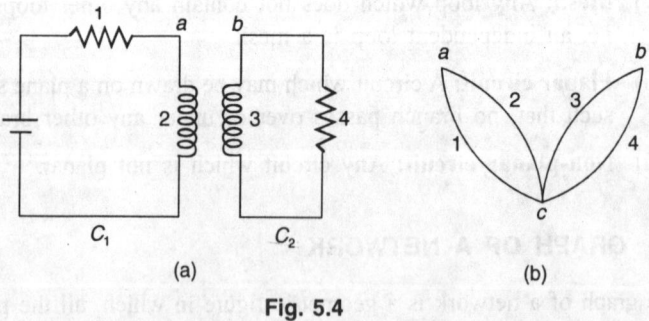

Fig. 5.4

From the study of the above networks and their graphs, we realise that, irrespective of their types and values, all the elements are represented by line segments, which are only recognised by their numbers, 1, 2, 3, ..., etc. The nodes are identified by letters a, b, c, ..., etc.

5.4 ORIENTED GRAPH

A graph is said to be oriented, when all its nodes are named, all its branches are numbered and arbitrary orientations are assigned to the branches. The arbitrary orientations in the branches incidentally indicate the directions of branch currents. The oriented graph of the network given in Fig. 5.5 (a) is as shown in Fig. 5.5 (b).

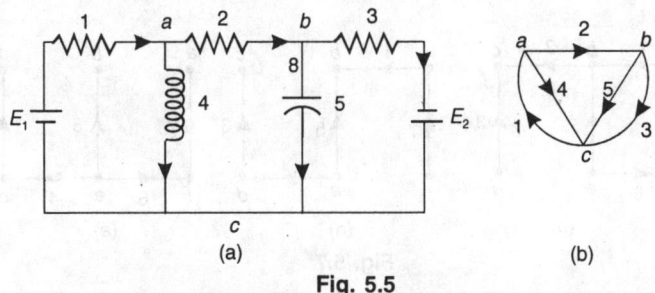

(a)   (b)

Fig. 5.5

5.5 TREE AND CO-TREE

A *tree* is a 'sub-graph' of a graph, which does not contain any loops. A network graph can have several trees, depending on the branches that are removed from the graph to form the trees. The number of possible trees for a given graph is discussed later in section 5.14.

Consider an oriented graph as shown in Fig. 5.6.

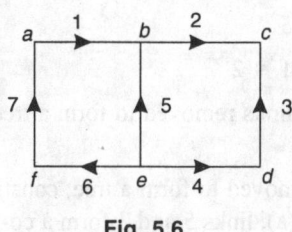

Fig. 5.6

The various trees of the graph are shown in Fig. 5.7 (a), (b), (c), (d) and (e).

The tree of a particular graph is formed by removing a few branches of the graph such that, it does not contain any loops. Branches 3 and 5 are removed to form the tree in Fig. 5.7 (a). Branches 7 and 3 are removed to form the tree in Fig. 5.7 (b) and so on.

The branches that are removed from a graph to form a particular tree are called **links** or **chords**. The branch of a tree is called **tree-branch** or **twig**.

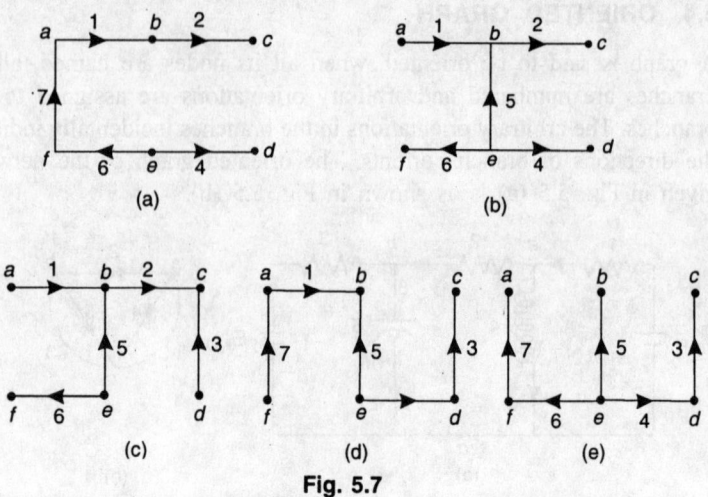

Fig. 5.7

Let n = number of nodes in a graph.

  b = number of branches in the graph.

The number of twigs t is given by, $t = n - 1$

The number of links is given by

$$l = b - t = b - (n - 1) = b - n + 1$$

For the graph discussed above and the trees drawn

  $b = 7$, $n = 6$ and therefore

  $t = 6 - 1 = 5$

  $l = 7 - 6 + 1 = 2$

It means that, two links removed to form a tree and the tree contains 5 branches.

The set of links removed to form a tree, constitute a **co-tree**. For the tree given in Fig. 5.7 (a), links 5 and 3 form a co-tree. For the tree given in Fig. 5.7 (b), links 7 and 3 form a co-tree and so on.

5.6 NETWORK VARIABLES

The solution of a network is nothing, but the calculation of currents flowing in the various branches and to find the voltages across the various branches. Hence, there are two types of network variables.

  (i) Current variables, and (ii) Voltage variables

The current variables may be either branch currents or loop currents. The voltage variables may be either node to datum voltages or node-pair voltages. In node voltage analysis, node to datum voltages are network variables. Depending on the method of analysis, these variables are assumed arbitrarily in the network and the network is analysed using appropriate method of analysis. The network variables are independent variables and all other quantities depend on their values.

Branch current: The current flowing in any branch of a network. Consider the network as shown in Fig. 5.8.

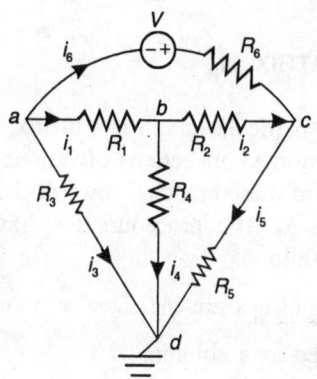

i_1, i_2 ... i_6 are branch currents.

Fig. 5.8

Branch Voltage

The voltage across any branch of the network. For Ex.: Consider the branch $a\ b$ of the above network. e_1 is the branch voltage. $e_1 = i_1\ R_1$.

As the current flows from node a to node b, point a is at a higher potential than b.

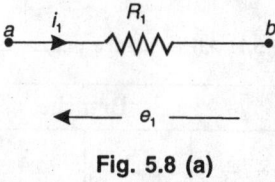

Fig. 5.8 (a)

e_1, e_2, e_3 ..., etc. represent the branch voltages in general.

Node-Pair Voltage

The voltage between any two nodes of a network is known as the node-pair voltage. In general all branch voltages are node-pair voltages. But, they are usually referred as tree-branch voltages in particular and are represented by, v_1, v_2, v_3 ..., etc.

Node to Datum Voltage

The potential of a particular node w.r.t. a reference node, whose potential is assumed to be zero, is known as the node to datum voltage. In the above Fig., the node d is the reference node. v_a, v_b and v_c are the node to datum voltages.

In the following sections, we discuss in detail, how these variables are assumed and networks are solved using topological approach.

5.7 ALL INCIDENCE MATRIX A_a

The all incidence matrix of a particular graph is a matrix, which gives complete information regarding the connections of various branches to the nodes and the orientation of these branches towards the nodes. The order of the matrix is $(n \times b)$. The procedure for drawing the all incidence matrix table is as follows.

1. The numbers of the branches are indicated as a row.

2. The nodes are indicated as a column.

3. If a particular branch is not connected to a particular node, the corresponding element in the matrix table is filled with 0.

4. If a particular branch is connected to a particular node and its orientation is away from the node, the corresponding element is filled as +1, and if the orientation is towards the node, it is filled as −1. The above assumption of +1 and −1 is only arbitrary.

Consider the oriented graph as shown in Fig. 5.9. The incidence matrix table for this oriented graph is as shown in Table 5.1.

Table 5.1: All incidence matrix table

Nodes	Branches					
	1	2	3	4	5	6
a	−1	+1	+1	0	0	0
b	0	−1	0	−1	+1	0
c	0	0	−1	+1	0	+1
d	+1	0	0	0	−1	−1

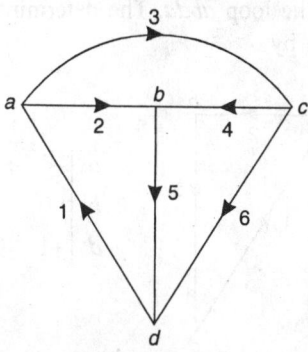

Fig. 5.9

The elements of the all incidence matrix table also represent the elements of the all incidence matrix A_a as shown in (5.1). The all incidence matrix is also sometimes called as 'node incidence matrix'.

$$
A_a = \begin{array}{c} \\ \text{Nodes} \\ a \\ b \\ c \\ d \end{array}
\begin{array}{c} \text{Branches} \\ \begin{array}{cccccc} 1 & 2 & 3 & 4 & 5 & 6 \end{array} \\
\begin{bmatrix} -1 & +1 & +1 & 0 & 0 & 0 \\ 0 & -1 & 0 & -1 & +1 & 0 \\ 0 & 0 & -1 & +1 & 0 & +1 \\ +1 & 0 & 0 & 0 & -1 & -1 \end{bmatrix}
\end{array}
\qquad (5.1)
$$

The first element of the matrix -1 indicates that, branch 1 is connected to node a and is oriented towards it. The last element of the matrix -1 indicates that, branch 6 is connected to node d and is oriented towards it. The last element of the first row 0 indicates that, branch 6 is not connected to node a. Thus, the elements of the incidence matrix gives complete information regarding the connection of the branches to the nodes and their orientations.

One of the important features of the all incidence matrix is that the sum of the elements of any column is zero. This is because, each column gives the information about a particular branch connected between any two nodes. For Ex: Branch 1 is connected between nodes a and d. The orientation of this branch w.r.t. node a is naturally opposite to its orientation w.r.t. d and hence, the truthfuness of the above fact.

Another feature of the all incidence matrix is that the determinant of the incidence matrix of a particular loop, which is called as a sub-matrix is always zero.

For Ex: Consider the loop *abda*. The determinent of the sub-matrix for this loop is given by

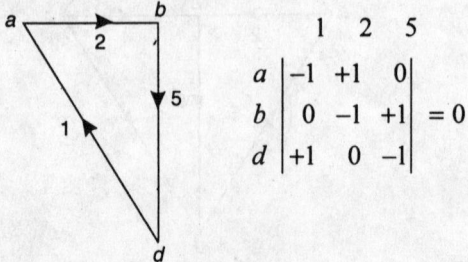

$$
\begin{array}{c c}
 & \begin{array}{c c c} 1 & 2 & 5 \end{array} \\
\begin{array}{c} a \\ b \\ d \end{array} &
\begin{vmatrix} -1 & +1 & 0 \\ 0 & -1 & +1 \\ +1 & 0 & -1 \end{vmatrix} = 0
\end{array}
$$

5.8 REDUCED INCIDENCE MATRIX A

If in the all incidence matrix, the information about a particular node is completely omitted, i.e. one of the rows is eliminated, then that matrix is called as "Incidence Matrix". But, it is usually referred as "Reduced Incidence Matrix". From the all incidence matrix $\mathbf{A_a}$ given in (5.1), if the information about node d is omitted or the fourth row is eliminated, it results in reduced incidence matrix as given in (5.2).

$$
\mathbf{A} = \begin{array}{c}
 & \text{Branches} \\
\begin{array}{c} \text{Nodes} \\ a \\ b \\ c \end{array} &
\begin{array}{c}
\begin{array}{c c c c c c} 1 & 2 & 3 & 4 & 5 & 6 \end{array} \\
\begin{bmatrix} -1 & +1 & +1 & 0 & 0 & 0 \\ 0 & -1 & 0 & -1 & +1 & 0 \\ 0 & 0 & -1 & +1 & 0 & +1 \end{bmatrix}
\end{array}
\end{array} \qquad (5.2)
$$

From the above matrix, we find that the sum of the elements of each column is not zero. Information about the missing node may be found, by writing another row of elements, satisfying the condition that the sum of the elements in each column must be zero. To satisfy this condition, the elements of the fourth row are: +1 0 0 0 −1 −1. These are the elements of node d. By adding this row, the matrix $\mathbf{A}$ becomes $\mathbf{A_a}$.

It is possible to write down the oriented graph from a particular all incidence matrix or reduced incidence matrix, which is clearly explained in the few examples worked out later in this chapter.

5.9 NODE TO DATUM VOLTAGES AND MATRIX A

As explained in section 5.8 in the reduced incidence matrix, the information about one node is missing, which is considered as reference node. As the information about node d is missing in matrix $\mathbf{A}$ given in

(5.2), it is taken as the reference node as shown in Fig. 5.10. From the **A** matrix, the various branch voltages may be expressed in terms of node to datum voltages.

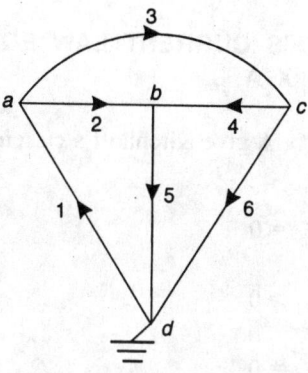

Fig. 5.10

$$e_1 = -v_a$$

$$e_2 = v_a - v_b$$

$$e_3 = v_a - v_c$$

$$e_4 = v_c - v_b$$

$$e_5 = v_b$$

$$e_6 = v_c \qquad\qquad (5.3)$$

The set of equations given in (5.3), may be written in matrix form as.

$$
\begin{bmatrix} e_1 \\ e_2 \\ e_3 \\ e_4 \\ e_5 \\ e_6 \end{bmatrix} =
\begin{bmatrix} -1 & 0 & 0 \\ +1 & -1 & 0 \\ +1 & 0 & -1 \\ 0 & -1 & +1 \\ 0 & +1 & 0 \\ 0 & 0 & +1 \end{bmatrix}
\begin{bmatrix} v_a \\ v_b \\ v_c \end{bmatrix} \qquad (5.4)
$$

i.e $$\mathbf{E_b} = \mathbf{A}^T \mathbf{V_n} \qquad\qquad (5.5)$$

Where, $\mathbf{E_b}$ is a column matrix of branch voltages and is of the order $(b \times 1)$. $\mathbf{A}^T$ is the transpose of matrix $\mathbf{A}$.

$\mathbf{V_n}$ is a column matrix of node to datum voltages and is of the order $\{(n - 1) \times 1\}$.

Equation (5.5) is known as the node transformation equation.

5.10 KIRCHHOFF'S CURRENT LAW EQUATIONS AND MATRIX A

The rows of the matrix $\mathbf{A}$ give Kirchhoff's current law equations, as in (5.6).

$$-i_1 + i_2 + i_3 = 0$$

$$-i_2 - i_4 + i_5 = 0$$

$$-i_3 + i_4 + i_6 = 0 \tag{5.6}$$

The set of Eqs (5.6) may be written in matrix form as in (5.7).

$$\begin{bmatrix} -1 & +1 & +1 & 0 & 0 & 0 \\ 0 & -1 & 0 & -1 & +1 & 0 \\ 0 & 0 & -1 & +1 & 0 & +1 \end{bmatrix} \begin{bmatrix} i_1 \\ i_2 \\ i_3 \\ i_4 \\ i_5 \\ i_6 \end{bmatrix} = 0 \tag{5.7}$$

(5.7) may be written as, $\mathbf{AI_b} = 0$ \hfill (5.8)

Where, $\mathbf{I_b}$ is a column matrix of branch currents and is of the order $(b \times 1)$.

Equations (5.5) and (5.8) are the basic relations required to derive equilibrium equations on node to datum voltages as independent variables.

5.11 EQUILIBRIUM EQUATIONS WITH NODE TO DATUM VOLTAGES AS VARIABLES

In this method of analysis, the node to datum voltages are used as independent variables.

Consider a general branch of a network as shown in Fig. 5.11.

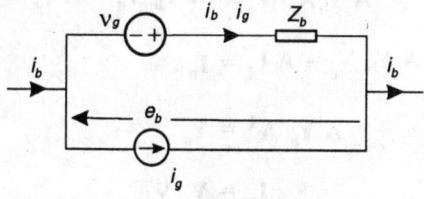

Fig. 5.11

Where, v_g = Total series voltage in the branch.

i_g = Total current source across the branch.

Z_b = Total impedance of the branch.

i_b = Branch current.

The voltage current relations for the branch can be written as

$$e_b = v_g + Z_b \left(i_b - i_g\right) \tag{5.9}$$

and $$i_b = i_g + Y_b \left(e_b - v_g\right) \tag{5.10}$$

Where, Y_b = total admittance of the branch. For a network with more number of branches, equations (5.9) and (5.10) may be written as

$$E_b = V_g + Z_b \left(I_b - I_g\right) \tag{5.11}$$

and $$I_b = I_g + Y_b \left(E_b - V_g\right) \tag{5.12}$$

Where, E_b, V_g, I_b and I_g are ($b \times 1$) matrices of branch voltages, source voltages in the branches, branch currents and source currents in the branches respectively. Z_b and Y_b are branch impedance and branch admittance matrices of the order ($b \times b$).

From Eq. (5.8), we know that

$$AI_b = 0 \tag{5.13}$$

Substituting (5.12) in (5.13), we get

$$A I_g + A Y_b \left(E_b - V_g\right) = 0$$

i.e. $$A Y_b V_g - A I_g = A Y_b E_b$$

but $$E_b = A^T V_n \qquad \text{(from 5.5)}$$

i.e. $$\mathbf{A} \; \mathbf{Y_b} \mathbf{A}^T \; \mathbf{V_n} = \mathbf{A} \; \mathbf{Y_b} \; \mathbf{V_g} - \mathbf{A} \; \mathbf{I_g} \qquad (5.14)$$

Let $$\mathbf{A} \; \mathbf{Y_b} \; \mathbf{V_g} - \mathbf{A} \; \mathbf{I_g} = \mathbf{I_n}$$

and $$\mathbf{A} \; \mathbf{Y_b} \; \mathbf{A}^T = \mathbf{Y_n}$$

Then $$\mathbf{I_n} = \mathbf{Y_n} \; \mathbf{V_n}$$

or $$\mathbf{V_n} = \mathbf{Y_n^{-1}} \; \mathbf{I_n} \qquad (5.15)$$

Where, $\mathbf{Y_n}$ is the node admittance matrix of the order $(n-1) \times (n-1)$ and is non-singular.

$\mathbf{V_n}$ is the column matrix of node to datum voltages of the order $\{(n-1) \times 1\}$ and $\mathbf{I_n}$ is the column matrix of the source currents to the nodes of the order $(n-1) \times 1$.

Equations (5.14) or (5.15) represents a set of $(n-1)$ equilibrium equations, the solution of which gives the node to datum voltages. Once, these voltages are found, the branch voltages and branch currents can be calculated.

$\mathbf{Y_n}$ and $\mathbf{I_n}$ can be written by mere inspection of the network.

Note: (i) v_g also includes initial capacitor voltages in loop analysis.

(ii) i_g also includes initial inductor currents in nodal analysis.

5.12 BRANCH ADMITTANCE MATRIX $\mathbf{Y_b}$

The branch admittance matrix $\mathbf{Y_b}$ is a matrix of the order $(b \times b)$, containing the branch admittances as diagonal elements and mutual admittances between the branches as off-diagonal elements. For a network which does not contain any transformers and there is no mutual coupling between the branches, the $\mathbf{Y_b}$ matrix contains only diagonal elements and all the off-diagonal elements are zero. The general form of $\mathbf{Y_b}$ for a network is written as

$$\mathbf{Y_b} = \begin{bmatrix} y_{11} & y_{12} & y_{13} & \cdots\cdots & y_{1b} \\ y_{21} & y_{22} & y_{23} & \cdots\cdots & y_{2b} \\ & \vdots & & & \\ y_{b1} & y_{b2} & y_{b3} & \cdots\cdots & y_{bb} \end{bmatrix} \qquad (5.16)$$

Where, $y_{11}, y_{22}, y_{33} \cdots y_{bb}$ represent the self admittances of branches 1, 2, 3 ... b respectively.

y_{12}, y_{23}, y_{b3} ... represent the mutual admittances between branches 1 and 2, 2 and 3, b and 3 ... respectively.

5.13 THE NODE ADMITTANCE MATRIX Y_n

The node admittance matrix $\mathbf{Y_n}$ is a matrix of the order $\{(n-1) \times (n-1)\}$, containing the node admittances as its elements. The general form of $\mathbf{Y_n}$ is written as

$$\mathbf{Y_n} = \begin{bmatrix} y_{aa} & y_{ab} & y_{ac} & \cdots\cdots & y_{ak} \\ y_{ba} & y_{bb} & y_{bc} & \cdots\cdots & y_{bk} \\ & \vdots & & & \\ y_{ka} & y_{kb} & y_{kc} & \cdots\cdots & y_{kk} \end{bmatrix} \qquad (5.17)$$

Where, k is $(n-1)^{\text{th}}$ node.

y_{aa}, y_{bb} ... y_{kk} are the self admittances of nodes a, b ... k respectively or the total admittances of the branches connected to nodes, a, b ... k respectively.

y_{ab}, y_{bc} ... y_{kb} are the admittances between nodes a and b, nodes b and c, and nodes k and b respectively.

For Ex: Consider the network as shown in Fig. 5.12. The node d is taken as the reference node. The ohmic values of the branches also indicate the branch numbers. Then, the branch admittance matrix is as in (5.18).

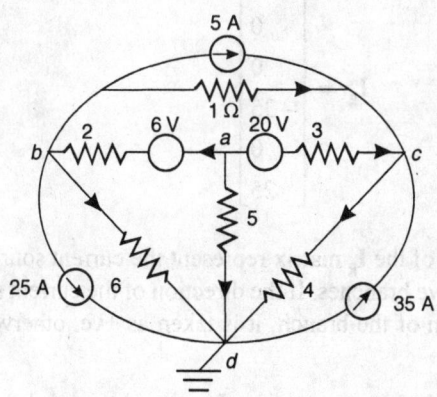

Fig. 5.12

The arrow marks indicate the arbitrary orientations of the branches.

$$Y_b = \begin{bmatrix} 1 & 0 & 0 & 0 & 0 & 0 \\ 0 & 1/2 & 0 & 0 & 0 & 0 \\ 0 & 0 & 1/3 & 0 & 0 & 0 \\ 0 & 0 & 0 & 1/4 & 0 & 0 \\ 0 & 0 & 0 & 0 & 1/5 & 0 \\ 0 & 0 & 0 & 0 & 0 & 1/6 \end{bmatrix}_{6 \times 6}$$ (5.18)

The V_g matrix is given as in (5.19).

$$V_g = \begin{bmatrix} 0 \\ -6 \\ -20 \\ 0 \\ 0 \\ 0 \end{bmatrix}$$ (5.19)

The elements of the V_g matrix represent the voltage sources in the respective branches. If the voltage source is a rise along the orientation of that branch, then it is taken as –ve, otherwise, it is +ve.

The I_g matrix is given as in (5.20).

$$I_g = \begin{bmatrix} 5 \\ 0 \\ 0 \\ -35 \\ 0 \\ 25 \end{bmatrix}$$ (5.20)

The elements of the I_g matrix represent the current sources associated with the respective branches. If the direction of the current source is same as the orientation of the branch, it is taken as +ve, otherwise it is taken as –ve.

The node admittance matrix Y_n is obtained by the singular transformation, $Y_n = A \, Y_b \, A^T$.

5.14 NUMBER OF POSSIBLE TREES OF A GRAPH

For a given network graph, it is possible to write several trees. The number of possible trees of a graph is given by

$$N = \text{determinent of } \mathbf{A} \, \mathbf{A}^{\mathbf{T}} \qquad (5.21)$$

Where, N = number of possible trees

                $\mathbf{A}$ = reduced incidence matrix

                $\mathbf{A}^{\mathbf{T}}$ = transpose of $\mathbf{A}$

Consider an oriented graph as shown in Fig. 5.13.

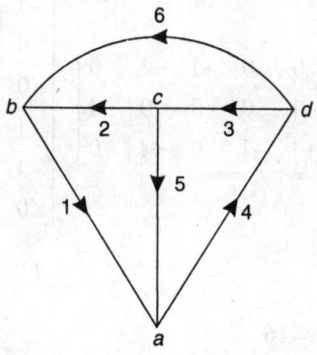

Fig. 5.13

The all incidence matrix of the above graph is

$$A_a = \begin{array}{c} \\ \text{Nodes} \\ a \\ b \\ c \\ d \end{array} \begin{array}{c} \text{Branches} \\ \begin{array}{cccccc} 1 & 2 & 3 & 4 & 5 & 6 \end{array} \\ \left[\begin{array}{cccccc} -1 & 0 & 0 & +1 & -1 & 0 \\ +1 & -1 & 0 & 0 & 0 & -1 \\ 0 & +1 & -1 & 0 & +1 & 0 \\ 0 & 0 & +1 & -1 & 0 & +1 \end{array} \right] \end{array}$$

The reduced incidence matrix is obtained by omitting one of the rows, say 4th row.

$$\therefore \quad \mathbf{A} = \begin{array}{c} a \\ b \\ c \end{array} \left[\begin{array}{cccccc} -1 & 0 & 0 & +1 & -1 & 0 \\ +1 & -1 & 0 & 0 & 0 & -1 \\ 0 & +1 & -1 & 0 & +1 & 0 \end{array} \right]$$

$$\mathbf{A^T} = \begin{bmatrix} -1 & +1 & 0 \\ 0 & -1 & +1 \\ 0 & 0 & -1 \\ +1 & 0 & 0 \\ -1 & 0 & +1 \\ 0 & -1 & 0 \end{bmatrix}$$

The number of possible trees for the given graph is given by

$$N = \det \mathbf{A} \, \mathbf{A^T}$$

$$= \det \text{ of } \begin{bmatrix} -1 & 0 & 0 & +1 & -1 & 0 \\ +1 & -1 & 0 & 0 & 0 & -1 \\ 0 & +1 & -1 & 0 & +1 & 0 \end{bmatrix} \begin{bmatrix} -1 & +1 & 0 \\ 0 & -1 & +1 \\ 0 & 0 & -1 \\ +1 & 0 & 0 \\ -1 & 0 & +1 \\ 0 & -1 & 0 \end{bmatrix}$$

$$= \begin{vmatrix} 3 & -1 & -1 \\ -1 & 3 & -1 \\ -1 & -1 & 3 \end{vmatrix} = 16$$

5.15 CUT-SET

A *cut-set* is a set of branches of a connected graph, whose removal causes the graph to become unconnected into exactly two connected sub-graphs. Restoring any one of the branches of the cut-set destroys the separation property of the two sub-graphs.

Consider the oriented graph of a network as shown in Fig. 5.14. The group of branches (1, 2, 3) represented as A is a cut-set. By removing these *branches*, the graph becomes unconnected and two sub-graphs are formed. One sub-graph is node a and the other sub-graph contains nodes, b, c, d and branches (4, 5, 6). A single isolated node is considered as a connected sub-graph. By replacing any one of the branches of the cut-set, the two sub-graphs get connected. Similarly, B, C and D, consisting of branches (1, 5, 6), (3, 4, 6), (2, 4, 5) are some of the other cut-sets. For a given graph, it is possible to identify a large number of cut-sets. But, for network solution, we must be able to identify the *fundamental cut-sets*, which are the minimum number of cut-sets that are required to be identified for the network solution. This is possible by identifying a possible tree for the given graph.

5.16 FUNDAMENTAL CUT-SET

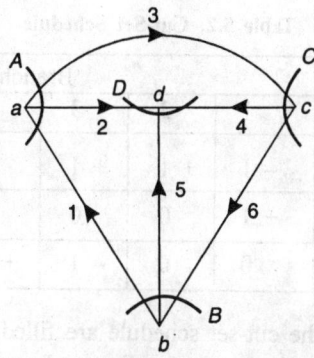

Fig. 5.14

For the oriented graph given in Fig. 5.14, one possible tree with branches 2, 4, and 5 is as shown in Fig. 5.15. The thick lines represent the tree-branches and the dotted lines represent the links. Branches 1, 3 and 6 are the links. A cut-set identified with any one of the tree branches is called as a fundamental cut-set. Thus, A (1, 2, 3), B (1, 5, 6), and C (3, 4, 6) are the fundamental cut-sets. The orientation of the fundamental cut-set is usually assumed to be the same as the orientation of the tree-branch it contains. The orientations of the cut-sets are marked as shown in Fig. 5.15. The number of fundamental cut-sets is equal to the number of tree-branches or $(n - 1)$, where n is the number of nodes in the graph.

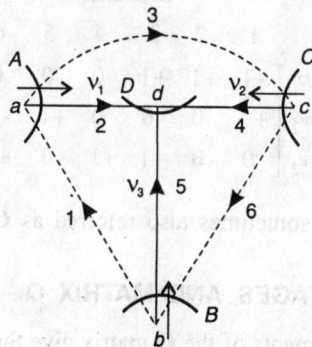

Fig. 5.15

Once the fundamental cut-sets are identified and their orientations are fixed, it is possible to write-down a schedule known as *cut-set schedule*, which gives the relation between the tree-branch voltages and all the other branch voltages. For the oriented graph shown in Fig. 5.15, the cut-set schedule is as shown in Table 5.2.

Table 5.2: Cut-Set Schedule

Tree branch voltages and f cutsets	Branches					
	1	2	3	4	5	6
v_1, A (1, 2, 3)	-1	$+1$	$+1$	0	0	0
v_2, B (1, 5, 6)	$+1$	0	0	0	$+1$	-1
v_3, C (3, 4, 6)	0	0	-1	$+1$	0	$+1$

The elements of the cut-set schedule are filled, taking into account the orientation of a particular cut-set. For ex: For cut-set A (1, 2, 3), the orientation of branch 1 is opposite to the orientation of cut-set A. Hence, first element of the first row is filled as -1. The orientation of the branch 2 is same as that of cut-set A. Hence, the second element of the first row is filled with $+1$. The orientation of branch 3 is same as the orientation of cut-set A and hence, the third element of the first row is filled as $+1$. The other three branches 4, 5 and 6 does not belong to cut-set A and hence, the remaining elements of the first row are filled as 0_s. Similarly, the elements of the other two rows are filled, taking into consideration, the other two cut-sets B and C, and their orientations w.r.t. the elements constituting them.

The elements of the cut-set schedule may be written in the form of a matrix known as *fundamental cut-set matrix*, represented by Q.

$$Q = \begin{matrix} & & \text{Branches} & & & & \\ & 1 & 2 & 3 & 4 & 5 & 6 \\ v_1 \\ v_2 \\ v_3 \end{matrix} \begin{bmatrix} -1 & +1 & +1 & 0 & 0 & 0 \\ +1 & 0 & 0 & 0 & +1 & -1 \\ 0 & 0 & -1 & +1 & 0 & +1 \end{bmatrix} \qquad (5.22)$$

The **Q** matrix is sometimes also referred as **C** matrix.

5.17 TWIG VOLTAGES AND MATRIX Q

The columnwise elements of the **Q** matrix give the relation between the branch voltages and the tree-branch voltages. (twig voltages)

$$e_1 = v_2 - v_1$$

$$e_2 = v_1$$

$$e_3 = v_1 - v_3$$

$$e_4 = v_3$$

$$e_5 = v_2$$

$$e_6 = v_3 - v_2 \tag{5.23}$$

The set of Eqs (5.23) may be written as

$$\begin{bmatrix} e_1 \\ e_2 \\ e_3 \\ e_4 \\ e_5 \\ e_6 \end{bmatrix} = \begin{bmatrix} -1 & +1 & 0 \\ +1 & 0 & 0 \\ +1 & 0 & -1 \\ 0 & 0 & +1 \\ 0 & +1 & 0 \\ 0 & -1 & 1 \end{bmatrix} \begin{bmatrix} v_1 \\ v_2 \\ v_3 \end{bmatrix} \tag{5.24}$$

or $$\mathbf{E_b} = \mathbf{Q^T V_t} \tag{5.25}$$

Equation (5.25) is known as node transformation equation.

$\mathbf{E_b}$ is a column matrix of branch voltages of the order ($b \times 1$).

$\mathbf{V_t}$ is the column matrix of tree-branch voltages of the order $\{(n - 1) \times 1\}$.

5.18 KIRCHHOFF'S CURRENT LAW AND MATRIX Q

The rows of the matrix **Q**, give the relation between branch currents, satisfying Kirchhoff's current law.

$$-i_1 + i_2 + i_3 = 0$$

$$i_1 + i_5 - i_6 = 0$$

$$-i_3 + i_4 + i_6 = 0 \tag{5.26}$$

The set of Eqs in (5.26) may be written in matrix form as

$$\begin{bmatrix} -1 & +1 & +1 & 0 & 0 & 0 \\ +1 & 0 & 0 & 0 & +1 & -1 \\ 0 & 0 & -1 & +1 & 0 & +1 \end{bmatrix} \begin{bmatrix} i_1 \\ i_2 \\ i_3 \\ i_4 \\ i_5 \\ i_6 \end{bmatrix} = 0 \tag{5.27}$$

In the matrix form, the set of Eqs in (5.27) may be written as

$$\mathbf{Q}\mathbf{I}_b = 0 \tag{5.28}$$

Where, $\mathbf{I}_b$ is a column matrix of branch currents and is of the order $(b \times 1)$.

Equations (5.25) and (5.28) are the basic equations, which are used to derive equilibrium equations with tree-branch voltages as variables for a given network.

5.19 EQUILIBRIUM EQUATIONS WITH TREE-BRANCH VOLTAGES AS VARIABLES

From Eqs (5.28), (5.25), and (5.12), we know that

$$\mathbf{Q}\mathbf{I}_b = 0 \tag{5.29}$$

$$\mathbf{E}_b = \mathbf{Q}^T \mathbf{V}_t \tag{5.30}$$

$$\mathbf{I}_b = \mathbf{I}_g + \mathbf{Y}_b \left(\mathbf{E}_b - \mathbf{V}_g \right) \tag{5.31}$$

Substituting (5.31) in (5.29), we get

$$\mathbf{Q}\mathbf{I}_g + \mathbf{Q}\mathbf{Y}_b \left(\mathbf{E}_b - \mathbf{V}_g \right) = 0$$

i.e. $\mathbf{Q}\mathbf{I}_g + \mathbf{Q}\mathbf{Y}_b \left(\mathbf{Q}^T \mathbf{V}_t - \mathbf{V}_g \right) = 0$ (from 5.30)

i.e. $\qquad\qquad \mathbf{Q}\mathbf{Y}_b\,\mathbf{Q}^T \mathbf{V}_t = \mathbf{Q}\mathbf{Y}_b\,\mathbf{V}_g - \mathbf{Q}\mathbf{I}_g. \tag{5.32}$

let, $\qquad\qquad \mathbf{Q}\mathbf{Y}_b\,\mathbf{V}_g - \mathbf{Q}\mathbf{I}_g = \mathbf{I}_n$

and $\qquad\qquad \mathbf{Q}\mathbf{Y}_b\,\mathbf{Q}^T = \mathbf{Y}_n$

Then, the Eq. (5.32) may be written as

$$\mathbf{I}_n = \mathbf{Y}_n\,\mathbf{V}_t$$

or $\qquad\qquad \mathbf{V}_t = \mathbf{Y}_n^{-1}\,\mathbf{I}_n \tag{5.33}$

Equations (5.32) or (5.33) gives a set of $(n-1)$ equilibrium equations with tree-branch voltages as variables. On solving these equations and knowing the tree-branch voltages, all the other branch voltages can be found. If the elements of the various branches of the network are known, then all the branch voltages can be found.

5.20 TIE-SET AND TIE-SET SCHEDULE

The node incidence matrix, which is explained in section 5.7, gives only the information about the connection of the various branches to the various nodes and their orientation, but does not give any information about the way in which the branches constitute the loops. This information can be conveniently given in the form of a matrix, known as *loop incidence matrix* or more popularly known as *tie-set matrix*.

A tie-set is defined as a collection of branches forming a loop. Normally, a large number of loops can be formed for a given graph. But, we must be able to identify the least number of loops required to solve the network, which are called as fundamental loops. Such fundamental loops are identified by selecting one possible tree of the given graph. A tree of the graph can be written by removing sufficient number of branches from the graph, such that, the graph does not contain any loops in it. The removed branches are called links. The branches of the tree are callled twigs.

By adding one link to the tree, a loop is formed. Similarly, by adding other links to the graph, remaining loops are also formed. They are called fundamental loops. Hence, the number of fundamental loops for a graph is equal to the number of links. All the branches of a fundamental loop constitute a fundamental tie-set. Hence, there will be as many fundamental tie-sets, as there are links. The loop currents are assumed in the same direction as the direction of currents in the corresponding links, constituting the loops, so that the orientation of the links and direction of loop currents are the same. Using this concept of tie-set, a matrix, which gives the relation between loop currents and branch currents may be written, which is known as the 'Tie-set Matrix' or 'Loop Incidence Matrix'. A schedule written to give the above relations is called the 'Tie-set Schedule'.

In the tie-set schedule, each element will be $+1$, -1 or 0. It is $+1$, if in a tie-set, the orientation of a branch is same as the direction of loop current, otherwise, it is -1. It is 0, if the branch does not belong to that particular tie-set.

Consider the oriented graph as shown in Fig. 5.16.

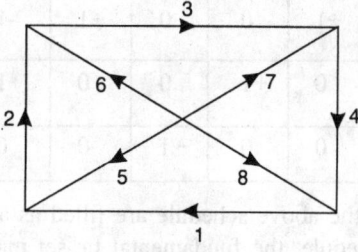

Fig. 5.16

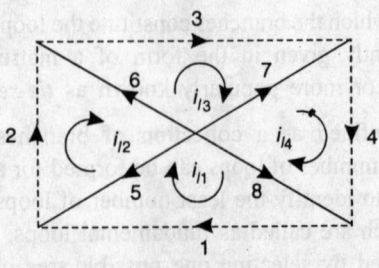

Fig. 5.16 (a)

One possible tree for the given oriented graph is as shown in Fig. 5.16 (a). The thick lines represent the tree-branches or twigs. The dotted lines represent the links. By replacing one link at a time, four fundamental loops are formed. i_1, i_2, ... i_8 are the branch currents, out of which i_1, i_2, i_3 and i_4 are link currents. i_5, i_6, i_7 and i_8 are tree-branch currents. The loop currents i_{l1}, i_{l2}, i_{l3} and i_{l4} are assumed in such a way that their directions coincide with the orientations of the links 1, 2, 3 and 4 respectively.

The three branches 1, 5 and 8 which constitute the loop of i_{l1} form the tie-set of i_{l1} and is written as i_{l1} (1, 5, 8). Similarly, the other tie-sets are: i_{l2} (2, 5, 6), i_{l3} (3, 6, 7) and i_{l4} (4, 7, 8).

The tie-set schedule of the oriented graph for the selected tree is given in Table 5.3.

Table 5.3 Tie-set schedule

Tie-sets	Branches							
	1	2	3	4	5	6	7	8
i_{l1} (1,5,8)	+1	0	0	0	−1	0	0	+1
i_{l2} (2,5,6)	0	+1	0	0	+1	−1	0	0
i_{l3} (3,6,7)	0	0	+1	0	0	+1	−1	0
i_{l4} (4,7,8)	0	0	0	+1	0	0	+1	−1

The elements of the above schedule are filled as already explained. From the above schedule, the fundamental tie-set matrix is written as under.

Branches

$$
\begin{array}{c}
\text{Loop currents}
\end{array}
\quad
\begin{array}{cccccccc}
1 & 2 & 3 & 4 & 5 & 6 & 7 & 8
\end{array}
$$

$$
\mathbf{B} =
\begin{array}{c}
i_{l1} \\ i_{l2} \\ i_{l3} \\ i_{l4}
\end{array}
\begin{bmatrix}
+1 & 0 & 0 & 0 & -1 & 0 & 0 & +1 \\
0 & +1 & 0 & 0 & +1 & -1 & 0 & 0 \\
0 & 0 & +1 & 0 & 0 & +1 & -1 & 0 \\
0 & 0 & 0 & +1 & 0 & 0 & +1 & -1
\end{bmatrix}
\tag{5.34}
$$

Matrix **B** is sometimes also referred as **M** matrix.

5.21 KIRCHHOFF'S CURRENT LAW EQUATIONS AND MATRIX B

The elements of a particular column of matrix **B** gives the relation between the branch currents and the loop currents.

i.e.

$$i_1 = i_{l1}$$

$$i_2 = i_{l2}$$

$$i_3 = i_{l3}$$

$$i_4 = i_{l4}$$

$$i_5 = i_{l2} - i_{l1}$$

$$i_6 = i_{l3} - i_{l2}$$

$$i_7 = i_{l4} - i_{l3}$$

$$i_8 = i_{l1} - i_{l4} \tag{5.35}$$

The set of Eq. in (5.35) can be expressed in the matrix form as

$$
\begin{bmatrix}
i_1 \\ i_2 \\ i_3 \\ i_4 \\ i_5 \\ i_6 \\ i_7 \\ i_8
\end{bmatrix}
=
\begin{bmatrix}
+1 & 0 & 0 & 0 \\
0 & +1 & 0 & 0 \\
0 & 0 & +1 & 0 \\
0 & 0 & 0 & +1 \\
-1 & +1 & 0 & 0 \\
0 & -1 & +1 & 0 \\
0 & 0 & -1 & +1 \\
+1 & 0 & 0 & -1
\end{bmatrix}
\begin{bmatrix}
i_{l1} \\ i_{l2} \\ i_{l3} \\ i_{l4}
\end{bmatrix}
\tag{5.36}
$$

i.e.

$$\mathbf{I_b} = \mathbf{B^T I_l} \tag{5.37}$$

Where, I_b is column matrix of branch currents of the order $(b \times 1)$.
B^T is the transpose of the fundamental tie-set matrix B.
I_l is the column matrix of loop currents of the order $(m \times 1)$, where, m is the number of independent loops.

5.22 BRANCH VOLTAGES AND MATRIX B

Each row of B, gives the voltage relations in the loop.

i.e.
$$e_1 - e_5 + e_8 = 0$$
$$e_2 + e_5 - e_6 = 0$$
$$e_3 + e_6 - e_7 = 0$$
$$e_4 + e_7 - e_8 = 0 \tag{5.38}$$

The set of Eqs (5.38) may be expressed in the matrix form as

$$
\begin{bmatrix}
+1 & 0 & 0 & 0 & -1 & 0 & 0 & +1 \\
0 & +1 & 0 & 0 & +1 & -1 & 0 & 0 \\
0 & 0 & +1 & 0 & 0 & +1 & -1 & 0 \\
0 & 0 & 0 & +1 & 0 & 0 & +1 & -1
\end{bmatrix}
\begin{bmatrix}
e_1 \\ e_2 \\ e_3 \\ e_4 \\ e_5 \\ e_6 \\ e_7 \\ e_8
\end{bmatrix} = 0 \tag{5.39}
$$

i.e.
$$B E_b = 0 \tag{5.40}$$

Where, E_b is a column matrix of branch voltages of the order $(b \times 1)$.

Equations (5.37), and (5.40) are the basic equations required to derive equilibrium equations on loop current basis.

5.23 EQUILIBRIUM EQUATIONS WITH LOOP CURRENTS AS VARIABLES

From Eqs (5.11), (5.37) and (5.40), we know that

$$E_b = V_g + Z_b \left(I_b - I_g \right) \tag{5.41}$$

$$I_b = B^T I_l \tag{5.42}$$

and
$$B E_b = 0 \tag{5.43}$$

Substituting (5.41) in (5.43), we get

$$BV_g + BZ_b \left(I_b - I_g\right) = 0$$

i.e.

$$BZ_b I_b = BZ_b I_g - BV_g$$

i.e.

$$BZ_b B^T I_l = BZ_b I_g - BV_g \qquad (5.44)$$

Let

$$BZ_b I_g - BV_g = V_l$$

and

$$BZ_b B^T = Z_l$$

Then, Eq. (5.44) may be written as

$$Z_l I_l = V_l$$

or

$$I_l = Z_l^{-1} V_l \qquad (5.45)$$

Equations (5.44) or (5.45) represents a set of equilibrium equations with loop currents as independent variables. On solving these equations, loop currents are obtained. Once the loop currents are known, all the branch currents can be found. If the elements of the branches are known, then, all the branch voltages also can be found.

5.24 THE BRANCH IMPEDANCE MATRIX Z_b

The branch impedance matrix Z_b is a matrix of the order ($b \times b$), containing branch impedances as diagonal elements and the mutual impedances between the branches as off-diagonal elements. For a network, which does not contain any transformers and there is no mutual coupling between the branches, the Z_b matrix contains only diagonal elements and all the off-diagonal elements are zero. The general form of Z_b for a network is written as in (5.46).

$$Z_b = \begin{bmatrix} Z_{11} & Z_{12} & Z_{13} & \cdots\cdots Z_{1b} \\ Z_{21} & Z_{22} & Z_{23} & \cdots\cdots Z_{2b} \\ & \vdots & & \\ Z_{b1} & Z_{b2} & Z_{b3} & \cdots\cdots Z_{bb} \end{bmatrix} \qquad (5.46)$$

Where, $Z_{11}, Z_{22}, Z_{33} \cdots Z_{bb}$ represent the self impedances of the branches 1, 2, 3 ... b respectively.

$Z_{12}, Z_{23}, Z_{b3} \cdots$ represent the mutual impedances between the branches 1 and 2, 2 and 3, and b and 3 ... respectively.

The Loop Impedance Matrix Z_l

The loop impedance matrix is of the order $(m \times m)$, where, m is the number of independent loops. The elements of this matrix are the loop impedances and mutual impedances between loops. The general form of Z_l is as in (5.47)

$$Z_l = \begin{bmatrix} Z_{l11} & Z_{l12} & Z_{l13} & \text{.......} & Z_{l1m} \\ Z_{l21} & Z_{l22} & Z_{l23} & \text{.......} & Z_{l2m} \\ & & \vdots & & \\ Z_{lm1} & Z_{lm2} & Z_{lm3} & \text{.......} & Z_{lmm} \end{bmatrix} \qquad (5.47)$$

Where, Z_{l11}, Z_{l22}, Z_{l33} ... Z_{lmm} are the self impedances of loop 1, loop 2, loop 3 and loop m respectively. They are the sum of the impedances of the branches in the respective loops.

Z_{l12}, Z_{l23} ... Z_{lm3} are the mutual impedances between loops 1 and 2, loops 2 and 3 and loops m and 3 respectively.

Z_l is obtained by $Z_l = B Z_b B^T$

For the circuit shown in Fig. 5.11, the Z_b matrix is as in (5.48).

$$Z_b = \begin{bmatrix} 1 & 0 & 0 & 0 & 0 & 0 \\ 0 & 2 & 0 & 0 & 0 & 0 \\ 0 & 0 & 3 & 0 & 0 & 0 \\ 0 & 0 & 0 & 4 & 0 & 0 \\ 0 & 0 & 0 & 0 & 5 & 0 \\ 0 & 0 & 0 & 0 & 0 & 6 \end{bmatrix} \qquad (5.48)$$

5.25 'E' SHIFT AND 'I' SHIFT

(a) The E shift: In complex networks, if any one of the branches contains an ideal voltage source, then, normal nodal analysis cannot be applied, because, the source current due to this branch to the particular node is indeterminate. In such cases, the ideal voltage source is shifted to the other branches that are connected in series with it, without changing the characteristics of the network. This is known as E shift.

Consider a complex network as shown in Fig. 5.17. V_g is the ideal voltage source connected between nodes e and a. The nodal analysis can't be applied to solve this network. The ideal voltage source V_g is shifted to the branches connected in series with it as shown in Fig. 5.18 (a) or as shown in Fig. 5.18 (b).

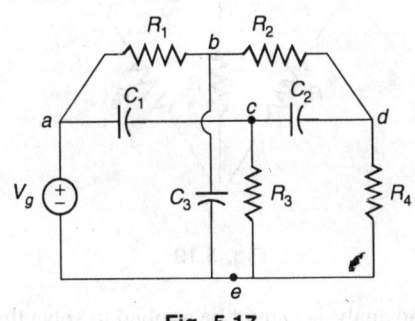

Fig. 5.17

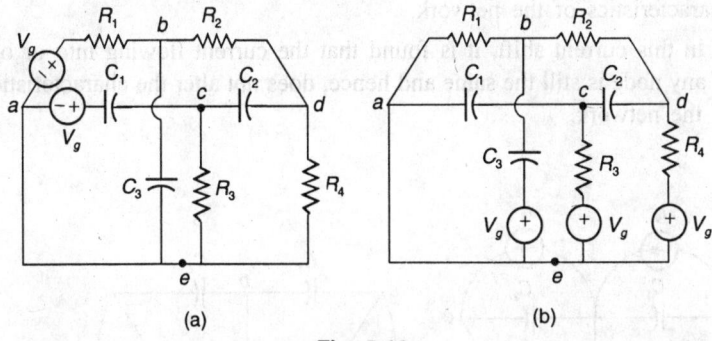

(a)   (b)

Fig. 5.18

In Fig. 5.18 (a), the voltage source V_g is shifted to the branches connected to node a and in Fig. 5.18 (b), it is shifted to the branches connected to node e. By these shifts, it can be shown that the loop equations will remain the same for the networks in Figs 5.17, 5.18 (a) and (b), indicating that, this shift does not alter the characteristics of the network.

(b) The *I* shift: When a network contains an ideal current source, then loop analysis can't be applied, because, the voltage drop in the branch containing the ideal current source is indeterminate. In such case, the ideal current source is shifted to be in parallel with each of the branches that form a closed loop with i_g. This is known as *I* shift.

Consider the network as shown in Fig. 5.19, containing an ideal current source i_g.

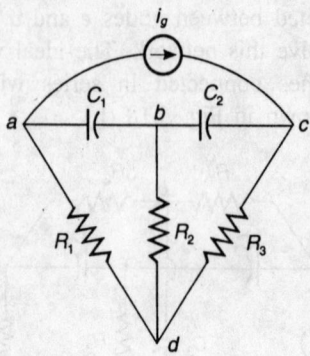

Fig. 5.19

The normal loop analysis cannot be applied to solve this network. The 'I' shift is done either as in Fig. 5.20 (a) or (b), without altering the characteristics of the network.

In this current shift, it is found that the current flowing into or out of any node is still the same and hence, does not alter the characteristics of the network.

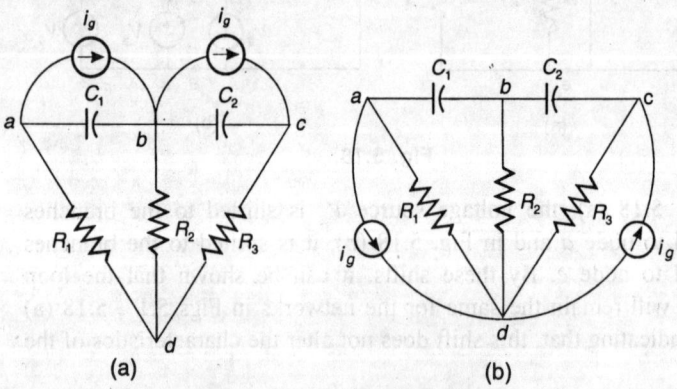

(a)   (b)

Fig. 5.20

Important equations

$$1. \quad E_b = V_g + Z_b \left(I_b - I_g\right)$$

$$2. \quad I_b = I_g + Y_b \left(E_b - V_g\right)$$

$$3. \quad E_b = A^T V_n, \quad A I_b = 0$$

and $\quad AY_b A^T V_n = AY_b V_g - AI_g$

4. $E_b = Q^T V_t, \quad QI_b = 0$

   and $\quad QY_b Q^T V_t = QY_b V_g - QI_g$

5. $I_b = B^T I_l, \quad BE_b = 0$

   and $\quad BZ_b B^T I_l = BZ_b I_g - BV_g$

WORKED EXAMPLES

5.1 Obtain the oriented graph from the given node incidence matrix given in (5.49). (Kuvempu University)

$$
\begin{array}{c}
\\
\\
A_a = \\
\\
\\
\end{array}
\quad
\begin{array}{c}
\text{Branches} \\
\begin{array}{c|cccccc}
\text{Nodes} & 1 & 2 & 3 & 4 & 5 & 6 \\
\hline
a & +1 & +1 & 0 & 0 & 0 & -1 \\
b & 0 & -1 & -1 & +1 & 0 & 0 \\
c & 0 & 0 & 0 & -1 & +1 & +1 \\
d & -1 & 0 & +1 & 0 & -1 & 0
\end{array}
\end{array}
\qquad (5.49)
$$

Solution: The given matrix is an all incidence matrix. The oriented graph contains 4 nodes *a, b, c* and *d*. The oriented graph is as shown in Fig. 5.21. The procedure is as follows:

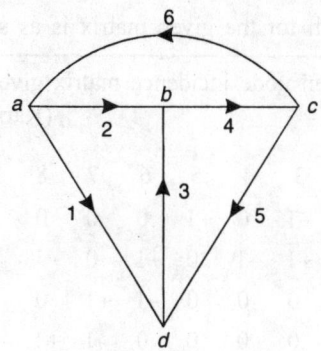

Fig. 5.21

Mark the four nodes *a, b, c* and *d* arbitrarily. Branch 1 is connected between nodes *a* and *d* and its orientation is away from *a*. Branch 2 is connected between nodes a and b and its orientation is away from a. Branch 6 is connected between nodes a and c and is oriented towards a. In a similar way branches 3, 4 and 5 are connected and their orientations are marked on the graph.

5.2 Sketch the oriented graph of the node incidence matrix shown in (5.50). (Mysore University)

$$
\mathbf{A} = \begin{array}{c} \\ \\ a \\ b \\ c \end{array}
\begin{array}{c} \text{Branches} \\ \begin{array}{ccccccc} 1 & 2 & 3 & 4 & 5 & 6 & 7 \end{array} \\ \left[\begin{array}{ccccccc} -1 & -1 & 0 & 0 & +1 & 0 & 0 \\ 0 & +1 & +1 & 0 & 0 & +1 & 0 \\ 0 & 0 & -1 & +1 & 0 & 0 & +1 \end{array}\right] \end{array} \quad (5.50)
$$

Nodes

Solution: The given matrix is a reduced incidene matrix. Hence, information about node d is missing. The elements of node d or fourth row are:

$$d\begin{bmatrix} +1 & 0 & 0 & -1 & -1 & -1 & -1 \end{bmatrix}$$

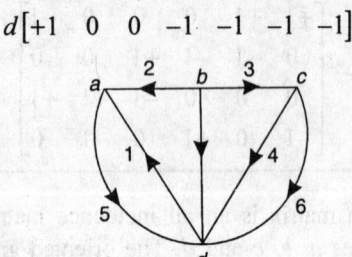

Fig. 5.22

The oriented graph for the given matrix is as shown in Fig. 5.22.

5.3 From the given node incidence matrix given in (5.51), obtain the oriented graph. (Karnataka University)

$$
\mathbf{A} = \begin{array}{c} a \\ b \\ c \\ d \end{array}
\begin{array}{c} \begin{array}{cccccccccc} 1 & 2 & 3 & 4 & 5 & 6 & 7 & 8 & 9 & 10 \end{array} \\ \left[\begin{array}{cccccccccc} +1 & 0 & -1 & 0 & +1 & 0 & 0 & 0 & 0 & -1 \\ 0 & 0 & +1 & -1 & 0 & +1 & 0 & -1 & 0 & 0 \\ 0 & -1 & 0 & 0 & 0 & -1 & +1 & 0 & 0 & 0 \\ 0 & 0 & 0 & 0 & 0 & 0 & -1 & +1 & -1 & +1 \end{array}\right] \end{array} \quad (5.51)
$$

Solution: The elements of node e are:

$$e\begin{bmatrix} -1 & +1 & 0 & +1 & -1 & 0 & 0 & 0 & +1 & 0 \end{bmatrix}$$

The oriented graph is as shown in Fig. 5.22 (a).

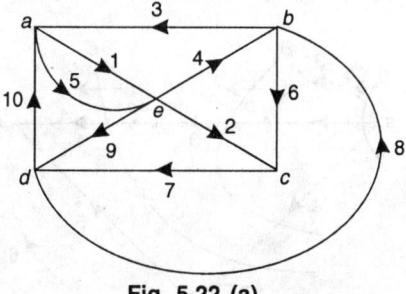

Fig. 5.22 (a)

5.4 The A matrix of a linear graph is given in (5.52). Draw the oriented graph. Select a tree with the branches 1, 3, 4 and 5 and construct cut-set matrix and tie-set matrix.

(Mysore University)

$$A = \begin{bmatrix} +1 & +1 & 0 & 0 & 0 & 0 & +1 \\ -1 & -1 & +1 & 0 & 0 & 0 & 0 \\ 0 & 0 & -1 & +1 & 0 & 0 & 0 \\ 0 & 0 & 0 & -1 & -1 & -1 & 0 \end{bmatrix} \qquad (5.52)$$

Solution: The given matrix is a reduced incidence matrix. The elements of the 5th node is added to make it an 'all incidence matrix'. The elements of 5th row are.

$$\begin{bmatrix} 0 & 0 & 0 & 0 & +1 & +1 & -1 \end{bmatrix}$$

The oriented graph is as shown in Fig. 5.23.

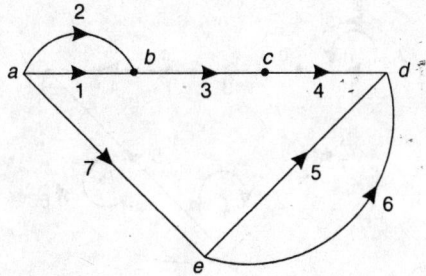

Fig. 5.23

The tree with branches 1, 3, 4 and 5 is as shown in Fig. 5.24.

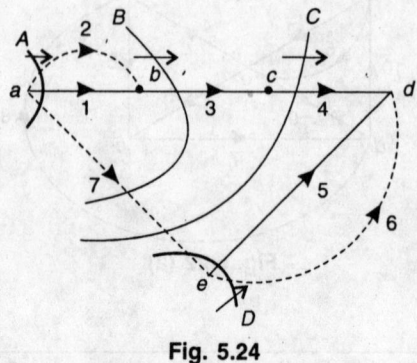

Fig. 5.24

To draw the cut-set matrix, 4 cut-sets are identified, one with respect to each twig. A (1, 2, 7), B (3, 7), C (4, 7) and D (5, 6, 7) are the 4 cut-sets and their orientations are marked to coincide with the orientations of the twigs. The cut-set matrix Q is as shown in (5.53).

$$
\mathbf{Q} = \begin{array}{c} \text{Cut-sets} \\[2pt] \begin{array}{c} A(1,2,7) \\ B(3,7) \\ C(4,7) \\ D(5,6,7) \end{array} \end{array}
\begin{array}{c} \overset{\textstyle \text{Branches}}{} \\ \begin{array}{ccccccc} 1 & 2 & 3 & 4 & 5 & 6 & 7 \end{array} \\ \left[\begin{array}{ccccccc} +1 & +1 & 0 & 0 & 0 & 0 & +1 \\ 0 & 0 & +1 & 0 & 0 & 0 & +1 \\ 0 & 0 & 0 & +1 & 0 & 0 & +1 \\ 0 & 0 & 0 & 0 & +1 & +1 & -1 \end{array}\right] \end{array}
\qquad (5.53)
$$

To draw the tie-set matrix, 3 tie-sets are identified with respect to each link as shown in Fig. 5.25.

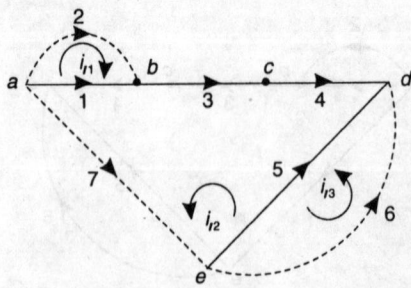

Fig. 5.25

i_{l1} (1, 2), i_{l2} (1, 3, 4, 5, 7) and i_{l3} (5, 6) are the tie-sets. The directions of i_{l1}, i_{l2} and i_{l3} are assumed to coincide with the orientations of the respecitve links.

The tie-set matrix **B** is as in (5.54).

$$
\begin{array}{c}
\text{Branches} \\[2pt]
\begin{array}{cc}
 & \begin{array}{ccccccc} 1 & 2 & 3 & 4 & 5 & 6 & 7 \end{array} \\
\mathbf{B} = \begin{array}{l} i_{l1}\,(1,2) \\ i_{l2}\,(1,3,4,5,7) \\ i_{l3}\,(5,6) \end{array} & \left[\begin{array}{ccccccc} -1 & +1 & 0 & 0 & 0 & 0 & 0 \\ -1 & 0 & -1 & -1 & +1 & 0 & +1 \\ 0 & 0 & 0 & 0 & -1 & +1 & 0 \end{array}\right]
\end{array}
\end{array}
\tag{5.54}
$$

5.5 For the graph shown in Fig. 5.26, select a tree and write the cut-set schedule. Obtain therefrom the equations giving the branch voltages in terms of tree-branch voltages.　　(Mysore University)

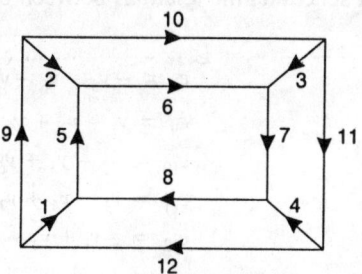

Fig. 5.26

Select a tree with branches, 1, 2, 3, 4, 5, 6, and 7 and form the cut-sets as shown in Fig. 5.27.

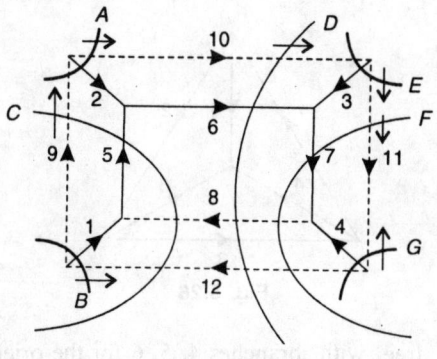

Fig. 5.27

Solution: The cut-set schedule is as in (5.55).

Twig voltages and cut-sets	Branches											
	1	2	3	4	5	6	7	8	9	10	11	12
v_1 B (1,9,12)	+1	0	0	0	0	0	0	0	+1	0	0	-1
v_2, A (2,9,10)	0	+1	0	0	0	0	0	0	-1	+1	0	0
v_3, E (3,10,11)	0	0	+1	0	0	0	0	0	0	-1	+1	0
v_4, G (4,11,12)	0	0	0	+1	0	0	0	0	0	0	-1	+1
v_5, C (5,8,9,12)	0	0	0	0	+1	0	0	-1	+1	0	0	-1
v_6, D (6,8,10,12)	0	0	0	0	0	+1	0	-1	0	+1	0	-1
v_7, F (7,8,11,12)	0	0	0	0	0	0	+1	-1	0	0	+1	-1

$$(5.55)$$

From the cut-set schedule, the relations between branch voltages and twig-voltages are:

$$e_1 = v_1 \qquad\qquad e_8 = -v_5 - v_6 - v_7$$
$$e_2 = v_2 \qquad\qquad e_9 = v_1 - v_2 + v_5$$
$$e_3 = v_3 \qquad\qquad e_{10} = v_2 - v_3 + v_6$$
$$e_4 = v_4 \qquad\qquad e_{11} = v_3 - v_4 + v_7$$
$$e_5 = v_5 \qquad\qquad e_{12} = -v_1 + v_4 - v_5 - v_6 - v_7$$
$$e_6 = v_6$$
$$e_7 = v_7$$

5.6 For the oriented graph given in Fig. 5.28, form a tree. Write the tie-set schedule and thereby write down the relations between the branch currents and loop currents. (Mangalore University)

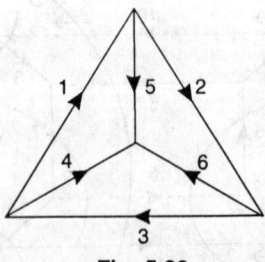

Fig. 5.28

One possible tree with branches 4, 5, 6 for the oriented graph in Fig. 5.28 is as shown in Fig. 5.29.

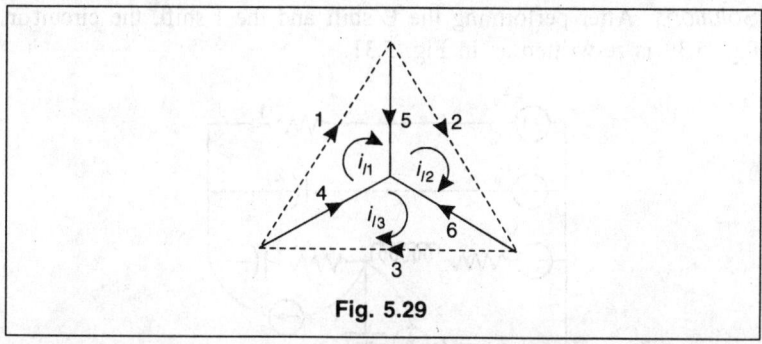

Fig. 5.29

Solution: Loop currents i_{l1}, i_{l2} and i_{l3} are assumed, such that, their directions coincide with the orientations of the corresponding links. The tie-set schedule is as in (5.58).

	Branches					
Tie-sets	1	2	3	4	5	6
i_{l1} (1, 4, 5)	+1	0	0	−1	+1	0
i_{l2} (2, 5, 6)	0	+1	0	0	−1	−1
i_{l3} (3, 4, 6)	0	0	+1	+1	0	+1

(5.56)

The relations between branch currents and loop currents are:

$$i_1 = i_{l1} \qquad\qquad i_4 = i_{l3} - i_{l1}$$
$$i_2 = i_{l2} \qquad\qquad i_5 = i_{l1} - i_{l2}$$
$$i_3 = i_{l3} \qquad\qquad i_6 = i_{l3} - i_{l2}$$

5.7 For the network shown and Fig. 5.30, after performing the E shift and I shift, draw the oriented graph. Select a tree corresponding to the branches 3 and 4. Write the tie-set and the cut-set matrices. The branch numbers are indicated in the circuit.

(Bangalore University)

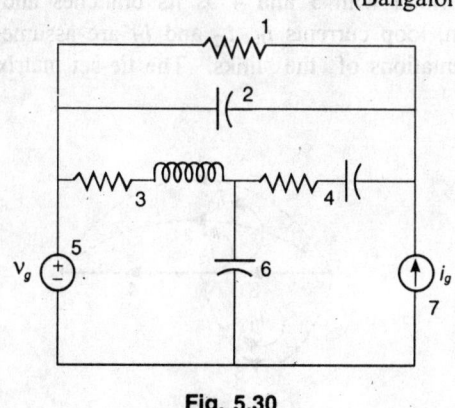

Fig. 5.30

Solution: After performing the E shift and the I shift, the circuit in Fig. 5.30 is re-written as in Fig. 5.31.

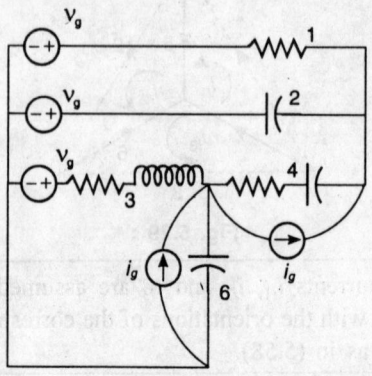

Fig. 5.31

The network graph for the circuit in Fig. 5.31 is as shown in Fig. 5.32. The orientations of the branches are arbitrarily assumed. The branches 5 and 7 does not form the branches of the graph.

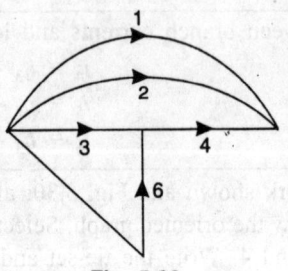

Fig. 5.32

A tree is formed with 3 and 4 as its branches and is shown in Fig. 5.33. Then, loop currents i_{l1}, i_{l2} and i_{l3} are assumed to coincide with the orientations of the links. The tie-set matrix is as shown in (5.57).

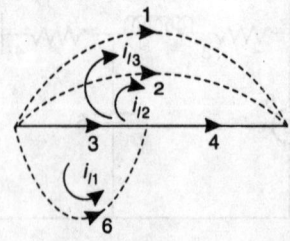

Fig. 5.33

Branches

$$
\begin{array}{c}
\text{Loop currents} \qquad \quad 1 \quad 2 \quad 3 \quad 4 \quad 6 \\[4pt]
\mathbf{B} = \begin{array}{c} i_{l1} \\ i_{l2} \\ i_{l3} \end{array}
\begin{bmatrix}
0 & 0 & -1 & 0 & +1 \\
0 & +1 & -1 & -1 & 0 \\
+1 & 0 & -1 & -1 & 0
\end{bmatrix}
\end{array}
\tag{5.57}
$$

For the same tree shown in Fig. 5.33, the cut-sets are identified as shown in Fig. 5.34.

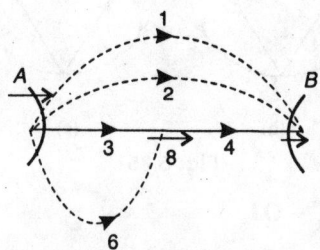

Fig. 5.34

The cut-set matrix is as shown in (5.58).

Branches

$$
\begin{array}{c}
\text{Cut-sets} \quad 1 \quad 2 \quad 3 \quad 4 \quad 6 \\[4pt]
\mathbf{Q} = \begin{array}{c} A \\ v \end{array}
\begin{bmatrix}
+1 & +1 & +1 & 0 & +1 \\
+1 & +1 & 0 & +1 & 0
\end{bmatrix}
\end{array}
\tag{5.58}
$$

5.8 For the network shown in Fig. 5.35 (a), draw the graph. Select a tree with branches 2 and 4. Draw the cut-set matrix. Write down the equilibrium equations, with node-pair voltages as variables. Solve these equations, and find the various branch voltages and currents. The integers indicate branch numbers.

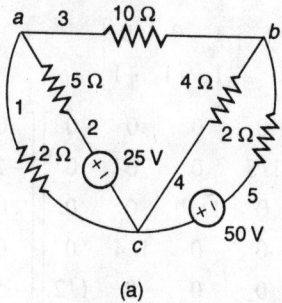

(a)

Fig. 5.35

Solution: The oriented graph of the given network is as shown in Fig. 5.35 (b). The orientations of the branches are arbitrarily assumed. The tree with branches 2 and 4 is as shown in Fig. 5.35 (c). The cut-sets A and B and their orientations are also marked. The set of equilibrium equations are given by

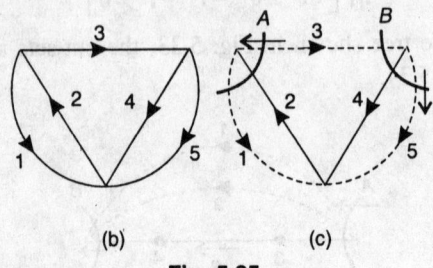

(b)   (c)

Fig. 5.35

$$Q\,Y_b\,Q^T\,V_t = Q\,Y_b\,V_g - Q\,I_g \qquad (1)$$

$$Q\,Y_b\,Q^T = \begin{bmatrix} -1 & +1 & -1 & 0 & 0 \\ 0 & 0 & -1 & +1 & +1 \end{bmatrix}$$

$$\begin{bmatrix} 1/2 & 0 & 0 & 0 & 0 \\ 0 & 1/5 & 0 & 0 & 0 \\ 0 & 0 & 1/10 & 0 & 0 \\ 0 & 0 & 0 & 1/4 & 0 \\ 0 & 0 & 0 & 0 & 1/2 \end{bmatrix} \begin{bmatrix} -1 & 0 \\ +1 & 0 \\ -1 & -1 \\ 0 & +1 \\ 0 & +1 \end{bmatrix}$$

$$= \begin{bmatrix} 0.8 & 0.1 \\ 0.1 & 0.85 \end{bmatrix} \qquad (5.59)$$

$$Q\,I_g = \begin{bmatrix} 0 \\ 0 \end{bmatrix} \qquad \because I_g \text{ is a null matrix.}$$

$$Q\,Y_b\,V_g = \begin{bmatrix} -1 & +1 & -1 & 0 & 0 \\ 0 & 0 & -1 & +1 & +1 \end{bmatrix}$$

$$\begin{bmatrix} 1/2 & 0 & 0 & 0 & 0 \\ 0 & 1/5 & 0 & 0 & 0 \\ 0 & 0 & 1/10 & 0 & 0 \\ 0 & 0 & 0 & 1/4 & 0 \\ 0 & 0 & 0 & 0 & 1/2 \end{bmatrix} \begin{bmatrix} 0 \\ -25 \\ 0 \\ 0 \\ -50 \end{bmatrix} = \begin{bmatrix} -5 \\ -25 \end{bmatrix} \qquad (5.60)$$

Then, Eq. (1), becomes

$$= \begin{bmatrix} 0.8 & 0.1 \\ 0.1 & 0.85 \end{bmatrix} \begin{bmatrix} v_2 \\ v_4 \end{bmatrix} = \begin{bmatrix} -5 \\ -25 \end{bmatrix} - \begin{bmatrix} 0 \\ 0 \end{bmatrix} \qquad (5.61)$$

The equilibrium equations are:

$$0.8\, v_2 + 0.1\, v_4 = -5$$

and $\qquad 0.1\, v_2 + 0.85\, v_4 = -25$

Solving these equations, we get

$$v_2 = -2.612 \text{ V} \qquad \text{and} \qquad v_4 = -29.104 \text{ V}$$

The various branch voltages are:

$$\mathbf{E_b} = \mathbf{Q^T V_t} = \begin{bmatrix} -1 & 0 \\ +1 & 0 \\ -1 & -1 \\ 0 & +1 \\ 0 & +1 \end{bmatrix} \begin{bmatrix} -2.612 \\ -29.104 \end{bmatrix} = \begin{bmatrix} 2.612 \\ -2.612 \\ 31.712 \\ -29.104 \\ -29.104 \end{bmatrix} \text{volts} \qquad (5.62)$$

∴ The voltages across the various branches are:

$e_1 = 2.612$ V, $e_2 = -2.612$ V, $e_3 = 31.712$ V,

$$e_4 = -29.104 \text{ V, and } e_5 = -29.104 \text{ V}$$

The various branch currents are given by

$$\mathbf{I_b} = \mathbf{I_g} + \mathbf{Y_b}\,(\mathbf{E_b} - \mathbf{V_g})$$

i.e.
$$\begin{bmatrix} i_1 \\ i_2 \\ i_3 \\ i_4 \\ i_5 \end{bmatrix} = \begin{bmatrix} 0 \\ 0 \\ 0 \\ 0 \\ 0 \end{bmatrix} + \begin{bmatrix} 1/2 & 0 & 0 & 0 & 0 \\ 0 & 1/5 & 0 & 0 & 0 \\ 0 & 0 & 1/10 & 0 & 0 \\ 0 & 0 & 0 & 1/4 & 0 \\ 0 & 0 & 0 & 0 & 1/2 \end{bmatrix} \begin{bmatrix} 2.612-0 \\ -2.612+25 \\ 31.712-0 \\ -29.104-0 \\ -29.104+50 \end{bmatrix}$$

$$= \begin{bmatrix} 1.306 \\ 4.478 \\ 31.712 \\ -7.276 \\ 10.448 \end{bmatrix}$$

∴ The various branch currents are:

$$i_1 = 1.306 \text{ A, } i_2 = 4.478 \text{ A, } i_3 = 3.1712 \text{ A,}$$

$$i_4 = -7.276 \text{ A, and } i_5 = 10.448 \text{ A}$$

The currents in the various branches of the network are marked as shown in Fig. 5.36.

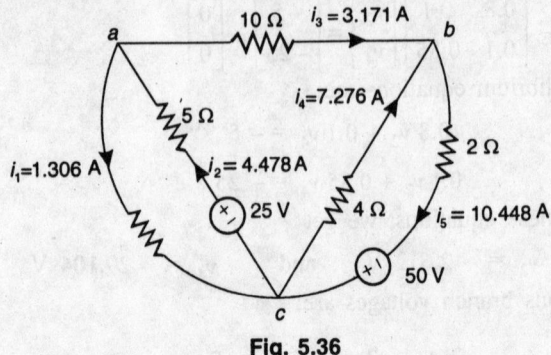

Fig. 5.36

5.9 For the network shown in Fig. 5.37 (a), draw the graph. Select 2 and 4 as tree branches. Draw the tie-set matrix. Write down the equilibrium equations with loop currents as variables. Solve these equations and find the various branch voltages and currents. The integers indicate branch numbers.

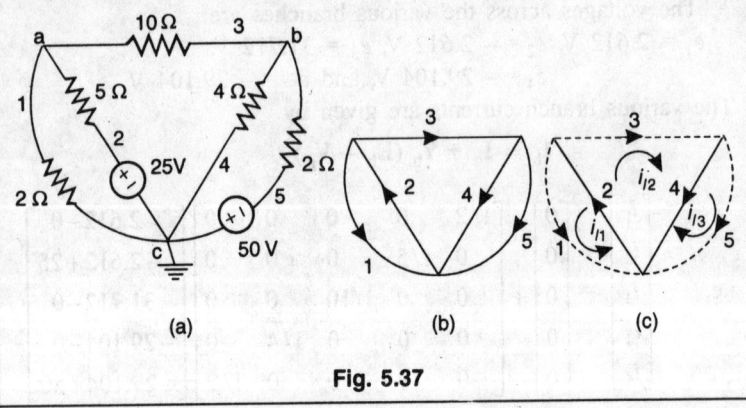

Fig. 5.37

Solution: The loop currents i_{l1}, i_{l2} and i_{l3} are assumed as shown in Fig. 5.37 (c), such that their directions coincide with the orientations of the links. The equilibrium equations with loop currents as variables are:

$$\mathbf{B Z_b B^T I_l} = \mathbf{B Z_b I_g} - \mathbf{B V_g} \tag{1}$$

$$\mathbf{B Z_b B^T} = \begin{bmatrix} +1 & +1 & 0 & 0 & 0 \\ 0 & +1 & +1 & +1 & 0 \\ 0 & 0 & 0 & -1 & +1 \end{bmatrix} \begin{bmatrix} 2 & 0 & 0 & 0 & 0 \\ 0 & 5 & 0 & 0 & 0 \\ 0 & 0 & 10 & 0 & 0 \\ 0 & 0 & 0 & 4 & 0 \\ 0 & 0 & 0 & 0 & 2 \end{bmatrix} \begin{bmatrix} +1 & 0 & 0 \\ +1 & +1 & 0 \\ 0 & +1 & 0 \\ 0 & +1 & -1 \\ 0 & 0 & +1 \end{bmatrix}$$

$$= \begin{bmatrix} 7 & 5 & 0 \\ 5 & 19 & -4 \\ 0 & -4 & 6 \end{bmatrix} \qquad (5.63)$$

$$\mathbf{BZ_b\,I_g} = \begin{bmatrix} 0 \\ 0 \\ 0 \end{bmatrix} \quad \because \mathbf{I_g} \text{ is a null matrix}$$

$$\mathbf{BV_g} = \begin{bmatrix} +1 & +1 & 0 & 0 & 0 \\ 0 & +1 & +1 & +1 & 0 \\ 0 & 0 & 0 & -1 & +1 \end{bmatrix} \begin{bmatrix} 0 \\ -25 \\ 0 \\ 0 \\ -50 \end{bmatrix} = \begin{bmatrix} -25 \\ -25 \\ -50 \end{bmatrix}$$

Substituting the above in equation (1), we get

$$= \begin{bmatrix} 7 & 5 & 0 \\ 5 & 19 & -4 \\ 0 & -4 & 6 \end{bmatrix} \begin{bmatrix} i_{l1} \\ i_{l2} \\ i_{l3} \end{bmatrix} = \begin{bmatrix} 0 \\ 0 \\ 0 \end{bmatrix} - \begin{bmatrix} -25 \\ -25 \\ -50 \end{bmatrix} = \begin{bmatrix} 25 \\ 25 \\ 50 \end{bmatrix}$$

Solving the above equations, we get

$$i_{l1} = 1.306 \text{ A}, \ i_{l2} = 3.171 \text{ A and } i_{l3} = 10.448 \text{ A}.$$

The various branch currents are given by

$$\mathbf{I_b} = \mathbf{B^T I_l} = \begin{bmatrix} +1 & 0 & 0 \\ +1 & +1 & 0 \\ 0 & +1 & 0 \\ 0 & +1 & -1 \\ 0 & 0 & +1 \end{bmatrix} \begin{bmatrix} 1.306 \\ 3.171 \\ 10.448 \end{bmatrix} = \begin{bmatrix} 1.306 \\ 4.478 \\ 3.171 \\ -7.276 \\ 10.448 \end{bmatrix}$$

The various branch currents are:

$$i_1 = 1.306 \text{ A}, \ i_2 = 4.478 \text{ A}, \ i_3 = 3.171 \text{ A},$$

$$i_4 = -7.276 \text{ A, and } i_5 = 10.448 \text{ A}$$

The various branch voltages are given by

$$\mathbf{E_b} = \mathbf{E_g} + \mathbf{Z_b}\,(\mathbf{I_b} - \mathbf{I_g})$$

$$\text{i.e. } \begin{bmatrix} e_1 \\ e_2 \\ e_3 \\ e_4 \\ e_5 \end{bmatrix} = \begin{bmatrix} 0 \\ -25 \\ 0 \\ 0 \\ -50 \end{bmatrix} + \begin{bmatrix} 2 & 0 & 0 & 0 & 0 \\ 0 & 5 & 0 & 0 & 0 \\ 0 & 0 & 10 & 0 & 0 \\ 0 & 0 & 0 & 4 & 0 \\ 0 & 0 & 0 & 0 & 2 \end{bmatrix} \begin{bmatrix} 1.306-0 \\ 4.478-0 \\ 3.171-0 \\ -7.276-0 \\ 10.448-0 \end{bmatrix} = \begin{bmatrix} 2.612 \\ -2.612 \\ 31.712 \\ -29.104 \\ -29.104 \end{bmatrix} \text{ volts}$$

5.10 For the network shown in Fig. 5.38 (a), draw the graph. Consider node c as datum node and write the **A** matrix. Obtain equilibrium equations with node to datum voltages as variables. Solve the equations and find all the branch currents and branch voltages. The integers indicate branch numbers.

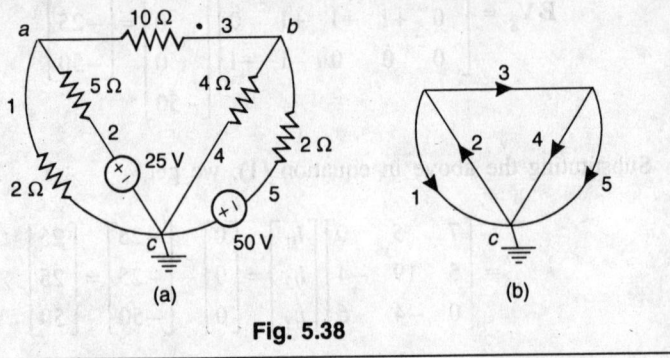

Fig. 5.38

Solution: The oriented graph of the network shown in Fig. 5.38 (a) is as shown in Fig. 5.38 (b). The orientations of the branches are arbitrarily assumed.

The equilibrium equations with node to datum voltages as variables are given by:

$$\mathbf{A Y_b A^T V_n} = \mathbf{A Y_b V_g} - \mathbf{A I_g} \tag{1}$$

$$\mathbf{A Y_b A^T} = \begin{bmatrix} +1 & -1 & +1 & 0 & 0 \\ 0 & 0 & -1 & +1 & +1 \end{bmatrix}$$

$$\begin{bmatrix} 1/2 & 0 & 0 & 0 & 0 \\ 0 & 1/5 & 0 & 0 & 0 \\ 0 & 0 & 1/10 & 0 & 0 \\ 0 & 0 & 0 & 1/4 & 0 \\ 0 & 0 & 0 & 0 & 1/2 \end{bmatrix} \begin{bmatrix} +1 & 0 \\ -1 & 0 \\ +1 & -1 \\ 0 & +1 \\ 0 & +1 \end{bmatrix} = \begin{bmatrix} 0.8 & -0.1 \\ -0.1 & 0.85 \end{bmatrix}$$

$$\mathbf{A}\,\mathbf{Y_b}\,\mathbf{V_g} = \begin{bmatrix} +1 & -1 & +1 & 0 & 0 \\ 0 & 0 & -1 & +1 & +1 \end{bmatrix}$$

$$\begin{bmatrix} 1/2 & 0 & 0 & 0 & 0 \\ 0 & 1/5 & 0 & 0 & 0 \\ 0 & 0 & 1/10 & 0 & 0 \\ 0 & 0 & 0 & 1/4 & 0 \\ 0 & 0 & 0 & 0 & 1/2 \end{bmatrix} \begin{bmatrix} 0 \\ -25 \\ 0 \\ 0 \\ -50 \end{bmatrix} = \begin{bmatrix} 5 \\ -25 \end{bmatrix}$$

$$\mathbf{A}\,\mathbf{I_g} = \begin{bmatrix} 0 \\ 0 \\ 0 \end{bmatrix} \quad \because \mathbf{I_g} \text{ is a null matrix}$$

$\therefore$ Equation (1) can be written as

$$= \begin{bmatrix} 0.8 & -0.1 \\ -0.1 & 0.85 \end{bmatrix} \begin{bmatrix} v_a \\ v_b \end{bmatrix} = \begin{bmatrix} 5 \\ -25 \end{bmatrix} - \begin{bmatrix} 0 \\ 0 \end{bmatrix} = \begin{bmatrix} 5 \\ -25 \end{bmatrix}$$

Solving the above equations, we get

$$v_a = 2.612 \text{ V}$$

and $\qquad v_b = -29.104 \text{ V}$

The various branch voltages are:

$$e_1 = v_a = 2.612 \text{ V}$$

$$e_2 = -v_a = -2.612 \text{ V}$$

$$e_3 = v_a - v_b = 2.612 + 29.104 = 31.712 \text{ V}$$

$$e_4 = v_b = -29.104 \text{ V}$$

and $\qquad e_5 = v_b = -29.104 \text{ V}$

The various branch currents are:

$$i_1 = \frac{v_a}{2} = \frac{2.612}{2} = 1.306 \text{ A,}$$

$$i_2 = \frac{25 - v_a}{5} = \frac{25 - 2.612}{5} = 4.477 \text{ A}$$

$$i_3 = \frac{v_a - v_b}{10} = \frac{2.612 + 29.104}{10} = 3.171 \text{ A}$$

$$i_4 = \frac{v_b}{4} = \frac{-29.104}{4} = -7.275 \text{ A}$$

and $$i_5 = \frac{v_a + 50}{2} = \frac{-29.104 + 50}{2} = 10.45 \text{ A}$$

5.11 For the circuit shown in Fig. 5.39 draw the graph. The ohmic values also represent the branch numbers. Form a tree with branches 4, 5 and 6. Take $e_a = v_5$, $e_b = v_6$, $e_c = v_4$ as the node-pair voltages. Obtain the network equilibrium equations with node-pair voltages as variables. Solve these equations to get the currents in the various branches of the network. (Mysore University)

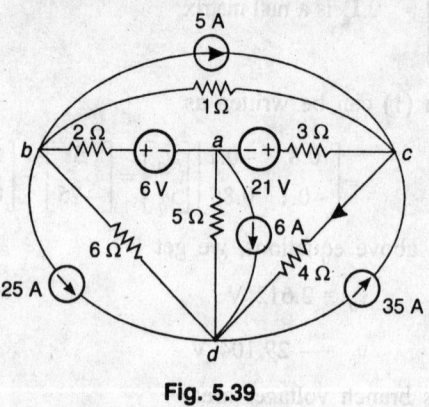

Fig. 5.39

Solution: The graph of the network is as shown in Fig. 5.40 (a) and the tree with branches 4, 5 and 6 is as shown in Fig. 5.40 (b). The orientations of the branches are assumed arbitrarily.

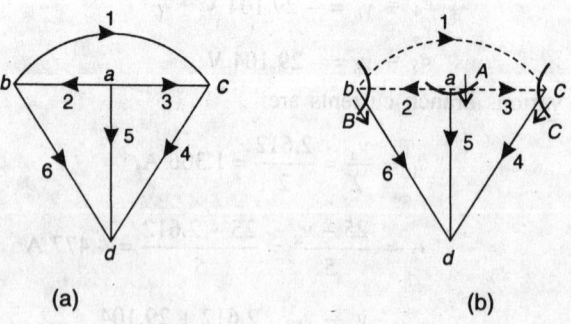

(a)   (b)

Fig. 5.40

$$\mathbf{Q}\,\mathbf{Y_b}\,\mathbf{Q^T} = \begin{bmatrix} 0 & +1 & +1 & 0 & +1 & 0 \\ +1 & -1 & 0 & 0 & 0 & +1 \\ -1 & 0 & -1 & +1 & 0 & 0 \end{bmatrix}$$

$$\begin{bmatrix} 1 & 0 & 0 & 0 & 0 & 0 \\ 0 & 1/2 & 0 & 0 & 0 & 0 \\ 0 & 0 & 1/3 & 0 & 0 & 0 \\ 0 & 0 & 0 & 1/4 & 0 & 0 \\ 0 & 0 & 0 & 0 & 1/5 & 0 \\ 0 & 0 & 0 & 0 & 0 & 1/6 \end{bmatrix} \begin{bmatrix} 0 & +1 & -1 \\ +1 & -1 & 0 \\ +1 & 0 & -1 \\ 0 & 0 & +1 \\ +1 & 0 & 0 \\ 0 & +1 & 0 \end{bmatrix}$$

$$= \begin{bmatrix} 31/30 & -1/2 & -1/3 \\ -1/2 & 20/12 & -1 \\ -1/3 & -1 & 19/12 \end{bmatrix}$$

$$\mathbf{Q}\,\mathbf{Y_b}\,\mathbf{V_g} = \begin{bmatrix} 0 & +1 & +1 & 0 & +1 & 0 \\ +1 & -1 & 0 & 0 & 0 & +1 \\ -1 & 0 & -1 & +1 & 0 & 0 \end{bmatrix}$$

$$\begin{bmatrix} 1 & 0 & 0 & 0 & 0 & 0 \\ 0 & 1/2 & 0 & 0 & 0 & 0 \\ 0 & 0 & 1/3 & 0 & 0 & 0 \\ 0 & 0 & 0 & 1/4 & 0 & 0 \\ 0 & 0 & 0 & 0 & 1/5 & 0 \\ 0 & 0 & 0 & 0 & 0 & 1/6 \end{bmatrix} \begin{bmatrix} 0 \\ -6 \\ -21 \\ 0 \\ 0 \\ 0 \end{bmatrix} = \begin{bmatrix} -10 \\ 3 \\ 7 \end{bmatrix}$$

$$\mathbf{Q}\,\mathbf{I_g} = \begin{bmatrix} 0 & +1 & +1 & 0 & +1 & 0 \\ +1 & -1 & 0 & 0 & 0 & +1 \\ -1 & 0 & -1 & +1 & 0 & 0 \end{bmatrix} \begin{bmatrix} 5 \\ 0 \\ 0 \\ -35 \\ 6 \\ 25 \end{bmatrix} = \begin{bmatrix} 6 \\ 30 \\ -40 \end{bmatrix}$$

The equilibrium equations are given by

$$Q Y_b Q^T V_t = Q Y_b Y_g - Q I_g$$

$$= \begin{bmatrix} 31/30 & -1/2 & -1/3 \\ -1/2 & 20/12 & -1 \\ -1/3 & -1 & 19/12 \end{bmatrix} \begin{bmatrix} v_5 \\ v_6 \\ v_4 \end{bmatrix} = \begin{bmatrix} -10 \\ 3 \\ 7 \end{bmatrix} - \begin{bmatrix} 6 \\ 30 \\ -40 \end{bmatrix} = \begin{bmatrix} -16 \\ -27 \\ 47 \end{bmatrix}$$

Solving the above set of equations, we get

$$v_5 = -8.968 \text{ V}, \quad v_6 = -3.562 \text{ V}, \quad v_4 = 25.552 \text{ V}.$$

The various branch voltages are given by

$$E_b = \begin{bmatrix} e_1 \\ e_2 \\ e_3 \\ e_4 \\ e_5 \\ e_6 \end{bmatrix} = Q^T V_b = \begin{bmatrix} 0 & +1 & -1 \\ 0 & -1 & 0 \\ +1 & 0 & -1 \\ 0 & 0 & +1 \\ +1 & 0 & 0 \\ 0 & +1 & 0 \end{bmatrix} \begin{bmatrix} -8.968 \\ -3.562 \\ 25.552 \end{bmatrix} = \begin{bmatrix} -29.114 \\ -5.406 \\ -34.52 \\ 25.552 \\ -8.968 \\ -3.562 \end{bmatrix} \text{volts}$$

The various branch currents are given by

$$I_b = I_g + Y_b (E_b - V_g)$$

i.e.

$$\begin{bmatrix} i_1 \\ i_2 \\ i_3 \\ i_4 \\ i_5 \\ i_6 \end{bmatrix} = \begin{bmatrix} 5 \\ 0 \\ 0 \\ -35 \\ 6 \\ 25 \end{bmatrix} + \begin{bmatrix} 1 & 0 & 0 & 0 & 0 & 0 \\ 0 & 1/2 & 0 & 0 & 0 & 0 \\ 0 & 0 & 1/3 & 0 & 0 & 0 \\ 0 & 0 & 0 & 1/4 & 0 & 0 \\ 0 & 0 & 0 & 0 & 1/5 & 0 \\ 0 & 0 & 0 & 0 & 0 & 1/6 \end{bmatrix} \begin{bmatrix} -29.114 - 0 \\ -5.406 + 6 \\ -34.52 + 21 \\ 25.552 + 0 \\ -8.968 + 0 \\ -3.562 + 0 \end{bmatrix}$$

$$= \begin{bmatrix} -24.114 \\ 0.297 \\ -4.5 \\ -28.6112 \\ 4.206 \\ 24.407 \end{bmatrix} \text{amps}$$

The current distribution in the various branches of the network is as shown in Fig. 5.41.

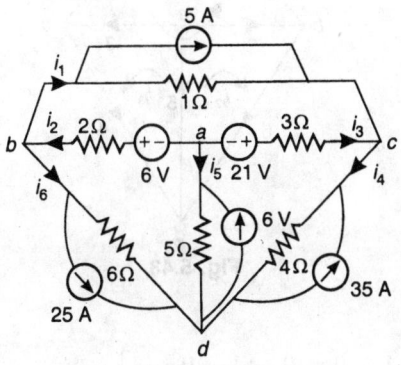

Fig. 5.41

5.12 For the circuit shown in Fig. 5.42, draw the graph. The ohmic values also represent the branch numbers. From a tree with branches 4, 5 and 6. Obtain the equilibrium equations with loop currents as variables. Solve these equations to get currents in the various branches of the network. *(Mysore University)*

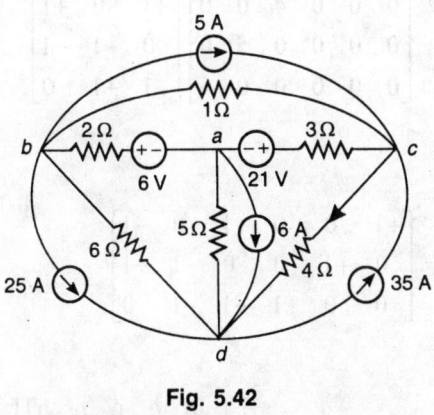

Fig. 5.42

Solution: For the network in Fig. 5.42, a tree with branches 4, 5 and 6 is formed as explained in example 5.11 and is shown in Fig. 5.43.

The loop currents i_{l1}, i_{l2} and i_{l3} are assumed such that, their directions coincide with the orientations of the corresponding links.

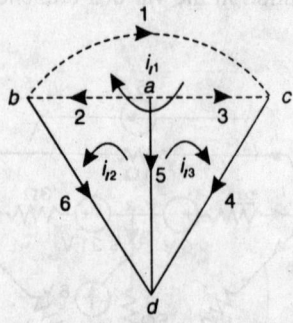

Fig. 5.43

$$\mathbf{B Z_b B^T} = \begin{bmatrix} +1 & 0 & 0 & +1 & 0 & -1 \\ 0 & +1 & 0 & 0 & -1 & +1 \\ 0 & 0 & +1 & +1 & -1 & 0 \end{bmatrix}$$

$$\begin{bmatrix} 1 & 0 & 0 & 0 & 0 & 0 \\ 0 & 2 & 0 & 0 & 0 & 0 \\ 0 & 0 & 3 & 0 & 0 & 0 \\ 0 & 0 & 0 & 4 & 0 & 0 \\ 0 & 0 & 0 & 0 & 5 & 0 \\ 0 & 0 & 0 & 0 & 0 & 6 \end{bmatrix} \begin{bmatrix} +1 & 0 & 0 \\ 0 & +1 & 0 \\ 0 & 0 & +1 \\ +1 & 0 & +1 \\ 0 & -1 & -1 \\ -1 & +1 & 0 \end{bmatrix} = \begin{bmatrix} 11 & -6 & 4 \\ -6 & 13 & 5 \\ 4 & 5 & 12 \end{bmatrix}$$

$$\mathbf{B Z_b I_g} = \begin{bmatrix} +1 & 0 & 0 & +1 & 0 & -1 \\ 0 & +1 & 0 & 0 & -1 & +1 \\ 0 & 0 & +1 & +1 & -1 & 0 \end{bmatrix}$$

$$\begin{bmatrix} 1 & 0 & 0 & 0 & 0 & 0 \\ 0 & 2 & 0 & 0 & 0 & 0 \\ 0 & 0 & 3 & 0 & 0 & 0 \\ 0 & 0 & 0 & 4 & 0 & 0 \\ 0 & 0 & 0 & 0 & 5 & 0 \\ 0 & 0 & 0 & 0 & 0 & 6 \end{bmatrix} \begin{bmatrix} 5 \\ 0 \\ 0 \\ -35 \\ 6 \\ 25 \end{bmatrix} = \begin{bmatrix} -285 \\ 120 \\ -170 \end{bmatrix}$$

$$
\mathbf{BV_g} =
\begin{bmatrix}
+1 & 0 & 0 & +1 & 0 & -1 \\
0 & +1 & 0 & 0 & -1 & +1 \\
0 & 0 & +1 & +1 & -1 & 0
\end{bmatrix}
\begin{bmatrix}
0 \\
-6 \\
-21 \\
0 \\
0 \\
0
\end{bmatrix}
=
\begin{bmatrix}
0 \\
-6 \\
-21
\end{bmatrix}
$$

The set of equilibrium equations are given by

$$
\mathbf{B Z_b B^T I_l} = \mathbf{B Z_b I_g} - \mathbf{B V_g}
$$

i.e.
$$
\begin{bmatrix}
11 & -6 & 4 \\
-6 & 13 & 5 \\
4 & 5 & 12
\end{bmatrix}
\begin{bmatrix}
i_{l1} \\
i_{l2} \\
i_{l3}
\end{bmatrix}
=
\begin{bmatrix}
-285 \\
120 \\
-170
\end{bmatrix}
-
\begin{bmatrix}
0 \\
-6 \\
-21
\end{bmatrix}
=
\begin{bmatrix}
-285 \\
126 \\
-149
\end{bmatrix}
$$

Solving the above equations, we get

$$
i_{l1} = -24.108 \text{ A}, \quad i_{l2} = 0.2977 \text{ A}, \quad \text{and} \quad i_{l3} = -4.504 \text{ A}
$$

The various branch currents are given by

$$
\mathbf{I_b} = \mathbf{B^T I_l} =
\begin{bmatrix}
+1 & 0 & 0 \\
0 & +1 & 0 \\
0 & 0 & +1 \\
+1 & 0 & +1 \\
0 & -1 & -1 \\
-1 & +1 & 0
\end{bmatrix}
\begin{bmatrix}
-24.108 \\
0.297 \\
-4.504
\end{bmatrix}
=
\begin{bmatrix}
-24.108 \\
0.292 \\
-4.505 \\
-28.613 \\
4.213 \\
24.4
\end{bmatrix}
\text{amps} =
\begin{bmatrix}
i_1 \\
i_2 \\
i_3 \\
i_4 \\
i_5 \\
i_6
\end{bmatrix}
$$

The various branch voltages are given by

$$
\mathbf{E_b} = \mathbf{V_g} + \mathbf{Z_b (I_b - I_g)}
$$

$$
=
\begin{bmatrix}
0 \\
-6 \\
-21 \\
0 \\
0 \\
0
\end{bmatrix}
+
\begin{bmatrix}
1 & 0 & 0 & 0 & 0 & 0 \\
0 & 2 & 0 & 0 & 0 & 0 \\
0 & 0 & 3 & 0 & 0 & 0 \\
0 & 0 & 0 & 4 & 0 & 0 \\
0 & 0 & 0 & 0 & 5 & 0 \\
0 & 0 & 0 & 0 & 0 & 6
\end{bmatrix}
\begin{bmatrix}
24.108 & -5 \\
0.292 & -0 \\
-4.505 & -0 \\
-28.613 & +35 \\
4.213 & -6 \\
24.4 & -25
\end{bmatrix}
$$

$$= \begin{bmatrix} -29.108 \\ -5.416 \\ -34.515 \\ +25.548 \\ -8.935 \\ -3.6 \end{bmatrix} \text{ volts} = \begin{bmatrix} e_1 \\ e_2 \\ e_3 \\ e_4 \\ e_5 \\ e_6 \end{bmatrix}$$

5.13 For the network shown in Fig. 5.44, draw the oriented graph. Construct a tree in which v_1 and v_2 are the branch voltage. Write the equilibrium equations and solve for v_1 and v_2.

(Gulbarga University)

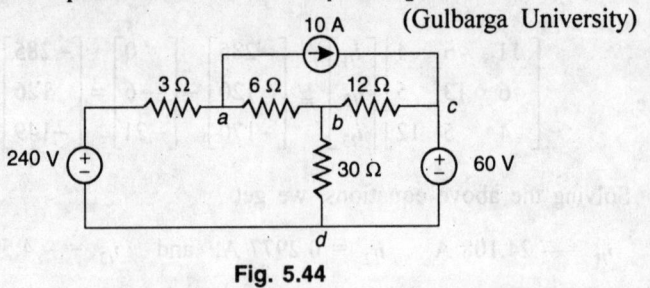

Fig. 5.44

Solution: The network in Fig. 5.44 is re-written as shown in Fig. 5.45 (a), after E shift and I shift.

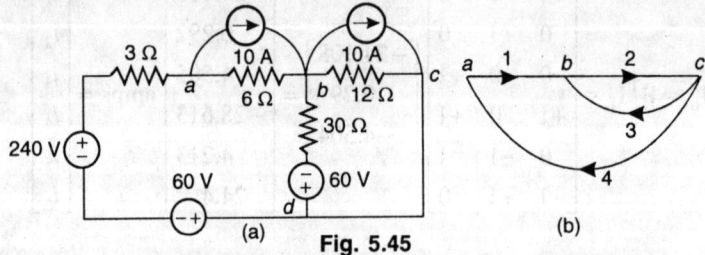

(a)   **Fig. 5.45**   (b)

The oriented graph of the network in Fig. 5.45 (a), is a shown in Fig. 5.45 (b). A tree with 1 and 2 as its branches is as shown in Fig. 5.46. The cut-sets A and B and their orientations are marked as shown.

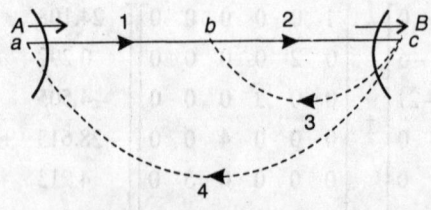

Fig. 5.46

Then

$$Q Y_b Q^T = \begin{bmatrix} +1 & 0 & 0 & -1 \\ 0 & +1 & -1 & -1 \end{bmatrix} \begin{bmatrix} 1/6 & 0 & 0 & 0 \\ 0 & 1/12 & 0 & 0 \\ 0 & 0 & 1/30 & 0 \\ 0 & 0 & 0 & 1/3 \end{bmatrix} \begin{bmatrix} +1 & 0 \\ 0 & +1 \\ 0 & -1 \\ -1 & -1 \end{bmatrix}$$

$$= \begin{bmatrix} 1/2 & 1/3 \\ 1/3 & 27/60 \end{bmatrix}$$

$$Q Y_b V_g = \begin{bmatrix} +1 & 0 & 0 & -1 \\ 0 & +1 & -1 & -1 \end{bmatrix} \begin{bmatrix} 1/6 & 0 & 0 & 0 \\ 0 & 1/12 & 0 & 0 \\ 0 & 0 & 1/30 & 0 \\ 0 & 0 & 0 & 1/3 \end{bmatrix} \begin{bmatrix} 0 \\ 0 \\ 60 \\ -180 \end{bmatrix} = \begin{bmatrix} 60 \\ 58 \end{bmatrix}$$

$$Q I_g = \begin{bmatrix} +1 & 0 & 0 & -1 \\ 0 & +1 & -1 & -1 \end{bmatrix} \begin{bmatrix} 10 \\ 10 \\ 0 \\ 0 \end{bmatrix} = \begin{bmatrix} 10 \\ 10 \end{bmatrix}$$

The equilibrium equations are given by

$$Q Y_b Q^T V_t = Q Y_b V_g - Q I_g$$

i.e. $$\begin{bmatrix} 1/2 & 1/3 \\ 1/3 & 27/60 \end{bmatrix} \begin{bmatrix} v_1 \\ v_2 \end{bmatrix} = \begin{bmatrix} 60 \\ 58 \end{bmatrix} - \begin{bmatrix} 10 \\ 10 \end{bmatrix} = \begin{bmatrix} 50 \\ 48 \end{bmatrix}$$

Solving the above equations, we get

$$v_1 = 57.067 \text{ V and } v_2 = 64.38 \text{ V}$$

The various branch voltages are given by

$$E_b = Q^T V_t = \begin{bmatrix} +1 & 0 \\ 0 & +1 \\ 0 & -1 \\ -1 & -1 \end{bmatrix} \begin{bmatrix} 57.067 \\ 64.38 \end{bmatrix} = \begin{bmatrix} 57.067 \\ 64.38 \\ -64.38 \\ -121.45 \end{bmatrix} \text{ volts} = \begin{bmatrix} e_1 \\ e_2 \\ e_3 \\ e_4 \end{bmatrix}$$

The various branch currents are given by

$$I_b = I_g + Y_b (E_b - V_g)$$

i.e.

$$\begin{bmatrix} i_1 \\ i_2 \\ i_3 \\ i_4 \end{bmatrix} = \begin{bmatrix} 10 \\ 10 \\ 0 \\ 0 \end{bmatrix} + \begin{bmatrix} 1/6 & 0 & 0 & 0 \\ 0 & 1/2 & 0 & 0 \\ 0 & 0 & 1/30 & 0 \\ 0 & 0 & 0 & 1/3 \end{bmatrix} \begin{bmatrix} 57.07 - 0 \\ 64.38 - 0 \\ -64.38 - 60 \\ -121.85 + 180 \end{bmatrix}$$

$$= \begin{bmatrix} 19.51 \\ 15.365 \\ -4.146 \\ 19.516 \end{bmatrix} \text{amps}$$

5.14 For the network shown in Fig. 5.47, draw the oriented graph. Select a tree with branches 1 and 3. Obtain equilibrium equations on loop current basis and find all the branch currents. The resistance values also indicate branch numbers. (Karnataka University)

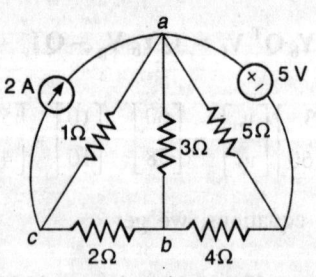

Fig. 5.47

Solution: The 5 Ω resistance in parallel with 5 V has no significance and can be neglected. The network in Fig. 5.47 is re-written as shown in Fig. 5.48 after shifting the 5 V voltage source.

The oriented graph of the network shown in Fig. 5.48 is as shown in Fig. 5.49. The orientations of the branches are assumed arbitrarily.

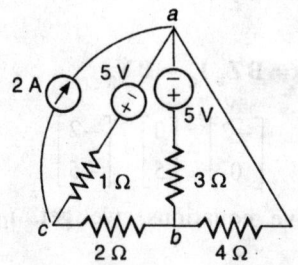

Fig. 5.48

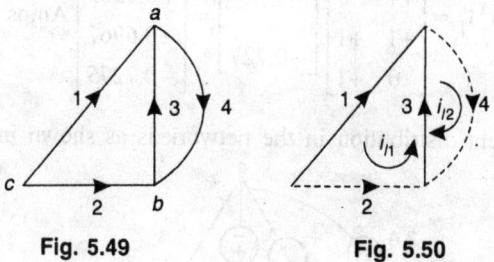

Fig. 5.49   **Fig. 5.50**

The tree with branches 1 and 3 is as shown in Fig. 5.50. The loop currents i_{l1} and i_{l2} are assumed as shown.

Then

$$\mathbf{B Z_b B^T} = \begin{bmatrix} -1 & +1 & +1 & 0 \\ 0 & 0 & +1 & +1 \end{bmatrix} \begin{bmatrix} 1 & 0 & 0 & 0 \\ 0 & 2 & 0 & 0 \\ 0 & 0 & 3 & 0 \\ 0 & 0 & 0 & 4 \end{bmatrix} \begin{bmatrix} -1 & 0 \\ +1 & 0 \\ +1 & +1 \\ 0 & +1 \end{bmatrix} = \begin{bmatrix} 6 & 3 \\ 3 & 7 \end{bmatrix}$$

$$\mathbf{B Z_b I_g} = \begin{bmatrix} -1 & +1 & +1 & 0 \\ 0 & 0 & +1 & +1 \end{bmatrix} \begin{bmatrix} 1 & 0 & 0 & 0 \\ 0 & 2 & 0 & 0 \\ 0 & 0 & 3 & 0 \\ 0 & 0 & 0 & 4 \end{bmatrix} \begin{bmatrix} 2 \\ 0 \\ 0 \\ 0 \end{bmatrix} = \begin{bmatrix} -2 \\ 0 \end{bmatrix}$$

$$\mathbf{B V_g} = \begin{bmatrix} -1 & +1 & +1 & 0 \\ 0 & 0 & +1 & +1 \end{bmatrix} \begin{bmatrix} 5 \\ 0 \\ 5 \\ 0 \end{bmatrix} = \begin{bmatrix} 0 \\ 5 \end{bmatrix}$$

The set of equilibrium equations with loop currents as variables are given by

$$BZ_bB^TI_l = BZ_bI_g - BV_g$$

i.e. $\begin{bmatrix} 6 & 3 \\ 3 & 7 \end{bmatrix}\begin{bmatrix} i_{l1} \\ i_{l2} \end{bmatrix} = \begin{bmatrix} -2 \\ 0 \end{bmatrix} - \begin{bmatrix} 0 \\ 5 \end{bmatrix} = \begin{bmatrix} -2 \\ -5 \end{bmatrix}$

Solving the above equations, we get, $i_{l1} = 0.0303$ A and $i_{l2} = -0.727$ A.

The various branch currents are:

$$I_b = B^TI_l = \begin{bmatrix} -1 & 0 \\ +1 & 0 \\ +1 & +1 \\ 0 & +1 \end{bmatrix}\begin{bmatrix} 0.0303 \\ -0.727 \end{bmatrix} = \begin{bmatrix} -0.0303 \\ +0.0303 \\ -0.6967 \\ -0.7275 \end{bmatrix} \text{Amps} = \begin{bmatrix} i_1 \\ i_2 \\ i_3 \\ i_4 \end{bmatrix}$$

The current distribution in the network is as shown in Fig. 5.51.

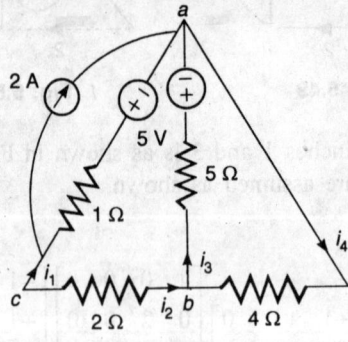

Fig. 5.51

The various branch voltages are given by

$$E_b = V_g + Z_b(I_b - I_g)$$

i.e. $\begin{bmatrix} e_1 \\ e_2 \\ e_3 \\ e_4 \end{bmatrix} = \begin{bmatrix} 5 \\ 0 \\ 5 \\ 0 \end{bmatrix} + \begin{bmatrix} 1 & 0 & 0 & 0 \\ 0 & 2 & 0 & 0 \\ 0 & 0 & 3 & 0 \\ 0 & 0 & 0 & 4 \end{bmatrix}\begin{bmatrix} -0.0303-2 \\ +0.0303-0 \\ -0.6967-0 \\ -0.7275-0 \end{bmatrix} = \begin{bmatrix} 2.9697 \\ 0.0606 \\ 2.91 \\ -2.908 \end{bmatrix}$ volts

5.15 For the network shown in Fig. 5.52, draw the oriented graph. Select a tree with branches 1 and 3. Obtain the equilibrium equations with tree-branch voltages as variables. Find all the branch currents and branch voltages. The resistance values also indicate branch numbers.

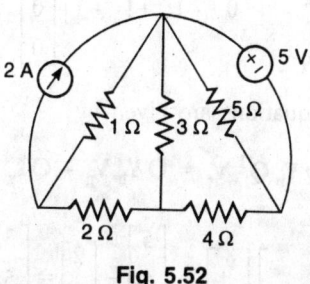

Fig. 5.52

Solution: For the network in Fig. 5.42, a tree with branches 4, 5 and 6 is formed as explained in example 5.11 and is shown in Fig. 5.43.

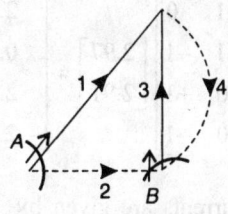

Fig. 5.53

Then

$$
Q Y_b Q^T = \begin{bmatrix} +1 & +1 & 0 & 0 \\ 0 & -1 & +1 & -1 \end{bmatrix} \begin{bmatrix} 1 & 0 & 0 & 0 \\ 0 & 1/2 & 0 & 0 \\ 0 & 0 & 1/3 & 0 \\ 0 & 0 & 0 & 1/4 \end{bmatrix} \begin{bmatrix} +1 & 0 \\ +1 & -1 \\ 0 & +1 \\ 0 & -1 \end{bmatrix}
$$

$$
= \begin{bmatrix} 1.5 & -0.5 \\ -0.5 & 1.0833 \end{bmatrix}
$$

$$
Q Y_b V_g = \begin{bmatrix} +1 & +1 & 0 & 0 \\ 0 & -1 & +1 & -1 \end{bmatrix} \begin{bmatrix} 1 & 0 & 0 & 0 \\ 0 & 1/2 & 0 & 0 \\ 0 & 0 & 1/3 & 0 \\ 0 & 0 & 0 & 1/4 \end{bmatrix} \begin{bmatrix} 5 \\ 0 \\ 5 \\ 0 \end{bmatrix} = \begin{bmatrix} 5 \\ \dfrac{5}{3} \end{bmatrix}
$$

$$\mathbf{Q I_g} = \begin{bmatrix} +1 & +1 & 0 & 0 \\ 0 & -1 & +1 & -1 \end{bmatrix} \begin{bmatrix} 2 \\ 0 \\ 0 \\ 0 \end{bmatrix} = \begin{bmatrix} 2 \\ 0 \end{bmatrix}$$

The equilibrium equations are given by

$$\mathbf{Q Y_b Q^T V_t = Q Y_b V_g - Q I_g}$$

i.e. $\begin{bmatrix} 1.5 & -0.5 \\ -0.5 & 1.0833 \end{bmatrix} \begin{bmatrix} v_1 \\ v_3 \end{bmatrix} = \begin{bmatrix} 5 \\ \dfrac{5}{3} \end{bmatrix} - \begin{bmatrix} 2 \\ 0 \end{bmatrix} = \begin{bmatrix} 3 \\ \dfrac{5}{3} \end{bmatrix}$

Solving the above equations, we get, $v_1 = 2.97$ V and $v_3 = 2.91$ V

The various branch voltages are given by

$$\mathbf{E_b = Q^T V_t} = \begin{bmatrix} +1 & 0 \\ +1 & -1 \\ 0 & +1 \\ 0 & -1 \end{bmatrix} \begin{bmatrix} 2.97 \\ 2.91 \end{bmatrix} = \begin{bmatrix} 2.97 \\ 0.06 \\ 2.91 \\ -2.91 \end{bmatrix} \text{ volts} = \begin{bmatrix} e_1 \\ e_2 \\ e_3 \\ e_4 \end{bmatrix}$$

The various branch currents are given by

$$\mathbf{I_b = I_g + Y_b (E_b - V_g)}$$

i.e. $\begin{bmatrix} i_1 \\ i_2 \\ i_3 \\ i_4 \end{bmatrix} = \begin{bmatrix} 2 \\ 0 \\ 0 \\ 0 \end{bmatrix} + \begin{bmatrix} 1 & 0 & 0 & 0 \\ 0 & 1/2 & 0 & 0 \\ 0 & 0 & 1/3 & 0 \\ 0 & 0 & 0 & 1/4 \end{bmatrix} \begin{bmatrix} 2.97-5 \\ 0.06-0 \\ 2.91-5 \\ -2.91-0 \end{bmatrix} = \begin{bmatrix} -0.0303 \\ +0.0303 \\ -0.6967 \\ -0.7275 \end{bmatrix}$ amps

The current distribution is as in Fig. 5.51.

5.16 For the network shown in Fig. 5.54, draw the oriented graph. Select the tree with branches 2 and 3. Draw the cut-set matrix. Write down the equilibrium equations with node-pair voltages as variables. Solve these equations and find the various branch voltages and currents. The figures in the parenthesis indicate branch numbers.

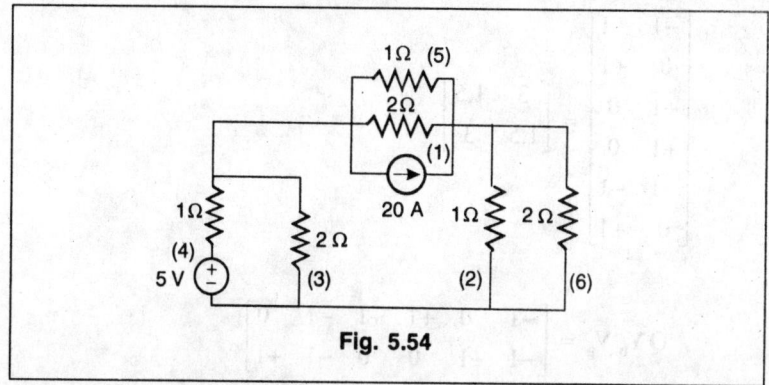

Fig. 5.54

Solution: The oriented graph for the network in Fig. 5.54 is as shown in Fig. 5.55 (a). The orientations of the branches are assumed arbitrarily.

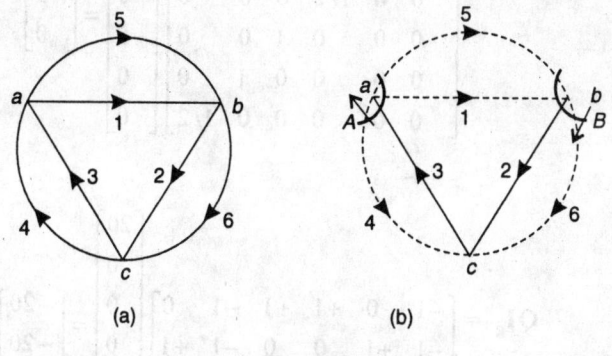

Fig. 5.55

The tree with branches 2 and 3 is as shown in Fig. 5.55 (b). The cut-set and their orientations are as shown.

Then

$$Q\,Y_b\,Q^T = \begin{bmatrix} -1 & 0 & +1 & +1 & -1 & 0 \\ -1 & +1 & 0 & 0 & -1 & +1 \end{bmatrix} \begin{bmatrix} 1/2 & 0 & 0 & 0 & 0 & 0 \\ 0 & 1 & 0 & 0 & 0 & 0 \\ 0 & 0 & 1/2 & 0 & 0 & 0 \\ 0 & 0 & 0 & 1 & 0 & 0 \\ 0 & 0 & 0 & 0 & 1 & 0 \\ 0 & 0 & 0 & 0 & 0 & 1/2 \end{bmatrix}$$

$$\begin{bmatrix} -1 & -1 \\ 0 & +1 \\ +1 & 0 \\ +1 & 0 \\ -1 & -1 \\ 0 & +1 \end{bmatrix} = \begin{bmatrix} 3 & 1.5 \\ 1.5 & 3 \end{bmatrix}$$

$$Q Y_b V_g = \begin{bmatrix} -1 & 0 & +1 & +1 & -1 & 0 \\ -1 & +1 & 0 & 0 & -1 & +1 \end{bmatrix}$$

$$\begin{bmatrix} 1/2 & 0 & 0 & 0 & 0 & 0 \\ 0 & 1 & 0 & 0 & 0 & 0 \\ 0 & 0 & 1/2 & 0 & 0 & 0 \\ 0 & 0 & 0 & 1 & 0 & 0 \\ 0 & 0 & 0 & 0 & 1 & 0 \\ 0 & 0 & 0 & 0 & 0 & 1/2 \end{bmatrix} \begin{bmatrix} 0 \\ 0 \\ 0 \\ -5 \\ 0 \\ 0 \end{bmatrix} = \begin{bmatrix} -5 \\ 0 \end{bmatrix}$$

$$Q I_g = \begin{bmatrix} -1 & 0 & +1 & +1 & -1 & 0 \\ -1 & +1 & 0 & 0 & -1 & +1 \end{bmatrix} \begin{bmatrix} 20 \\ 0 \\ 0 \\ 0 \\ 0 \\ 0 \end{bmatrix} = \begin{bmatrix} -20 \\ -20 \end{bmatrix}$$

The equilibrium equations are given by

$$Q Y_b Q^T V_t = Q Y_b V_g - Q I_g$$

i.e. $$\begin{bmatrix} 3 & 1.5 \\ 1.5 & 3 \end{bmatrix} \begin{bmatrix} v_3 \\ v_2 \end{bmatrix} = \begin{bmatrix} -5 \\ 0 \end{bmatrix} - \begin{bmatrix} -20 \\ -20 \end{bmatrix} = \begin{bmatrix} 15 \\ 20 \end{bmatrix}$$

Solving the above equations, we get

$$v_3 = 2.22 \text{ V}$$

and $$v_2 = 5.5 \text{ V}$$

The various branch voltages are given by

$$\mathbf{E_b} = \mathbf{Q^T V_t} = \begin{bmatrix} -1 & -1 \\ 0 & +1 \\ +1 & 0 \\ +1 & 0 \\ -1 & -1 \\ 0 & +1 \end{bmatrix} \begin{bmatrix} 2.22 \\ 5.5 \end{bmatrix} = \begin{bmatrix} -7.77 \\ 5.55 \\ 2.22 \\ 2.22 \\ -7.77 \\ 5.55 \end{bmatrix} \text{ volts} = \begin{bmatrix} e_1 \\ e_2 \\ e_3 \\ e_4 \\ e_5 \\ e_6 \end{bmatrix}$$

The various branch currents are given by $\mathbf{I_b} = \mathbf{I_g} + \mathbf{Y_b}\,(\mathbf{E_b} - \mathbf{V_g})$

i.e.
$$\begin{bmatrix} i_1 \\ i_2 \\ i_3 \\ i_4 \\ i_5 \\ i_6 \end{bmatrix} = \begin{bmatrix} 20 \\ 0 \\ 0 \\ 0 \\ 0 \\ 0 \end{bmatrix} + \begin{bmatrix} 1/2 & 0 & 0 & 0 & 0 & 0 \\ 0 & 1 & 0 & 0 & 0 & 0 \\ 0 & 0 & 1/2 & 0 & 0 & 0 \\ 0 & 0 & 0 & 1 & 0 & 0 \\ 0 & 0 & 0 & 0 & 1 & 0 \\ 0 & 0 & 0 & 0 & 0 & 1/2 \end{bmatrix} \begin{bmatrix} -7.77-0 \\ 5.55-0 \\ 2.22-0 \\ 2.22+5 \\ -7.77-0 \\ 5.55-0 \end{bmatrix}$$

$$= \begin{bmatrix} 16.115 \\ 5.55 \\ 1.11 \\ 7.22 \\ -7.77 \\ 2.775 \end{bmatrix} \text{ amps}$$

The current distribution in the network is as shown in Fig. 5.56.

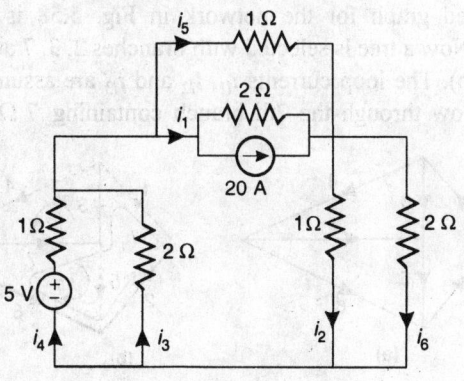

Fig. 5.56

5.17 For the network shown in Fig. 5.57, construct a tree such that, all the currents flow through 7 Ω. Write the equilibrium equations on current basis and find the current i_5. The resistance values also indicate branch numbers.

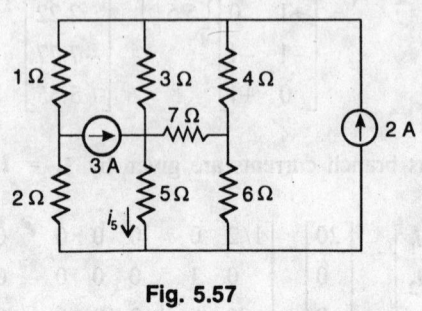

Fig. 5.57

Solution: After the *I* shift, the network in Fig. 5.57, is re-written as in Fig. 5.58.

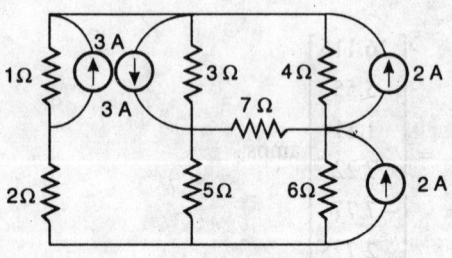

Fig. 5.58

The oriented graph for the network in Fig. 5.58 is as shown in Fig. 5.59 (a). Now a tree is selected with branches 2, 3, 7 and 6 as shown in Fig. 5.59 (b). The loop currents i_{l1}, i_{l2} and i_{l3} are assumed such that, all of them flow through the 7th branch containing 7 Ω.

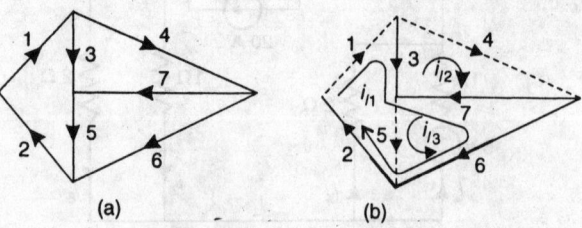

Fig. 5.59

The equilibrium equations on loop basis are given by

$$\mathbf{B Z_b B^T V_l} = \mathbf{B Z_b I_g} - \mathbf{B V_g} \tag{1}$$

$$\mathbf{B Z_b B^T} = \begin{bmatrix} +1 & +1 & +1 & 0 & 0 & +1 & -1 \\ 0 & 0 & -1 & +1 & 0 & 0 & +1 \\ 0 & 0 & 0 & 0 & +1 & -1 & +1 \end{bmatrix} \begin{bmatrix} 1 & 0 & 0 & 0 & 0 & 0 & 0 \\ 0 & 2 & 0 & 0 & 0 & 0 & 0 \\ 0 & 0 & 3 & 0 & 0 & 0 & 0 \\ 0 & 0 & 0 & 4 & 0 & 0 & 0 \\ 0 & 0 & 0 & 0 & 5 & 0 & 0 \\ 0 & 0 & 0 & 0 & 0 & 6 & 0 \\ 0 & 0 & 0 & 0 & 0 & 0 & 7 \end{bmatrix}$$

$$\begin{bmatrix} +1 & 0 & 0 \\ +1 & 0 & 0 \\ +1 & -1 & 0 \\ 0 & +1 & 0 \\ 0 & 0 & +1 \\ +1 & 0 & -1 \\ -1 & +1 & +1 \end{bmatrix} = \begin{bmatrix} +19 & -10 & -13 \\ -10 & +14 & +7 \\ -13 & +7 & +18 \end{bmatrix}$$

$$\mathbf{B Z_b I_g} = \begin{bmatrix} +1 & +1 & +1 & 0 & 0 & +1 & -1 \\ 0 & 0 & -1 & +1 & 0 & 0 & +1 \\ 0 & 0 & 0 & 0 & +1 & -1 & +1 \end{bmatrix} \begin{bmatrix} 1 & 0 & 0 & 0 & 0 & 0 & 0 \\ 0 & 2 & 0 & 0 & 0 & 0 & 0 \\ 0 & 0 & 3 & 0 & 0 & 0 & 0 \\ 0 & 0 & 0 & 4 & 0 & 0 & 0 \\ 0 & 0 & 0 & 0 & 5 & 0 & 0 \\ 0 & 0 & 0 & 0 & 0 & 6 & 0 \\ 0 & 0 & 0 & 0 & 0 & 0 & 7 \end{bmatrix}$$

$$\begin{bmatrix} 3 \\ 0 \\ 3 \\ -2 \\ 0 \\ -2 \\ 0 \end{bmatrix} = \begin{bmatrix} 0 \\ -17 \\ 12 \end{bmatrix}$$

$$\mathbf{B V_g} = \begin{bmatrix} 0 \\ 0 \\ 0 \end{bmatrix} \quad \because \mathbf{V_g} \text{ is a null matrix.}$$

Substituting the above in equation (1), we get

$$\begin{bmatrix} +19 & -10 & -13 \\ -10 & +14 & +7 \\ -13 & +7 & +18 \end{bmatrix} \begin{bmatrix} i_{l1} \\ i_{l2} \\ i_{l3} \end{bmatrix} = \begin{bmatrix} 0 \\ -17 \\ 12 \end{bmatrix} - \begin{bmatrix} 0 \\ 0 \\ 0 \end{bmatrix} = \begin{bmatrix} 0 \\ -17 \\ 12 \end{bmatrix}$$

Solving the above equations, we get

$$i_{l3} = i_5 = 1.352 \text{ A}$$

5.18 For the network shown in Fig. 5.60, write the oriented graph. Write the equilibrium equations on loop basis.

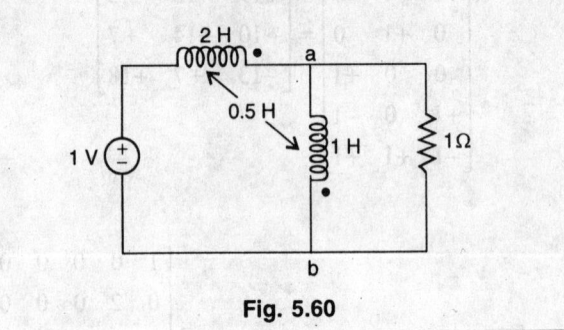

Fig. 5.60

Solution: The oriented graph and the tree are as shown in Figs 5.61 (a) and (b) respectively.

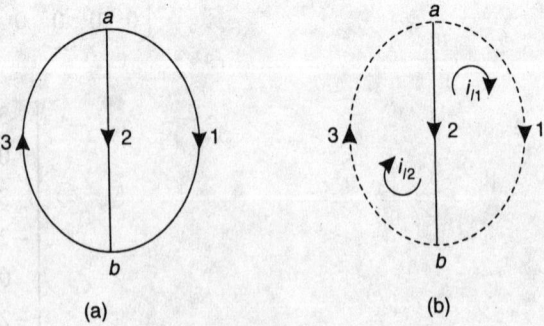

(a)   (b)

Fig. 5.61

The equilibrium equations are given by

$$BZ_b B^T V_l = BZ_b I_g - BV_g \qquad (1)$$

$$\begin{bmatrix} +1 & -1 & 0 \\ 0 & +1 & +1 \end{bmatrix} \begin{bmatrix} +1 & 0 & 0 \\ 0 & 5 & 0.5s \\ 0 & 0.5s & 2s \end{bmatrix} \begin{bmatrix} +1 & 0 \\ -1 & +1 \\ 0 & +1 \end{bmatrix} \begin{bmatrix} i_{l1} \\ i_{l2} \end{bmatrix}$$

$$= \begin{bmatrix} 0 \\ 0 \\ 0 \end{bmatrix} - \begin{bmatrix} +1 & -1 & 0 \\ 0 & +1 & +1 \end{bmatrix} \begin{bmatrix} 0 \\ 0 \\ -1 \end{bmatrix}$$

i.e. $\begin{bmatrix} 1+s & -1.5s \\ -1.5s & 4s \end{bmatrix} \begin{bmatrix} i_{l1} \\ i_{l2} \end{bmatrix} = \begin{bmatrix} 0 \\ 1 \end{bmatrix}$

The equilibrium equations are

$$(1 + s) i_{l1} - 1.5 s i_{l2} = 0$$

$$- 1.5 s i_{l1} + 4s i_{l2} = 1$$

5.19 For the network shown in Fig. 5.62, write down the equilibrium equations on loop basis. $v_c (0) = 5$ V, is the initial voltage on the capacitor. The suffixes of the elements also indicates branch numbers.

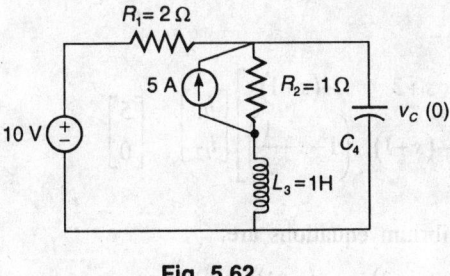

Fig. 5.62

Solution: The oriented graph and the tree are as shown in Figs 5.63 (a) and (b).

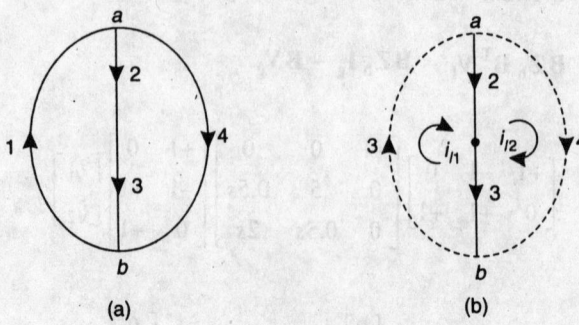

Fig. 5.63

The equilibrium equations are given by

$$\mathbf{B Z_b B^T I_l} = \mathbf{B Z_b I_g} - \mathbf{B V_g}$$

i.e.

$$= \begin{bmatrix} +1 & +1 & +1 & 0 \\ 0 & -1 & -1 & +1 \end{bmatrix} \begin{bmatrix} 2 & 0 & 0 & 0 \\ 0 & 1 & 0 & 0 \\ 0 & 0 & s & 0 \\ 0 & 0 & 0 & 1/s \end{bmatrix} \begin{bmatrix} +1 & 0 \\ +1 & -1 \\ +1 & -1 \\ 0 & +1 \end{bmatrix} \begin{bmatrix} i_{l1} \\ i_{l2} \end{bmatrix}$$

$$= \begin{bmatrix} +1 & +1 & +1 & 0 \\ 0 & -1 & -1 & +1 \end{bmatrix} \begin{bmatrix} 2 & 0 & 0 & 0 \\ 0 & 1 & 0 & 0 \\ 0 & 0 & s & 0 \\ 0 & 0 & 0 & 1/s \end{bmatrix} \begin{bmatrix} 0 \\ -5 \\ 0 \\ 0 \end{bmatrix}$$

$$- \begin{bmatrix} +1 & +1 & +1 & 0 \\ 0 & -1 & -1 & +1 \end{bmatrix} \begin{bmatrix} -10 \\ 0 \\ 0 \\ 5 \end{bmatrix}$$

i.e.

$$\begin{bmatrix} s+3 & -(s+1) \\ -(s+1) & \left(1+s+\dfrac{1}{s}\right) \end{bmatrix} \begin{bmatrix} i_{l1} \\ i_{l2} \end{bmatrix} = \begin{bmatrix} 5 \\ 0 \end{bmatrix}$$

The equilibrium equations are:

$$(s+3)i_{l1} - (s+1)i_{l2} = 5$$

and

$$-(s+1)i_{l1} + \left(1+s+\frac{1}{s}\right)i_{l2} = 0$$

DUALITY

5.26 DUALITY

We observe several analogous situations in the study of electrical networks. The statements of Kirchhoff's current law is similar to the statement of Kirchhoff's voltage law, when current is replaced by voltage and a node is replaced by a mesh. The integro-differential equations written by the application of the two laws also appear similar. This similarity between two quantities or concepts is known as *duality.*

Consider the two circuits as shown in Figs 5.64 (a) and (b), which are completely different in physical appearance.

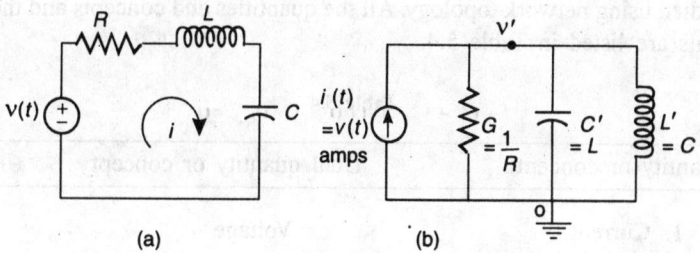

Fig. 5.64

The integro-differential equations for the two circuits are:

$$Ri + L\frac{di}{dt} + \frac{1}{C}\int i\,dt = v(t) \tag{5.70}$$

$$Gv + C\frac{dv}{dt} + \frac{1}{L}\int v\,dt = i(t) \tag{5.71}$$

Both the Eqs (5.70) and (5.71) appear to be similar. Hence, equation (5.71) is the dual of equation (5.70). The network in Fig. 5.64 (b) is the dual of the network in Fig. 5.64 (a). Hence, two networks are said to be 'dual networks', if the mesh current equations of one are similar to the node voltage equations of the other. The two networks are duals when

$$G = \frac{1}{R}, \ C' = L, L' = C \text{ and } i'(t) = v(t).$$

From Eqs (5.70) and (5.71), the dual quantities that can be identified are: (with primes omitted)

(i) Ri and Gv

(ii) $L\dfrac{di}{dt}$ and $C\dfrac{dv}{dt}$

(iii) $\dfrac{1}{C}\int i\, dt$ and $\dfrac{1}{L}\int v\, dt$

The pairs of dual quantities identified are:

(i) R and G

(ii) L and C

(iii) Loop current i and node voltage v.

There are dual quantities which can be identified, when networks are studied using network topology. All the quantities and concepts and their duals are listed in Table 5.4.

Table 5.4

Quantity or concept	Dual quantity or concept
1. Current	Voltage
2. Branch current	Branch voltage
3. Mesh	Node
4. Loop	Node-pair
5. Loop current	Node-pair voltage
6. Mesh current	Node voltage
7. Number of loops	Number of nodes
8. Link	Twig
9. Tie-set	Cut-set
10. Short-circuit	Open-circuit
11. Series circuit	Parallel circuit
12. Inductance	Capacitance
13. Resistance	Conductance
14. Thevenin's circuit	Norton's circuit
15. Kirchhoff's current law	Kirchhoff's voltage law
16. Closing switch	Opening switch

5.27 CONSTRUCTION OF DUAL NETWORKS

The procedure for constructing a dual network for a given network is as follows.

1. Inside each loop of the given network, place a node and name it by a letter such as *a*, for convenience. Place an extra node 0, which is the datum node, external to the network.
2. Draw lines from node to node through the elements in the original network, traversing only one element at a time.
3. For drawing the dual network, arrange the nodes marked in the original network in a separate space on the paper.
4. For each element traversed in the original network, connect its dual element between corresponding nodes. The network constructed in this manner is the dual network of the original network. The procedure is best illustrated in Figs 5.65 (a) and (b).

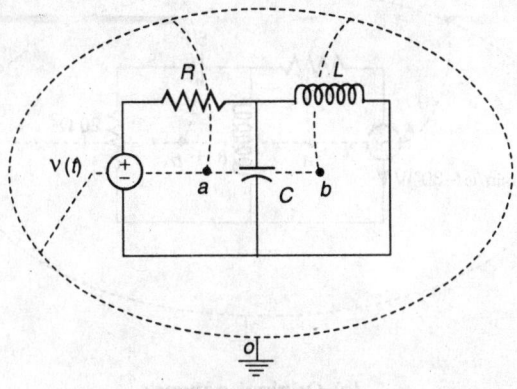

(a) Original network

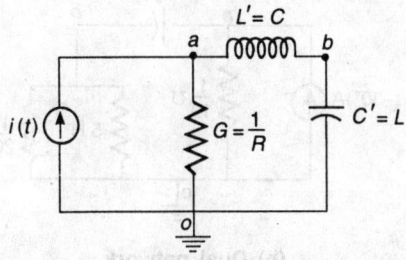

(b) Dual network

Fig. 5.65

Example: Obtain the dual of the network shown in Fig. 5.66.

(Mysore University)

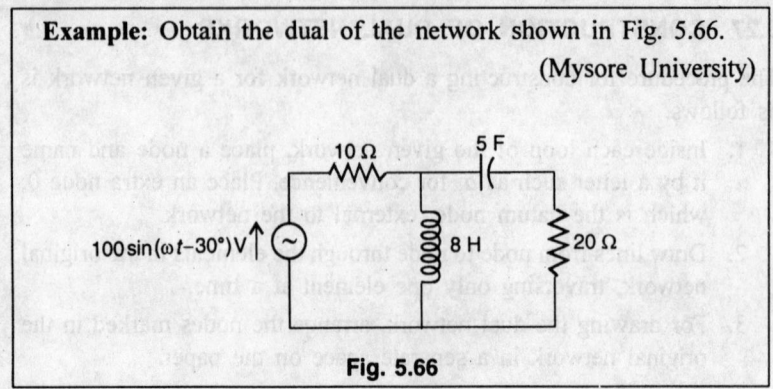

Fig. 5.66

Solution: The nodes are marked for the given network as shown in Fig. 5.67 (a), and its dual network is as shown in Fig. 5.67 (b).

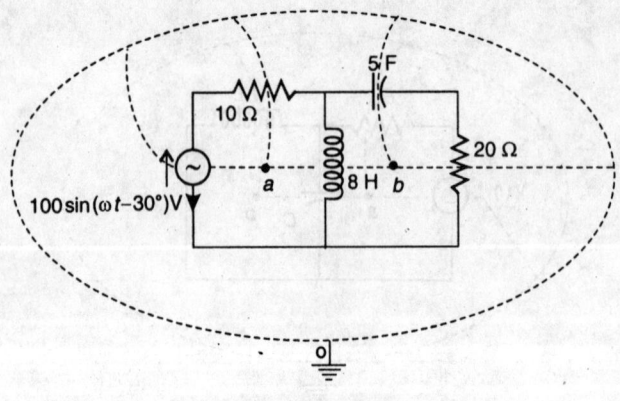

(a) Original network

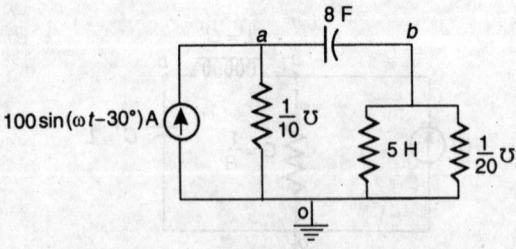

(b) Dual network

Fig. 5.67

Example: Obtain the dual of the given network shown in Fig. 5.68.

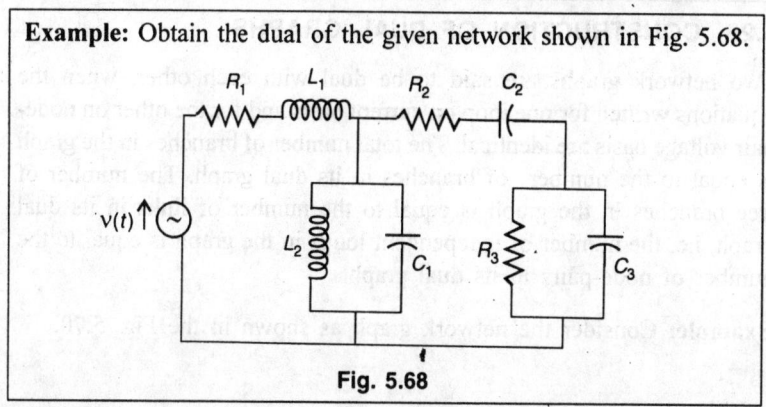

Fig. 5.68

Solution: The nodes are marked in the original network and all the elements are traversed as shown in Fig. 5.69 (a) and the dual network is as shown in Fig. 5.69 (b).

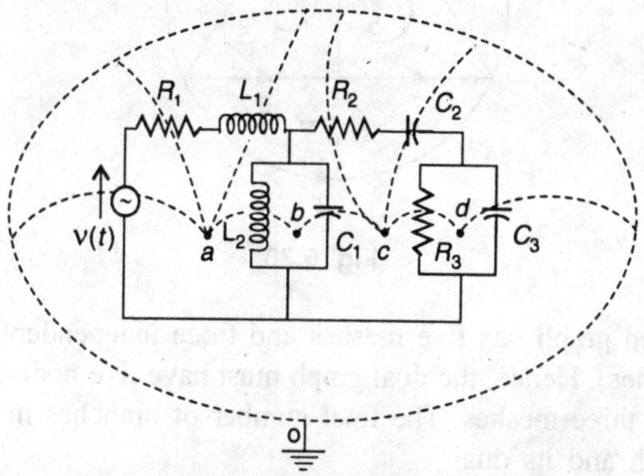

(a) Original network

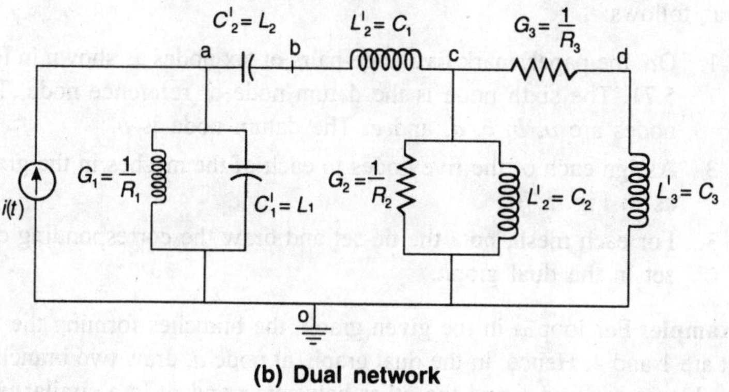

(b) Dual network

Fig. 5.69

5.28 CONSTRUCTION OF DUAL GRAPHS

Two network graphs are said to be dual with each other, when the equations written for one loop on current basis and for the other on node-pair voltage basis are identical. The total number of branches in the graph is equal to the number of branches in its dual graph. The number of tree branches in the graph is equal to the number of links in its dual graph, i.e. the number of independent loops in the graph is equal to the number of node-pairs in its dual graph.

Example: Consider the network graph as shown in the Fig. 5.70.

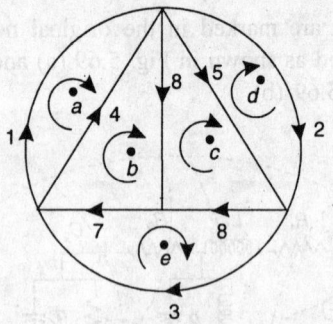

Fig. 5.70

The given graph has five meshes and three independent node pairs (or four nodes). Hence, the dual graph must have five node-pairs (or six nodes) and three meshes. The total number of branches must be same in the graph and its dual.

The procedure for drawing the dual of the graph given in Fig. 5.70 is as follows:

1. On the paper, mark five node-pairs or six nodes as shown in Fig. 5.71. The sixth node is the datum node or reference node. The nodes are a, b, c, d, and e. The datum node is o.
2. Assign each of the five nodes to each of the meshes in the graph as in Fig. 5.70.
3. For each mesh, note the tie-set and draw the corresponding cut-set in the dual graph.

Example: For loop a in the given graph, the branches forming the tie-set are 1 and 4. Hence, in the dual graph, at node a, draw two branches, one between a and b and the other between a and o. In a similar way, the other branches are connected in the dual graph.

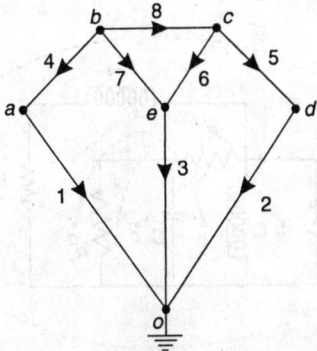

Fig. 5.71

4. The orientations of the branches in the dual graph are marked as follows. When mesh *a* is traced in clockwise direction, the orientation of branch 1 is divergent from node *a* and the orientation of branch 4 is convergent towards node *a* and accordingly, the orientations of branches 1 and 4 are marked on the dual graph. In a similar way, the orientations of the other branches are marked.

WORKED EXAMPLES

5.20 Draw the dual of the given network shown in Fig. 5.72.
(Mysore University)

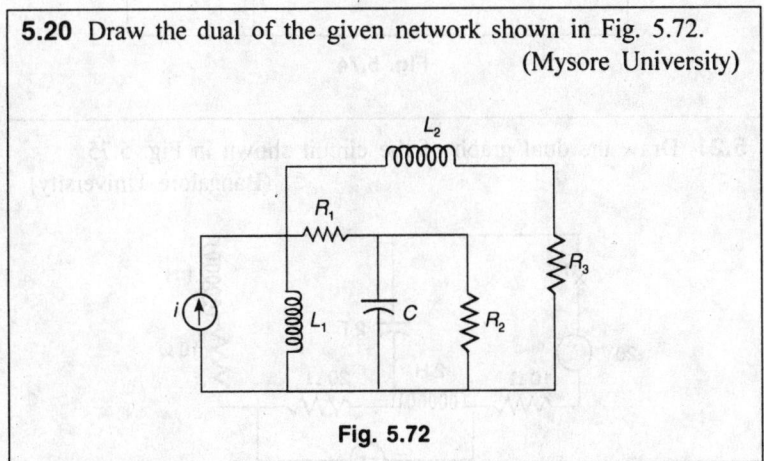

Fig. 5.72

Solution: The nodes *a*, *b*, *c*, *d* and reference node o are marked as shown in Fig. 5.73.

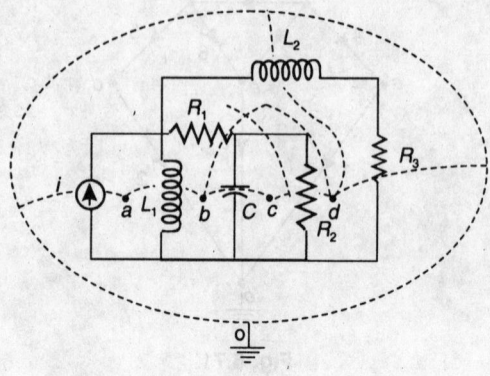

Fig. 5.73

All the nodes are joined through elements as indicated. The dual network is as shown in Fig. 5.74.

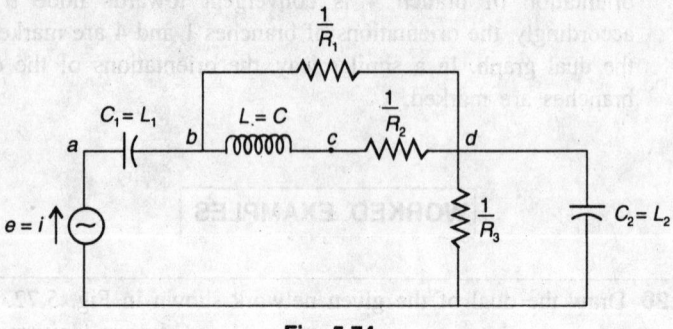

Fig. 5.74

5.21 Draw the dual graph of the circuit shown in Fig. 5.75.
(Bangalore University)

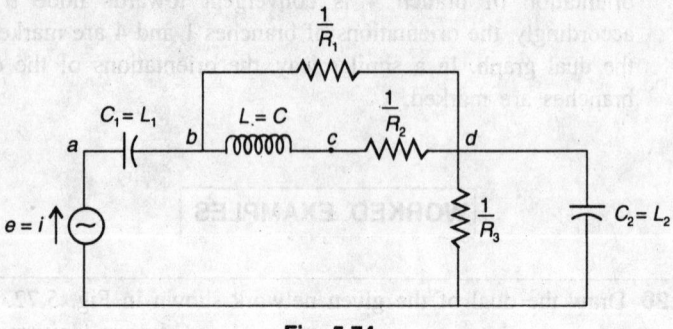

Fig. 5.75

Solution: The nodes *a*, *b*, *c* and reference node *o* are marked as shown in Fig. 5.76.

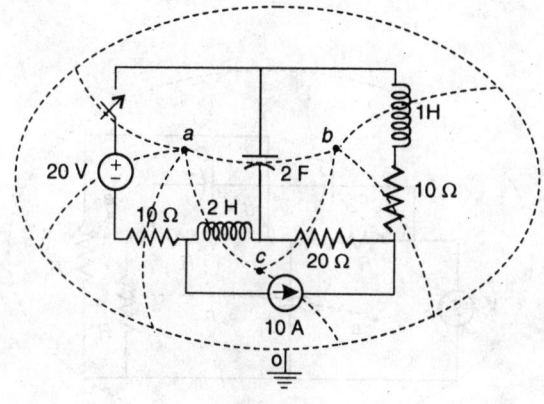

Fig. 5.76

The dual network is as shown in Fig. 5.77.

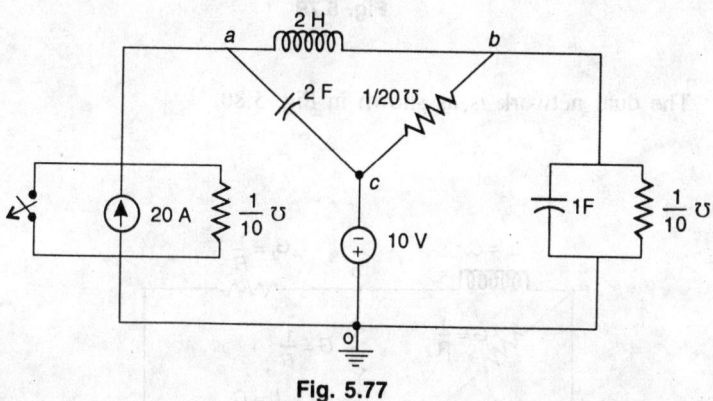

Fig. 5.77

5.22 Draw the dual network of the circuit shown in Fig. 5.78.

(Mangalore University)

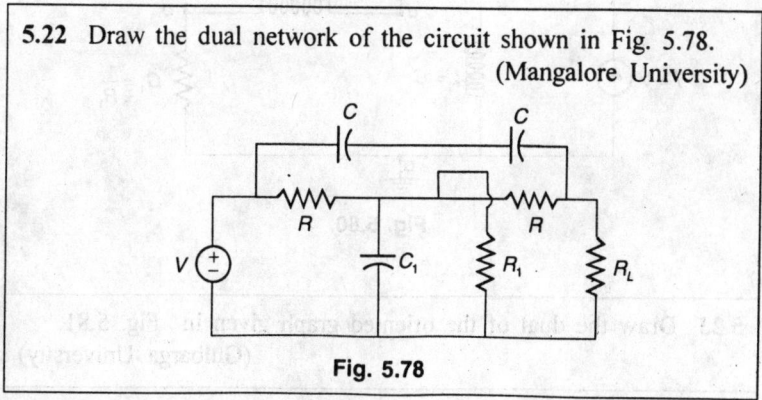

Fig. 5.78

Solution: The circuit in Fig. 5.78 is modified as in Fig. 5.79. The nodes *a*, *b*, *c*, *d*, *e*, and reference node *o* are marked.

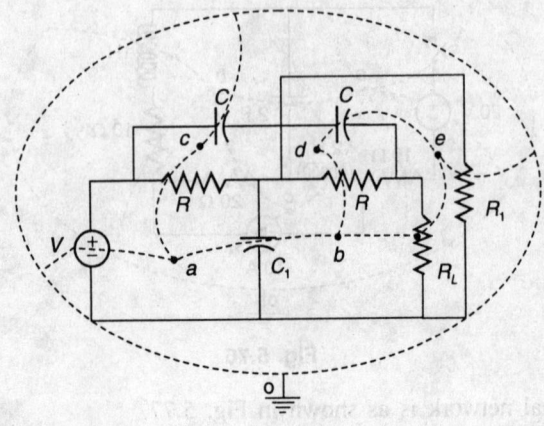

Fig. 5.79

The dual network is as shown in Fig. 5.80.

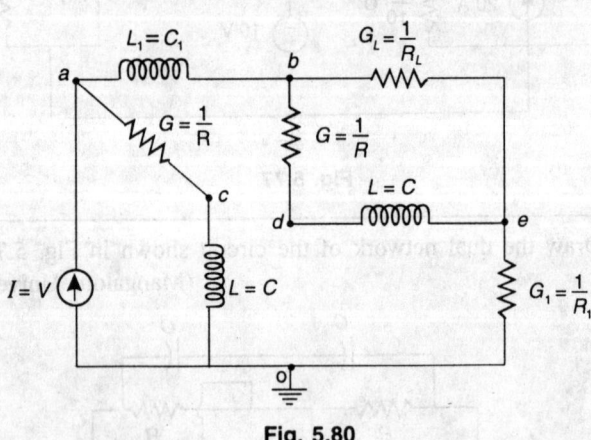

Fig. 5.80

5.23 Draw the dual of the oriented graph given in Fig. 5.81.
(Gulbarga University)

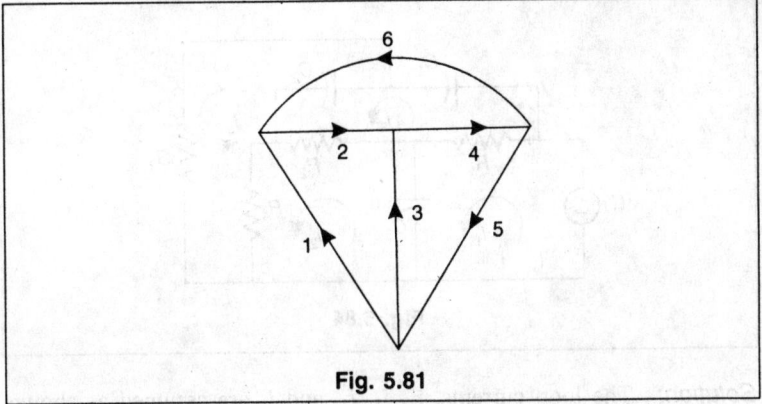

Fig. 5.81

Solution: The dual of the above graph is drawn as follows:

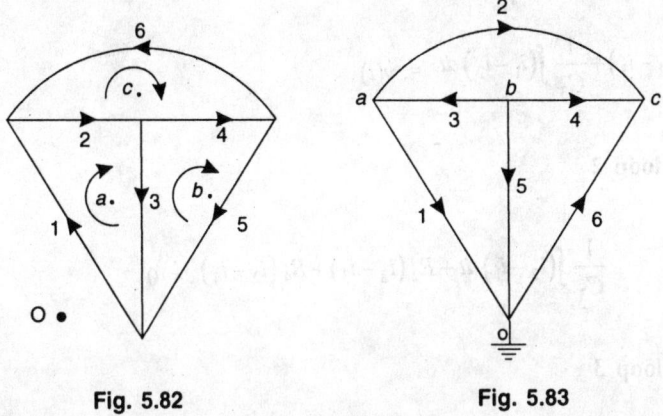

Fig. 5.82   **Fig. 5.83**

The graph has 3 meshes and 3 independent node-pairs (or 4 nodes). Hence, the dual graph must have three node-pairs (or 4 nodes) and 3 meshes. The total number of branches is six in both the graph and dual graph. Place the nodes *a*, *b* and *c* in the there meshes of the graph and the datum node *o* outside the graph as shown in Fig. 5.82. The dual graph is drawn as explained in section 5.28 and is as shown in Fig. 5.83.

5.24 Draw the dual of the network shown in Fig. 5.84. Show that the loop equations of the network are the duals of the nodal equations of the dual network.   (Bangalore University)

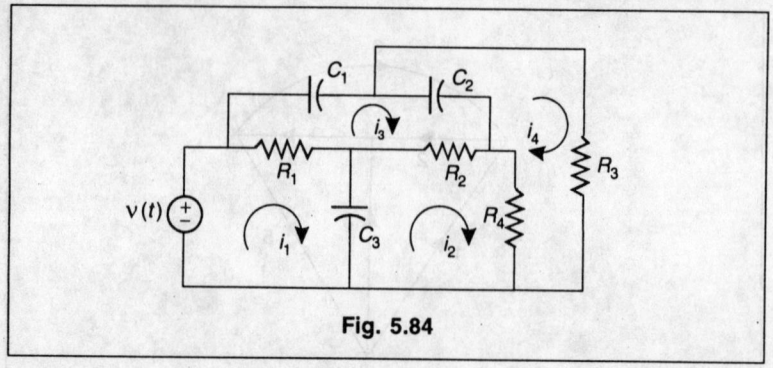

Fig. 5.84

Solution: The loop currents, i_1, i_2, i_3, and i_4 are assumed as shown. The loop equations for the network are:

For loop 1

$$R_1(i_1 - i_3) + \frac{1}{C_3}\int(i_1 - i_2)\,dt = v(t) \tag{1}$$

For loop 2

$$\frac{1}{C_3}\int(i_2 - i_1)\,dt + R_2(i_2 - i_3) + R_4(i_2 - i_4) = 0 \tag{2}$$

For loop 3

$$R_1(i_3 - i_1) + \frac{1}{C_1}\int i_3\,dt + \frac{1}{C_2}\int(i_3 - i_4)\,dt + R_2(i_3 - i_2) = 0 \tag{3}$$

For loop 4

$$R_3 i_4 + R_4(i_4 - i_2) + \frac{1}{C_2}\int(i_4 - i_3)\,dt = 0 \tag{4}$$

The dual of the network given is obtained as shown in Fig. 5.85. The nodes 1, 2, 3, 4, and the reference node *o* are marked. The various elements between the nodes are joined.

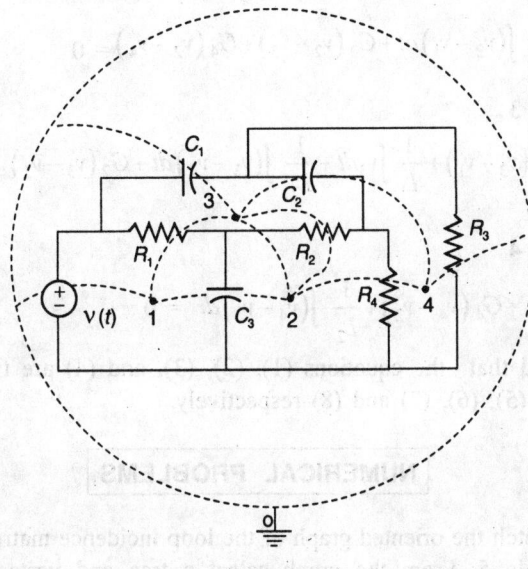

Fig. 5.85

The dual of this network is as shown in Fig. 5.86.

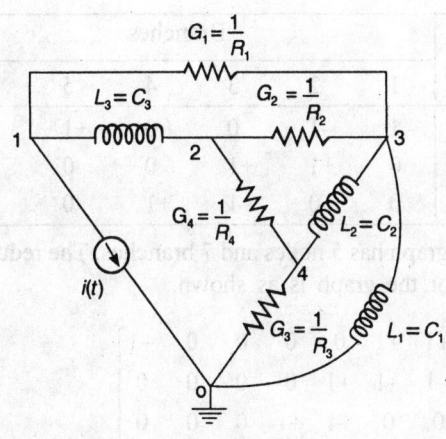

Fig. 5.86

The nodal equations are:

For node 1

$$G_1(v_1 - v_3) + \frac{1}{L_3}\int(v_1 - v_2)\,dt = i(t) \qquad (5)$$

For node 2

$$\frac{1}{L_3} \int (v_2 - v_1) dt + G_2 (v_2 - v_3) + G_4 (v_2 - v_4) = 0 \qquad (6)$$

For node 3

$$G_1 (v_3 - v_1) + \frac{1}{L_1} \int v_3 dt + \frac{1}{L_2} \int (v_3 - v_4) dt + G_2 (v_3 - v_2) = 0 \qquad (7)$$

For node 4

$$G_3 v_4 + G_4 (v_4 - v_2) + \frac{1}{L_2} \int (v_4 - v_3) dt = 0 \qquad (8)$$

We find that the equations (1), (2), (3), and (4) are the duals of equations (5), (6), (7) and (8) respectively.

NUMERICAL PROBLEMS

5.1 Sketch the oriented graph of the loop incidence matrix shown in Table 5. From the graph select a tree and write the cut-set schedule. (Bangalore University)

Table 5.5

Loops	Branches						
	1	2	3	4	5	6	7
1	−1	−1	0	0	+1	0	0
2	0	+1	+1	0	0	+1	0
3	0	0	−1	+1	0	0	+1

5.2 A linear graph has 5 nodes and 7 branches. The reduced incidence matrix for the graph is as shown.

$$\begin{bmatrix} +1 & +1 & 0 & 0 & 0 & 0 & +1 \\ -1 & -1 & +1 & 0 & 0 & 0 & 0 \\ 0 & 0 & -1 & +1 & 0 & 0 & 0 \\ 0 & 0 & 0 & -1 & -1 & -1 & 0 \end{bmatrix}$$

Draw the oriented graph. Verify, whether the branches (1, 3, 4, 5) constitute a tree. If so, far this tree write (i) the tie-set matrix and (ii) the cut-set matrix. (Mysore University)

5.3 From the given node incidence matrix, obtain the oriented graph. (Kuvempu University)

$$
A_a = \begin{array}{c} \\ 1 \\ 2 \\ 3 \\ 4 \\ 5 \end{array}
\begin{array}{cccccccc}
1 & 2 & 3 & 4 & 5 & 6 & 7 & 8 \\
\left[\begin{array}{cccccccc}
+1 & +1 & 0 & 0 & 0 & 0 & 0 & -1 \\
0 & -1 & +1 & +1 & 0 & 0 & 0 & 0 \\
0 & 0 & 0 & -1 & +1 & +1 & 0 & 0 \\
0 & 0 & 0 & 0 & 0 & -1 & +1 & +1 \\
-1 & 0 & -1 & 0 & -1 & 0 & -1 & 0
\end{array}\right]
\end{array}
$$

5.4 From the given reduced incidence matrix, obtain the oriented graph. (Mangalore University)

$$
A = \begin{array}{c} \text{Nodes} \\ \\ 1 \\ 2 \\ 3 \\ 4 \end{array}
\begin{array}{cccccccccc}
\multicolumn{10}{c}{\text{Branches}} \\
1 & 2 & 3 & 4 & 5 & 6 & 7 & 8 & 9 & 10 \\
\left[\begin{array}{cccccccccc}
+1 & 0 & -1 & 0 & +1 & 0 & 0 & 0 & 0 & -1 \\
0 & 0 & +1 & -1 & 0 & +1 & 0 & -1 & 0 & 0 \\
0 & -1 & 0 & 0 & 0 & -1 & +1 & 0 & 0 & 0 \\
0 & 0 & 0 & 0 & 0 & 0 & -1 & +1 & -1 & +1
\end{array}\right]
\end{array}
$$

5.5 Obtain the cut-set matrix and branch admittance matrix for the network shown in Fig. 5.87. The suffixes indicate branch numbers. Assume the orientations of the branches as indicated.

(Karnataka University)

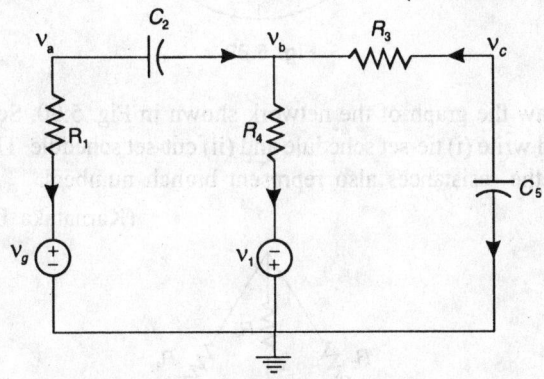

Fig. 5.87

5.6 For the oriented graph shown in Fig. 5.88, choose the tree (1, 5, 4). (i) Draw the tie-sets and from these, deduce the tie-set matrix, and (ii) Draw the cut-sets and from these, deduce the cut-set matrix. (Gulbarga University)

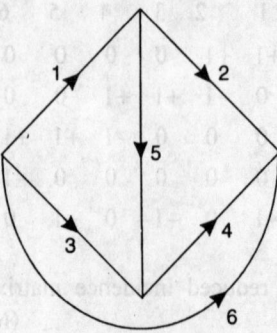

Fig. 5.88

5.7 Select a tree for the network graph shown in Fig. 5.89, and write (i) tie-set schedule and (ii) cut-set schedule.

(Gulbarga University)

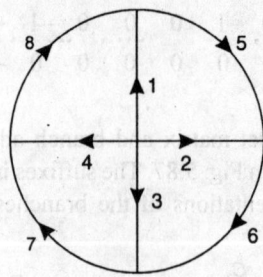

Fig. 5.89

5.8 Draw the graph of the network shown in Fig. 5.90. Select a tree and write (i) tie-set schedule and (ii) cut-set schedule. The suffixes of the resistances also represent branch numbers.

(Karnataka University)

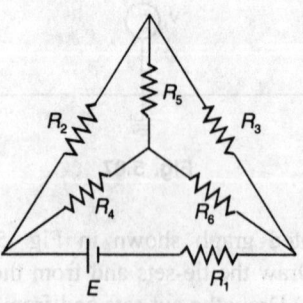

Fig. 5.90

5.9 Draw the oriented graph for the network shown in Fig. 5.91. Take tree (3, 4) and write node admittance matrix and loop impedance matrix. The suffixes of the elements also represent branch numbers. (Mysore University)

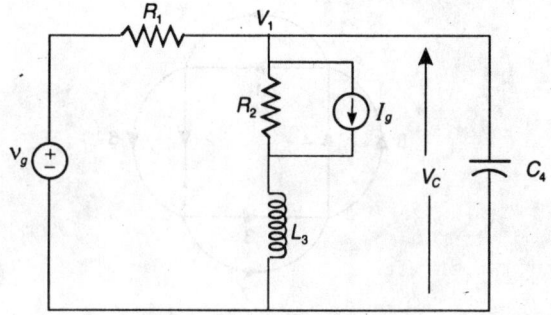

Fig. 5.91

5.10 For the circuit shown in Fig. 5.92, draw the oriented graph. Select the tree (5, 6, 7, 8) and draw (i) tie-set schedule and (ii) cut-set schedule. The suffixes of the resistances also indicate branch numbers. (Bangalore University)

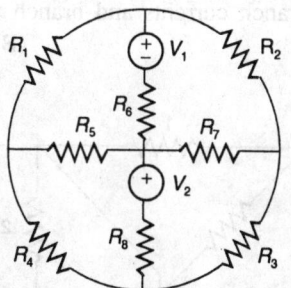

Fig. 5.92

5.11 For the all incidence matrix given, draw the oriented graph. Selecting a tree with branches 1, 2, 4 and 7, write the cut-set matrix. (Mysore University)

		Branches						
	Nodes	1	2	3	4	5	6	7
	a	-1	0	0	0	$+1$	0	0
	b	0	-1	$+1$	0	-1	$+1$	0
$\mathbf{A_a} =$	c	0	0	0	-1	0	-1	0
	d	0	0	-1	$+1$	0	0	-1
	e	$+1$	$+1$	0	0	0	0	$+1$

5.12 For the oriented graph given in Fig. 5.93, write the tie-set and cut-set schedule, choosing a tree consisting of branches 2, 3 and 4. (Kuvempu University)

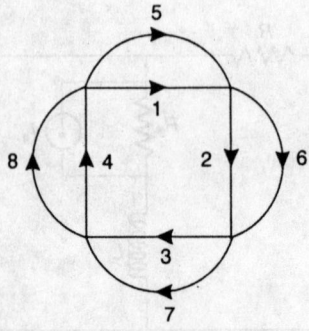

Fig. 5.93

5.13 For the network shown in Fig. 5.94, the branch numbers may be regarded as also indicating the branch conductance values in mhos. Construct the cut-set schedule for the choice of node-pair voltages, $e_1 = v_1$, $e_2 = v_2$ and $e_3 = v_3$. Therefrom, obtain equilibrium equations on node basis. Solve these equations and obtain all the branch currents and branch voltages.

(Bangalore University)

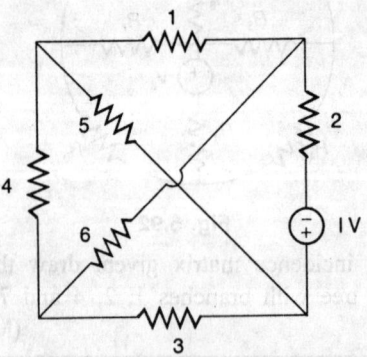

Fig. 5.94

5.14 For the network shown in Fig. 5.95, draw the graph. Construct a tree with branches 1, 3 and 4. Write the tie-set schedule. Obtain the equilibrium equations on loop basis and find all the branch currents and branch voltages. The branch numbers also indicate the resistance values in ohms. (Mysore University)

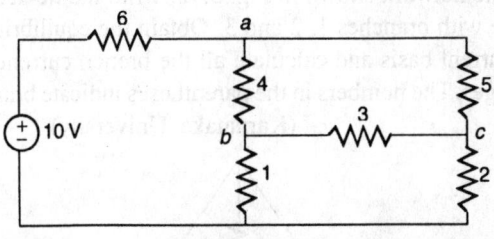

Fig. 5.95

5.15 Find the currents in the various branches, by constructing a tie-set schedule, for the network shown in Fig. 5.96. Take branch numbers same as those of branch resistances. Choose branches 1, 2 and 3 as tree-branches. (Mangalore University)

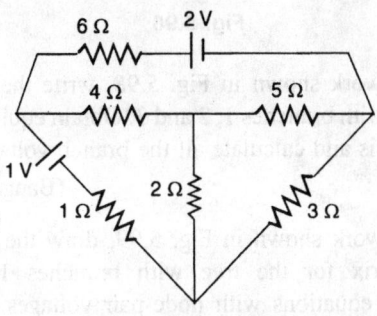

Fig. 5.96

5.16 For the network shown in Fig. 5.97, considering c as the reference node, draw the reduced incidence matrix **A**. Obtain equilibrium equations with node to datum voltages as variables. Determine all the branch voltages and currents.

(Mangalore University)

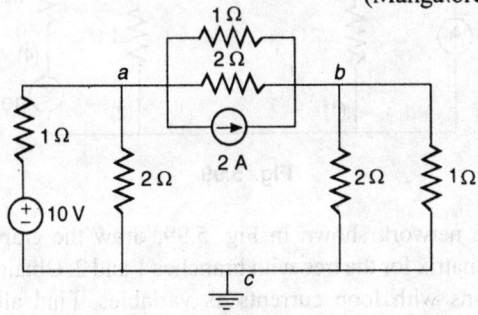

Fig. 5.97

5.17 For the network shown in Fig. 5.98, write the tie-set schedule for a tree with branches 1, 2 and 3. Obtain the equilibrium equations on current basis and calculate all the branch currents and branch voltages. The numbers in the parentheses indicate branch numbers.

(Karnataka University)

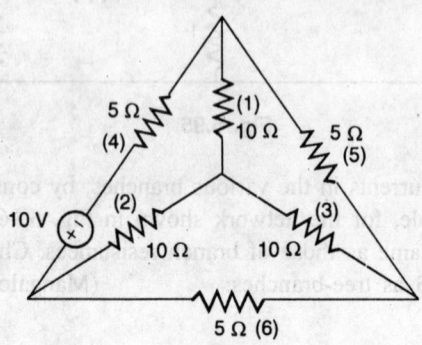

Fig. 5.98

5.18 For the network shown in Fig. 5.98, write the cut-set schedule for the tree with branches 1, 2 and 3. Obtain equilibrium equations on node basis and calculate all the branch voltages and currents.

(Bangalore University)

5.19 For the network shown in Fig. 5.99, draw the graph. Write the cut-set matrix for the tree with branches 1 and 2. Obtain equilibrium equations with node-pair voltages as variables and find all the branch voltages and currents. The numbers in the parentheses indicate branch numbers.   (Kuvempu University)

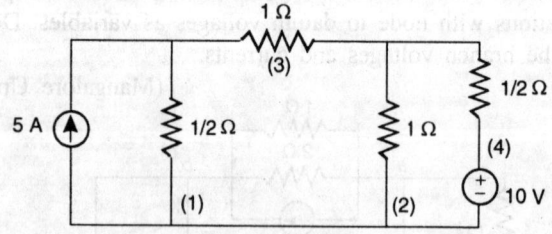

Fig. 5.99

5.20 For the network shown in Fig. 5.99, draw the graph. Write the tie-set matrix for the tree with branches 1 and 2. Obtain equilibrium equations with loop currents as variables. Find all the branch currents and branch voltages.   (Mysore University)

5.21 Write the loop equations for the network shown in Fig. 5.100, on tie-set basis, (Mysore University)

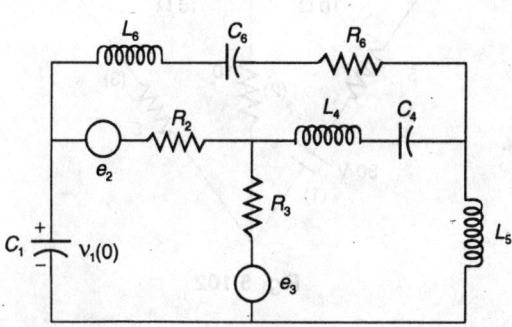

Fig. 5.100

5.22 Draw a tree for the circuit shown in Fig. 5.101, in which all loop currents flow through $1\,\Omega$ resistor. Obtain for this tree, the tie-set matrix and therefrom the equilibrium equations.

(Mysore University)

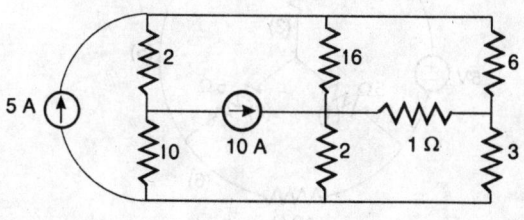

Fig. 5.101

5.23 For the network shown in Fig. 5.102, write the tie-set schedule and obtain equilibrium equations on current basis. Obtain all the branch currents and branch voltages. The numbers in the parentheses indicate branch numbers. Take the tree with branches 1, 2 and 3. (Gulbarga University)

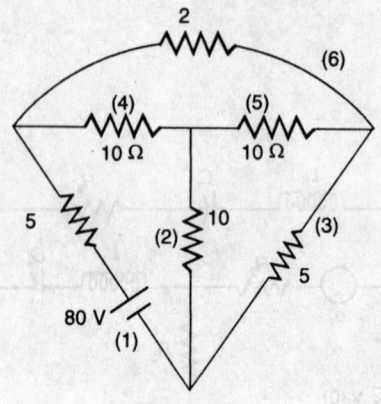

Fig. 5.102

5.24 For the network shown in Fig. 5.103, write the cut-set schedule. Obtain equilibrium equations on node basis. Obtain all the branch voltages and branch currents. The numbers in the parentheses indicate branch numbers. Take the tree with branches 1, 2 and 3. (Karnataka University)

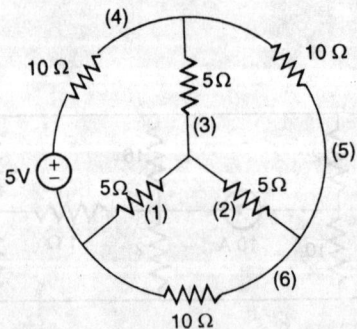

Fig. 5.103

5.25 For the network shown in Fig. 5.104, form equilibrium equations, choosing elements *oa*, *bc* and *ac* as links. Obtain all the branch currents and voltages, using tree-branch voltages as variables.

(Bangalore University)

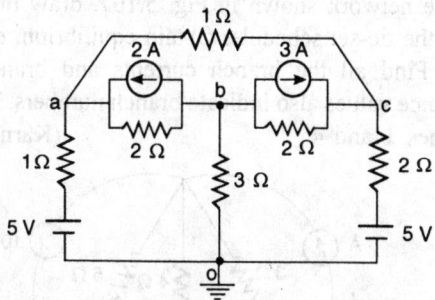

Fig. 5.104

5.26 For the circuit shown in Fig. 5.105, construct a tree with branches 4, 5 and 6. Write the tie-set schedule and hence, obtain the branch currents and branch voltages. The numbers in the parentheses indicate branch numbers. (Mangalore University)

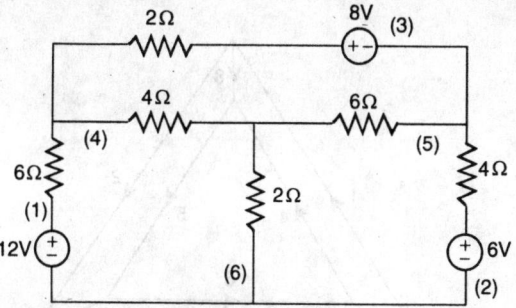

Fig. 5.105

5.27 For the network shown in Fig. 5.106, after performing *E* shift and *I* shift, draw the oriented graph. Select a tree corresponding to branches 3, 4 and 6. Write tie-set and cut-set matrices. The branch numbers are indicated in the figure.

(Mysore University)

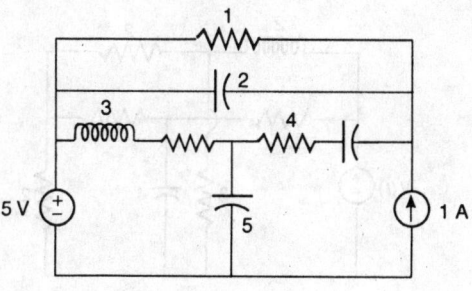

Fig. 5.106

5.28 For the network shown in Fig. 5.107, draw the oriented graph. Write the tie-set schedule. Obtain equilibrium equations on loop basis. Find all the branch currents and branch voltages. The resistance values also indicate branch numbers. Take the tree with branches 1 and 4. (Karnataka University)

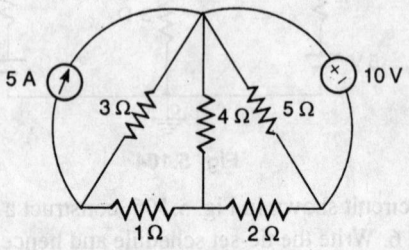

Fig. 5.107

5.29 Obtain the dual of the graph shown in Fig. 5.108. The numbers indicate branch numbers. (Kuvempu University)

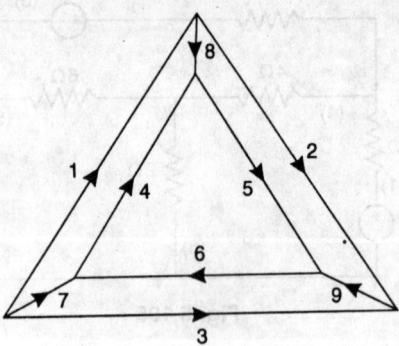

Fig. 5.108

5.30 Construct the dual of the network shown in Fig. 5.109. Write loop equations for the given network and node equations for the dual network such that both the sets form dual equations. The numbers indicate the values of *R, L* and *C.* (Bangalore University)

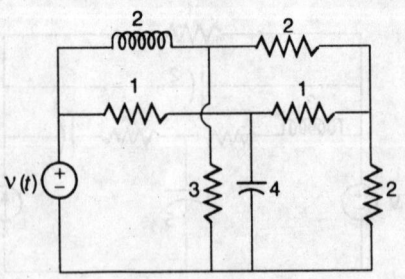

Fig. 5.109

5.31 Draw the dual network of the network shown in Fig. 5.110.

(Karnataka University)

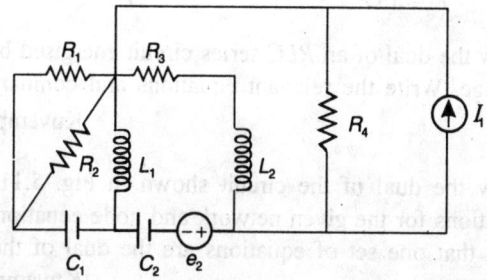

Fig. 5.110

5.32 Draw the dual of the network shown in Fig. 5.111. Write loop equations for the network given and the node equations for the dual network and show that they form a set of dual equations.

(Gulbarga University)

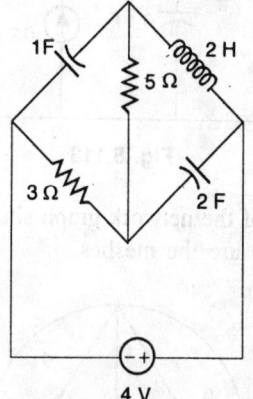

Fig. 5.111

5.33 Construct the dual of the network shown in Fig. 5.112. Write loop equations for the given network and node equations for the dual.

(Mangalore University)

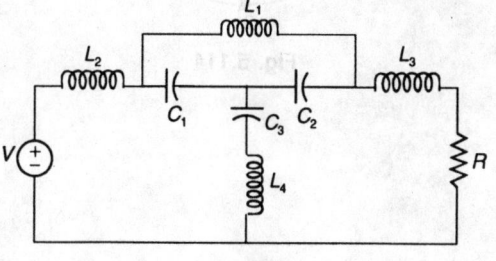

Fig. 5.112

$$L_1 = L_2 = L_3 = L_4 = 1\text{H} \qquad C_1 = C_2 = C_3 = \frac{1}{2}\text{ F}$$

$$R = 1\,\Omega \qquad\qquad V = 1\text{ V}$$

5.34 Draw the dual of an *RLC* series circuit energised by a sinusoidal voltage. Write the relevant equations and compare them.

(Kuvempu University)

5.35 Draw the dual of the circuit shown in Fig. 5.113. Write loop equations for the given network and node equations for the dual such that one set of equations are the dual of the other.

(Kuvempu University)

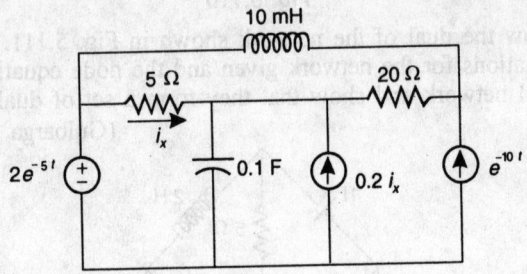

Fig. 5.113

5.36 Draw the dual of the network graph shown in Fig. 5.114, where *a*, *b*, *c*, *d* and *e* are the meshes.   (Mysore University)

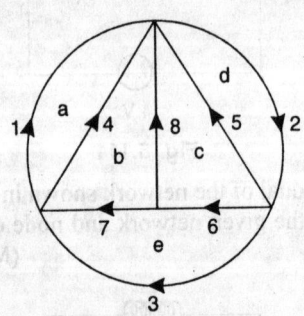

Fig. 5.114

Locus Diagrams

6.1 INTRODUCTION

Locus diagrams are the graphical representations of the way in which the responses of electrical circuits vary, when one or more parameters are continuously changing. They help us to study the way in which (i) current and/or p.f. vary, when voltage is kept constant, (ii) voltage and/or p.f. vary, when current is kept constant, when one of the parameters of the circuit (whether series or parallel) is varied. The variations usually take the forms of circles or straight lines. The locus diagrams yield such important informations as I_{max}, I_{min}, V_{max}, V_{min} and the p.f.s at which they occur. In some parallel circuits, they will also indicate whether or not, a condition for resonance is possible.

6.2 R-X_L SERIES CIRCUIT

Consider an R-X_L series circuit as shown in Fig. 6.1, across which a constant voltage E is applied. By varying R or X_L, a wide range of currents and p.f.s can be obtained. R can be varied by the rheostatic adjustment and X_L can be varied by using a variable inductor or by applying a variable frequency source. When the variations are uniform and lie between 0 and ∞, the resulting locus diagrams are circles.

Case (i) When R is varied

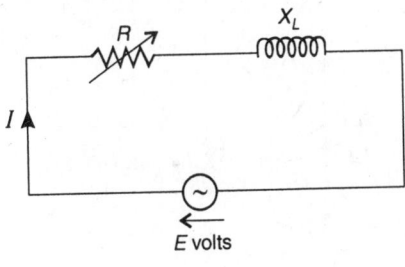

Fig. 6.1

Figure 6.1 represents an R-X_L series circuit across which, a constant voltage, constant frequency source is applied. I is the current flowing through the circuit.

When $R = 0$, the current is maximum and is given by

$$I_{max} = \frac{E}{X_L}$$

and lags E by 90°. The p.f. is zero.

When, $R = \infty$, the current is minimum and is given by

$$I_{min} = 0, \quad \phi = 0$$

and   p.f. = 1

For any other value of R, the current lags the voltage by an angle $\phi = \tan^{-1} X_L / R$.

The general expression for the current is

$$I = \frac{E}{\sqrt{R^2 + X_L^2}} = \frac{E}{Z} \times \frac{X_L}{X_L} = \frac{E}{X_L} \times \frac{X_L}{Z} = \frac{E}{X_L} \sin\phi \qquad (6.1)$$

Equation (6.1) is the equation of a circle in the polar form, E/X_L is the diameter of the circle. The locus diagram of current, i.e. the way in which, the current varies in the circuit, as R is varied from 0 to ∞ is shown in Fig. 6.2, which is a semi-circle.

This variation of current can be analysed in another way also. From Fig. 6.2.

$$E = IZ = \left(I_x - jI_y\right)\left(R + jX_L\right)$$
$$= I_x R - jI_y R + jI_x X_L + I_y X_L \qquad (6.2)$$

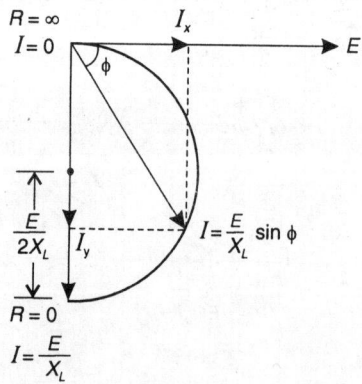

Fig. 6.2

Equating real and imaginery parts of equation (6.2), we get

i.e. $\qquad I_x R + I_y X_L = E$ $\hfill$ (6.3)

and $\qquad I_x X_L - I_y R = 0$ $\hfill$ (6.4)

multiplying equation (6.3) by I_y and (6.4) by I_x, to eliminate the variable quantity, we get

$$I_x I_y R + I_y^2 X_L = E I_y \hfill (6.5)$$

$$-I_x I_y R + I_x^2 X_L = 0 \hfill (6.6)$$

Solving Eqs (6.5) and (6.6), we get

$$X_L \left(I_x^2 + I_y^2 \right) = E I_y$$

or $\qquad I_x^2 + I_y^2 - \dfrac{E}{X_L} I_y = 0$ $\qquad$ Completing the squares, we get

$$I_x^2 + \left(I_y - \frac{E}{2X_L} \right)^2 = \left(\frac{E}{2X_L} \right)^2 \hfill (6.7)$$

Equation (6.7) is the equation of a circle, whose radius is $\dfrac{E}{2X_L}$, and whose centre is $\left(0, \dfrac{E}{2X_L} \right)$, which is as shown in Fig. 6.2.

Case (ii) When X_L is varied

Figure 6.3 represents and R-L series circuit, across which a constant voltage, constant frequency source is applied. I is the current flowing through the circuit.

When $X_L = 0$, the current is maximum and is given by, $I_m = E/R$ and is in phase with E. The p.f. is unity.

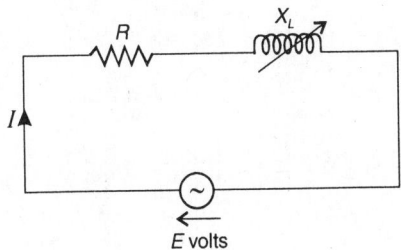

Fig. 6.3

When, $X_L = \infty$, the current is zero. p.f. is zero and $\phi = 90°$.

For any other value of X_L, the current lags the voltage by an angle $\phi = \tan^{-1} X_L/R$.

The general expression for current is

$$I = \frac{E}{\sqrt{R^2 + X^2}} = \frac{E}{Z} \times \frac{R}{R} = \frac{E}{R}\frac{R}{Z} = \frac{E}{R}\cos\phi \qquad (6.8)$$

Equation (6.8) is the equation of a circle in the polar form, where E/R is the diameter of the circle. The locus diagram of the current is as shown in Fig. 6.4, which is a semi-circle.

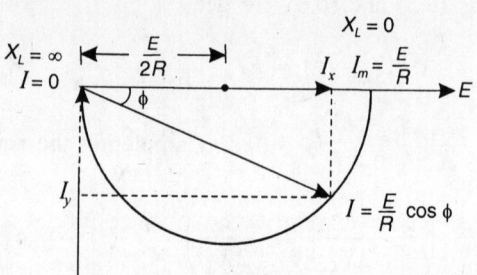

Fig. 6.4

This variation of current can be analysed in another way also.
From Fig. 6.4.

$E = IZ = (I_x - jI_y)(R + jX_L)$ Equating real and imaginary parts, we get,

$$I_x R + I_y X_L = E$$

and $\qquad -I_y R + I_x X_L = 0$

i.e. $\qquad I_x^2 R + I_x I_y X_L = EI_x \qquad (6.9)$

and $\qquad -I_y^2 R + I_x I_y X_L = 0 \qquad (6.10)$

Solving Eqs (6.9) and (6.10), we get

$$\left(I_x^2 + I_y^2\right)R = EI_x$$

or $\qquad I_x^2 + I_y^2 - \dfrac{E}{R}I_x = 0 \qquad$ Completing the squares, we get

$$I_x^2 + I_y^2 - \frac{E}{R}I_x + \left(\frac{E}{2R}\right)^2 = \left(\frac{E}{2R}\right)^2$$

i.e. $\qquad I_y^2 + \left(I_x - \dfrac{E}{2R}\right)^2 = \left(\dfrac{E}{2R}\right)^2 \qquad (6.11)$

Equation (6.11) is the equation of a circle, whose radius is $E/2R$ and whose centre is $(E/2R, 0)$, which is as shown in Fig. 6.4.

6.3 *R-X$_C$* SERIES CIRCUIT

Case (i) When *R* is varied

Figure 6.5 represents and *R-X$_C$* series circuit across which a constant voltage, constant frequency source is applied. *I* is the current flowing through the circuit.

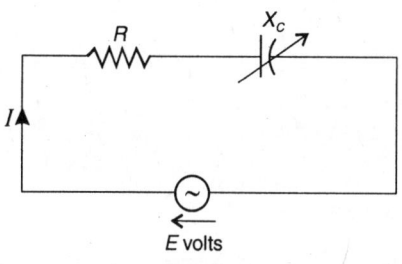

Fig. 6.5

When $R = 0$, the current is maximum and is given by, $I_{max} = E/X_C$, which leads the voltage by 90°. The p.f. is zero.

When $R = \infty$, the current is zero. The p.f. is unity and $\phi = 0$.

For any other value of R, the current leads the voltage by an angle $\phi = \tan^{-1} X_C/R$.

The general expression for the current is

$$I = \frac{E}{\sqrt{R^2 + X_C^2}} = \frac{E}{Z} \times \frac{X_C}{X_C} = \frac{E}{X_C} \frac{X_C}{Z} = \frac{E}{X_C} \sin\phi \qquad (6.12)$$

Equation (6.12) is the equation of a circle in the polar form, where E/X_C is the diameter of the circle. The locus diagram of the current is as shown in Fig. 6.6.

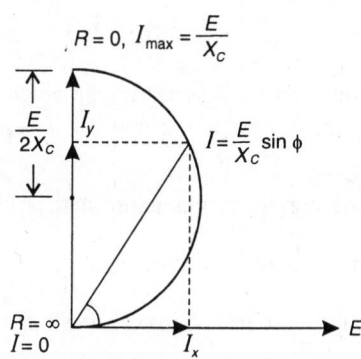

Fig. 6.6

The variation of current can be analysed in another way also.

$$E = IZ = (I_x + jI_y)(R - jX_C) = I_x R + jI_y R - jI_x X_C + I_y X_C$$

Equating real and imaginary parts on both sides, we get

$$I_x R + I_y X_C = E \quad \text{and} \quad I_y R - I_x X_C = 0$$

i.e. $$I_x I_y R + I_y^2 X_C = EI_y \quad \text{and} \quad I_x I_y R - I_x^2 X_C = 0$$

i.e. $$\left(I_x^2 + I_y^2\right) X_C = EI_y \quad \text{i.e.} \quad I_x^2 + I_y^2 - \frac{E}{X_C} I_y = 0$$

Completing the squares, we get

$$I_x + \left(I_y - \frac{E}{2X_C}\right)^2 = \left(\frac{E}{2X_C}\right)^2 \tag{6.13}$$

Equation (6.13) is the equation of a circle of radius $E/2X_C$ and centre $(0, E/2X_C)$, which is as shown in Fig. 6.6.

Case (ii) When X_C is varied

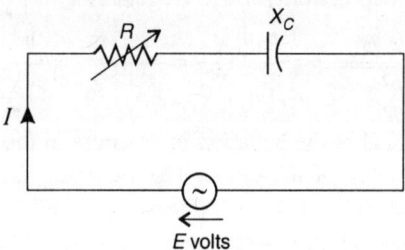

Fig. 6.7

Figure 6.7 represents an R-X_C series circuit across which, a constant voltage, constant frequency source is applied. I is the current flowing through the circuit.

When $X_C = 0$, the current is maximum and is given by

$I_{max} = \dfrac{E}{R}$, which is in phase with E. The p.f. is unity and $\phi = 0°$.

When $X_C = \infty$, the current is zero. The p.f. is 0 and $\phi = 90°$.

For any other value of X_C, the current leads the voltage by an angle $\phi = \tan^{-1} X_C/R$.

The general equation for the current is

$$I = \frac{E}{Z} = \frac{E}{Z} \times \frac{R}{R} = \frac{E}{R} \times \frac{R}{Z} = \frac{E}{R} \cos \phi \qquad (6.14)$$

Equation (6.14) is the equation of a circle in the polar form, where E/R is the diameter of the circle. The locus diagram of the current is as shown in Fig. 6.8.

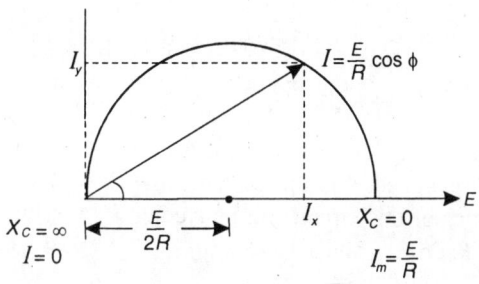

Fig. 6.8

The variation of current can be analysed in another way also.

$$E = IZ = (I_x + jI_y)(R - jX_C)$$

Equating real and imaginary parts, we get

$$I_x R + I_y X_C = E \qquad \text{and} \qquad I_y R - I_x X_C = 0$$

i.e. $I_x^2 R + I_x I_y X_C = EI_x$ and $I_y^2 R - I_x I_y X_C = 0$

i.e. $(I_x^2 + I_y^2) R = EI_x$ and $I_x^2 + I_y^2 - \frac{E}{R} I_x = 0$

Completing the squares, we get

$$I_x^2 + I_y^2 - \frac{E}{R} I_x + \left(\frac{E}{2R}\right)^2 = \left(\frac{E}{2R}\right)^2$$

i.e. $$I_y^2 + \left(I_x - \frac{E}{2R}\right)^2 = \left(\frac{E}{2R}\right)^2 \qquad (6.15)$$

This is the equation of a circle of radius $E/2R$ and centre $(E/2R, 0)$, which is as shown in Fig. 6.8.

6.4 R-X_L-X_C SERIES CIRCUIT

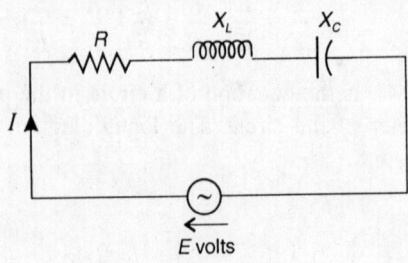

Fig. 6.9

Figure 6.9 represents an R-X_L-X_C series circuit across which, a constant voltage source is applied. I is the current flowing through the circuit. The characteristics of this circuit can be studied by varying any one of the parameters R, X_L, X_C and f.

Case (i) When R is varied

When R is varied and the other three parameters are constant, the locus diagrams of current are similar to those of (a) an R-X_L series circuit, if $X_L > X_C$ as shown in Fig. 6.2 and (b) an R-X_C series circuit, if $X_C > X_L$ as shown in Fig. 6.6. The only difference would be, the resulting reactance is either X_L-X_C or X_C-X_L as the case may be.

Case (ii) When X_L is varied

When $X_L = 0$, the circuit behaves as an R-X_C series circuit and the current is given by

$$I = \frac{E}{\sqrt{R^2 + X_C^2}}$$

and
$$\phi = \tan^{-1}\frac{X_C}{R} \text{ (leading)}$$

When $X_L = X_C$, the circuit behaves as a pure resistance circuit. The current is maximum and is given by

$$I_{max} = \frac{E}{R}$$

and $\phi = 0$.   The p.f. is unity.

When $X_L > X_C$, the circuit behaves as an R-X_L series circuit and the current is given by

$$I = \frac{E}{\sqrt{R^2 + (X_L - X_C)^2}}$$

and

$$\phi = \tan^{-1} \frac{X_L - X_C}{R} \text{ (lagging)}$$

When $X_L \stackrel{=}{=} \infty$, $I = 0$

For any value of X_L lying between X_C and ∞, the locus of current is a semi-circle of radius $E/2R$. The complete locus diagram of current as X_L varied from 0 to ∞ is as shown in Fig. 6.10.

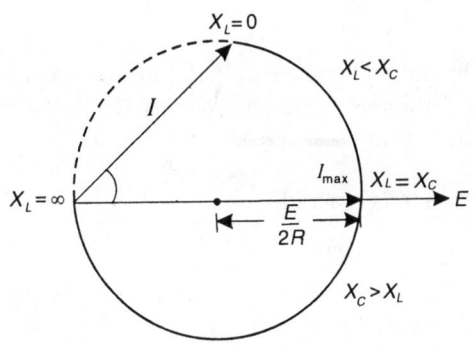

Fig. 6.10

Case (iii) When X_C is varied

When $X_C = 0$, the circuit behaves as an $R\text{-}X_L$ series circuit and the current is given by

$$I = \frac{E}{\sqrt{R^2 + X_L^2}} \quad \text{and} \quad \phi = \tan^{-1} \frac{X_L}{R} \text{ (lagging)}$$

When $X_C = X_L$, the circuit behaves as a pure resistance circuit. The current is maximum and is given by, $I_{max} = E/R$, $\phi = 0$, and p.f. is unity. When $X_C > X_L$, the circuit behaves as an $R\text{–}X_C$ series circuit and the current is given by

$$I = \frac{E}{\sqrt{R^2 + (X_C - X_L)^2}} \quad \text{and} \quad \phi = \tan^{-1} \frac{X_C - X_L}{R} \text{ (leading)}$$

When $X_C = \infty$, $I = 0$.

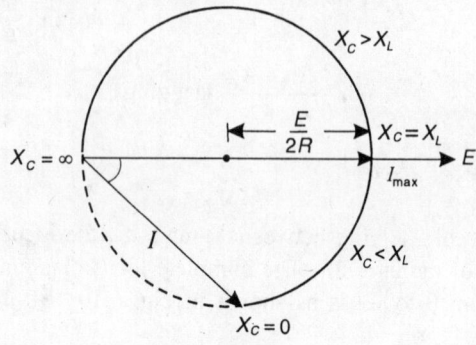

Fig. 6.11

For any value of X_C lying between X_L and ∞, the locus of current is a semi-circle of radius $E/2R$. The complete locus diagram of current as X_C varies from 0 to ∞ is as shown in Fig. 6.11.

Case (iv) When *f* is varied

When $f = 0$, $X_C = \infty$, Hence, $I = 0$.

For values of f, for which $X_C > X_L$, the circuit behaves as an R-X_C series circuit and the locus diagram of current lies in the upper half of the X-Y plane and is a semi-circle with $E/2R$ as radius.

For the value of f at which $X_C = X_L$, the current is maximum and is given by, $I_{max} = E/R$, $\phi = 0$ and p.f. is unity.

For values of f, for which $X_C < X_L$, the circuit behaves as an R-X_L series circuit and the locus diagram of current lies in the lower half of the X-Y plane and is a semi-circle of radius $E/2R$.

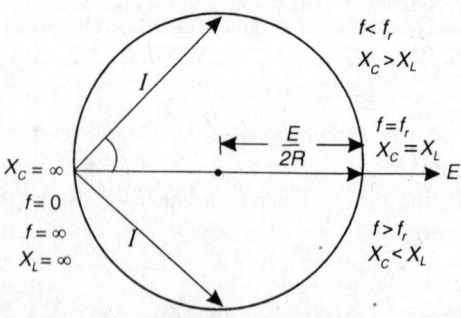

Fig. 6.12

For $f = \infty$, $X_L = \infty$ and hence, $I = 0$. The complete locus diagram of current as f varies from 0 to ∞ is as shown in Fig. 6.12.

6.5 LOCUS DIAGRAMS OF PARALLEL CIRCUITS

When a constant voltage, constant frequency source is applied across a parallel circuit and any one parameter, in one of the parallel branches is varied, current varies only in that branch and the total current locus is got by adding the variable current locus with the constant current, flowing in the other branch. Various such parallel circuits can be considered.

Case (i) R and X_L in parallel and R varying

Consider a parallel circuit as shown in Fig. 6.13, across which a constant voltage, constant frequency source is applied.

$$\vec{I} = \vec{I}_L + \vec{I}_R$$

As X_L is constant, I_L is constant.

As R is variable, I_R is variable.

When $R = \infty$, $I_R = 0$ and $I = I_L$ which lags E by 90°.

For any other values of $R = R_1$ the current I_L remains constant, but $I_{R1} = E/R_1$ and is in phase with E.

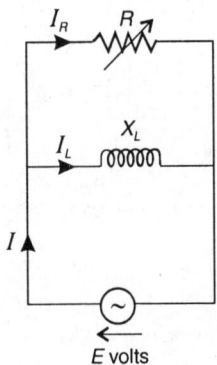

Fig. 6.13

The total current is given by

$$\vec{I} = \vec{I}_L + \vec{I}_R$$

Similarly for other values of $R = R_2$, R_3, etc. I_{R_2}, I_{R_3} etc. and I_1, I_2, etc. can be found and plotted. The locus of the total current is plotted and is as shown in Fig. 6.14.

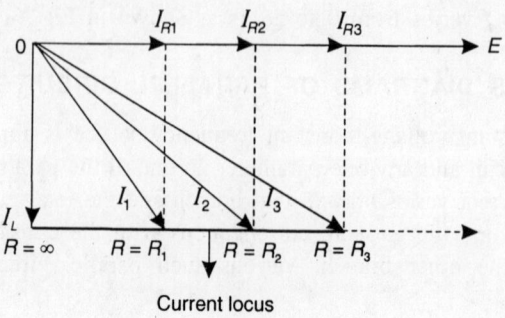

Current locus

Fig. 6.14

Case (ii) R_C-X_C in parallel with R and R varying

Consider a parallel circuit consisting of R_C-X_C branch in parallel with R as shown in Fig. 6.15.

$$\vec{I} = \vec{I}_C + \vec{I}_R$$

As R_C and X_C are constant, I_C remains constant and is given by

$$I_C = \frac{E}{\sqrt{R_C^2 + X_C^2}}$$

and

$$\phi_C = \tan^{-1}\frac{X_C}{R} \text{ (leading)}$$

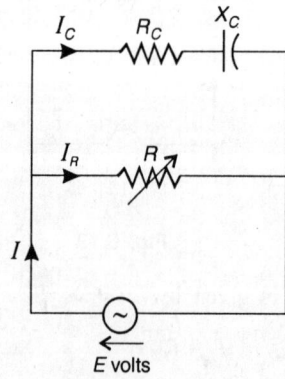

Fig. 6.15

As R is variable, I_R is also variable.

When $R = \infty$, $I_R = 0$ and hence, $I = I_C$.

For any other value of $R = R_1$, I_C remains constant, but $I_{R1} = E/R_1$ and is in phase with E. The total current is given by

$$\vec{I_1} = \vec{I_C} + \vec{I_{R1}}$$

Similarly for other values of $R = R_2$, R_3 etc. I_{R2}, I_{R3}, etc. and I_2, I_3, etc. can be plotted. The locus of the total current is as shown in Fig. 6.16.

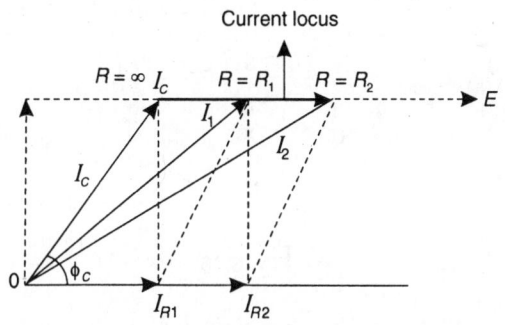

Fig. 6.16

Case (iii) R_L-X_L in parallel with R_C-X_C, R_C varying

Consider a parallel circuit as shown in Fig. 6.17 in which R_C is variable. The total current is given by

$$\vec{I} = \vec{I_L} + \vec{I_C}$$

As R_L and X_L are constant, I_L is constant and is given by

$$I_L = \frac{E}{\sqrt{R_L^2 + X_L^2}} \quad \text{and} \quad \phi_L = \tan^{-1}\frac{X_L}{R}\,(\text{lagging})$$

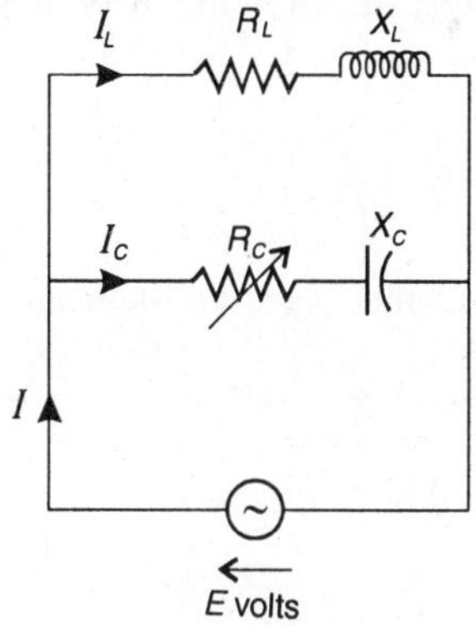

Fig. 6.17

The locus of current I_C is a semi-circle as R_C is varied from 0 to ∞ as explained in section 6.3, case (i). The locus of the total current is the sum of the constant current I_L and the locus of I_C which is as shown in Fig. 6.18.

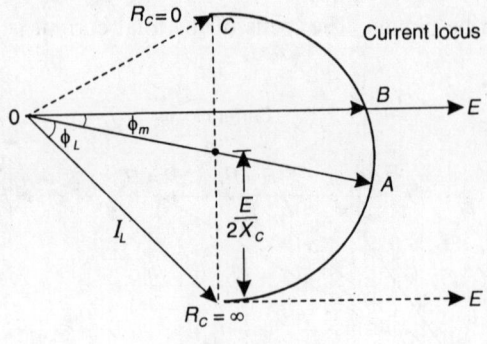

Fig. 6.18

The vector *OA* drawn from the origin, through the centre of the circle gives the maximum current drawn from the supply. $\cos\phi_m$ gives the p.f. at which the current is maximum. *OB* gives the current at resonance. *OC* represents the minimum current that flows through the circuit and its p.f. also can be found, knowing the angle between this vector and the reference voltage. The locus diagram may be drawn to scale and all the relevant quantities can be found from the diagram.

Case (iv) R_C-X_C in parallel with R_L-X_L, R_L varying

Consider a parallel circuit as shown in Fig. 6.19 in which R_L is variable. The total current is given by

$$\vec{I} = \vec{I}_C + \vec{I}_L$$

As R_C and X_C are constants, I_C is constant and is given by

$$I_C = \frac{E}{\sqrt{R_C^2 + X_C^2}}$$

and $$\phi_C = \tan^{-1} \frac{X_C}{R} \text{ (leading)}$$

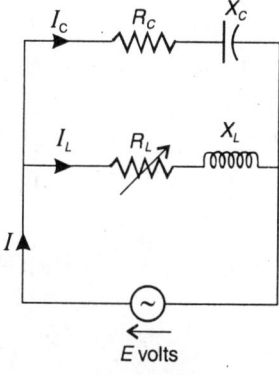

Fig. 6.19

The locus of current I_L is a semi-circle as R_L is varied from 0 to ∞ as explained in section 6.2, case (i). The locus of the total current is the sum of the constant current I_C and the locus of I_L, which is as shown in Fig. 6.20.

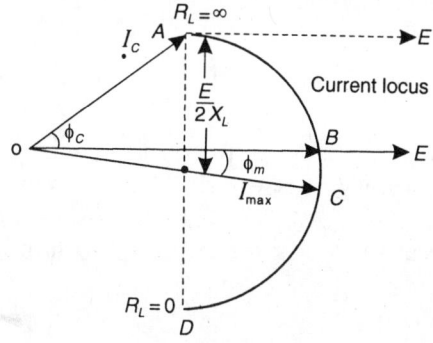

Fig. 6.20

OC represents the maximum current drawn from the supply and $\cos \phi_m$, gives the corresponding p.f. OB gives the current at resonance. $OA = I_C$ gives the minimum current and $\cos \phi_C$ gives the corresponding p.f.

Case (v) $R_C\text{-}X_C$ in parallel with $R_L\text{-}X_L$, X_L varying

Consider a parallel circuit as shown in Fig. 6.21 in which, X_L is variable. The total current is given by

$$\vec{I} = \vec{I}_C + \vec{I}_L$$

As R_C and X_C are constants, I_C is constant and is given by

$$I_C = \frac{E}{\sqrt{R_C^2 + X_C^2}}$$

and $$\phi_C = \tan^{-1} \frac{X_C}{R} \text{ (leading)}$$

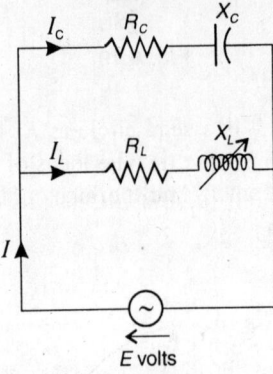

Fig. 6.21

The locus of current I_L is a semi-circle as X_L is varied from 0 to ∞ as explained in section 6.2, case (ii). The locus of the total current is the sum of the current I_C and the locus of I_L, which is as shown in Fig. 6.21 (a).

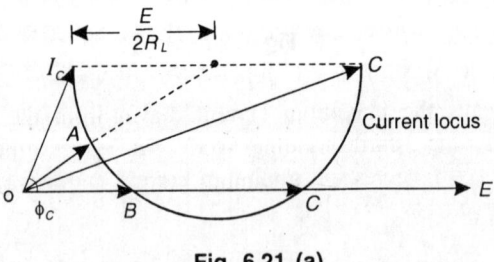

Fig. 6.21 (a)

OA gives the minimum current and the corresponding p.f. can be found. *OC* gives the maximum current and the corresponding p.f. can be found. *OB* and *OC* are the two currents at resonance.

Case (vi) R_L-X_L in parallel with R_C-X_C, X_C varying

Consider a parallel circuit as shown in Fig. 6.22 in which, X_C is variable. The total current is given by

$$\vec{I} = \vec{I}_L + \vec{I}_C$$

As R_L and X_L are constants, I_L is constant and is given by

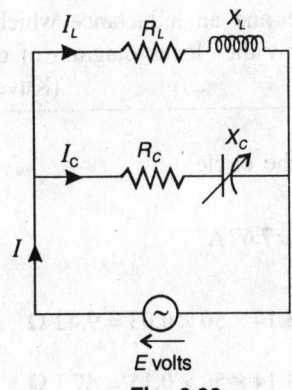

Fig. 6.22

$$I_L = \frac{E}{\sqrt{R_L^2 + X_L^2}}$$

and $$\phi_L = \tan^{-1}\frac{X_L}{R_L}\,(\text{lagging})$$

The locus of current I_C is a semi-circle as X_C is varied from 0 to ∞ as explained in section 6.3, case (ii). The locus of the total current is the sum of the current I_L and the locus of I_C, which is shown in Fig. 6.23.

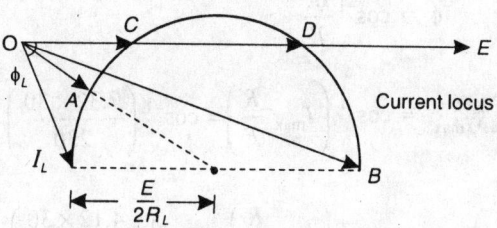

Fig. 6.23

OA represents the minimum current and the corresponding p.f. can be found. *OB* represents the maximum current and the corresponding p.f. can be found. *OC* and *OD* represents the two currents at resonance.

WORKED EXAMPLES

6.1 A 230 V, 50 Hz source is connected to a series circuit consisting of a resistance of 30 Ω and an inductance which varies between 0.03 H and 0.15 H. Draw the locus diagram of current.

(Kuvempu University)

Solution: Diameter of the circle

$$= \frac{E}{R} = \frac{230}{30} = 7.67 \, A$$

$$X_{min} = 2 \times 3.14 \times 50 \times 0.03 = 9.42 \, \Omega$$

$$X_{max} = 2 \times 3.14 \times 50 \times 0.15 = 47.1 \, \Omega$$

$$I_{max} = \frac{230}{\sqrt{30^2 + 9.42^2}} = 7.32 \, A$$

$$I_{min} = \frac{230}{\sqrt{30^2 + 47.1^2}} = 4.12 \, A$$

We know that the equation of the current locus is

$$I = \frac{E}{R} \cos \phi$$

or

$$\phi = \cos^{-1} \frac{IR}{E}$$

$$\therefore \quad \phi_{I_{max}} = \cos^{-1}\left(I_{max} \frac{R}{E}\right) = \cos^{-1}\left(\frac{7.32 \times 30}{230}\right) = 17.3°$$

$$\phi_{I_{min}} = \cos^{-1}\left(I_{min} \frac{R}{E}\right) = \cos^{-1}\left(\frac{4.12 \times 30}{230}\right) = 57.49°$$

The locus diagram is as shown in Fig. 6.24.

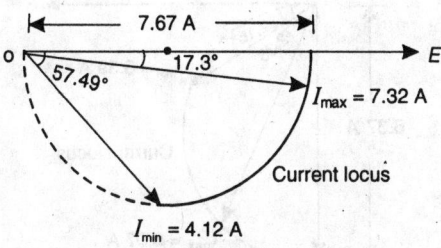

Fig. 6.24

6.2 A 200 V, 50 Hz source is connected to a series circuit consisting of an inductance of 0.1 H and a resistance, which is variable between 10 and 50 Ω. Draw the current locus and indicate in the locus diagram, the circle diameter, maximum and minimum currents and their phase angles. (Mysore University)

Solution

$$X_L = 2 \times 3.14 \times 50 \times 0.1 = 31.4\ \Omega$$

Diameter of the circle $= \dfrac{E}{X_L} = \dfrac{200}{31.4} = 6.37$ A

$$I_{max} = \frac{200}{\sqrt{10^2 + 31.4^2}} = 6.07 \text{ A}$$

$$I_{min} = \frac{200}{\sqrt{50^2 + 31.4^2}} = 3.39 \text{ A}$$

The equation for the current locus is $I = \dfrac{E}{X_L} \sin \phi$

$$\therefore \quad \phi_{I_{max}} = \sin^{-1}\left(I_{max}\,\frac{X_L}{E}\right) = \sin^{-1}\left(\frac{6.07 \times 31.4}{200}\right) = 72.36°$$

$$\phi_{I_{min}} = \sin^{-1}\left(I_{min}\,\frac{X_L}{E}\right) = \sin^{-1}\left(\frac{3.39 \times 31.4}{200}\right) = 32.16°$$

The locus diagram is as shown in Fig. 6.25.

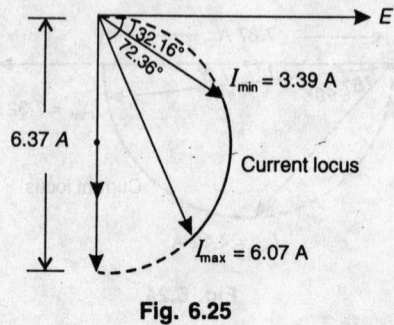

Fig. 6.25

6.3 An impedance coil having a resistance of 5 Ω and an inductive reactance of 12 Ω is connected in series with a variable resistance which is adjustable from 0 to 30 Ω. Assuming a constant impressed e.m.f. of 156 V, draw the current locus for the circuit, indicating there on (a) the circle diameter, (b) I_{max}, I_{min}, (c) $\phi_{I_{max}}$, $\phi_{I_{min}}$.

(Bangalore University)

Solution

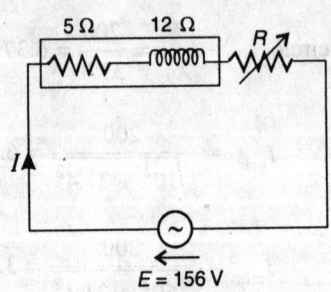

Fig. 6.26

When $R = 0$, $I_{max} = \dfrac{156}{\sqrt{5^2 + 12^2}} = 12\,\text{A}$

$\phi_{I_{max}} = \tan^{-1}\dfrac{12}{5} = 67.3° \ (\text{lagging})$

When $R = 30\Omega$, $I_{min} = \dfrac{156}{\sqrt{35^2 + 12^2}} = 4.21\,\text{A}$

$$\phi_{I_{min}} = \tan^{-1}\frac{12}{35} = 18.9° \text{ (lagging)}$$

diameter of the circle$= \dfrac{E}{X_L} = \dfrac{156}{12} = 13 \text{ A}$

The locus diagram of the current is as shown in Fig. 6.27.

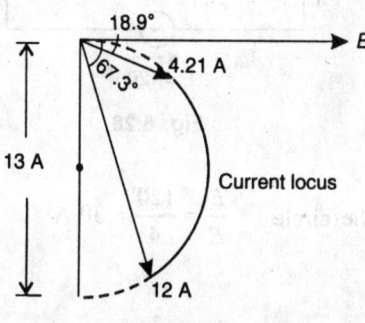

Fig. 6.27

6.4 An impedance coil having a resistance of 4 Ω and inductive reactance of 2 Ω is connected in series with a reactor that is variable between 1 and 11.4 Ω. Assuming a constant applied voltage of 120 V, draw the current locus for the circuit indicating thereon (a) the circle diameter, (b) I_{max}, I_{min} and (c) $\phi_{I_{max}}$, $\phi_{I_{min}}$.

(Karnataka University)

Solution

When $X_L = 1 \ \Omega$

$$Z_1 = \sqrt{4^2 + 3^2} = 5\Omega$$

$$I_{max} = \frac{120}{5} = 24 \text{ A}$$

When $X_L = 11.4 \ \Omega,$

$$Z_2 = \sqrt{4^2 + 13.4^2} = 13.98 \ \Omega$$

$$I_{min} = \frac{120}{13.98} = 8.57 \text{ A}$$

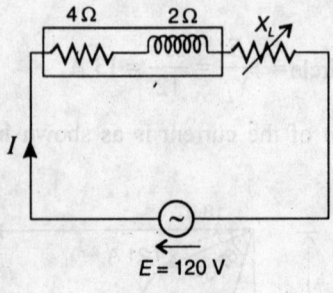

Fig. 6.28

diameter of the circle $= \dfrac{E}{R} = \dfrac{120}{4} = 30$ A

$$\phi_{I_{max}} = \tan^{-1} \dfrac{3}{4} = 36.87° \text{ (lagging)}$$

$$\phi_{I_{min}} = \tan^{-1} \dfrac{13.4}{4} = 73.38° \text{ (lagging)}$$

The locus diagram is as shown in Fig. 6.29.

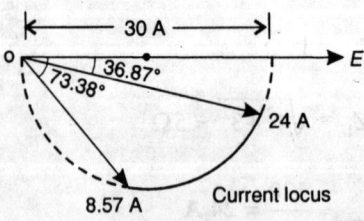

Fig. 6.29

6.5 In the problem 6.4, current is maintained constant at 9 A by adjusting the impressed e.m.f., as the inductive reactance is varied between 1 and 11.4 Ω. Using current as horizontal reference vector, sketch the voltage locus. (Gulbarga University)

Solution

When $\quad X_L = 1\,\Omega$

$$Z_1 = \sqrt{4^2 + 3^2} = 5\,\Omega$$

$$E_{min} = 9 \times 5 = 45\ \text{V}$$

$$\phi_{E_{min}} = \tan^{-1}\frac{3}{4} = 36.87°\ (\text{lagging})$$

When $\quad X_L = 11.4\,\Omega$

$$Z_2 = \sqrt{4^2 + 13.4^2} = 13.98\,\Omega$$

$$E_{max} = 9 \times 13.98 = 125.82\ \text{V}$$

$$\phi_{E_{max}} = \tan^{-1}\frac{13.4}{4} = 73.38°\ (\text{lagging})$$

The locus diagram of the voltage is as shown in Fig. 6.30.

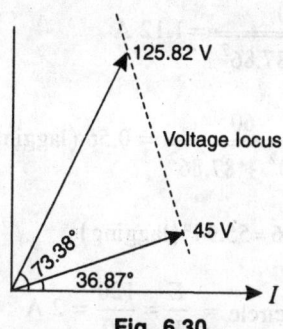

Fig. 6.30

The voltage locus is a straight line.

6.6 A series circuit consisting of a resistance of 60 Ω, an inductance of 0.4 H and a capacitance of 17.6 μF is connected to a variable frequency source, the potential of which is maintained constant at 120 V. If the frequency, is varied through a range of 40 to 80 Hz. Calculate (a) the resonant frequency, (b) current and p.f. at 40 Hz and 80 Hz. Draw the complete current locus for the problem.

(Mangalore University)

Solution

(a) $\quad f_r = \dfrac{1}{2\pi\sqrt{LC}} = \dfrac{1}{2 \times 3.14\ \sqrt{0.4 \times 17.6 \times 10^{-6}}} = 60\ \text{Hz}$

(b) $\quad X_{40\text{Hz}} = (2\times3.14\times40\times0.4)-\left(\dfrac{1}{2\times3.14\times40\times17.6\times10^{-6}}\right)$

$\qquad\quad = -126\ \Omega\ (\text{capacitive})$

$$I = \frac{120}{\sqrt{60^2+126^2}} = 0.86\ \text{A (leading)}$$

$$\cos\phi = \frac{R}{Z} = \frac{60}{\sqrt{60^2+126^2}} = 0.43\ \text{(leading)}$$

$$\phi = \cos^{-1}0.43 = 64.5°\ \text{(leading)}$$

$$X_{80\text{Hz}} = (2\times3.14\times80\times0.4)-\left(\frac{1}{2\times3.14\times80\times17.6\times10^{-6}}\right)$$

$\qquad\quad = 87.86\Omega\ (\text{inductive})$

$$I = \frac{120}{\sqrt{60^2+87.86^2}} = 1.12\ \text{A}$$

$$\cos\phi = \frac{R}{Z} = \frac{60}{\sqrt{60^2+87.86^2}} = 0.56\ \text{(lagging)}$$

$$\phi = \cos^{-1}0.56 = 55.94°\ \text{(lagging)}$$

$$\text{Dia. of the circle} = \frac{E}{R} = \frac{120}{60} = 2\ \text{A}$$

The current locus is as shown in Fig. 6.31.

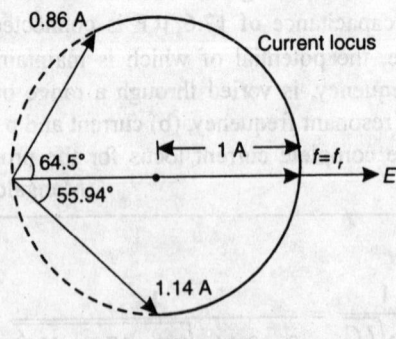

Fig. 6.31

6.7 In the circuit shown in Fig. 6.32, R is varied from 0 to ∞. Draw the total current locus. Find (a) the u.p.f. current, (b) the minimum p.f. and the corresponding current and (c) the maximum current and the corresponding p.f. (Mysore University)

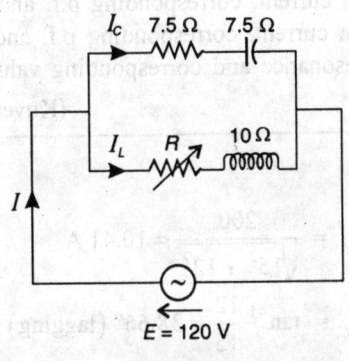

Fig. 6.32

Solution

$$I_C = \frac{120}{7.5 - j\,7.5} = \frac{120}{10.61\,\angle -45°} = 11.31\,\angle\,45°\ \text{A}$$

For the I_L locus, dia. of the circle $= \dfrac{E}{X_L} = \dfrac{120}{10} = 12$ A

The current locus is as shown in Fig. 6.33.

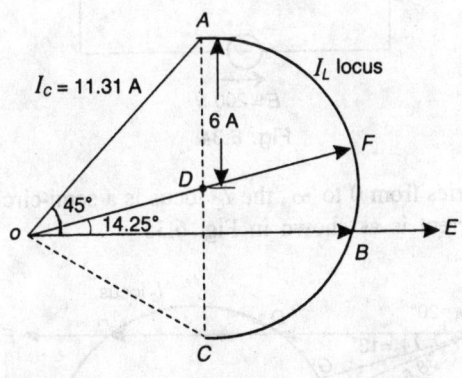

Fig. 6.33

from the locus diagram:

(a) u.p.f. current $= OB = 13.8$ A.
(b) minimum p.f. $= \cos 45° = 0.707$ leading and corresponding current $= OA = 11.31$ A.

(c) $I_{max} = OF = 15.8$ A, and the corresponding p.f. is $\cos 14.25°$ = 0.969 leading.

6.8 For the parallel circuit shown in Fig. 6.34, draw the current locus and thereby find.
(i) The minimum current, corresponding p.f. and X_C.
(ii) The maximum current, corresponding p.f. and X_C.
(iii) Currents at resonance and corresponding values of X_C.

(Kuvempu University)

Solution

$$I_L = \frac{200}{\sqrt{15^2 + 12^2}} = 10.41 \text{ A}$$

$$\phi_L = \tan^{-1} \frac{12}{15} = 38.66° \text{ (lagging)}$$

For I_C locus, diameter of the circle $= \frac{200}{10} = 20$ A

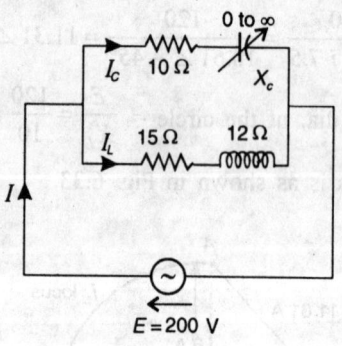

Fig. 6.34

When X_C varies from 0 to ∞, the I_C locus is a semi-circle. The locus of the total current is as shown in Fig. 6.35.

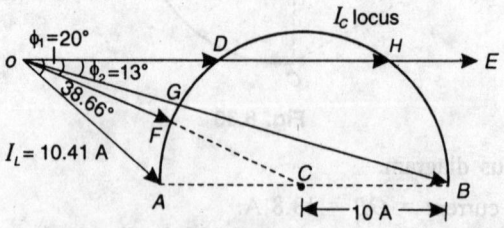

Fig. 6.35

(i) Minimum current = OF = 9.07 A

Corresponding p.f. = $\cos \phi_1 = \cos 20° = 0.94$ (lagging)

Corresponding $I_C = AF$ = 3.4 A

Corresponding $Z_C = \dfrac{200}{3.4} = 58.82 \ \Omega$

Corresponding $X_C = \sqrt{58.82^2 - 10^2} = 57.96 \ \Omega$

(ii) Maximum current = OB = 28.4 A

Corresponding p.f. = $\cos \phi_2 = \cos 13° = 0.97$ (lagging)

Corresponding $I_C = AB$ = 20 A

Corresponding $Z_C = \dfrac{200}{20} = 10 \ \Omega$

Corresponding $X_C = \sqrt{10^2 - 10^2} = 0 \ \Omega$

(iii) Currents are resonance are:

$$I_{r1} \ OD = 10.2 \text{ A} \quad \text{and} \quad I_{r2} = OH = 25.2 \text{ A}$$

Corresponding I_C, Z_C and X_C are given by:

$$I_C = AD = 6.72 \text{ A}$$

$$Z_C = \frac{200}{6.72} = 29.76 \text{ A}$$

$$X_C = \sqrt{29.76^2 - 10^2} = 28.03 \ \Omega$$

and $\quad I_{r2} = OH$ = 25.2 A

Corresponding I_C, Z_C and X_C are given by:

$$I_C = AH = 18.45 \text{ A}$$

$$Z_C = \frac{200}{18.45} = 10.84 \ \Omega$$

$$X_C = \sqrt{10.84^2 - 10^2} = 4.18 \ \Omega$$

6.9 Draw the locus of the total current for the circuit shown in Fig. 6.36 and find (i) the values of X_L for which the entire circuit is at resonance, (ii) find also the minimum and maximum currents and the corresponding power factors. (Bangalore University)

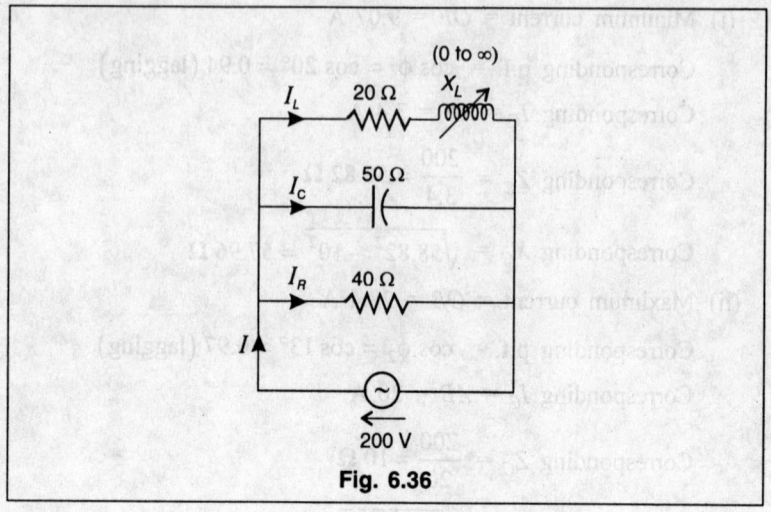

Fig. 6.36

Solution: $I_R = \dfrac{200}{40} = 5$ A, in phase with E.

$$I_C = \frac{200}{50} = 4 \text{ A, leading } E \text{ by } 90°.$$

Diameter of I_L locus $= \dfrac{200}{20} = 10$ A

The locus of the total current is as shown in Fig. 6.37.

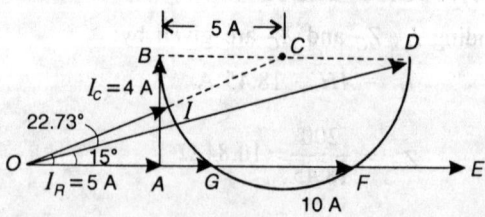

Fig. 6.37

From the locus diagram, the currents at resonance are:

$$I_{r1} = OG = 7 \text{ A} \quad \text{and} \quad I_{r2} = OF = 12.8 \text{ A}$$

Corresponding $I_L = BG = 4.53$ A $\qquad I_L = BF = 8.84$ A

Corresponding $Z_L \doteq \dfrac{200}{4.53} = 44.15\ \Omega$, $\quad Z_L = \dfrac{200}{8.84} = 22.62$ A

Corresponding $X_L = \sqrt{44.15^2 - 20^2} = 39.36\ \Omega$,

$$X_L = \sqrt{22.62^2 - 20^2} = 10.58\ \Omega$$

(i) Minimum current = OI = 5.75 A

   Corresponding p.f. = cos 22.73° = 0.922 (leading)

   Maximum current = OD = 15.43 A

   Corresponding p.f. = cos 15° = 0.966 (leading)

6.10 Determine the circuit constants for the locus diagram of the total current shown in Fig. 6.38.   (Mangalore University)

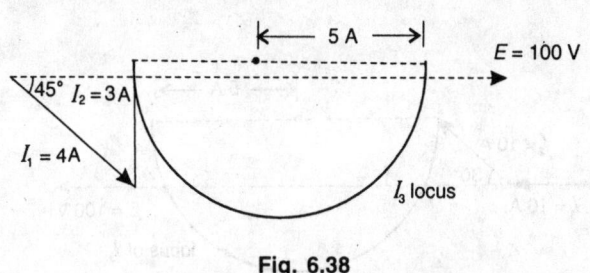

Fig. 6.38

Solution

First branch: $Z_1 = \dfrac{100}{4} = 25\ \Omega$

$$R_1 = Z_1 \cos \phi = 25 \times 0.707 = 17.675\ \Omega$$

$$X_1 = \sqrt{25^2 - 17.675^2} = 17.675\ \Omega \ \text{(inductive)}$$

Second branch: $X_C = \dfrac{100}{3} = 33.33\ \Omega$

Third branch: $\dfrac{E}{R_3} = 10 = \dfrac{100}{R_3}$

$$\therefore \qquad R_3 = 10 \ \Omega$$

$$X_3 = 0 \text{ to } \infty \text{ (inductive)}$$

The circuit is as shown in Fig. 6.39.

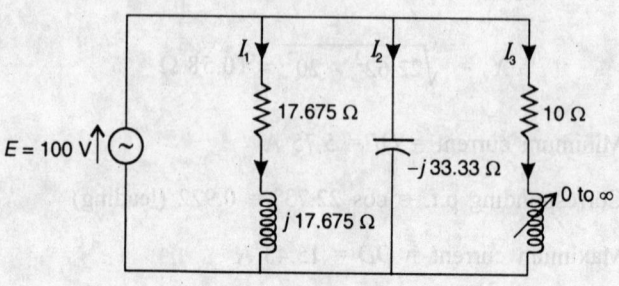

Fig. 6.39

6.11 For the locus diagram shown in Fig. 6.40, determine the circuit elements. (Gulbarga University)

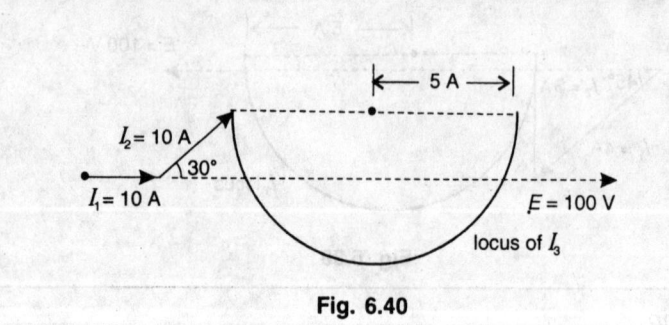

Fig. 6.40

Solution

First branch: $\qquad R_1 = \dfrac{200}{10} = 20 \ \Omega$

Second branch: $\qquad Z_2 = \dfrac{200}{10} = 20 \ \Omega$

$$R_2 = Z_2 \cos 30° = 20 \cos 30° = 17.32 \ \Omega$$

$$X_2 = Z_2 \sin 30° = 20 \sin 30° = 10 \ \Omega \text{ (capacitive)}$$

Third branch: $\qquad \dfrac{E}{R_3} = 10 = \dfrac{200}{R_3}$

$$\therefore \qquad R_3 = 20 \ \Omega$$

$$X_3 = 0 \text{ to } \infty \text{ (inductive)}$$

The circuit is as shown in Fig. 6.41.

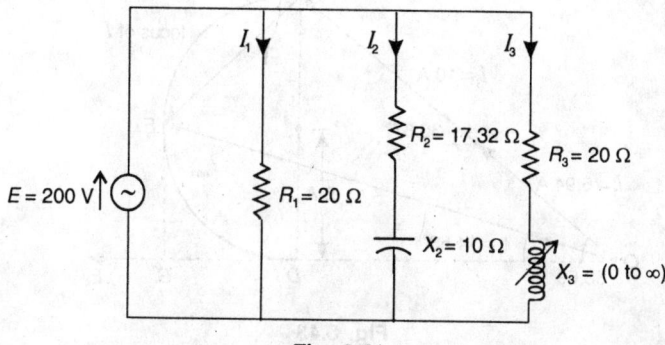

Fig. 6.41

6.12 For the given circuit shown in Fig. 6.42, draw the locus diagram of current. Find the value of R_2 which results in maximum power and the corresponding current and power factor.

(Karnataka University)

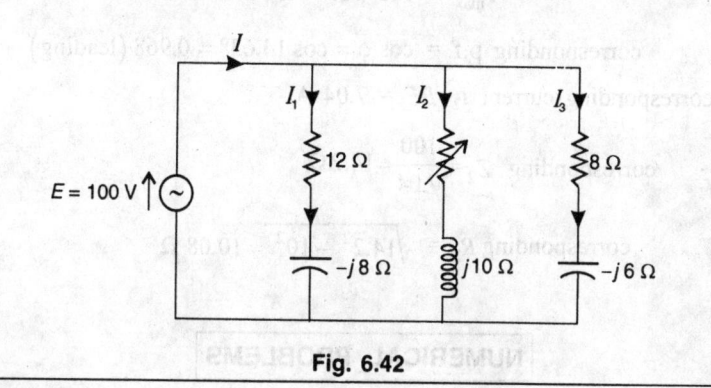

Fig. 6.42

Solution

$$I_1 = \frac{100}{12 - j8} = \frac{100}{14.42 \angle -33.69°}$$

$$= 6.94 \angle 33.69° \text{ A}$$

$$I_3 = \frac{100}{8 - j6} = \frac{100}{10 \angle -36.87°} = 10 \angle 36.87° \text{ A}$$

Diameter of I_2 locus $= \dfrac{100}{10} = 10$ A

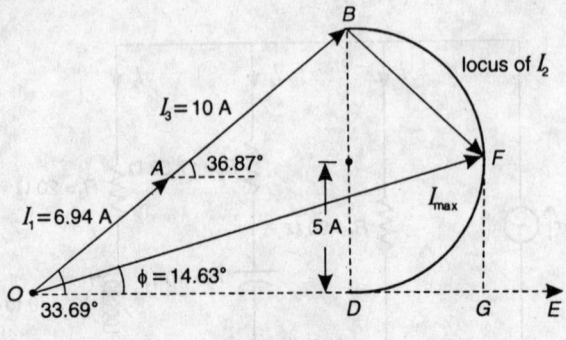

Fig. 6.43

The locus diagram is as shown in Fig. 6.43.

Maximum power is absorbed from the supply due to the component of maximum current, which is in phase with E, i.e. OG.

$$OG = 19 \text{ A}$$

$$\therefore \quad P_{max} = 100 \times 19 = 1900 \text{ watts}$$

corresponding p.f. $= \cos \phi = \cos 14.63° = 0.968 \text{ (leading)}$

corresponding current $I_2 = BF = 7.04 \text{ A}$

$$\therefore \quad \text{corresponding } Z_2 = \frac{100}{7.04} = 14.2 \,\Omega$$

$$\text{corresponding } R_2 = \sqrt{14.2^2 - 10^2} = 10.08 \,\Omega$$

NUMERICAL PROBLEMS

6.1 A 200 V supply is connected to a series circuit consisting of a resistance, which varies from 10 to 20 Ω and an inductive reactance of 30 Ω. Draw the locus diagram of the current.

(Kuvempu University)

6.2 A 200 V, 50 Hz supply is connected to a series circuit consisting of a resistance of 20 Ω and a capacitance, which varies from 150 μF to 300 μF Draw the locus diagram of current.

(Mysore University)

6.3 An impedance coil having a resistance of 10 Ω and an inductive reactance of 15 Ω is connected in series with a variable resistance which is varied from 4 to 8 Ω. If the supply voltage is 100 V, draw the locus diagram of the current. (Bangalore University)

6.4 An impedance coil having a resistance of 20 Ω and an inductive reactance of 10 Ω is connected in series with a capacitive reactance, which varies from 16 to 24 Ω. Draw the locus diagram of the current.(Karnataka University)

6.5 A series circuit consists of $R = 20$ Ω, $L = 0.1$ H and $C = 100$ μ F and is connected to a variable frequency source of 100 V. If the frequency is varied through a range of 50 to 75 Hz, calculate the (i) resonant frequency, (ii) currents and p.f.s at 50 Hz and 75 Hz. Draw the complete current locus.

(Gulbarga University)

6.6 A parallel circuit consists of an inductive coil of resistance 10 Ω and inductive reactance of 15 Ω connected in parallel with another branch of resistance 20 Ω and a capacitive reactance which is variable between 0 to ∞. Draw the current locus and thereby find (i) minimum and maximum currents and the corresponding power factors and (ii) the u.p.f. currents and the capacitive reactances, at which these currents occur. The applied voltage is 230 V. (Mangalore University)

6.7 A parallel circuit consists of two branches. One branch consists of a resistance of 12 Ω and an inductive reactance, which is variable between 0 to ∞. The other branch consists of a resistance of 20 Ω and a capacitive reactance of 10 Ω. If the applied voltage is 100 V, draw the locus diagram of current and thereby find (i) minimum and maximum currents and the corresponding power factors and (ii) u.p.f. currents and the corresponding inductive reactances.(Bangalore University)

6.8 A pure inductive reactance of X_L ohms is connected in parallel with a branch consisting of a resistance R, which is variable from 0 to ∞ and a capacitive reactance $X_C = X_L/2$. If the applied voltage is E volts, sketch the current locus. (Kuvempu University)

6.9 The following information is given in connection with two branch *RC-RL* parallel circuit. The applied voltage $E = 150$ V, $R_C = 7.5$ Ω, $X_C = 8$ Ω, $X_L = 10$ Ω and R_L varies from 0 to ∞.

Draw the total current locus and find (i) u.p.f. currents, (ii) minimum p.f. and the corresponding current and (iii) the maximum current and p.f.(Mysore University)

6.10 Two impedances $Z_1 = 20 + jX_L$ and $Z_2 = -j25$ Ω are connected in parallel across 200 V, 50 Hz single phase supply. If X_L is variable from 0 to ∞, draw the current locus and determine the maximum power factor and the corresponding current.

(Bangalore University)

6.11 A three branch parallel circuit consists of a pure resistance of 10 Ω in one branch, a pure capacitive reactance of 20 Ω in another branch. The third branch consists of an inductive reactance of 8 Ω and a resistance which is variable between 0 to ∞. The applied voltage is 150 V. Draw the current locus. Determine the value of the variable resistance in the third branch, which results in maximum power. Also find the corresponding current and p.f.

(Karnataka University)

6.12 For the locus diagram given in Fig. 6.44, find the configuration of the circuit and insert the values of the circuit elements.

(Gulbarga University)

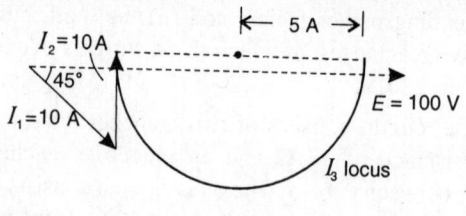

Fig. 6.44

6.13 For the locus diagram shown in Fig. 6.45, find the configuration of the circuit and insert the circuit elements.

(Mangalore University)

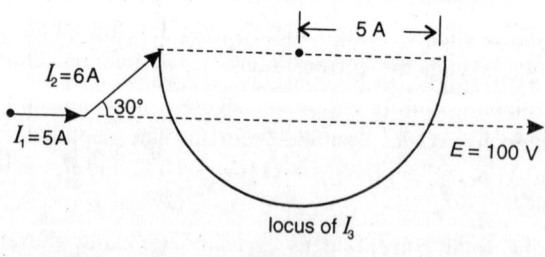

Fig. 6.45

Also find the two values of X_L in the third branch, at which resonance occurs.

6.14 For the locus diagram shown in Fig. 6.46, find the configuration of the circuit and insert the values of the circuit elements.

(Bangalore University)

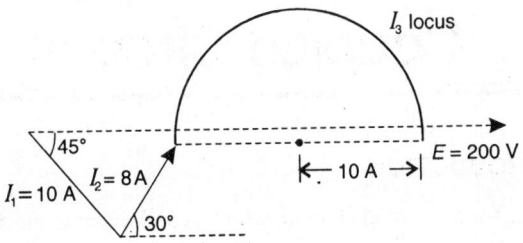

Fig. 6.46

Also find the values of X_C in the third branch for which resonance occurs.

7

Coupled Circuits

7.1 INTRODUCTION

A pure inductor is a circuit element which is defined in terms of the voltage across it and the time rate of change of the current through it. Figure 7.1 represents an inductive coil, across which an alternating voltage $e(t)$ is applied, due to which, an alternating current $i(t)$ flows through it. This alternating current produces an alternating flux which links the coil. Hence, an e.m.f. $e'(t)$ is induced in the coil, which is given by

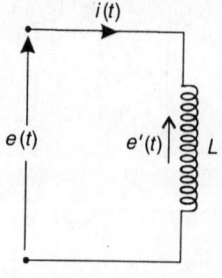

Fig. 7.1

$$e'(t) = -L\frac{di(t)}{dt} = -e(t) \qquad \therefore e(t) = L\frac{di(t)}{dt} \qquad (7.1)$$

Where, L is known as the self-inductance or simply inductance of the coil in henrys.

When two inductive coils are placed physically close to each other, then the flux produced by one coil links the other coil also. Therefore, eventhough the second coil is not energised, due to the mutual flux between the two coils, an e.m.f. is induced in the second coil, which is known as the mutually induced e.m.f. Mutual inductance results due to the presence of the mutual flux between the two coils. Eventhough, a circuit element called 'mutual inductor' does not exist, mutual inductance exists between two coils, which are physically very close to each other.

There exists an inductive coupling between the two coils, because of the mutual flux between them. The physical device whose operation is based inherently on mutual inductance is the transformer.

7.2 MUTUAL INDUCTANCE (*M*)

Two coils which are placed close to each other are said to be mutually coupled, when a part of the alternating flux produced in one coil links the other coil. As the flux is of alternating type, e.m.f. is induced in both the coils. The e.m.f. induced in the first coil, where the flux is produced, is called as *self induced e.m.f.* and the e.m.f. induced in the second coil, which links a part of the flux produced in the first coil, is known as *mutually induced e.m.f.*

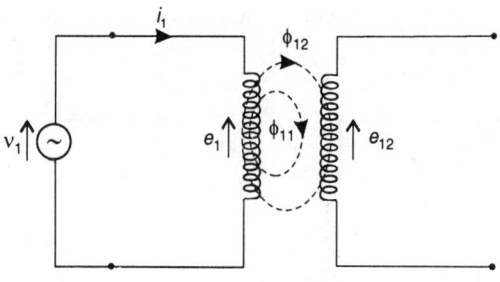

Fig. 7.2

Consider two coils of turns N_1 and N_2, which are placed very close to each other, as shown in Fig. 7.2. When an alternating voltage v_1 is applied to the first coil, an alternating current i_1 flows through it, producing an alternating flux ϕ_1. This flux ϕ_1, links coil 1 and hence an e.m.f. e_1 is induced in it, which is given by

$$e_1 = -N_1 \frac{d\phi_1}{dt} \tag{7.2}$$

This is known as the self induced e.m.f. in coil 1.

A part of the flux ϕ_1, produced in coil 1, links the coil 2 also. This flux ϕ_{12} which links both coil 1 and coil 2, is called as the mutual flux between the two coils. The flux ϕ_{11} links only coil 1. Hence, the flux ϕ_1 is the sum of the two fluxes ϕ_{11} and ϕ_{12}.

$$\phi = \phi_{11} + \phi_{12} \tag{7.3}$$

The mutual flux ϕ_{12}, linking coil 2, induces an e.m.f. e_{12} in that coil. This e.m.f. is known as the mutually induced e.m.f. and is given by

$$e_{12} = -N_2 \frac{d\phi_{12}}{dt} \tag{7.4}$$

The equation for e_{12} may also be written as

$$e_{12} = -M_{12}.\frac{di_1}{dt} \tag{7.5}$$

e_{12} = e.m.f. induced in coil 2 due to the current flowing in coil 1. M_{12} is known as the mutual inductance between coil 1 and coil 2. The equation for the mutual inductance M_{12} may be written as

$$M_{12} = N_2 \frac{d\phi_{12}}{di_1} \tag{7.6}$$

Similar equations can be written, when coil 2 is energised by an alternating current i_2, producing a total flux ϕ_2 in it, as shown in Fig. 7.3.

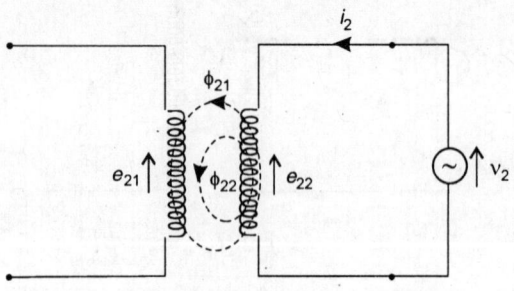

Fig. 7.3

$$\phi_2 = \phi_{22} + \phi_{21} \tag{7.7}$$

Where ϕ_2 = total flux produced in coil 2.

ϕ_{22} = flux that links only coil 2.

ϕ_{21} = flux that links both coil 2 and coil 1.

The self induced e.m.f. in coil 2 is given by

$$e_2 = -N_2 \frac{d\phi_2}{dt} \tag{7.8}$$

The mutually induced e.m.f. in coil 1 is given by

$$e_{21} = -N_1 \frac{d\phi_{21}}{dt} = -M_{21} \frac{di_2}{dt} \tag{7.9}$$

Where $\qquad M_{21} = N_1 \dfrac{d\phi_{21}}{di_2}$ $\hspace{2cm}$ (7.10)

M_{21} is the mutual inductance between coil 2 and coil 1.

As the coupling between the two coils is bilateral, which means that the coupled circuit has the same characteristics in both directions.

$$M_{12} = M_{21} = M \hspace{2cm} (7.11)$$

Hence, *the mutual inductance between any two coils, which are placed close to each other may be defined as the ability of one coil to induce an e.m.f. in the other coil, when an alternating current flows through one of the coils.*

$$M = N_2 \dfrac{d\phi_{12}}{di_1} = N_1 \dfrac{d\phi_{21}}{di_2} \hspace{2cm} (7.12)$$

7.3 COEFFICIENT OF COUPLING (K)

The coefficient of coupling is the ratio of the mutual flux to the total flux.

$$\therefore \qquad K_{12} = \dfrac{\phi_{12}}{\phi_1}$$

and $\qquad K_{21} = \dfrac{\phi_{21}}{\phi_2}$ $\hspace{2cm}$ (7.13)

As the coupling is bilateral

$$K_{12} = K_{21} = K \hspace{2cm} (7.14)$$

$$\therefore \qquad \phi_{12} = K\phi_1$$

and $\qquad \phi_{21} = K\phi_2$ $\hspace{2cm}$ (7.15)

from Eq. (7.11)

$$M^2 = M_{12}.M_{21} = N_2 \dfrac{d\phi_{12}}{di_1} \times N_1 \dfrac{d\phi_{21}}{di_2} = N_1 N_2 \dfrac{d(K\phi_1)}{di_1} \times \dfrac{d(K\phi_2)}{di_2}$$

$$= K^2 N_1 \dfrac{d\phi_1}{di_1} \times N_2 \dfrac{d\phi_2}{di_2} = K^2 L_1 L_2$$

$$\therefore \qquad K = \frac{M}{\sqrt{L_1 L_2}} \qquad\qquad (7.16)$$

7.4 MUTUAL INDUCTANCE BETWEEN TWO COILS WHICH ARE CONNECTED ON THE SAME MAGNETIC MATERIAL

Consider two coils of N_1 and N_2 turns, wound on the same magnetic material. Let a current I_1 flowing through the first coil produce a flux ϕ_1 in it. Then

$$\phi_1 = \frac{N_1 I_1}{l / \mu_0 \mu_r a}$$

Where $\qquad\qquad l$ = length of the electromagnet.

$\qquad\qquad\qquad a$ = area of cross section of the electromagnet.

As the coils are wound on the same magnetic material, the entire flux ϕ_1 produced by coil 1, links coil 2 also. The mutual inductance between the coils is given by

$$M = \frac{N_2 \phi_1}{I_1} = N_2 \frac{N_1 I_1}{l / \mu_0 \mu_r a} \frac{1}{I_1} = \frac{\mu_0 \mu_r a N_1 N_2}{l}$$

7.5 DOT CONVENTION

Dot convention is used to find the relative polarities of the voltages due to mutual inductance with respect to the voltages of self inductance in coupled coils.

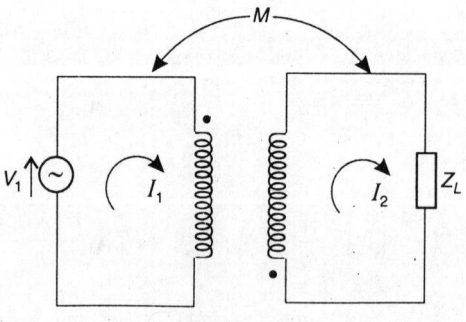

Fig. 7.4 (a)

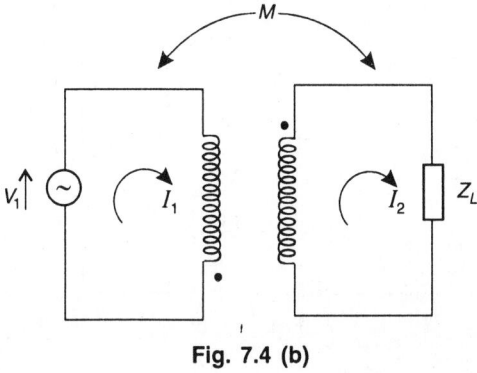

Fig. 7.4 (b)

Let the dots be assigned for the two mutually coupled coils as shown in Figs 7.4 (a) and (b) respectively. In the first case, both the assumed loop currents enter the dotted terminals of the coils. In the second case, both the currents leave the dotted ends of the coils. In both cases, the polarity or sign of the mutually induced e.m.f. is the same as the sign of the self induced e.m.f. The loop equations in both the cases are:

$$j\omega L_1 I_1 + j\omega M\, I_2 = V_1 \tag{7.17}$$

$$j\omega M\, I_1 + (j\omega L_2 + Z_L) I_2 = 0 \tag{7.18}$$

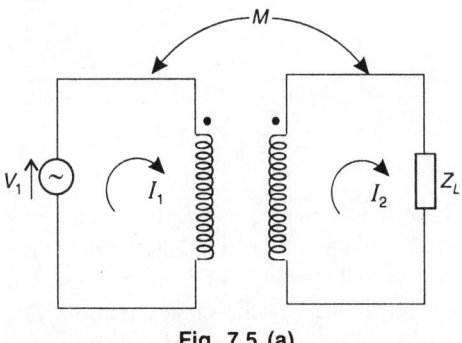

Fig. 7.5 (a)

In the circuit shown in Figs 7.5 (a) and (b), one current enters the dotted terminal and the other current leaves the dotted terminal. In both cases, the polarity or sign of the mutually induced e.m.f. is opposite to the sign of the self induced e.m.f. The loop equations in both the cases are

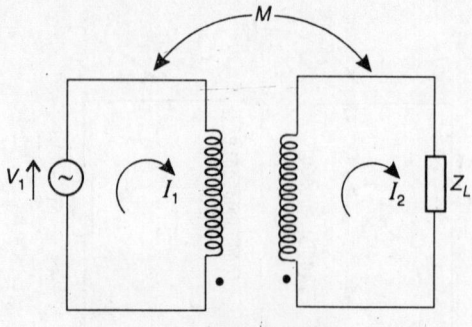

Fig. 7.5 (b)

$$j\omega L_1 I_1 - j\omega M\, I_2 \;=\; V_1 \tag{7.19}$$

$$-j\omega M\, I_1 + \left(j\omega L_2 + Z_L\right)I_2 \;=\; 0 \tag{7.20}$$

7.6 COUPLED COILS IN SERIES

(a) Coupled coils connected in series aiding

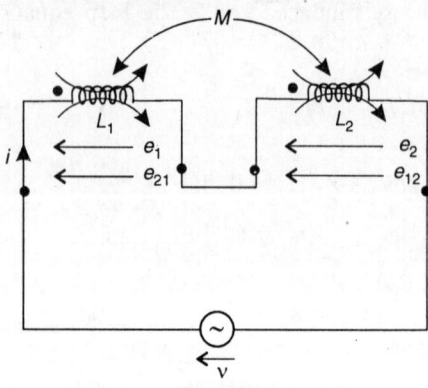

Fig. 7.6

Let two coils of inductance L_1 and L_2 be connected in series across an alternating voltage v due to which an alternating current i flows. The current i enters the dotted terminals of both the coils. Hence, the fluxes produced in both the coils are in the same direction. The e.m.f.s of self induction e_1, e_2 and the e.m.f.s of mutual induction e_{21} and e_{12} are in the same direction as shown in Fig. 7.6. The total induced e.m.f. is given by

$$e = e_1 + e_{21} + e_2 + e_{21}$$

$$-L\frac{di}{dt} \;=\; -L_1\frac{di}{dt} - M\frac{di}{dt} - L_2\frac{di}{dt} - M\frac{di}{dt}$$

$\therefore$ $$L = L_1 + L_2 + 2M \qquad (7.21)$$

Where L is equivalent inductance.

(b) Coupled coils connected in series opposition

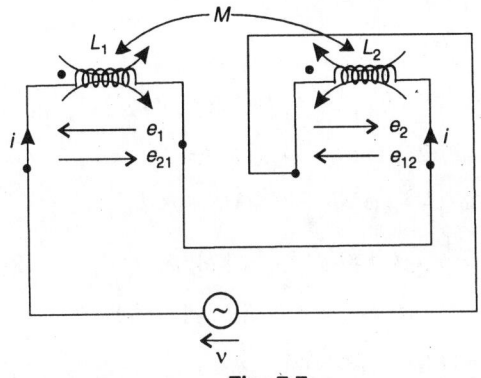

Fig. 7.7

In this type of connection, it is found that the current i enters the dot in one coil and leaves the dot in the other coil. Hence, the direction of the mutually induced e.m.f.s and the self-induced e.m.f.s are in the opposite direction. The total induced e.m.f is given by

$$e = e_1 + e_2 - e_{21} - e_{12}$$

$$-L\frac{di}{dt} = -L_1\frac{di}{dt} - L_2\frac{di}{dt} + M\frac{di}{dt} + M\frac{di}{dt}$$

$\therefore$ $\quad L = L_1 + L_2 - 2M =$ equivalent inductance. $\qquad (7.22)$

7.7 COUPLED COILS IN PARALLEL

(a) Coupled coils in parallel aiding

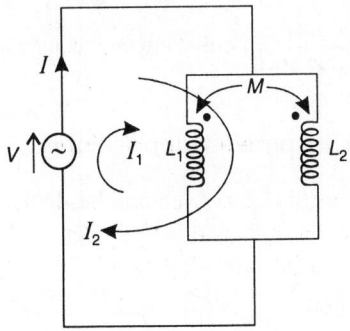

Fig. 7.8

In the circuit shown in Fig. 7.8, both the currents I_1 and I_2 enter at the dots and hence, the fluxes produced by them are in the same direction. Then the two coils are in parallel aiding. The directions of self induced e.m.f.s and the mutually induced e.m.f.s are same. The loop equations are

$$V = j\omega L_1 I_1 + j\omega M I_2 \qquad (7.23)$$

$$V = j\omega M I_1 + j\omega L_2 I_2 \qquad (7.24)$$

Equating Eqs (7.23) and (7.24), we get

$$j\omega L_1 I_1 + j\omega M I_2 = j\omega M I_1 + j\omega L_2 I_2$$

i.e. $\qquad\qquad L_1 I_1 + M I_2 = M I_1 + L_2 I_2 \qquad (7.25)$

or $\qquad\qquad L_1 I_1 + M(I - I_1) = M I_1 + L_2 (I - I_1)$

On simplification, we get

$$I_1 = \frac{L_2 - M}{L_1 + L_2 - 2M} I$$

Similarly, by putting $I_1 = I - I_2$ in Eq. (7.25), we get

$$I_2 = \frac{L_1 - M}{L_1 + L_2 - 2M} I$$

Substituting I_1 and I_2 in equation (7.23), we get

$$V = j\omega L_1 \left(\frac{L_2 - M}{L_1 + L_2 - 2M} \right) I + j\omega M \left(\frac{L_1 - M}{L_1 + L_2 - 2M} \right) I$$

$$I = j\omega I \left(\frac{L_1 L_2 - M^2}{L_1 + L_2 - 2M} \right) = j\omega L I$$

$$\therefore \qquad L = \frac{L_1 L_2 - M^2}{L_1 + L_2 - 2M} = \text{equivalent inductance} \qquad (7.26)$$

(b) Coupled coils in parallel opposition

In the circuit shown in Fig. 7.9, I_1 enters the dotted terminal of coil 1 and I_2 leaves the dotted terminals of coil 2. Hence, the directions of self induced e.m.f.s are opposite to the directions of mutually induced e.m.f.s. The loop equations are:

$$V = j\omega L_1 I_1 - j\omega M I_2 \qquad (7.27)$$

$$V = -j\omega M\,I_1 + j\omega L_2\,I_2 \qquad (7.28)$$

$$\therefore \qquad j\omega L_1 I_1 - j\omega M\,I_2 = -j\omega M\,I_1 - j\omega L_2\,I_2$$

or $\qquad L_1\,I_1 - M\,I_2 = -M\,I_1 + L_2\,I_2 \qquad (7.29)$

Putting $I_2 = I - I_1$ in Eq. (7.29), we get

or $\qquad L_1 I_1 - M\left(I - I_1\right) = -MI_1 + L_2\left(I - I_1\right)$

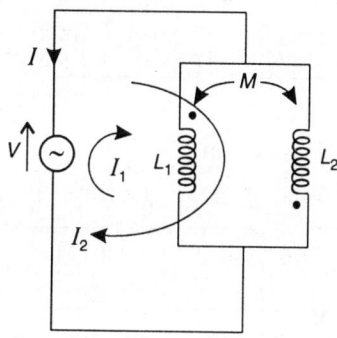

Fig. 7.9

On simplification, we get

$$I_1 = \frac{L_2 + M}{L_1 + L_2 + 2M}\,I$$

Putting $I_1 = I - I_2$ in equation (7.27), we get

$$I_2 = \frac{L_1 + M}{L_1 + L_2 + 2M}\,I$$

Substituting I_1 and I_2 in equation (7.28), we get

$$V = j\omega L_1\left(\frac{L_2 + M}{L_1 + L_2 + 2M}\right)I - j\omega M\left(\frac{L_1 + M}{L_1 + L_2 + 2M}\right)I = j\omega L I$$

$$\therefore \qquad L = \left(\frac{L_1 L_2 - M^2}{L_1 + L_2 + 2M}\right) \qquad (7.30)$$

= equivalent inductance

WORKED EXAMPLES

7.1 Two 1000-turn air-cored coils, 100 cm long, having a cross sectional area of 5 cm², are placed side by side. The mutual inductance between them is 25 mH. Find the self inductances of the coils and the coefficient of coupling. (Kuvempu University)

Solution: As the two coils are same.

$$L_1 = L_2 = \frac{\mu_0 \mu_r a N^2}{l} = \frac{4\pi \times 10^{-7} \times 1 \times 5 \times 10^{-2} \times 1000^2}{100 \times 10^{-2}}$$

$$= 0.0628 \text{ H} = 62.8 \text{ mH}$$

$$K = \frac{M}{\sqrt{L_1 L_2}} = \frac{25 \times 10^{-3}}{\sqrt{62.8 \times 10^{-3} \times 62.8 \times 10^{-3}}} = 0.3981$$

7.2 Two coils having 1000 turns and 1600 turns respectively are placed close to each other such that, 60% of the flux produced by one coil links the other. If a current of 10 A, flowing in the first coil, produces a flux of 0.5 mWb, find the inductance of the second coil. (Mysore University)

Solution

$$L_1 = \frac{N_1 \phi_1}{I_1} = \frac{1000 \times 0.5 \times 10^{-3}}{10} = 0.05 \text{ H}$$

$$K = 60\% = 0.6$$

$$M = N_2 \frac{\phi_{12}}{I_1} = \frac{1600 \times \left(0.5 \times 10^{-3} \times 0.6\right)}{10} = 0.048 \text{ H}$$

$$K = \frac{M}{\sqrt{L_1 L_2}} \quad \text{or} \quad L_1 L_2 = \frac{M^2}{K^2}$$

or $$L_2 = \frac{M^2}{K^2} \frac{1}{L_1} = \frac{0.048^2}{0.6^2} \times \frac{1}{0.05} = 0.128 \text{ H}$$

7.3 Two indentical coils of 1200 turns each, are placed side by side such that, 60% of the flux produced by one coil links the other. A current of 10 A in the first coil, sets up a flux of 0.12 mWb. If the current in the first coil changes from + 10 A to − 10 A in 20 mSec, find (a) the self inductances of the coils, (b) the e.m.f.s and induced in both the coils.

(Bangalore University)

Solution

(a) $\quad L_1 = L_2 = \dfrac{N_1\phi_1}{I_1} = \dfrac{1200 \times 0.12 \times 10^{-3}}{10} = 0.0144\,\mathrm{H}$

(b) $\quad e_1 = -L_1 \dfrac{di_1}{dt} = -0.0144 \times \dfrac{10-(-10)}{20 \times 10^{-3}} = -14.4\,\mathrm{V}$

$\qquad M = N_2 \dfrac{\phi_{12}}{I_1} = \dfrac{1200 \times (0.6 \times 0.12)10^{-3}}{10} = 0.00864\,\mathrm{H}$

$\qquad e_{12} = -M \dfrac{di_1}{dt} = -0.008664 \times \dfrac{10-(-10)}{20 \times 10^{-3}} = -8.64\ \text{volts.}$

7.4 Two coupled coils of self inductances 0.8 H and 0.2 H, have a coefficient of coupling 0.9. Find the mutual inductance and turns ratio.

(Karnatak University)

Solution

$$M = K\sqrt{L_1 L_2} = 0.9\sqrt{0.8 \times 0.2} = 0.36\,\mathrm{H}$$

$$M = N_2 \dfrac{K\phi_1}{I_1} = KN_2 \dfrac{L_1}{N_1} \quad \left[\because L_1 = \dfrac{N_1\phi_1}{I_1} \right]$$

$$\therefore \qquad \dfrac{N_1}{N_2} = \dfrac{KL_1}{M} = \dfrac{0.9 \times 0.8}{0.36} = 2$$

7.5 An air cored solenoid consists of 1500 turns of wire wound on a length of 60 cm. A search coil of 500 turns, enclosing a mean area of 20 cm², is placed centrally in the solenoid. Find (a) the mutual inductance of the arrangement and (b) the e.m.f induced in the search coil, when the current in the solenoid is changing uniformly at the rate of 250 A/Sec.

(Gulbarga University)

Solution

$$L = \frac{\mu_0 \mu_r a N_1 N_2}{l} = \frac{4\pi \times 10^{-7} \times 1 \times 20 \times 10^{-4} \times 1500 \times 500}{60 \times 10^{-2}}$$

$$= 0.00314 \text{ H} = 3.14 \text{ mH}$$

$$e_{12} = -M \frac{di_1}{dt} = -0.00314 \times 250 = -0.785 \text{V}$$

7.6 Find the voltage across 5 W resistor in the circuit shown in Fig. 7.10, (a) for dots as given, (b) reversing the polarty of one coil.

(Kuvempu University)

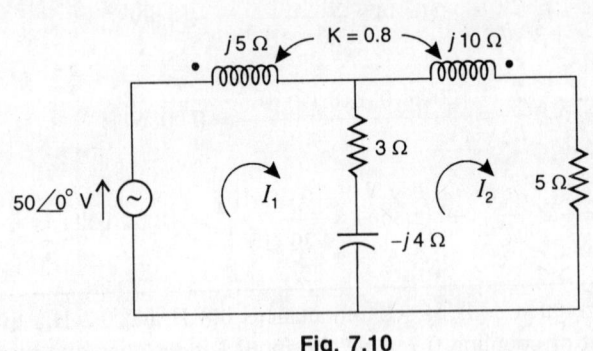

Fig. 7.10

Solution

(a)

$$M = K\sqrt{L_1 L_2}$$

$$j\omega M = K\sqrt{j\omega L_1 \times j\omega L_2} = j0.8\sqrt{5 \times 10}$$

$$jX_m = j5.66 \ \Omega$$

Loop current I_1 and I_2 are assumed as shown. The loop equations are:

$$(3 + j1)I_1 - (3 - j4)I_2 - j5.66 I_2 = 50\angle 0°$$

i.e. $$(3 + j1)I_1 + (-3 - j1.66)I_2 = 50\angle 0° \quad (1)$$

$$(-3 - j1.66)I_1 + (8 + j6)I_2 = 0 \quad (2)$$

Solving (1) and (2), we get

$$I_2 = 8.6\angle -24.8° \text{ A}$$

∴ $$V_{5\Omega} = 5I_2 = 5 \times 8.6\angle -24.8°$$

$$= 43\angle -24.8° \text{ V}$$

(b) When the polarity of the second coil is reversed, the dot is put at the beginning of the second coil. In that case, both currents I_1 and I_2 enter the dots and hence, the sign of voltage drops due to self inductance and mutual inductance are same.

Then, the loop equations are:

$$(3+j1)I_1-(3-j4)I_2+j5.66I_2 = 50\angle 0°$$

i.e $$(3+j1)I_1+(-3+j9.66)I_2 = 50\angle 0° \qquad (1)$$

$$(-3+j9.66)I_1+(8+j6)I_2 = 0 \qquad (2)$$

Solving (1) and (2), we get

$$I_2 = 3.82\angle-112.11°\,A$$

$$\therefore \quad V_{5\Omega} = 5I_2 = 5\times 3.82\angle-112.11°$$

$$= 19.1\angle-112.11°\,V$$

7.7 Draw the conductively coupled circuit (or T equivalent circuit) for the mutually coupled circuit shown in Fig. 7.11.

(Bangalore University)

Solution: The loop equations are

$$(R_1+j\omega L_1)I_1-j\omega M I_2 = V_1 \qquad (1)$$

$$-j\omega M I_1+(R_2+j\omega L_2)I_2 = V_2 \qquad (2)$$

In the above equations

$$Z_{11} = R_1+j\omega L_1$$

$$Z_{12} = -j\omega M \quad \text{and} \quad Z_{22} = (R_2+j\omega L_2)$$

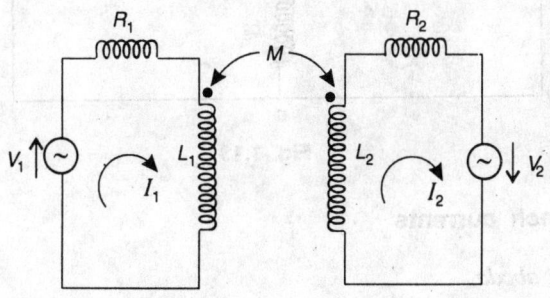

Fig. 7.11

The conductively coupled circuit which satisfies the above equations is shown in Fig. 7.12.

Where
$$Z_1 = R_1 + j\omega L_1 - j\omega M = R_1 + j\omega(L_1 - M)$$
$$Z_2 = R_2 + j\omega L_2 - j\omega M = R_2 + j\omega(L_2 - M)$$
$$Z_m = j\omega M$$

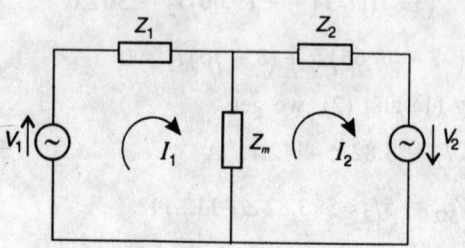

Fig. 7.12

7.8 Write down loop equations for the network shown in Fig. 7.13 in terms of (a) branch currents and (b) loop currents.

(Mysore University)

Solution

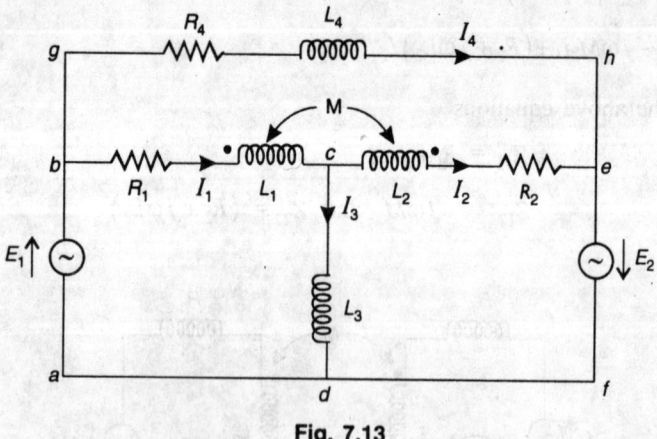

Fig. 7.13

(a) Branch currents

For loop $abcda$,

$$E_1 - I_1 R_1 - j\omega L_1 I_1 - j\omega L_3 I_3 + j\omega M I_2 = 0$$

i.e. $\qquad E_1 - (R_1 + j\omega L_1)I_1 + j\omega M\,I_2 - j\omega L_3\,I_3 = 0 \qquad\qquad (1)$

For loop *dcefd*

$$+ j\omega L_3 I_3 - j\omega L_2 I_2 - R_2 I_2 + E_2 + j\omega\,MI_1 = 0$$

i.e. $\qquad E_2 + j\omega MI_1 - (R_2 + j\omega L_2)I_2 + j\omega L_3 I_3 = 0 \qquad\qquad (2)$

For loop *bghecb*

$$-R_4 I_4 - j\omega L_4\,I_4 + R_2\,L_2 + j\omega L_2\,I_2 + j\omega L_1 I_1 + I_1 R_1 - J\omega MI_1 - J\omega MI_2 = 0$$

i.e. $(R_1 + j\omega L_1 - j\omega M)I_1 + (R_2 + j\omega L_2 - j\omega M)I_2 - (R_4 + j\omega L_4)I_4 = 0 \quad (3)$

(b) Loop currents

The loop currents are assumed as shown in Fig. 7.14.

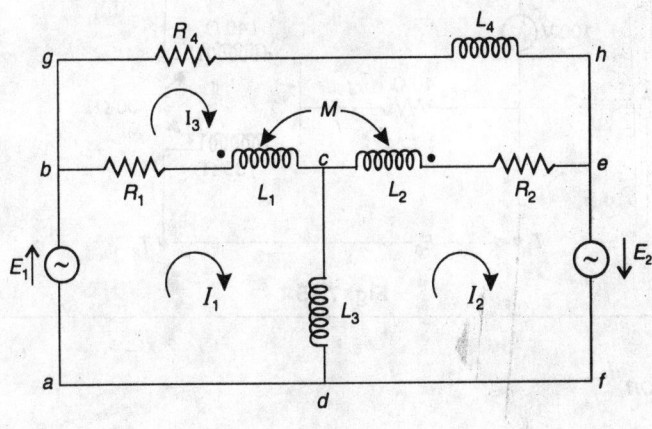

Fig. 7.14

The loop equations are
For loop 1

$$\{R_1 + j\omega(L_1 + L_3)\}I_1 - j\omega L_3 I_2 - j\omega M(I_2 - I_3) - (R_1 + j\omega L_1)I_3 = E_1$$

i.e. $\{R_1 + j\omega(L_1 + L_3)\}I_1 - (j\omega L_3 + j\omega M)I_2 - (R_1 + j\omega L_1 - j\omega M)I_3 = E_1$

$$(1)$$

For loop 2

$$-j\omega L_3 I_1 + j\omega M(I_3 - I_1) + \{R_2 + j\omega(L_2 + L_3)\}I_2 - (R_2 + j\omega L_2)I_3 = E_2$$

i.e. $\qquad -(j\omega L_3 + j\omega M)I_1 + \{R_2 + j\omega(L_2 + L_3)\}I_2$

$$-(R_2 + j\omega L_2 - j\omega M)I_3 = E_2 \qquad (2)$$

For loop 3

$$-(R_1 + j\omega L_1)I_1 - (R_2 + j\omega L_2)I_2 + \{R_1 + R_2 + R_4 +$$

$$j\omega(L_1 + L_2 + l_4)\}I_3 = 0 \qquad (3)$$

7.9 Obtain Thevenin's equivalent circuit for the circuit shown in Fig. 7.15 as seen from the terminals T_1 and T_2.

(Gulbarga University)

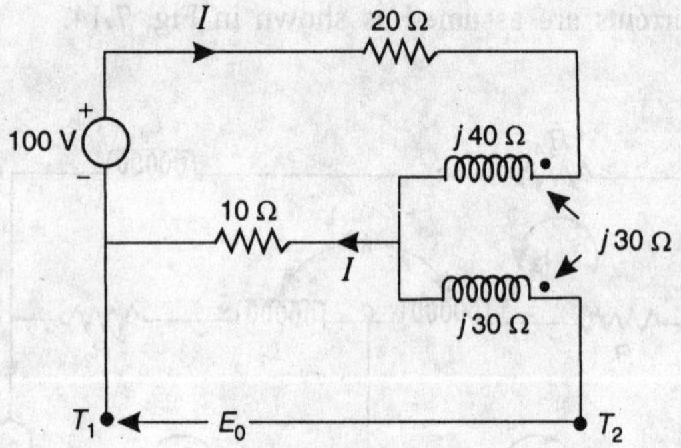

Fig. 7.15

Solution

To find E_0

Let I be the current flowing in the circuit, No current flows through $j\,30\,\Omega$.

$$\therefore \qquad I = \frac{100}{30 + j40} = 2\angle -53.13° \, \text{A}$$

$$\therefore \quad E_0 + 10I - j30I = 0$$

$$\therefore \qquad E_0 = (-10 + j30)I = (-10 + j30)2\angle -53.13°$$

$$= 63.24 \angle -124.7° \, \text{V}$$

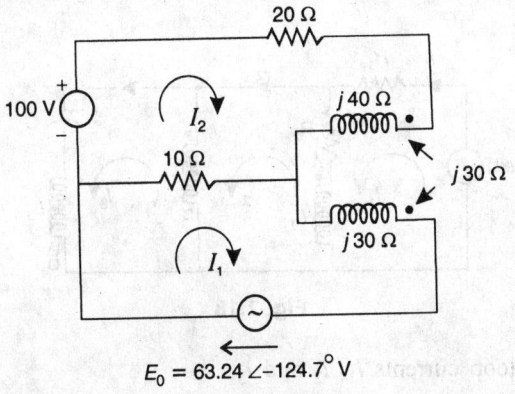

Fig. 7.16

The loop equations are: (from Fig. 7.26)

$$(10+j30)I_1-(10+j30)I_2 = 63.24\angle-124.7° \tag{1}$$

$$-(10+j30)I_1+(30+j40)I_2 = 100 \tag{2}$$

Solving (1) and (2), we get

$$I_1 = 4.4715\angle-169.7°$$

$$Z_0 = \frac{E_0}{I_1}=\frac{63.24\angle-124.7°}{4.4715\angle-169.7°} =14.1429\angle-45°\,\Omega$$

The Thevenin's equivalent circuit is shown in Fig. 7.17.

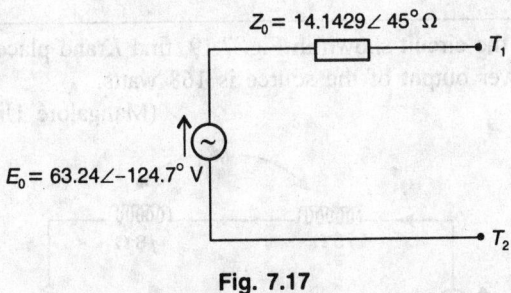

Fig. 7.17

7.10 For the coupled circuit shown in Fig. 7.18, there is coupling between every two coils. Taking M_{12}, M_{23} and M_{31} as mutual inductances, write the circuit equations using KVL. Δ, • and □ are used for dot conventions. $e = 141.4 \sin \omega t$. (Karnataka University)

Solution

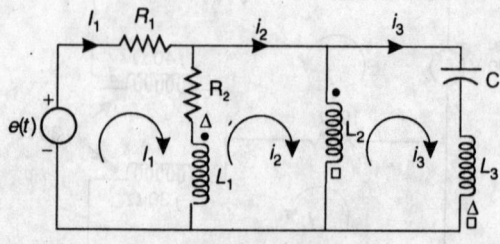

Fig. 7.18

Assume loop currents i_1, i_2 and i_3.
The loop equations are,
For loop 1

$$(R_1 + R_2 + j\omega L_1)i_1 - (R_2 + j\omega L_1)i_2 + j\omega M_{12}(i_2 - i_3) - j\omega M_{13}i_3$$
$$= 141.4 \sin \omega t$$

i.e. $(R_1 + R_2 + j\omega L_1)i_1 - (R_2 + j\omega L_1 - j\omega M_{12})i_2 - (j\omega M_{12} + j\omega M_{13})i_3$
$$= 141.4 \sin \omega t \tag{1}$$

For loop 2
$$-(R_2 + j\omega L_1 - j\omega M_{12})i_1 + (R_2 + j\omega L_1 + j\omega L_2)i_2 - (j\omega L_2 - j\omega M_{23})i_3 = 0 \tag{2}$$

For loop 3
$$-(j\omega M_{12} + j\omega M_{13})i_1 - (j\omega L_2 - j\omega M_{23})i_2 + (j\omega L_2 + j\omega L_3 - j/\omega C)i_3 = 0 \tag{3}$$

7.11 For the circuit shown in Fig. 7.19, find K and place the dots so that power output of the source is 168 watts.

(Mangalore University)

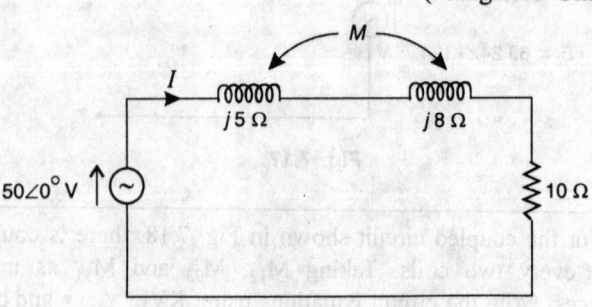

Fig. 7.19

Solution: Only 10 Ω consumes power.

$$\therefore \quad |I| = \sqrt{\frac{168}{10}} = 4.1\text{A}$$

$$P = EI\cos\phi$$

$$\therefore \quad \cos\phi = \frac{168}{50\times4.1} = 0.8195 \text{ lagging}$$

$$\phi = 35° = \tan^{-1}\frac{X}{R} = \tan^{-1}\frac{X}{10}$$

$$\therefore \quad X = 7\,\Omega = \text{total reactance.}$$

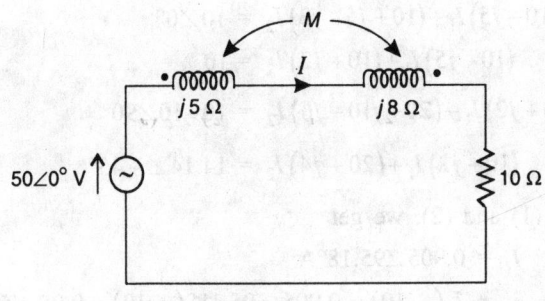

Fig. 7.19 (a)

Dots must be placed in such a way that

$$X = X_{L_1} + X_{L_2} - 2X_m \qquad \left[\because X < X_{L_1} + X_{L_2}\right]$$

The dots placed are as shown, in Fig. 7.19 (a)

$$2X_m = X_{L_1} + X_{L_2} - X = 13 - 7 = 6\,\Omega$$

$$\therefore \quad X_m = 3\,\Omega$$

$$X_m = 3 = K\sqrt{X_{L_1}X_{L_2}}$$

$$\therefore \quad K = \frac{X_m}{\sqrt{X_{L_1}X_{L_2}}} = \frac{3}{\sqrt{5\times8}} = 0.474$$

7.12 Find the voltage across $-j10\,\Omega$ in the circuit shown in Fig. 7.20. (Mysore University)

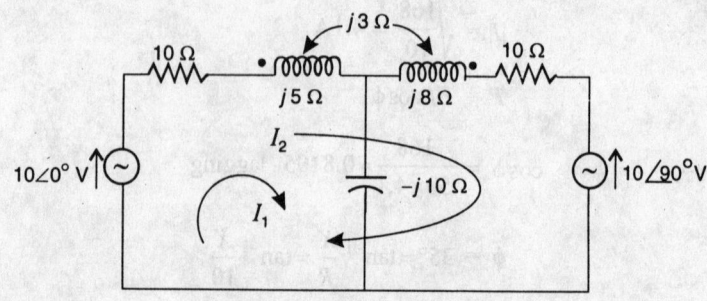

Fig. 7.20

Solution: The loop current equations are:

$$(10 - j5)I_1 + (10 + j5 - j3)I_2 = 10\angle 0°$$

i.e. $$(10 - j5)I_1 + (10 + j2)I_2 = 10 \qquad (1)$$

$$(10 + j2)I_1 + (20 + j10 - j6)I_2 = 10 - 10\angle 90°$$

i.e. $$(10 + j2)I_1 + (20 + j4)I_2 = 14.14\angle -45° \qquad (2)$$

Solving (1) and (2), we get

$$I_1 = 0.905\angle 95.18°\text{A}$$

$$V_{-j10\Omega} = I_1(-j10) = 0.905\angle 95.18°(-j10) = 9.05\angle 5.18°\text{V}$$

7.13 Replace the coupled circuit shown in Fig. 7.21 by Thevenin's equivalent circuit at the terminals T_1 and T_2.

(Mangalore University)

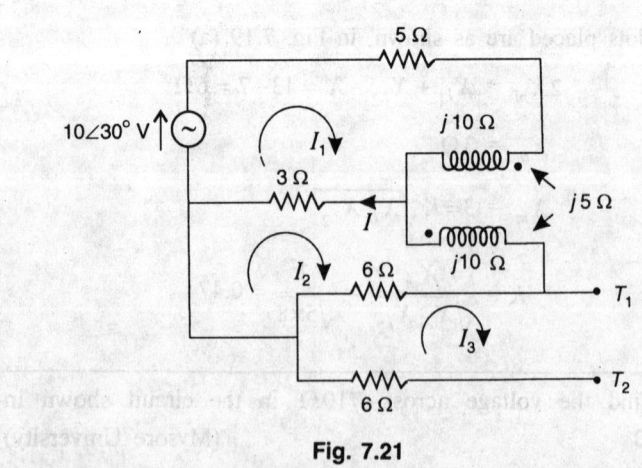

Fig. 7.21

Solution

To find E_0

Assume loop currents I_1, I_2 and I_3 as shown. As there is o.c. between T_1 and T_2, $I_3 = 0$. The loop equations for I_1 and I_2 are:

$$(8+j10)I_1 +(-3+j5)I_2 = 10\angle 30° \tag{1}$$

$$(-3+j5)I_1 +(9+j10)I_2 = 0 \tag{2}$$

Solving (1) and (2), we get

$$I_2 = 0.291\angle -122.47°A$$

$$\therefore \qquad E_0 = I_2 \times 6 = 0.291\angle -122.47°\times 6 = 1.746\angle -122.47° \text{ V}$$

The Thevenin's impedance between T_1 and T_2 is obtained using the circuit shown in Fig. 7.22.

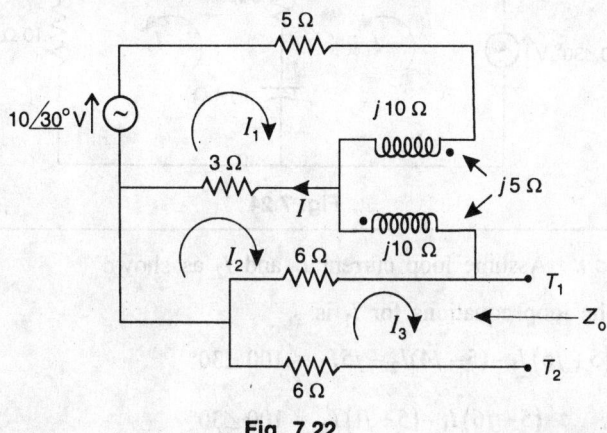

Fig. 7.22

$$Z_0 = Z_{\text{input 3}} = \frac{\Delta}{\Delta_{33}} = \frac{\begin{vmatrix} Z_{11} & Z_{12} & Z_{13} \\ Z_{21} & Z_{22} & Z_{23} \\ Z_{31} & Z_{32} & Z_{33} \end{vmatrix}}{\begin{vmatrix} Z_{11} & Z_{12} \\ Z_{21} & Z_{22} \end{vmatrix}} = \frac{\begin{vmatrix} 8+j10 & (-3+j5) & 0 \\ -3+j5 & (9+j10) & -6 \\ 0 & -6 & 12 \end{vmatrix}}{\begin{vmatrix} 8+j10 & -3+j5 \\ -3+j5 & 9+j10 \end{vmatrix}}$$

$$= 10.4\angle 8.52° \text{ } \Omega$$

The Thevenin's equivalent circuit is shown in Fig. 8.23.

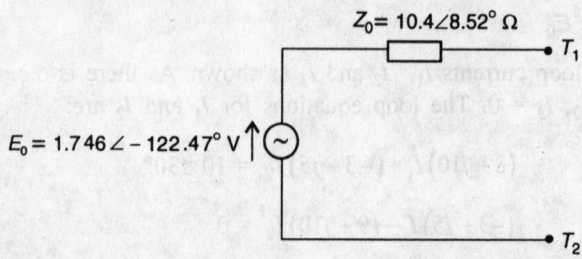

Fig. 7.23

7.14 Obtain conductively coupled equivalent circuit for the mutually coupled circuit shown in Fig. 7.24. (Kuvempu University)

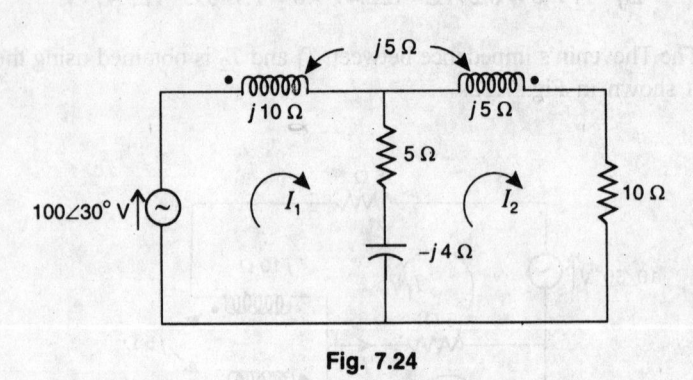

Fig. 7.24

Solution: Assume loop current I_1 and I_2 as shown

The loop equations for I_1 is

$$(5+j6)I_1-(5-j4)I_2-j5I_2 = 100\angle30°$$

i.e. $$(5+j6)I_1-(5+j1)I_2 = 100\angle30°$$

$\therefore$ $$Z_{11} = (5+j6)\Omega, \quad Z_{12}=(5+j1)\,\Omega$$

and $$Z_{22} = (15+j1)\,\Omega$$

The conductively equivalent circuit is shown in Fig. 7.25.

$$Z_m = (5+j1)\Omega$$

$$Z_1 = j5\Omega$$

and $$Z_2 = 10\Omega$$

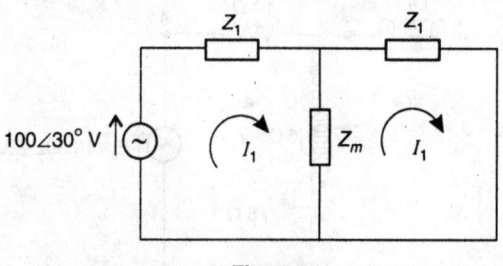

Fig. 7.25

7.15 Obtain Thevenin's equivalent circuit for the network shown in Fig. 7.26. (Karnataka University)

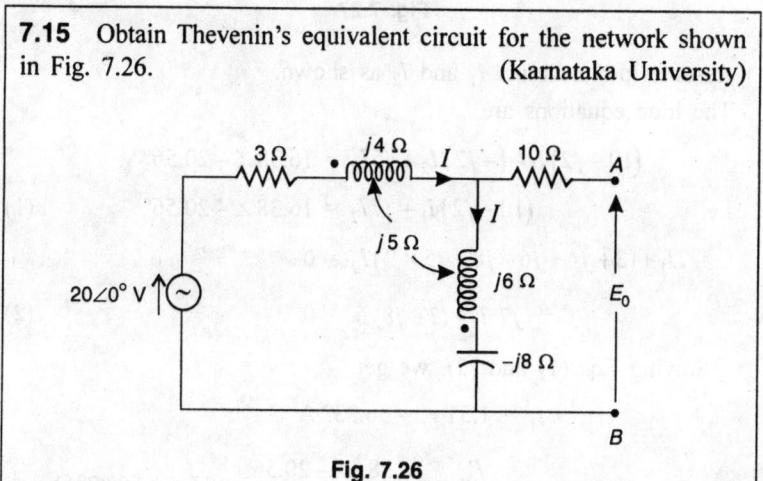

Fig. 7.26

Solution

To find E_0

Let I be the current flowing through the circuit.

$$(3+j2-j10)I = 20\angle 0°$$

$\therefore$
$$I = 2.34\angle 69.44° \text{ A}$$

$\therefore$
$$E_0 = I(j6-j8)-j5I$$

$$=I(-j7) = 2.34\angle 69.44°(-j7)=16.38\angle -20.56° \text{ V}$$

To find Z_0

Consider the circuit as in Fig. 7.27, short circuiting the voltage source.

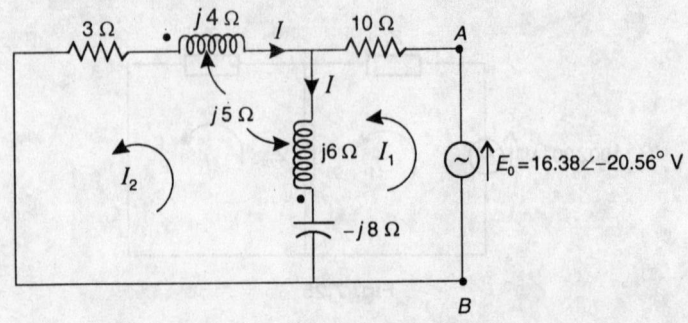

Fig. 7.27

Assume loop currents I_1 and I_2 as shown.

The loop equations are

$$(10-j2)I_1 - (-j2)I_2 + j5I_2 = 16.38\angle - 20.56° \text{ V}$$

i.e. $$(10-j2)I_1 + j7I_2 = 16.38\angle - 20.56° \qquad (1)$$

$$j7I_1 + (3+j4+j6-j8-j5\times2)I_2 = 0$$

i.e. $$j7\,I_1 + (3-j8)I_2 = 0 \qquad (2)$$

Solving Eqs (1) and (2), we get

$$I_1 = 1.31 \angle - 36.23° \text{ A}$$

$$\therefore \qquad Z_0 = \frac{E_o}{I_1} = \frac{16.38 \angle - 20.56°}{1.31 \angle - 36.23°} = 12.5\angle 15.67° \Omega$$

The Thevenin's equivalent circuit is as shown in Fig. 7.28.

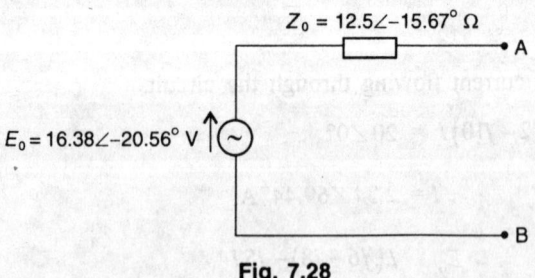

Fig. 7.28

NUMERICAL PROBLEMS

7.1 Two magnetically coupled coils have a coefficient of coupling 0.6. When they are connected on series aiding, the total inductance is 200 mH and when they are connected in series opposing, the total inductance is 100 mH. Find L_1, L_2 and M.

(Kuvempu University)

7.2 Two coils with terminals AB and CD respectively are inductively coupled. The inductance measured between the terminals AB is 300 μH and that between terminals CD is 640 μH. With B joined to C, the inductance measured between the terminals AD is 1600 μH Calculate (a) the mutual inductance between the coils and (b) the inductance between the terminals AC, when B is connected to D.

(Mysore University)

7.3 An e.m.f. of 15 V is induced in a coil, when the current in the adjacent coil varies at the rate of 100 A/Sec. What is the value of the mutual inductance between the two coils?

(Bangalore University)

7.4 Two 2000-turn air-cored coils, 200 cm long having a cross sectional area of 8 cm², are placed side by side. The mutual inductance between them is 0.5 mH. Find L_1, L_2 and K.

(Mangaloroe University)

7.5 Two similar coils, having 1600 turns each are placed side by side, so that, 60% of the flux produced by one coil links the other. If a current of 5 A, flowing in one coil, produces a flux of 0.1 mWb, find the mutual inductance between the two coils.

(Karnataka University)

7.6 Find the voltage across $(3+j4)\Omega$ in the circuit shown in Fig. 7.29.

(Gulbarga University)

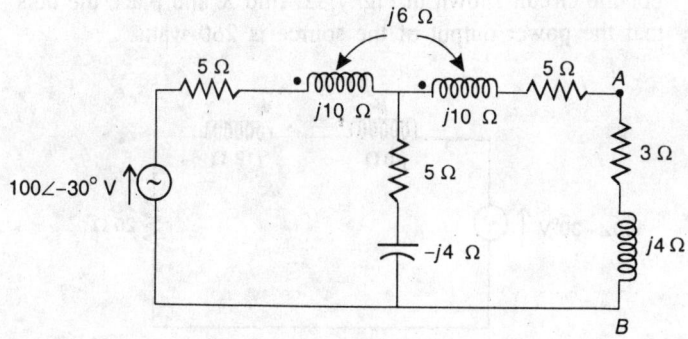

Fig. 7.29

7.7 Draw the conductively coupled equivalent circuit for the mutually coupled circuit shown in Fig. 7.30. (Mysore University)

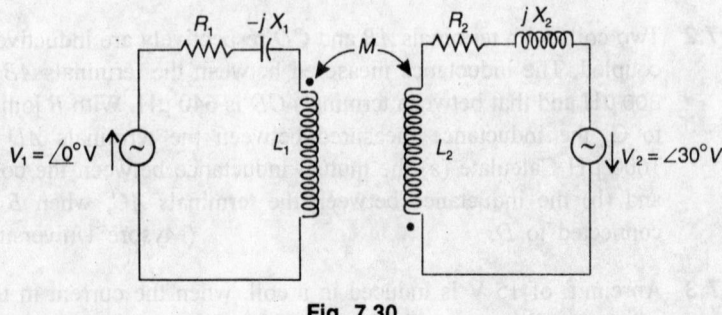

Fig. 7.30

7.8 Find the Thevenin's equivalent circuit across the terminals AB of the circuit shown in Fig. 7.31.

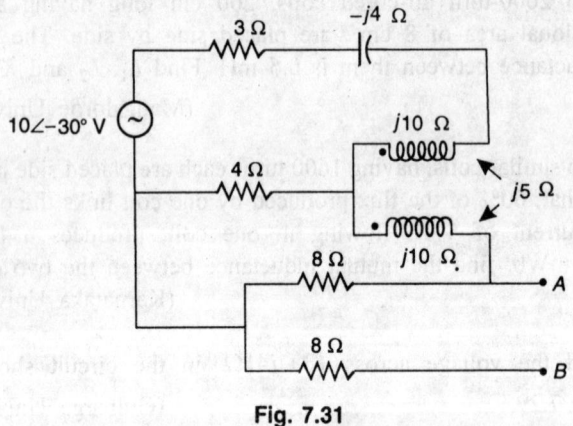

Fig. 7.31

7.9 For the circuit shown in Fig. 7.32, find K and place the dots so that the power output of the source is 260 watts.

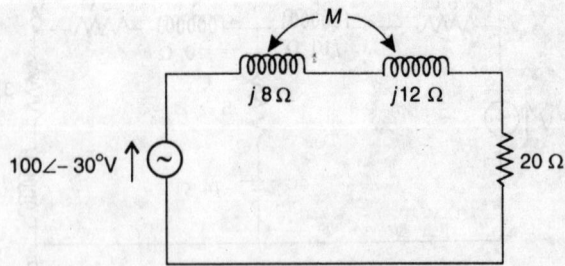

Fig. 7.32

7.10 Find the source current in the circuit shown in Fig. 7.33.

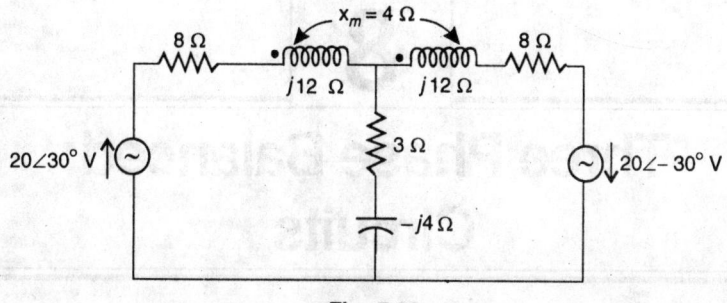

Fig. 7.33

7.11 Obtain the conductively coupled equivalent circuit for the magnetically coupled circuit shown in Fig. 7.34.

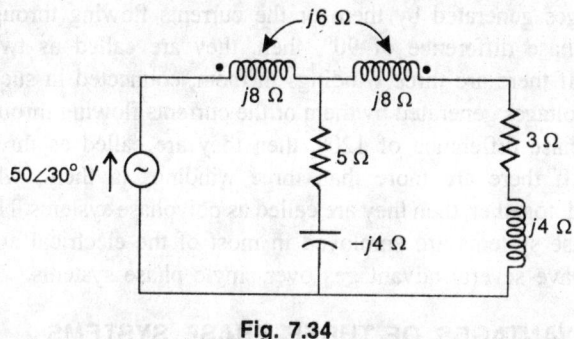

Fig. 7.34

7.12 Using Thevenin's theorem, find the current flowing through $(3+j4)\,\Omega$ in the network shown in Fig. 7.35.

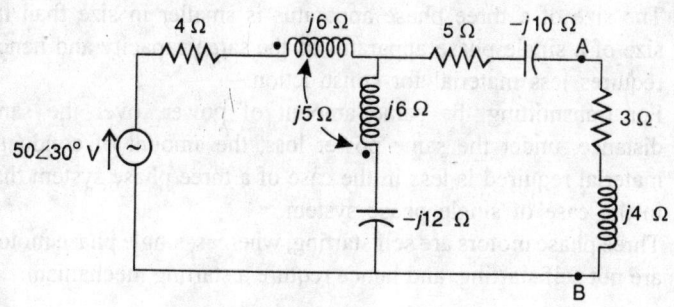

Fig. 7.35

Three Phase Balanced Circuits

8.1 INTRODUCTION

Any electrical apparatus such as a generator, motor, transformer or rectifier having only one winding is called a single phase system. If there are two windings in the above apparatus, connected in such a way that the voltages generated by them or the currents flowing through them have a phase difference of 90°, then, they are called as two phase systems. If there are three windings in them, connected in such a way that the voltages generated by them or the currents flowing through them have a phase difference of 120°, then they are called as three phase systems. If there are more than three windings in them, which are connected together, then they are called as polyphase systems. Normally, three phase systems are employed in most of the electrical apparatus, as they have several advantages over single phase systems.

8.2 ADVANTAGES OF THREE PHASE SYSTEMS

1. A three phase apparatus is more efficient than a single phase apparatus.
2. For the same capacity, a three phase apparatus costs less than a single phase apparatus.
3. The size of a three phase apparatus is smaller in size than the size of a single phase apparatus of the same capacity and hence, requires less material for construction.
4. For transmitting the same amount of power, over the same distance, under the same power loss, the amount of conductor material required is less in the case of a three phase system than in the case of single phase system.
5. Three phase motors are self starting, whereas, single phase motors are not self starting, and hence require a starting mechanism.

6. Three phase motors produce uniform torque whereas, the torque produced by single phase motors is pulsating.

7. The connection of single phase generators in parallel gives rise to harmonics, whereas, three phase generators can be conveniently connected in parallel without giving rise to the generation of harmonics.

8. In the case of a three phase star system, two different voltages can be obtained, one between lines and the other between the line and phase, whereas, only one voltage can be obtained in a single phase system.

8.3 GENERATION OF THREE PHASE VOLTAGES

The electrical machine which generates three phase voltages is called an alternator. A three phase alternator shown in Fig. 8.1, mainly consists of a stator and a rotor. The stator is stationary and the rotor rotates. The stator is cylindrical in shape and has uniform slots on its inner periphery. The conductors which form the windings of the alternator are connected together in such a way that, the e.m.f.s induced in them are additive, forming one winding. In a similar way, the remaining conductors are connected together to form two separate windings. These windings are connected together either as a star winding or as a delta winding which are discussed later in section 8.4 of this chapter.

The rotor which is the rotating part of the alternator is represented by a magnet of two poles N and S, only for the sake of explaining the generation of three phase voltages.

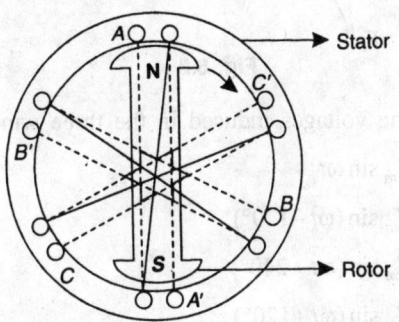

Fig. 8.1

AA', BB' and CC' are three independent coils, which are electrically displaced by 120° with respect to one another. When the rotor rotates in the clockwise direction with a particular speed N_s, the flux produced by it sweeps across the stator conductors and hence, e.m.f.s are induced in all the three phases, which have a phase displacement of 120° with respect to one another. The waveforms of the voltages generated are sinusoidal in nature and can be represented by the wave forms as shown in Fig. 8.2.

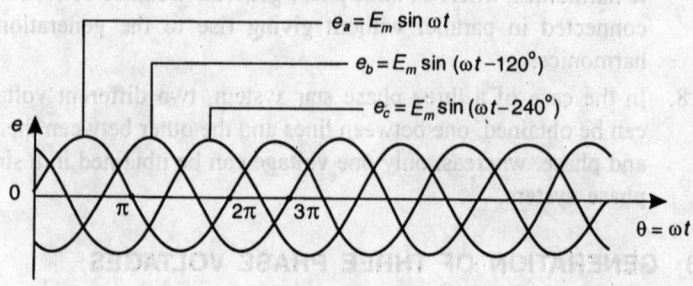

Fig. 8.2

As the number of conductors in each winding are same, the maximum values of the e.m.f.s induced in each of the windings and hence, their r.m.s. values are the same. Vectorially, the r.m.s. values of the voltages induced in the three windings are represented as shown in Fig. 8.3.

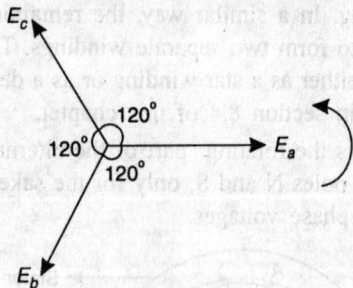

Fig. 8.3

Equations for the voltages induced in the three windings are

$$e_a = E_m \sin \omega t \qquad (8.1)$$

$$e_b = E_m \sin (\omega t - 120°) \qquad (8.2)$$

$$e_c = E_m \sin (\omega t - 240°)$$

$$= E_m \sin (\omega t + 120°) \qquad (8.3)$$

The three vectors represented in Fig. 8.3 are arbitrarily assumed to rotate in the anti-clockwise direction, which is taken as the positive direction of rotation of vectors. Hence, E_b lags E_a by 120° and E_c lags E_b by 120°.

From the wave diagram shown in Fig. 8.2, we observe that at any instant

$$e_a + e_b + e_c = 0 \qquad (8.4)$$

It can be proved mathematically also.

$$
\begin{aligned}
e_a + e_b + e_c &= E_m\left[\sin\omega t + \sin(\omega t - 120°) + \sin(\omega t + 120°)\right] \\
&= E_m[\sin\omega t + \sin\omega t \cos 120° - \cos\omega t \sin 120° \\
&\qquad + \sin\omega t \cos 120° + \cos\omega t \sin 120°] \\
&= E_m\left[\sin\omega t + 2\sin\omega t \cos 120°\right] \\
&= E_m\left[\sin\omega t + 2\sin\omega t (-1/2)\right] = 0
\end{aligned}
$$

8.4 THREE PHASE CONNECTIONS

Three are two types of three phase connections (i) Star connection (Y) and (ii) Delta connection (Δ).

(i) Star Connection (Y)

A star connection is formed, when the ends of the three coils are joined together at point n, the other three ends being free as shown in Fig. 8.4. The point n is known as the neutral point. E_{an}, E_{bn} and E_{cn} are the phase voltages and each one of them is equal to E_{ph}, E_{ab}, E_{bc} and E_{ca} are the line voltages and each of them is equal to E_l.

From the diagram shown in Fig. 8.4., we observe that the currents flowing through the lines are the same as the currents flowing through the phases.

Hence, line current = phase current

i.e. $$I_l = I_{ph} \qquad (8.5)$$

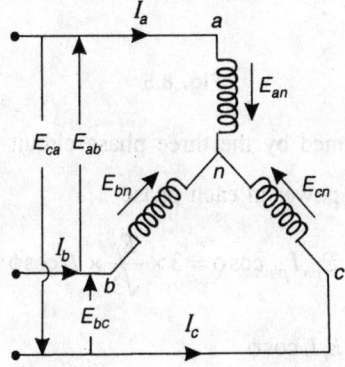

Fig. 8.4

The vector diagram of line voltages and phase voltages for the star connection is shown in Fig. 8.5.

The line voltage E_{ab} is given by

$$E_{ab} = E_{an} + E_{nb} = E_{an} - E_{bn}$$

The vector sum of E_{an} and E_{nb} gives E_{ab} as shown in the vector diagram of Fig. 8.5.

Draw a perpendicular AC on OB, $\angle AOC = 30°$

From triangle OAC

$$\cos 30° = \frac{OC}{OA} = \frac{OB/2}{OA} = \frac{E_{ab}/2}{E_{an}}$$

$\therefore$
$$E_{ab} = 2 E_{an} \cos 30° = 2 E_{an} \frac{\sqrt{3}}{2} = \sqrt{3} E_{an}$$

i.e.
$$E_l = \sqrt{3} E_{ph} \qquad (8.6)$$

$\therefore$ Line voltage = $\sqrt{3}$ phase voltage

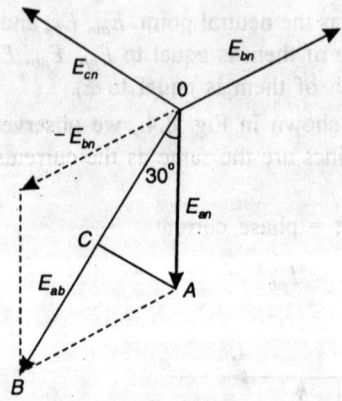

Fig. 8.5

The power consumed by the three phase circuit is given by

$$P = 3 \times \text{power in each phase}$$

$$= 3 \times E_{ph} I_{ph} \cos\phi = 3 \times \frac{E_l}{\sqrt{3}} \times I_l \cos\phi$$

$$= \sqrt{3} E_l I_l \cos\phi \qquad (8.7)$$

Note: ϕ is the angle between E_{ph} and I_{ph} and not E_l and I_l.

(ii) Delta Connection (Δ)

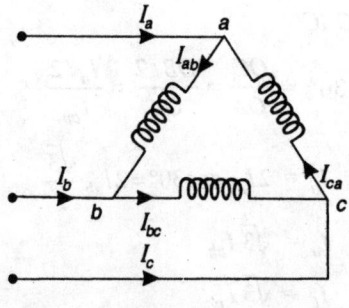

Fig. 8.6

For forming a delta connection, the three coils are connected, end to end, as shown in Fig. 8.6. I_a, I_b and I_c are the line currents and each is equal to I_l. I_{ab}, I_{bc} and I_{ca} are the phase currents and each is equal to I_{ph}. From the diagram shown in Fig. 8.6, we observe that the voltages between the lines are the same as the voltages between the phases.

Hence, $$E_l = E_{ph} \tag{8.8}$$

The vector diagram of phase currents and line currents is as shown in Fig. 8.7.

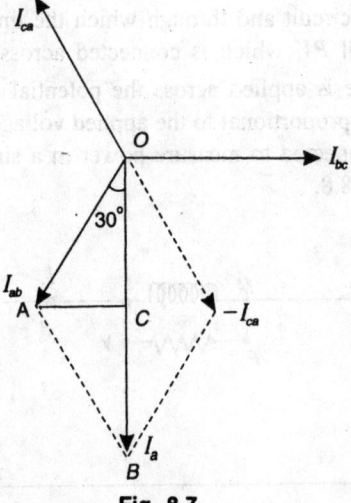

Fig. 8.7

Applying Kirchhoff's current law to point a

$$I_a + I_{ca} = I_{ab}$$

$$I_a = I_{ab} - I_{ca}$$

I_a is given by the vector sum of I_{ab} and $-I_{ca}$.
Draw AC perpendicular to OB. $\angle AOC = 30°$
From triangle OAC

$$\cos 30° = \frac{OC}{OA} = \frac{OB/2}{OA} = \frac{I_a/2}{I_{ab}}$$

$$\therefore \qquad I_a = 2I_{ab}\cos 30° = 2I_{ab}\frac{\sqrt{3}}{2}$$

$$I_a = \sqrt{3}\,I_{ab}$$

i.e. $\qquad I_l = \sqrt{3}\,I_{ph}$ \hfill (8.9)

$\therefore$ \qquad Line current $= \sqrt{3}$ phase current

The three phase power is given by

$$P = 3E_{ph}I_{ph}\cos\phi = 3 \times E_l\frac{I_l}{\sqrt{3}}\cos\phi = \sqrt{3}\,E_lI_l\cos\phi \qquad (8.10)$$

8.5 MEASUREMENT OF POWER IN A THREE PHASE CIRCUIT

Wattmeter is the instrument, which is used to measure power in an electrical circuit. It consists of (i) a current coil ML, which is connected in series with the circuit and through which the line current flows and (ii) a potential coil PV, which is connected across the circuit.

The full voltage is applied across the potential coil and it carries a very small current proportional to the applied voltage. The way in which a wattmeter is connected to measure power in a single phase circuit is as shown in Fig. 8.8.

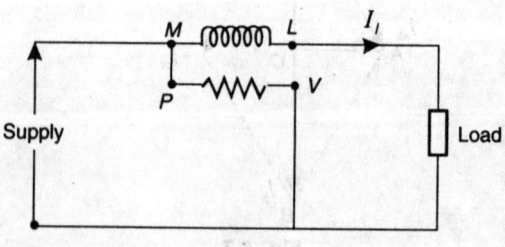

Fig. 8.8

Three single phase wattmeters may be connected in each phase and the algebraic sum of their readings gives the total power consumed by the three phase circuit. But, it can be proved that only two wattmeters are sufficient to measure power in a three phase circuit.

8.6 TWO WATTMETERS METHOD

(a) Balanced or Unbalanced Load

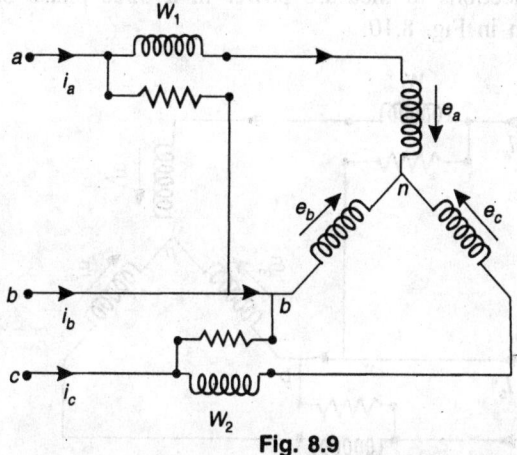

Fig. 8.9

The way in which two wattmeters are connected to measure three phase power is shown in Fig. 8.9. The instantaneous current through $W_1 = i_a$. The instantaneous voltage across the potential coil of $W_1 = e_{ab}$.

Hence at any instant, $W_1 = e_{ab} i_a$, but $e_{ab} = e_a - e_b$, where e_a and e_b are the phase voltages across phase a and phase b respectively.

$$\therefore \qquad W_1 = (e_a - e_b)i_a$$

Similarly

$$W_2 = e_{cb} i_c = (e_c - e_b)i_c$$

$$\begin{aligned} W_1 + W_2 &= (e_a - e_b)i_a + (e_c - e_b)i_c \\ &= e_a i_a + e_c i_c - e_b(i_a + i_c) \end{aligned}$$

But, we know that

$$i_a + i_b + i_c = 0$$

$$\therefore \qquad i_a + i_c = -i_b$$

$$W_1 + W_2 = e_a i_a + e_b i_b + e_c i_c = \text{Three phase power} \qquad (8.11)$$

Hence, the sum of the readings of the two wattmeters gives the three phase power.

(b) Balanced Load

The load is said to be balanced, when the impedances of the three phases are equal, otherwise, they are said to be unbalanced. The supply is said to be balanced, if the three voltages are equal and are displaced by 120° with respect to one another. When a balanced supply is given to a balanced load, the currents flowing through the three phases will be equal in magnitude and are displaced by 120° with respect to each other. The wattmeter connections to measure power in a three phase balanced circuit is shown in Fig. 8.10.

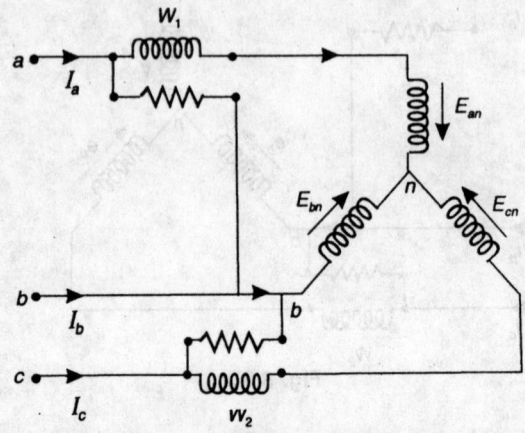

Fig. 8.10

The wattmeter reading W_1 is given by

W_1 = Voltage across its potential coil × current through its current coil × cosine of the angle between the voltage and current.

$$= E_{ab}I_a \cos \angle E_{ab} \text{ and } I_a \qquad (8.12)$$

Similarly $\qquad W_2 = E_{cb}I_c \cos \angle E_{cb} \text{ and } I_c \qquad (8.13)$

The angles between E_{ab} and I_a and E_{cb} and I_c are found by the vector diagram as shown in Fig. 8.11.

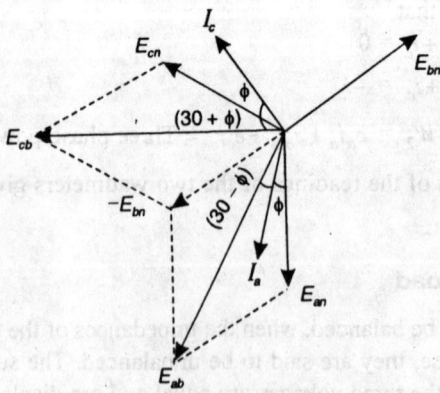

Fig. 8.11

$$E_{ab} = E_{an} + E_{nb} = E_{an} - E_{bn}$$

Assuming the load to be inductive, I_a lags E_{an} by an angle ϕ. Hence, the angle between E_{ab} and I_a is $(30 - \phi)$.

$$\therefore \quad W_1 = E_{ab} I_a \cos(30 - \phi)$$

$$= E_l I_l \cos(30 - \phi) \tag{8.14}$$

$$E_{cb} = E_{cn} + E_{nb} = E_{cn} - E_{bn}$$

I_c lags E_{cb} by an angle ϕ.

Hence, the angle between E_{cb} and I_c is $(30 + \phi)$

$$\therefore \quad W_2 = E_{cb} I_c \cos(30 + \phi)$$

$$= E_l I_l \cos(30 + \phi) \tag{8.15}$$

Adding equations (8.14) and (8.15), we get

$$W_1 + W_2 = E_l I_l \left[\cos(30 - \phi) + \cos(30 + \phi) \right]$$

$$= E_l I_l \, 2\cos 30° \cos\phi$$

$$= \sqrt{3} \, E_l I_l \cos\phi = \text{Three phase power} \tag{8.16}$$

Thus, it is shown that, two wattmeters are sufficient to measure power in a three phase circuit.

(c) Expression for p.f.

We get from Eqs (8.14) and (8.15)

$$W_1 - W_2 = E_l I_l \left[\cos(30 - \phi) - \cos(30 + \phi) \right]$$

$$= E_l I_l \, 2\sin 30° \sin\phi = E_l I_l \sin\phi \tag{8.17}$$

From Eqs (8.16) and (8.17), we get

$$\frac{W_1 - W_2}{W_1 + W_2} = \frac{\tan\phi}{\sqrt{3}} \quad \text{or} \quad \tan\phi = \frac{\sqrt{3}(W_1 - W_2)}{W_1 + W_2}$$

$$\text{or} \qquad \phi = \tan^{-1}\left\{ \frac{\sqrt{3}(W_1 - W_2)}{W_1 + W_2} \right\}$$

The p.f. is given by

$$\cos\phi = \cos\left[\tan^{-1}\left\{ \frac{\sqrt{3}(W_1 - W_2)}{W_1 + W_2} \right\} \right] \tag{8.18}$$

(d) Effect of p.f. on W_1 and W_2

(i) When p.f. = 1, $\phi = 0°$

$$W_1 = E_l I_l \cos 30° = \frac{\sqrt{3}}{2} E_l I_l$$

$$W_2 = E_l I_l \cos 30° = \frac{\sqrt{3}}{2} E_l I_l$$

The two wattmeter readings are positive and equal.

(ii) When p.f. = 0.5, $\phi = 60°$

$$W_1 = E_l I_l \cos(30-60°) = \frac{\sqrt{3}}{2} E_l I_l$$

$$W_2 = E_l I_l \cos(30+60°) = 0$$

One of the Wattmeters reads zero.

(iii) When p.f. = 0, $\phi = 90°$

$$W_1 = E_l I_l \cos(30-90°) = \frac{1}{2} E_l I_l$$

$$W_2 = E_l I_l \cos(30+90°) = -\frac{1}{2} E_l I_l$$

One of the wattmeters reads –ve. The pointer of this wattmeter kicks back and hence the reading can not be taken. Then, either the current coil connections or potential coil connections are interchanged, then the pointer moves in the forward direction and the reading can be taken. But this reading has to be considered as –ve.

From the above discussion, we can conclude that, for p.f.s lying between 0 to 0.5, one of the wattmeters, reads negative. When p.f.= 0.5, one wattmeter reads zero. When p.f. lies between 0.5 to 1.0, both wattmeter readings are +ve. When p.f. = 1, the readings of both wattmeters are equal.

WORKED EXAMPLES

8.1 When three balanced impedances are connected in star, across a 3 phase, 415 V, 50 Hz supply the line current drawn is 20 A, at a lagging p.f. of 0.4. Determine the parameters of the impedance in each phase.

As the p.f. is lagging, the impedance consists of resistance and inductance.

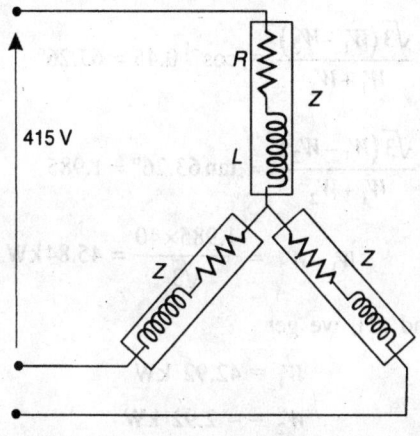

415 V

Fig. 8.12

Solution

$$V_{ph} = \frac{V_l}{\sqrt{3}} = \frac{415}{\sqrt{3}} = 239.6\,V \qquad\qquad I_l = I_{ph} = 20\,A$$

$$Z = \frac{V_{ph}}{I_{ph}} = \frac{239.6}{20} = 11.98\,\Omega$$

$$\cos\phi = \frac{R}{Z}, \quad \text{i.e.} \quad 0.4 = \frac{R}{11.98}, \quad \therefore R = 4.792\,\Omega$$

$$X_L = \sqrt{Z^2 - R^2} = \sqrt{11.98^2 - 4.792^2} = 10.98\,\Omega$$

$$L = \frac{X_L}{2\pi F} = \frac{10.98}{2\times3.14\times50} = 0.035\,H = 35\,mH$$

8.2 A 3 phase, 400 V, motor takes an input of 40 kW at 0.45 p.f. lag. Find the reading of each of the two single phase wattmeters connected to measure the input.

Solution

$$W_1 + W_2 = 40 \text{ kW} \tag{1}$$

$$\cos\phi = \cos\left\{ \tan^{-1} \frac{\sqrt{3}\left(W_1 - W_2\right)}{W_1 + W_2} \right\}$$

$$0.45 = \cos \left\{ \tan^{-1} \frac{\sqrt{3}(W_1 - W_2)}{W_1 + W_2} \right\}$$

$$\therefore \qquad \tan^{-1} \frac{\sqrt{3}(W_1 - W_2)}{W_1 + W_2} = \cos^{-1} 0.45 = 63.26°$$

$$\frac{\sqrt{3}(W_1 - W_2)}{W_1 + W_2} = \tan 63.26° = 1.985$$

$$W_1 - W_2 = \frac{1.985 \times 40}{\sqrt{3}} = 45.84 \, \text{kW} \tag{2}$$

Solving (1) and (2), we get

$$W_1 = 42.92 \ \text{kW}$$

$$W_2 = -2.92 \ \text{kW}$$

8.3 If the readings on two wattmeters in a 3 phase balanced load are 836 and 224 W, the latter reading being obtained after the reversal of the current coil connections, calculate the power and p.f. of the load.

Solution:

$$W_1 = 836 \ \text{W}$$

$$W_2 = -224 \ \text{W}$$

$$P = W_1 + W_2 = 836 - 224 \ \text{W} = 612 \ \text{W}$$

$$\text{P.f.} = \cos\phi = \cos\left\{ \tan^{-1} \frac{\sqrt{3}(W_1 - W_2)}{W_1 + W_2} \right\}$$

$$= \cos\left\{ \tan^{-1} \frac{\sqrt{3}(836 + 224)}{836 - 224} \right\} = 0.316$$

8.4 A 3 phase star connected load draws a line current of 25 A. The load kVA and kW are 20 and 16 respectively. Find the readings on each of the two wattmeters used to measure the 3 phase power.

$$W_1 + W_2 = 16 \ \text{kW} \tag{1}$$

$$\text{P.f.} = \frac{\text{kW}}{\text{kVA}} = \frac{16}{20} = 0.8$$

$$\text{p.f.} = 0.8 = \cos\left\{\tan^{-1}\frac{\sqrt{3}\left(W_1 - W_2\right)}{W_1 + W_2}\right\}$$

$$\therefore \quad \tan^{-1}\frac{\sqrt{3}\left(W_1 - W_2\right)}{W_1 + W_2} = \cos^{-1}0.8 = 36.87°$$

$$\frac{\sqrt{3}\left(W_1 - W_2\right)}{W_1 + W_2} = \tan 36.87° = 0.75$$

$$W_1 - W_2 = \frac{0.75 \times 16}{\sqrt{3}} = 6.928 \text{ kW} \tag{2}$$

Solving (1) and (2), we get

$$W_1 = 11.464 \text{ kW}$$

$$W_2 = 4.536 \text{ kW}$$

8.5 A star connected load consists of 6 Ω resistance and 8 Ω inductive reactance in each phase. A supply of 440 V at 50 Hz is applied to the load. Find the line current, power factor and power consumed by the load.

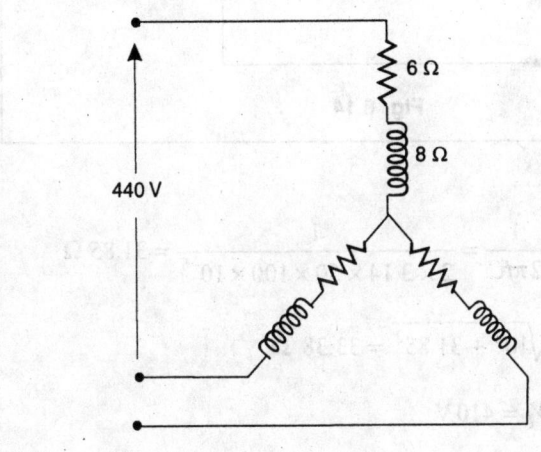

Fig. 8.13

Solution

$$V_{ph} = \frac{V_l}{\sqrt{3}} = \frac{440}{\sqrt{3}} = 254 \text{ V}$$

$$Z_{ph} = \sqrt{6^2 + 8^2} = 10 \ \Omega$$

$$I_{ph} = \frac{V_{ph}}{Z_{ph}} = \frac{254}{10} = 25.4\,\text{A} = I_l$$

$$\text{P.f.} = \frac{R}{Z} = \frac{6}{10} = 0.6 \text{ lagging}$$

$$\text{Power} = \sqrt{3}\,E_l I_l' \cos\phi = \sqrt{3} \times 440 \times 25.4 \times 0.6$$

$$= 11,614.44 \text{ watts.}$$

8.6 A delta connected load consists of a resistance of 10 Ω and a capacitance of 100 μF in each phase. A supply of 410 V at 50 Hz is applied to the load. Find the line current, power factor and power consumed by the load.

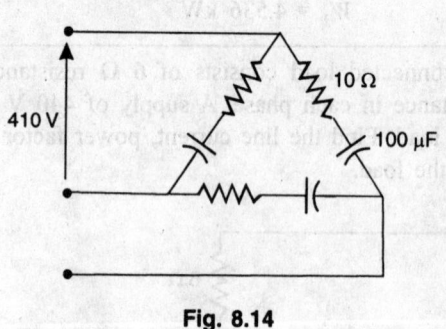

410 V

10 Ω

100 μF

Fig. 8.14

Solution

$$X_C = \frac{1}{2\pi f C} = \frac{1}{2 \times 3.14 \times 50 \times 100 \times 10^{-6}} = 31.85\,\Omega$$

$$Z_{ph} = \sqrt{10^2 + 31.85^2} = 33.38\,\Omega$$

$$V_{ph} = V_l = 410\,\text{V}$$

$$I_{ph} = \frac{V_{ph}}{Z_{ph}} = \frac{410}{33.38} = 12.28\,\text{A}$$

$$I_l = \sqrt{3}\,I_{ph} = \sqrt{3} \times 12.28 = 21.27\,\text{A}$$

$$\text{P.f.} = \frac{R}{Z} = \frac{10}{33.38} = 0.3 \text{ leading}$$

$$\text{Power} = \sqrt{3}\, E_l I_l \cos\phi = \sqrt{3} \times 410 \times 21.27 \times 0.3$$
$$= 4,531.41 \text{ watts.}$$

8.7 Two wattmeters are connected to measure the input to a 3 phase, 12 H.P. 50 Hz, induction motor, which works at a full load efficiency of 85% and a power factor of 0.8. Find the readings of the two wattmeters.

Solution: Output of I.M. = 12 H.P. = 12 × 735.5 = 8,826 watts

$$\text{Input} = \frac{8826}{0.85} = 10,383.53 \text{ watts}$$

∴ $\qquad\qquad W_1 + W_2 = 10,383.53 \text{ watts}$ $\qquad$ (1)

$$\cos\phi = \cos\left\{ \tan^{-1} \frac{\sqrt{3}\,(W_1 - W_2)}{W_1 + W_2} \right\}$$

i.e. $\qquad\qquad 0.8 = \cos\left\{ \tan^{-1} \frac{\sqrt{3}\,(W_1 - W_2)}{W_1 + W_2} \right\}$

∴ $\qquad \tan^{-1} \dfrac{\sqrt{3}\,(W_1 - W_2)}{W_1 + W_2} = \cos^{-1} 0.8 = 36.87°$

∴ $\qquad \dfrac{\sqrt{3}\,(W_1 - W_2)}{W_1 + W_2} = \tan 36.87° = 0.75$

∴ $\qquad W_1 - W_2 = \dfrac{0.75 \times 10,383.53}{\sqrt{3}} = 4,496.2 \text{ watts}$ $\qquad$ (2)

Solving (1) and (2), we get

$$W_1 = 7,439.87 \text{ watts}$$
$$W_2 = 2,943.66 \text{ watts}$$

NUMERICAL PROBLEMS

8.1 Three identical impedances are connected in delta and are supplied from a 400 V, 3 phase line. Calculate the phase and line currents and the power consumed. Each impedance is $(20 + j15)\ \Omega$.

8.2 Three identical coils having a resistance of 10 Ω and an inductance of 0.05 H each, are connected in star, across a 3 phase, 400 V, 50 Hz balanced supply. Calculate the line current and the power consumed. What will be the readings of the two wattmeters connected to measure the total power.

8.3 Three impedances, each consisting of 20 Ω resistance and 15 Ω inductive reactance in series, are connected in star, across 400 V, 3 phase supply. Calculate (i) line current, (ii) phase current, (iii) total power consumed and (iv) the p.f. of the load.

8.4 Two wattmeters connected to measure the power in a three phase balanced load read W_1 = 2000 W, W_2 = 1000 W. Calculate (i) total power and (ii) power factor. When does one of the two wattmeters read the power negative?

8.5 A balanced, 3 phase star connected load of 150 kW takes a leading current of 100 A, at a line voltage of 1100 V at 50 Hz. Find the circuit constants per phase.

8.6 When the balanced impedances are connected in delta across a 3 phase, 400 V, 50 Hz supply, the line current drawn is 20 A, at a lagging p.f. of 0.3. Determine the value of the impedance connected in each phase.

8.7 The power input to a 3 phase load connected to a three phase, 440 V, 50 Hz supply is measured by two wattmeter method. The readings are 40 kW and 10 kW. Calculate the power input, the p.f. and line current.

8.8 Each of the two wattmeters connected to measure the input to a 3 phase circuit, reads 20 kW. What does each instrument reads, when the load p.f. is 0.866 lagging, the total 3 phase power remaining unchanged?

8.9 Two wattmeters measure the total power in a 3 phase circuit and are correctly connected. One wattmeter reads 4800 W and the other reads backwards. On reversing the connections of the latter, it reads 400 watts. What is the total power and p.f.?

8.10 Two wattmeters are connected to measure the input of a 15 H.P., 50 Hz, 3 phase Induction motor at full load. The full load efficiency and p.f. are 0.9 and 0.8 lagging respectively. Find the readings of the two wattmeters.

Three Phase Unbalanced Circuits

9.1 INTRODUCTION

A three phase supply is said to be balanced, if all the line voltages or phase voltages are equal in magnitude and differ by 120°e in phase from one another. If the magnitude of any one line voltage or phase voltage is different in magnitude from others, then the supply is said to be unbalanced. Even if their magnitudes are equal, but the phase difference between one another is not 120°e, then also the supply is said to be unbalanced. Normally the supply voltage is assumed to be balanced.

A three phase load is said to be balanced, if the impedances of all the three phases are identical, both in magnitude and phase. If one of the three impedances differ either in magnitude or in phase, then the load is said to be unbalanced. Either way, whether the supply is unbalanced or the load is unbalanced or both are unbalanced, the three phase circuit is said to be unbalanced. The same methods of solution are applicable to a three phase unbalanced circuit, whether the unbalance is due to supply or load. Usually, the supply is assumed to be balanced and the load as unbalanced and the solutions are worked out.

9.2 PHASE SEQUENCE

The phase sequence of a three phase supply is the order in which the maximum values of the generated voltages occur. There can be only two phase sequences for a three phase supply (i) sequence *abc* and (ii) sequence *acb*. In some textbooks the phase sequences are referred as *ABC* or *ACB*, sometimes also as *RYB* or *RBY*. Some books refer the phase sequences as *CAB* which is nothing but *ABC*. The phase sequence *CBA* is nothing but phase sequence *ACB*.

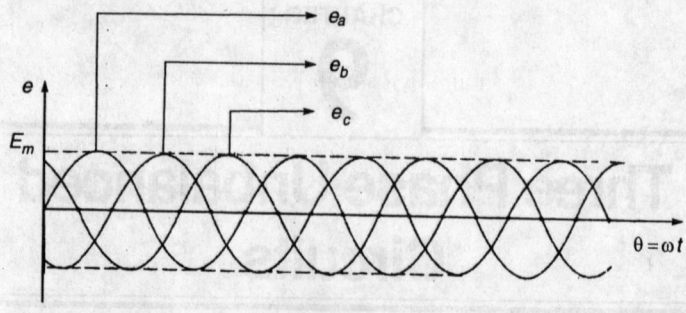

Fig. 9.1

Figure 9.1 represents the waveforms of three phase balanced voltages e_a, e_b and e_c, whose phase sequence is *abc*, and whose equations are.

$$e_a = E_m \sin \omega t$$

$$e_b = E_m \sin(\omega t - 120°)$$

and

$$e_c = E_m \sin(\omega t - 240°) = E_m \sin(\omega t + 120°)$$

The vector diagram of line and phase voltages and the triangle of voltages are as shown in Figs. 9.2 (a) and (b), respectively.

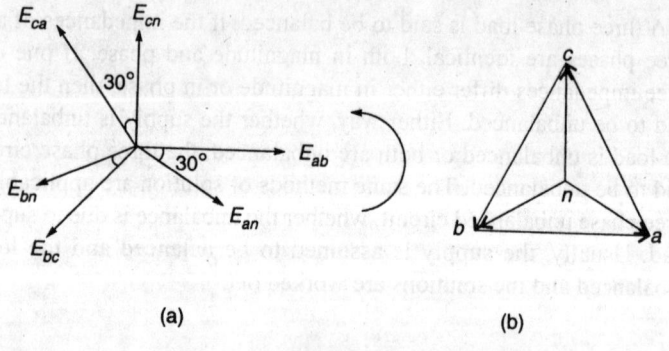

(a)   (b)

Fig. 9.2

E_{ab}, E_{bc} and E_{ca} are the r.m.s. values of line voltages.

E_{an}, E_{bn} and E_{cn} are the r.m.s. values of phase voltages.

It should be noted that for any generated voltage vector, at a particular instant, the point at arrow end is at a higher potential than the point at tail end.

As the supply is assumed to be balanced

$$|E_{ab}| = |E_{bc}| = |E_{ca}| = |E_l|$$

and $\qquad |E_{an}| = |E_{bn}| = |E_{cn}| = |E_{ph}|$

From the vector diagram or triangle of voltages, the following relations hold good for line voltages and phase voltages, in the case of a star connection, for the phase sequence *abc*.

$$E_{ab} = E_l \angle 0° \text{ V} \qquad\qquad E_{an} = \frac{E_l}{\sqrt{3}} \angle -30° \text{ V}$$

$$E_{bc} = E_l \angle -120° \text{ V} \qquad\qquad E_{bn} = \frac{E_l}{\sqrt{3}} \angle -150° \text{ V}$$

$$E_{ca} = E_l \angle +120° \text{ V} \qquad\qquad E_{cn} = \frac{E_l}{\sqrt{3}} \angle 90° \text{ V} \qquad (9.1)$$

Figure 9.3 represents the waveforms of three phase balanced voltages, whose phase sequence is *acb* and whose equations are

$$e_a = E_m \sin \omega t \qquad\qquad e_b = E_m \sin(\omega t - 240°)$$
$$= E_m \sin(\omega t + 120°) \qquad \text{and} \qquad e_c = E_m \sin(\omega t - 120°)$$

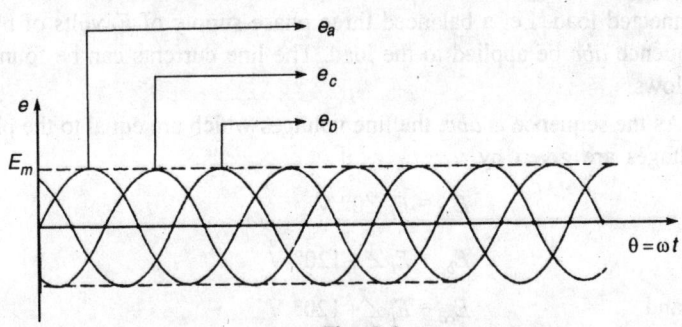

Fig. 9.3

The vector diagram of line and phase voltages and the triangle of voltages are as shown in Figs 9.4 (a) and (b) respectively.

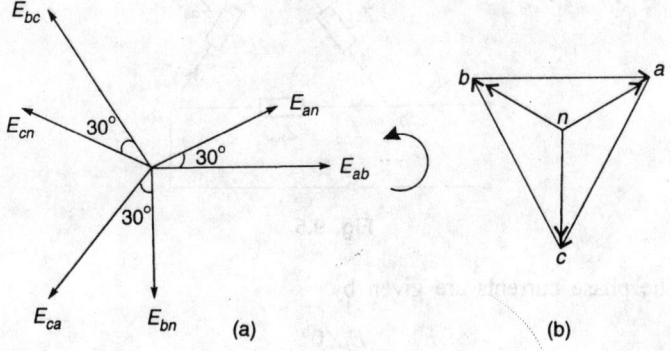

Fig. 9.4

From the above vector diagram or triangle of voltages, the following relations hold good for line voltages and phase voltages, in the case of a star connection, for the sequence *acb*.

$$E_{ab} = E_l \angle 0° \text{ V} \qquad\qquad E_{an} = \frac{E_l}{\sqrt{3}} \angle 30° \text{ V}$$

$$E_{bc} = E_l \angle + 120° \text{ V} \qquad\qquad E_{bn} = \frac{E_l}{\sqrt{3}} \angle 150° \text{ V}$$

$$E_{ca} = E_l \angle - 120° \text{ V} \qquad\qquad E_{cn} = \frac{E_l}{\sqrt{3}} \angle - 90° \text{ V} \qquad (9.2)$$

When the sequence of the supply changes, the direction of rotation of the motor changes and the currents also change.

9.3 UNBALANCED DELTA CONNECTED LOAD

Z_{ab}, Z_{bc} and Z_{ca} are the three impedances of an unbalanced delta connected load. Let a balanced three phase supply of E_l volts of phase sequence *abc* be applied to the load. The line currents can be found as follows.

As the sequence is *abc*, the line voltages which are equal to the phase voltages are given by

$$E_{ab} = E_l \angle 0° \text{ V}$$

$$E_{bc} = E_l \angle - 120° \text{ V}$$

and $\qquad\qquad E_{ca} = E_l \angle + 120° \text{ V}$

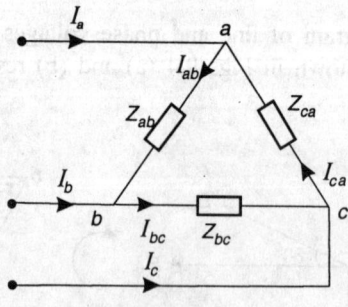

Fig. 9.5

The phase currents are given by

$$I_{ab} = \frac{E_{ab}}{Z_{ab}} = \frac{E_l \angle 0°}{Z_{ab}}$$

$$I_{bc} = \frac{E_{bc}}{Z_{bc}} = \frac{E_l\angle-120°}{Z_{bc}}$$

and
$$I_{ca} = \frac{E_{ca}}{Z_{ca}} = \frac{E_l\angle+120°}{Z_{ca}}$$

The line currents can be found by

$$I_a = I_{ab} - I_{ca}$$
$$I_b = I_{bc} - I_{ab}$$

and
$$I_c = I_{ca} - I_{bc}$$

The same procedure is followed, when the phase sequence is *acb*, where in the voltages are given by

$$E_{ab} = E_l\angle0° \text{ V}$$
$$E_{bc} = E_l\angle+120° \text{ V}$$

and
$$E_{ca} = E_l\angle-120° \text{ V}$$

WORKED EXAMPLES

9.1 A three phase, 3 wire, 400 V delta connected load has impedances $Z_{ab} = 10\angle0°\,\Omega$, $Z_{bc} = 10\angle-30°\,\Omega$ and $Z_{ca} = 10\angle30°\,\Omega$. The phase sequence is *abc*. Obtain the three line currents and draw the phasor diagram.

(Mysore University)

Solution

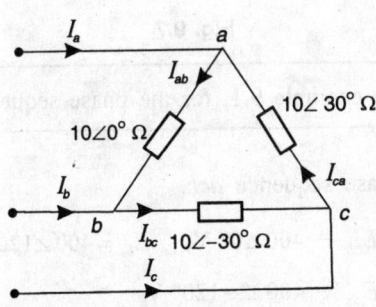

Fig. 9.6

For phase sequence *abc*

$$E_{ab} = 400\angle0° \text{ V}$$
$$E_{bc} = 400\angle-120° \text{ V}$$

and
$$E_{ca} = 400\angle+120° \text{ V}$$

The phase currents are given by

$$I_{ab} = \frac{E_{ab}}{Z_{ab}} = \frac{400\angle 0°}{10\angle 0°} = 40\,\text{A}$$

$$I_{bc} = \frac{E_{bc}}{Z_{bc}} = \frac{400\angle -120°}{10\angle -30°} = 40\angle -90°\,\text{A}$$

$$I_{ca} = \frac{E_{ca}}{Z_{ca}} = \frac{400\angle +120°}{10\angle 30°} = 40\angle 90°\,\text{A}$$

The line currents are:

$$I_a = I_{ab} - I_{ca} = 40 - 40\angle 90° = 56.57\angle -45°\,\text{A}$$
$$I_b = I_{bc} - I_{ab} = 40\angle -90° - 40\angle 0° = 56.57\angle -135°\,\text{A}$$
$$I_c = I_{ca} - I_{bc} = 40\angle 90° - 40\angle -90° = 80\angle 90°\,\text{A}$$

The phasor diagram is as shown in Fig. 9.7.

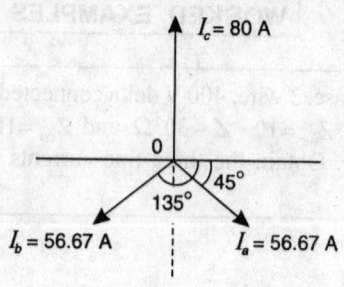

Fig. 9.7

9.2 Repeat the example 9.1, for the phase sequence *acb*.

Solution: For phase sequence *acb*.

$$E_{ab} = 400\angle 0°\,\text{V}, \quad E_{bc} = 400\angle 120°\,\text{V}$$

and

$$E_{ca} = 400\angle -120°\,\text{V}$$

The phase currents are:

$$I_{ab} = \frac{E_{ab}}{Z_{ab}} = \frac{400\angle 0°}{10\angle 0°} = 40\,\text{A}$$

$$I_{bc} = \frac{E_{bc}}{Z_{bc}} = \frac{400\angle 120°}{10\angle -30°} = 40\angle 150°\,\text{A}$$

and $$I_{ca} = \frac{E_{ca}}{Z_{ca}} = \frac{400\angle -120°}{10\angle 30°} = 40\angle -150°\,\text{A}$$

The line currents are:

$$I_a = I_{ab} - I_{ca} = 40 - 40\angle -150° = 77.27\angle 15°\,\text{A}$$

$$I_b = I_{bc} - I_{ab} = 40\angle 150° - 40 = 77.27\angle 165°\,\text{A}$$

$$I_c = I_{ca} - I_{bc} = 40\angle -150° - 40\angle 150° = -j\,40\,\text{A} = 40\angle -90°\,\text{A}$$

The phasor diagram is as shown in Fig. 9.8.

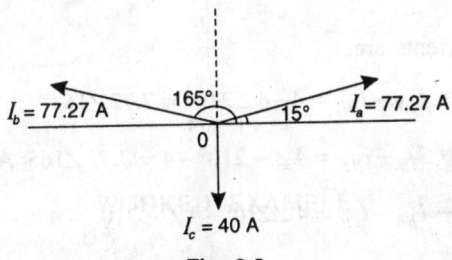

$I_b = 77.27$ A   165°   15° $I_a = 77.27$ A

0

$I_c = 40$ A

Fig. 9.8

9.3 A 400 V, 50 Hz, three phase supply is connected to a load having 100 Ω between R and Y, 318 mH between Y and B and 31.8 μF between B and R. Find (a) line currents for phase sequences, (i) *RYB* and (ii) *RBY* (b) Star connected balanced resistors for the same power. (Gulbarga University)

Solution: (i) When phase sequence is *RYB*, the line voltages are:

$$V_{RY} = 400\angle 0°\,\text{V}, \; V_{BY} = 400\angle -120°\,\text{V and}\; V_{YB} = 400\angle +120°\,\text{V}$$

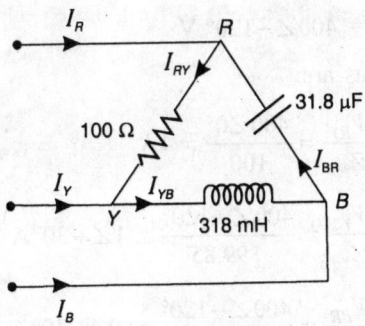

Fig. 9.9

The phase currents are:

$$I_{RY} = \frac{V_{RY}}{Z_{RY}} = \frac{400\angle 0°}{100} = 4\,A$$

$$I_{YB} = \frac{V_{YB}}{Z_{YB}} = \frac{400\angle -120°}{j2\times 3.14\times 50\times 318\times 10^{-3}} = \frac{400\angle -120°}{j99.85} \cong 4\angle -210°\,A$$

$$I_{BR} = \frac{V_{BR}}{Z_{BR}} = \frac{400\angle +120°}{-j\dfrac{1}{2\times 3.14\times 50\times 31.8\times 10^{-6}}} = \frac{400\angle +120°}{-j100.15} \cong 4\angle 210°\,A$$

The line currents are:

$$I_R = I_{RY} - I_{BR} = 4 - 4\angle 210° = 7.72\angle 15°\,A$$

$$I_Y = I_{YB} - I_{RY} = 4\angle -210° - 4 = 7.72\angle 165°\,A$$

$$I_B = I_{BR} - I_{YB} = 4\angle 210° - 4\angle -210° = 4\angle -90°\,A$$

Power is consumed only by phase *RY* and is given by

$$P = I_{RY}^2\,R = 4^2\times 100 = 1600 \text{ watts}$$

Power consumed per phase = 1600/3 watts.

If *R* is the resistance in each phase of a star connection then

$$\text{power per phase} = \frac{E_{Ph}^2}{R} = \frac{\left(400/\sqrt{3}\right)^2}{R} = \frac{1600}{3},\ \therefore\ R = 100\ \Omega$$

(ii) When the phase sequence is *RBY*, the line voltages are:

$$V_{RY} = 400\angle 0°\ V, \qquad\qquad V_{YB} = 400\angle +120°\ V$$

and $$V_{BR} = 400\angle -120°\ V$$

The phase currents are:

$$I_{RY} = \frac{V_{RY}}{Z_{RY}} = \frac{400\angle 0°}{100} = 4\,A$$

$$I_{YB} = \frac{V_{YB}}{Z_{YB}} = \frac{400\angle +120°}{j99.85} \cong 4\angle +30°\,A$$

$$I_{BR} = \frac{V_{BR}}{Z_{BR}} = \frac{400\angle -120°}{-j100.15} \cong 4\angle -30°\,A$$

The phase currents are:

$$I_R = I_{RY} - I_{BR} = 4 - 4\angle - 30° = 2.071\angle + 75° \text{ A}$$

$$I_Y = I_{YB} - I_{RY} = 4\angle + 30° - 4 = 2.071\angle 105° \text{ A}$$

$$I_B = I_{BR} - I_{YB} = 4\angle - 30° - 4\angle + 30° = 4\angle - 90° \text{ A}$$

The total power consumed in delta connection

$$= I_{RY}^2 \, R = 4^2 \times 100 = 1600 \text{ watts}$$

Power consumed per phase = 1600/3 watts

The equivalent resistance in star connection is 100 Ω as calculated earlier.

9.4 A resistor of 300 Ω and a capacitor of 8 μF are connected in series across lines *A* and *B* of a 50 Hz, 400 V, 3 phase system. Determine the voltage between the junction of the resistor and the capacitor and the line *C*. Assume the phase sequence as *ACB*.

(Kuvempu University)

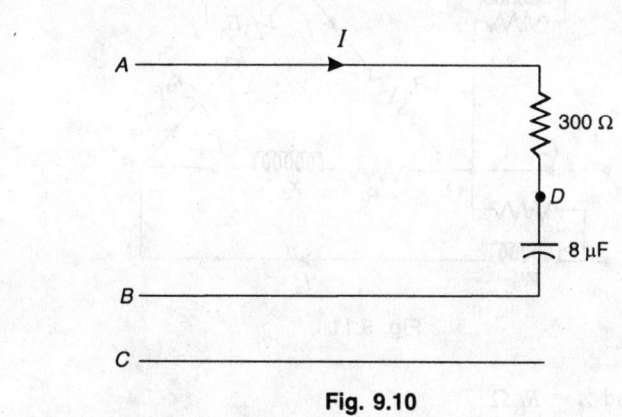

Fig. 9.10

Solution: The phase sequence is *ACB*.

The line voltages are:

$$V_{AB} = 400\angle 0° \text{ V}, \qquad V_{BC} = 400\angle + 120° \text{ V}$$

and $V_{CA} = 400\angle - 120° \text{ V}$

Let *D* be the junction point.

$$X_C = \frac{1}{2 \times 3.14 \times 50 \times 8 \times 10^{-6}} = 398 \text{ Ω}$$

$$I = \frac{V_{AB}}{300 - j398} = \frac{400\angle 0°}{498.4\angle - 53°} = 0.8\angle 53° \text{ A}$$

$$-300\,I - V_{DC} - V_{CA} = 0$$

$$\therefore \qquad -V_{DC} = 300\,I + V_{CA} = 300 \times 0.8\angle 53° + 400\angle -120°$$

$$= -55.56 - j154.74$$

$$\therefore \qquad V_{DC} = 55.56 + j154.74 = 164.41\angle 70.25°\,\text{V}$$

9.5 Three loads are delta connected to a symmetrical, 3 phase, 440 V system. Load *A* takes 25 kW at u.p.f., load *B* takes 40 kVA at 0.9 p.f. leading and load *C* takes 45 kVA at 0.7 p.f. lagging. Calculate the line currents and the readings of two wattmeters connected to measure the input, when the current coils of the wattmeters are in phases *R* and *B*. The phases sequence is *RYB*.

(Kuvempu University)

Solution: As per the given data, the elements of the delta connected network are as shown in Fig. 9.11.

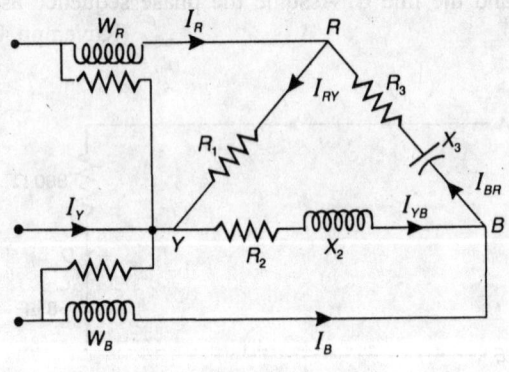

Fig. 9.11

load $A = R_1\ \Omega$

load $B = (R_3 - jX_3)\ \Omega$

load $C = (R_2 + jX_2)\ \Omega$

$$R_1 = \frac{E^2}{P} = \frac{440^2}{25 \times 10^3} = 7.744\ \Omega$$

$$I_{RY} = \frac{V_{RY}}{Z_{RY}} = \frac{440}{7.744}$$

$$= 56.82\,\text{A, in phase with } V_{RY}$$

$$I_{YB} = \frac{45 \times 10^3}{440} = 102.27 \text{ A, lagging by an angle}$$

$$\cos^{-1} 0.7 = 45.57° \text{ w.r.t. } V_{YB}$$

$$\therefore \quad Z_{YB} = \frac{440 \angle 45.57°}{102.27} = 4.3 \angle 45.57° \ \Omega = (3.01 + j3.07) \ \Omega$$

$$I_{BR} = \frac{40 \times 10^3}{440} = 90.91 \text{ A, leading by an angle}$$

$$\cos^{-1} 0.9 = 25.84° \text{ w.r.t. } V_{BR}$$

$$Z_{BR} = \frac{440 \angle -25.84°}{90.91} = 4.84 \angle -25.84° \ \Omega$$

$$= (4.36 - j2.11) \ \Omega$$

As the phase sequence is *RYB*,

$$V_{RY} = 440 \angle 0° \text{ V}, \quad V_{YB} = 440 \angle -120° \text{ V}$$

and $\qquad V_{BR} = 440 \angle +120° \text{ V}$

Then $\qquad I_{RY} = \dfrac{V_{RY}}{Z_{RY}} = \dfrac{440 \angle 0°}{7.744} = 56.82 \angle 0° \text{ A}$

$$I_{YB} = \frac{V_{YB}}{Z_{YB}} = \frac{440 \angle -120°}{4.3 \angle 45.57°} = 102.33 \angle -165.57° \text{ A}$$

$$= (-99.1 - j25.5) \text{ A}$$

$$I_{BR} = \frac{V_{BR}}{Z_{BR}} = \frac{440 \angle +120°}{4.84 \angle -25.84°} = 90.91 \angle 145.84° \text{ A}$$

$$= (-75.23 + j51.05) \text{ A}$$

The currents in the lines *R* and *B* are given by

$$I_R = I_{RY} - I_{BR} = 56.82 - (-75.23 + j51.05) = 141.57 \angle -21.14° \text{ A}$$

$$I_B = I_{BR} - I_{YB} = -75.23 + j51.05 - (-99.1 - j25.5) = 80.19 \angle 72.68° \text{ A}$$

$$W_R = V_{RY} I_R \cos \angle V_{RY} \text{ and}$$

$$I_R = 440 \times 141.57 \cos 21.14° = 58,098.75 \text{ watts}$$

and $\quad W_B = V_{BY} I_B \cos \angle V_{BY}$ and

$$I_B = 440 \times 80.19 \cos(-300° - 72.68°) = 34,423.08 \text{ watts}$$

9.6 A 3 phase voltage is applied to a load consisting of two equal resistors in series, phase B being connected to the junction. Find the ratio of the currents in the three lines and their relative phase positions. The phase sequence is RYB. (Bangalore University)

Solution

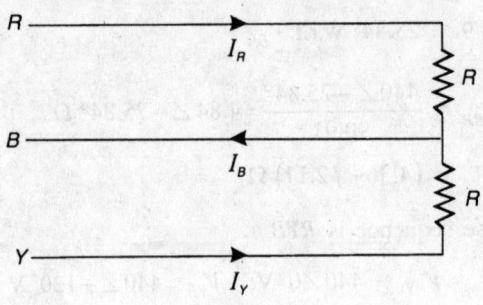

Fig. 9.12

Let the voltages be

$$V_{RY} = V \angle 0° \text{ V}, \quad V_{YB} = V \angle -120° \text{ V}$$

and $\quad V_{BR} = V \angle +120° \text{ V}$

Then $\quad I_R = \dfrac{V_{RB}}{R} = \dfrac{V \angle 300°}{R}$

$$I_Y = \dfrac{V_{YB}}{R} = \dfrac{V \angle -120°}{R}$$

and $\quad I_B = -(I_R + I_Y)$

$$= -\left(\dfrac{V \angle 300°}{R} + \dfrac{V \angle -120°}{R} \right) = 1.732 \dfrac{V \angle 90°}{R}$$

$\therefore \quad I_R : I_Y : I_B = 1 : 1 : 1.732$

If 60° is added to the phase position of I_R, then, it becomes reference. Then, the relative phase position of the currents are: $I_R \angle 0°$, $I_Y \angle -60°$ and $I_B \angle 150°$.

9.7 On a symmetrical, 3 phase system, a capacitive reactance of 8 Ω is across YB and a coil of $(R+jX)\,\Omega$ is across RY. Find R and X, such that $I_Y = 0$. The phase sequence is RYB.

(Karnataka University)

Solution

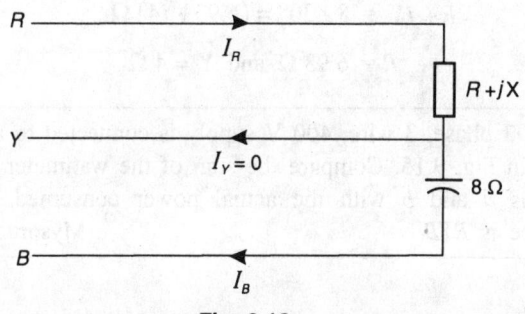

Fig. 9.13

As the sequence is RYB, the line voltages are

$$E_{RY} = E\angle 0°\,\text{V}, \quad E_{YB} = E\angle -120°\,\text{V}$$

and $\qquad E_{BR} = E\angle +120°\,\text{V}$

If $\qquad I_Y = 0$

Then $\qquad I_R = I_B$

$\qquad I_B$ leads E_{YB} by 90°

The vector diagram is as shown in Fig. 9.14.

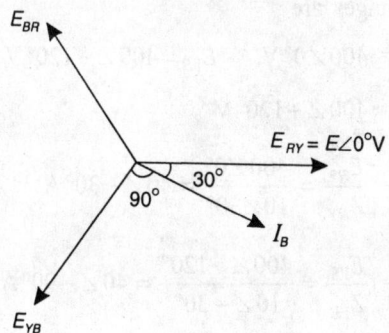

Fig. 9.14

$$I_B = \frac{E}{8}\angle -30° = \frac{E}{8\angle 30°}\,\text{A}$$

$$I_R = \frac{E_{RY}}{R+jX} = \frac{E}{R+jX}$$

But $\qquad I_B = I_R$

$\therefore \qquad \dfrac{E}{8\angle 30°} = \dfrac{E}{R+jX}$

$\therefore \qquad R+jX = 8\angle 30° = (6.93+j4)\,\Omega$

i.e. $\qquad R = 6.93\,\Omega$ and $X = 4\,\Omega$

9.8 A 3 phase, 3 wire, 400 V supply is connected to a delta load shown in Fig. 9.15. Compare the sum of the wattmeter readings in positions *a* and *b* with the actual power consumed. The phase sequence is *RYB*. (Mysore University)

Solution

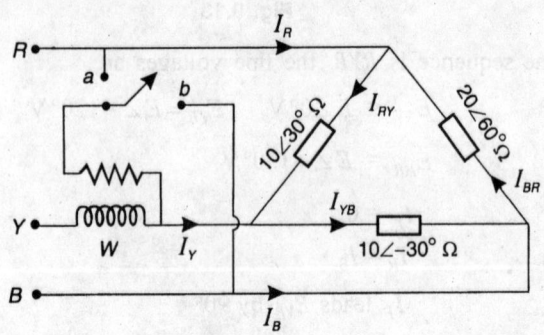

Fig. 9.15

The line voltages are

$$E_{RY} = 400\angle 0°\,\text{V}, \quad E_{YB} = 400\angle -120°\,\text{V}$$

and $\quad E_{BR} = 400\angle +120°\,\text{V}$

$$I_{RY} = \frac{E_{RY}}{Z_{RY}} = \frac{400\angle 0°}{10\angle 30°} = 40\angle -30°\,\text{A}$$

$$I_{YB} = \frac{E_{YB}}{Z_{YB}} = \frac{400\angle -120°}{10\angle -30°} = 40\angle -90°\,\text{A}$$

$$I_{BR} = \frac{E_{BR}}{Z_{BR}} = \frac{400\angle +120°}{20\angle 60°} = 20\angle 60°\,\text{A}$$

$$I_Y = I_{YB} - I_{RY} = 40\angle -90° - 40\angle -30° = 40\angle -150°\,\text{A}$$

$$W_a = E_{YR}I_Y \cos \angle E_{YR} \text{ and } I_Y$$

$$= 400 \times 40 \cos(180° + 150°) = 13856.4 \text{ W}$$

$$W_b = E_{YB}I_Y \cos \angle E_{YB} \text{ and } I_Y$$

$$= 400 \times 40 \cos(-120° + 150°) = 13,856.4 \text{ W}$$

$$\therefore \quad W_a + W_b = 27,712.8 \text{ watts}$$

Actual power consumed

$$= I_{RY}^2 R_{RY} + I_{YB}^2 R_{YB} + I_{BR}^2 R_{BR}$$

$$= 40^2 \times 10 \cos 30° + 40^2 \times 10 \cos(-30°) + 20^2 \times 20 \cos 60°$$

$$= 31,712.8 \text{ W}$$

Hence, the actual power consumed is 4,000 watts more than $(W_a + W_b)$.

9.9 Find the readings of the two wattmeters connected to the network as shown in Fig. 9.16, when it is connected to a symmetrical 400 V, 3 phase supply. The phase sequence is *RYB*.

(Kuvempu University)

Solution

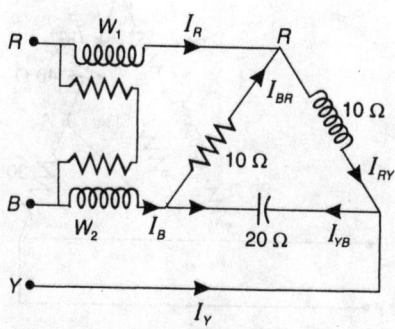

Fig. 9.16

The line voltages are

$$V_{RY} = 400 \angle 0° \text{ V}, \quad V_{YB} = 400 \angle -120° \text{ V}$$

and $$V_{BR} = 400 \angle +120° \text{ V}$$

The phase currents are given by

$$I_{RY} = \frac{V_{RY}}{Z_{RY}} = \frac{400 \angle 0°}{j10} = 40 \angle -90° \text{ A}$$

$$I_{YB} = \frac{V_{YB}}{Z_{YB}} = \frac{400 \angle -120°}{-j20} = 20 \angle -30° \text{ A}$$

$$I_{BR} = \frac{V_{BR}}{Z_{BR}} = \frac{400 \angle +120°}{10} = 40 \angle +120° \text{ A}$$

$$I_R = I_{RY} - I_{BR} = 40 \angle -90° - 40 \angle +120° = 77.27 \angle -75° \text{ A}$$

$$I_B = I_{BR} - I_{YB} = 40 \angle +120° - 20 \angle -30° = 58.185 \angle 129.9° \text{ A}$$

$$\therefore \quad W_1 = \frac{V_{RB}}{2} \times I_R \cos \angle V_{RB} \text{ and } I_R$$

$$= 200 \times 77.27 \cos(300 + 75°) = 14,927.418 \text{ watts}$$

$$\therefore \quad W_2 = \frac{V_{BR}}{2} \times I_B \cos \angle V_{BR} \text{ and } I_B$$

$$= 200 \times 58.185 \cos(120° - 129.9°) = 11,463.717 \text{ watts}$$

9.10 Find the reading of the wattmeter, when the network shown in Fig. 9.17 is connected to a balanced 400 V, 3 phase supply of phase sequence *RYB*. Neglect instrument losses. (Kuvempu University)

Solution

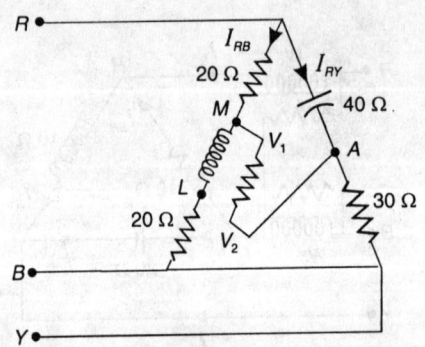

Fig. 9.17

Given: $V_{RY} = 400 \angle 0° \text{ V}, \quad V_{YB} = 400 \angle -120° \text{ V}$

and $\quad V_{BR} = 400 \angle +120° \text{ V}$

$$I_{RB} = \frac{V_{RB}}{Z_{RB}} = \frac{400 \angle 300°}{40} = 10 \angle 300° \text{ A}$$

$$I_{RY} = \frac{V_{RY}}{Z_{RY}} = \frac{400 \angle 0°}{30 - j40} = 8 \angle 53.13° \text{ A}$$

$$-V_{RM} - V_{v_1v_2} + V_{AR} = 0 \quad (\because \text{Current must flow from } V_1 \text{ to } V_2)$$

$$-20 \times 10\angle 300° - V_{v_1v_2} + (-j40)8\angle 53.13° = 0$$

$$\therefore \qquad V_{v_1v_2} = -157\angle 173.13° \text{ volts.}$$

$$W = V_{v_1v_2} I_{RB} \cos \angle V_{v_1v_2} \text{ and } I_{RB}$$

$$= -157 \times 10\cos(173.13 - 300) = 942.78 \text{ watts}$$

9.11 Find the reading of the wattmeter, when the network shown in Fig. 9.18 is connected to a balanced 400 V, 3 phase supply. The phase sequence is *RYB*. Neglect instrument losses.

(Mangalore University)

Solution

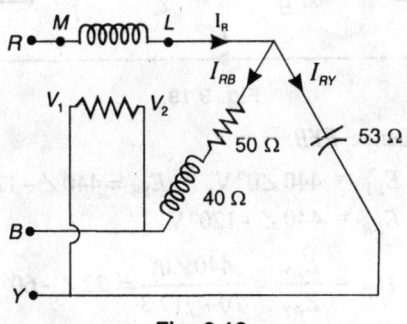

Fig. 9.18

Given: $\quad V_{RY} = 400\angle 0° \text{ V}, \quad V_{YB} = 400\angle -120° \text{ V}$

and $\qquad V_{BR} = 400\angle +120° \text{ V}$

$$I_{RB} = \frac{V_{RB}}{Z_{RB}} = \frac{400\angle 300°}{50 + j40} = 6.87\angle 261.35° \text{ A}$$

$$I_{RY} = \frac{V_{RY}}{Z_{RY}} = \frac{400\angle 0°}{-j53} = 7.55\angle 90° \text{ A}$$

$$I_R = I_{RB} + I_{RY} = 6.87\angle 261.35° + 7.55\angle 90°$$

$$= 1.662\angle 124.48° \text{ A}$$

$$W = V_{YB} I_R \cos \angle V_{YB} \text{ and } I_R$$

$$= 400 \times 1.662 \times \cos(-120 - 124.48) = -286.41 \text{ watts}$$

9.12 Three delta connected load impedances $Z_{RY} = (10 + j17.3) \,\Omega$, $Z_{YB} = (10 + j0)\,\Omega$ and $Z_{BR} = (15 - j15)\,\Omega$ are jointed to the delta connected secondary winding of a transformer having a symmetrical voltage of 440 V per phase. Find (a) the line currents to the load, (b) the current in each secondary phase of the transformer. Neglect line and transformer impedances. The phase sequence is *RYB*.

(Mangalore University)

Solution

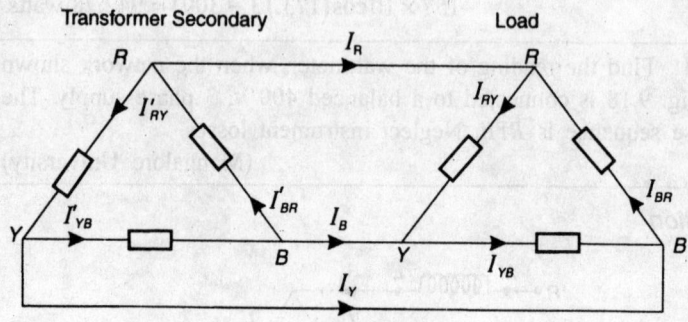

Fig. 9.19

For phase sequence *RYB*:

$$E_{RY} = 440 \angle 0° \text{V}, \quad E_{YB} = 440 \angle -120° \text{V}$$

and

$$E_{BR} = 440 \angle +120° \text{V}$$

$$I_{RY} = \frac{E_{RY}}{Z_{RY}} = \frac{440 \angle 0°}{10 + j17.3} = 22 \angle -60° \text{A}$$

$$I_{YB} = \frac{E_{YB}}{Z_{YB}} = \frac{440 \angle -120°}{10 \angle 0°} = 44 \angle -120° \text{A}$$

$$I_{BR} = \frac{E_{BR}}{Z_{BR}} = \frac{440 \angle +120°}{15 - j15} = 20.74 \angle -165° \text{A}$$

The line currents are given by

$$I_R = I_{RY} - I_{BR} = 22 \angle -60° - 20.74 \angle 165° = 39.46 \angle -38.23° \text{A}$$
$$I_B = I_{YB} - I_{RY} = 44 \angle -120° - 22 \angle -60° = 38.11 \angle -150° \text{A}$$
$$I_Y = I_{BR} - I_{YB} = 20.74 \angle 165° - 44 \angle -120° = 43.55 \angle 87.37° \text{A}$$

Secondary Phase Currents

Let, *A, B* and *C* be the tips of the currents I_{RY}, I_{YB} and I_{BR} respectively. If *G* is the centroid of the triangle *ABC*, then

$$OG = \frac{OA + OB + CC}{3}$$

$$= \frac{22 \angle -60° + 44 \angle -120° + 20.74 \angle 165°}{3} = -10.33 - j17.26$$

The currents in the secondary windings of the transformer are

$$I'_{RY} = GA = OA - OG = I_{RY} - OG = 22\angle -60° - (-10.33 - j17.26)$$
$$= 20.41\angle -5.03° \text{ A}$$

$$I'_{YB} = GB = OB - OG = I_{YB} - OG = 44\angle -120° - (-10.33 - j17.26)$$
$$= 23.89\angle -119.24° \text{ A}$$

$$I'_{BR} = GC = OC - OG = I_{BR} - OG = 20.74\angle 165° - (-10.33 - j17.26)$$
$$= 24.61\angle 113.14° \text{ A}$$

9.4 UNBALANCED STAR CONNECTED LOADS

There are two types of unbalanced star connected systems. (a) Three phase, three wire unbalanced system and (b) three phase, four wire unbalanced system.

9.5 THREE PHASE, THREE WIRE UNBALANCED SYSTEM

There are three methods of solving an unbalanced, three phase, three wire star connected system (i) Star-delta conversion method, (ii) Using Maxwell's loop current analysis and (iii) Displacement neutral method, We shall consider each one of these methods.

9.6 STAR-DELTA CONVERSION METHOD

Consider an unbalanced star connected load as shown in Fig. 9.20. Z_a, Z_b and Z_c are the star impedances. It is required to find the line currents and phase voltages, when a balanced, 3 phase supply is given to the load.

Let E_l be the applied voltage. If the phase sequence is abc, then the three line voltages are:

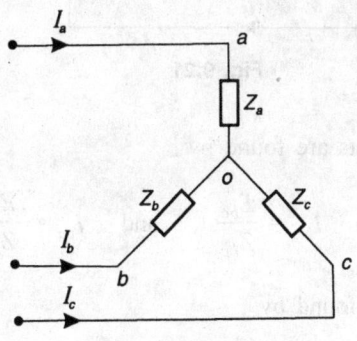

Fig. 9.20

$$E_{ab} = E_l \angle 0° \text{ V}, \quad E_{bc} = E_l \angle -120° \text{ V}$$

and $$E_{ca} = E_l = \angle +120° \text{ V}$$

If the phase sequence is *acb*, then the three line voltages are:

$$E_{ab} = E_l \angle 0° \text{ V}, \qquad E_{bc} = E_l \angle +120° \text{ V}$$

and $$E_{ca} = E_l = \angle -120° \text{ V}$$

As the phase impedances are not balanced, the phase voltages are also not balanced. In order to find the line currents and phase voltages, convert the given star load into delta load, which is as shown in Fig. 9.21. The formulae used for the conversion are:

$$Z_{ab} = Z_a + Z_b + \frac{Z_a Z_b}{Z_c}$$

$$Z_{bc} = Z_b + Z_c + \frac{Z_b Z_c}{Z_a}$$

$$Z_{ca} = Z_c + Z_a + \frac{Z_c Z_a}{Z_b}$$

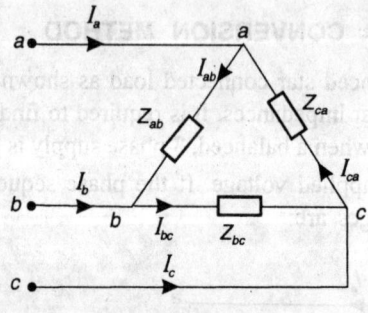

Fig. 9.21

The phase currents are found by

$$I_{ab} = \frac{E_{ab}}{Z_{ab}}, \qquad I_{bc} = \frac{E_{bc}}{Z_{bc}} \quad \text{and} \quad I_{ca} = \frac{E_{ca}}{Z_{ca}}$$

The line currents are found by

$$I_a = I_{ab} - I_{ca}, \ I_b = I_{bc} - I_{ab} \quad \text{and} \quad I_c = I_{ca} - I_{bc}$$

The phase voltages are found by

$$E_{ao} = I_a Z_a, \qquad E_{bo} = I_b Z_b$$

and $\qquad E_{co} = I_c Z_c$ (Refer Fig. 9.20)

Note that the junction point of an unbalanced star connected load is marked as o and not n, as in balanced star connected load. o is called the floating neutral, which possesses certain potential. In a balanced system, the neutral is n, which is at zero potential.

9.7 MAXWELL'S LOOP CURRENT ANALYSIS

Assume loop currents I_1 and I_2 as shown in Fig. 9.22 and write loop equations.

$$(Z_a + Z_b)I_1 - Z_b I_2 = E_{ab} \tag{1}$$

$$-Z_b I_1 + (Z_b + Z_c)I_2 = E_{bc} \tag{2}$$

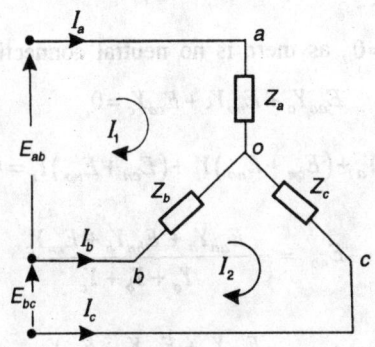

Fig. 9.22

Solving Eqs (1) and (2)

I_1 and I_2 can be found.

Then, the line currents are given by

$$I_a = I_1, \qquad I_b = I_2 - I_1 \quad \text{and} \quad I_c = -I_2$$

The phase voltages are given by

$$I_{ao} = I_a Z_a, \qquad I_{bo} = I_b Z_b \quad \text{and} \quad I_{co} = I_c Z_c$$

9.8 DISPLACEMENT NEUTRAL METHOD

As the phase voltages are not balanced, the neutral n shifts its position to o as shown in Figs 9.23 (a) and (b) for sequences abc and acb respectively.

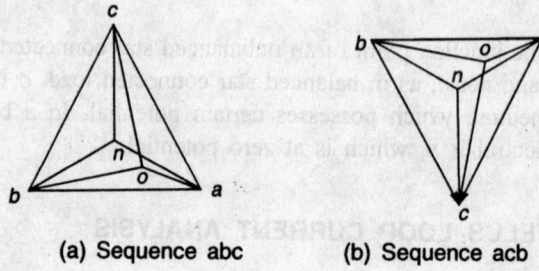

(a) Sequence abc   (b) Sequence acb

Fig. 9.23

From Fig. 9.22

$$I_a = \frac{E_{ao}}{Z_a} = E_{ao}Y_a, \qquad I_b = \frac{E_{bo}}{Z_b} = E_{bo}Y_b$$

and $\qquad I_c = \dfrac{E_{co}}{Z_c} = E_{co}Y_c$

But, $I_a + I_b + I_c = 0$, as there is no neutral connection.

$\therefore \qquad\qquad E_{ao}Y_a + E_{bo}Y_b + E_{co}Y_c = 0$

i.e. $(E_{an} + E_{no})Y_a + (E_{bn} + E_{no})Y_b + (E_{cn} + E_{no})Y_c = 0$

or $\qquad\qquad E_{no} = -\dfrac{E_{an}Y_a + E_{bn}Y_b + E_{cn}Y_c}{Y_a + Y_b + Y_c}$

$$E_{on} = \frac{E_{an}Y_a + E_{bn}Y_b + E_{cn}Y_c}{Y_a + Y_b + Y_c} \qquad (9.3)$$

E_{on} is known as the displacement neutral voltage or floating neutral voltage. It can be found, if E_{an}, E_{bn} and E_{cn} are known. E_{an}, E_{bn} and E_{cn} can be known by knowing the phase sequence and the line voltages as explained in section 9.2.

The phase voltages of the unbalanced system can be found by the equations

$$E_{ao} = E_{an} + E_{no} = E_{an} - E_{on}, \qquad E_{bo} = E_{bn} - E_{on}$$

and $\qquad E_{co} = E_{cn} - E_{on} \qquad\qquad\qquad\qquad (9.4)$

The line currents are calculated using the equations

$$I_a = E_{ao}Y_a, \qquad I_b = E_{bo}Y_b \qquad \text{and} \qquad I_c = E_{co}Y_c$$

9.9 UNBALANCED 4 WIRE STAR CONNECTED SYSTEM

In this type of star connected load as shown in Fig. 9.24, the neutral of the load is connected to the neutral of the supply. Hence, if the line voltages of a supply are balanced, the phase voltages of the load are also balanced. If the line voltages and the phase sequence are known, the phase voltages E_{an}, E_{bn} and E_{cn} can be found using the relations (9.1) or (9.2).

The unbalanced currents are given by

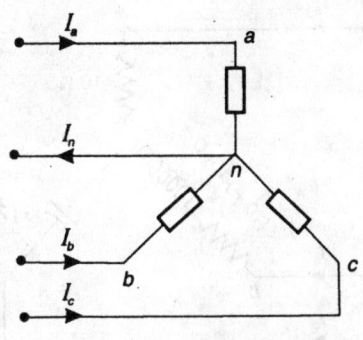

Fig. 9.24

$$I_a = \frac{E_{an}}{Z_a}, \qquad I_b = \frac{E_{bn}}{Z_b}, \qquad \text{and} \qquad I_c = \frac{E_{cn}}{Z_c}$$

The current through the neutral is given by

$$I_n = I_a + I_b + I_c$$

WORKED EXAMPLE

9.13 A balanced set of three phase voltages is connected to an unbalanced set of Y connected impedances.

$$V_{ab} = 212\angle 90° \text{ V}, \qquad V_{bc} = 212\angle -150° \text{ V}$$

and $\quad V_{ca} = 212\angle -30° \text{ V}, \qquad Z_a = (10+j0)\ \Omega$

$\quad\quad Z_b = (10+j10)\ \Omega, \qquad Z_c = (0-j20)\ \Omega$

Find (a) the line currents, (b) phase voltages, and (c) power dissipated in each phase. (Kuvempu University)

Solution

(a) Star-Delta Conversion Method

The star impedances load in Fig. 9.25 is converted into a delta impedances load, which is as shown in Fig. 9.26.

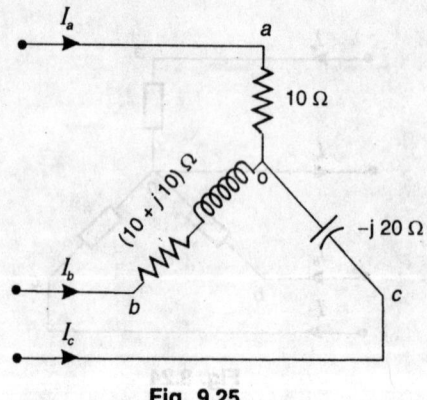

Fig. 9.25

$$Z_{ab} = Z_a + Z_b + \frac{Z_a Z_b}{Z_c}$$

$$= 10 + (10+j10) + \frac{10(10+j10)}{-j20}$$

$$= 21.2\angle 45°\ \Omega$$

$$Z_{bc} = Z_b + Z_c + \frac{Z_b Z_c}{Z_a}$$

$$= (10+j10) - j20 + \frac{(10+j10)(-j20)}{10}$$

$$= 42.4\angle -45°\ \Omega$$

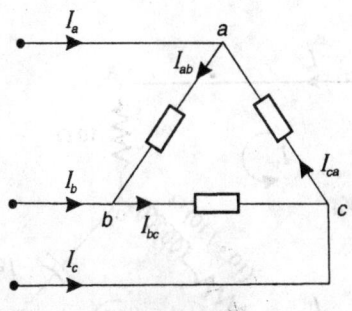

Fig. 9.26

$$Z_{ca} = Z_c + Z_a + \frac{Z_c Z_a}{Z_b} = -j20 + 10 + \frac{(-j20)(10)}{10 + j10} = 30\angle - 90° \ \Omega$$

The phase currents in the delta load are given by

$$I_{ab} = \frac{V_{ab}}{Z_{ab}} = \frac{212\angle 90°}{21.2\angle 45°} = 10\angle 45° \text{ A}$$

$$I_{bc} = \frac{V_{bc}}{Z_{bc}} = \frac{212\angle - 150°}{42.4\angle - 45°} = 5\angle - 105° \text{ A}$$

$$I_{ca} = \frac{V_{ca}}{Z_{ca}} = \frac{212\angle - 30°}{30\angle - 90°} = 7.07\angle 60° \text{ A}$$

(a) The line currents are given by

$$I_a = I_{ab} - I_{ca} = 10\angle 45° - 7.07\angle 60° = 3.66\angle 15° \text{ A}$$
$$I_b = I_{bc} - I_{ab} = 5\angle - 105° - 10\angle 45° = 14.56\angle - 125.1° \text{ A}$$
$$I_c = I_{ca} - I_{bc} = 7.07\angle 60° - 5\angle - 105° = 11.98\angle 66.2° \text{ A}$$

(b) The phase voltages are given by

$$V_{ao} = I_a Z_a = 3.66\angle 15° \times 10\angle 0° = 36.66\angle 15° \text{ V}$$

$$V_{bo} = I_b Z_b = 14.56\angle - 125.1° \times (10 + j10) = 205.91\angle - 80.1° \text{ V}$$

$$V_{co} = I_c Z_c = 11.98\angle 66.2° (-j20) = 239.6\angle - 23.8° \text{ V}$$

(c) The power dissipated in the phases are given by

$$P_a = I_a^2 R_a = 3.66^2 \times 10 = 134 \text{ watts}$$

$$P_b = I_b^2 R_b = 14.56^2 \times 10 = 2120 \text{ watts}$$

$$P_c = I_c^2 R_c = 0$$

(b) Maxwell's Loop Current Analysis

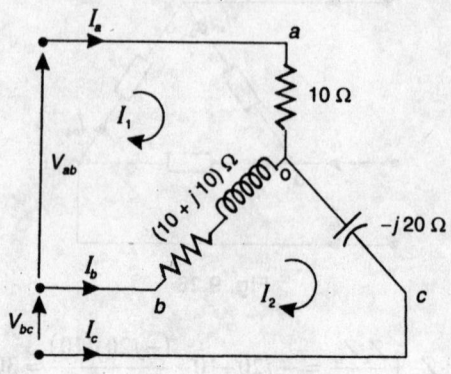

Fig. 9.27

The loop currents I_1 and I_2 are assumed as shown in Fig. 9.27. The loop equations are:

$$Z_{11}I_1 + Z_{12}I_2 = E_1$$

$$(20 + j10)I_1 - (10 + j10)I_2 = V_{ab}$$

i.e. $\quad 22.36\angle 26.57° I_1 - 14.14\angle 45° I_2 = 212\angle 90° \qquad (1)$

$$Z_{21}I_1 + Z_{22}I_2 = E_2$$

$$-14.14\angle 45° I_1 + (10 - j10)I_2 = V_{bc}$$

i.e $\quad -14.14\angle 45° I_1 + 14.14\angle -45° I_2 = 212\angle -150° \qquad (2)$

Solving equations (1) and (2), we get

$$I_1 = 3.66\angle 15° \text{A}$$

and $\qquad I_2 = -11.98\angle 66.2° \text{A}$

Hence, the line currents are

$$I_a = I_1 = 3.66\angle 15° \text{A}$$

$$I_b = I_1 - I_2 = 3.66\angle 15° + 11.98\angle 66.2°$$

$$= 14.56\angle -125.1° \text{A}$$

and $\qquad I_c = -I_2 = 11.98\angle 66.2° \text{A}$

The phase voltages and power dissipated in each phase are calculated as in method (a) discussed earlier.

(c) Displacement Neutral Method

The line voltages are rewritten, taking V_{ab}, as reference by subtracting $90°$ from the phase of these voltages.

$$V_{ab} = 212\angle 0° \text{ V}$$

$$V_{bc} = 212\angle -240° = 212\angle 120° \text{ V}$$

$$V_{ca} = 212\angle -120° \text{ V}$$

The phase sequence of these voltages is *acb*.

Then, for phase sequence *acb*

$$V_{an} = \frac{212}{\sqrt{3}}\angle +30° = 122.4\angle +30° \text{ V}$$

$$V_{bn} = \frac{212}{\sqrt{3}}\angle +150° = 122.4\angle +150° \text{ V}$$

$$V_{cn} = \frac{212}{\sqrt{3}}\angle -90° = 122.4\angle -90° \text{ V}$$

The displacement neutral voltage is given by

$$V_{on} = \frac{V_{an}Y_a + V_{bn}Y_b + V_{cn}Y_c}{Y_a + Y_b + Y_c}$$

i.e. $V_{on} = \dfrac{122.4\angle 30°\times\dfrac{1}{10} + 122.4\angle 150°\times\dfrac{1}{10+j10} + 122.4\angle -90°\times\dfrac{1}{-j20}}{\dfrac{1}{10} + \dfrac{1}{10+j10} + \dfrac{1}{-j20}}$

$$= 136.52\angle 45° \text{ V}$$

Then, the phase voltages are given by

$$V_{ao} = V_{an} - V_{on} = 122.4\angle +30° - 136.52\angle 45° = 36.6\angle -75° \text{ V}$$

$$V_{bo} = V_{bn} - V_{on} = 122.4\angle 150° - 136.52\angle 45° = 205.91\angle -170.1° \text{ V}$$

$$V_{co} = V_{cn} - V_{on} = 122.4\angle -90° - 136.52\angle 45° = 239.6\angle -113.8° \text{ V}$$

We must add $90°$ to get the actual phases of these voltages.

$$\therefore \qquad V_{ao} = 36.6\angle 15° \text{ V}$$

$$V_{bo} = 205.91\angle -80.1° \text{ V}$$

and $\qquad V_{co} = 239.6\angle -23.8° \text{ V}$

The line currents are given by

$$I_a = \frac{V_{ao}}{Z_a} = \frac{36.6\angle +15°}{10} = 3.66\angle 15° \, A$$

$$I_b = \frac{V_{bo}}{Z_b} = \frac{205.91\angle -80.1°}{10+j10} = 14.56\angle -125.1° \, A$$

$$I_c = \frac{V_{co}}{Z_c} = \frac{239.6\angle -23.8°}{-j20} = 11.98\angle 66.2° \, A$$

The power dissipated in each phase is calculated as earlier.

WORKED EXAMPLES

9.14 An unbalanced star connected load is fed from a symmetrical three phase system. The phase voltages across two of the arms of the load are $V_B = 295\angle 97.5°$ V and $V_R = 206\angle -25°$ V. Calculate the voltage between the star point and the supply neutral.

(Karnataka University)

Solution

$$V_{BO} = 295\angle 97.5° \text{ V} \quad \text{and} \quad V_{RO} = 206\angle -25° \text{ V}$$

$$V_{BR} = V_{BO} + V_{OR} = V_{BO} - V_{RO}$$

$$= 295\angle 97.5° - 206\angle -25° = 440\angle 120°$$

∴ The phase sequence is *RYB*.

∴
$$V_{RN} = \frac{440}{\sqrt{3}}\angle -30° = 254\angle -30° \text{ V}$$

∴
$$V_{ON} = V_{OR} + V_{RN} = V_{RN} - V_{RO}$$

$$= 254\angle -30° - 206\angle -25° = 51.98\angle -50.21° \text{ V}$$

9.15 Calculate the readings of the wattmeters W_1 and W_2 connected as shown in Fig. 9.28. Is $W_1 + W_2$ equal to the total power drawn by the load? Explain the error, if any. Supply voltage is 400 V. The phase sequence is *RYB*.

(Mangalore University)

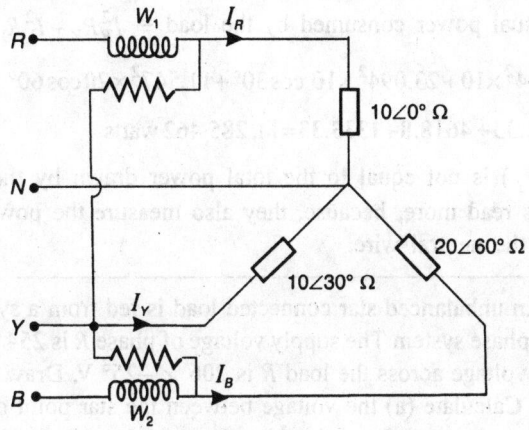

Fig. 9.28

Solution: The line voltages are

$$V_{RY} = 400\angle 0° \text{ V}, \qquad V_{YB} = 400\angle -120° \text{ V}$$

and $\qquad V_{BR} = 400\angle +120° \text{ V}$

The phase voltages are also balanced, because the neutral of the load is connected to the neutral of the supply. For sequence *RYB*.

$$\therefore \qquad V_{RN} = \frac{400}{\sqrt{3}}\angle -30° = 230.94\angle -30° \text{ V}$$

$$\therefore \qquad V_{YN} = 230.94\angle -150° \text{ V} \quad \text{and} \quad V_{BN} = 230.94\angle +90° \text{ V}$$

$$\therefore \qquad I_R = \frac{V_{RN}}{Z_R} = \frac{230.94\angle -30°}{10\angle 0°} = 23.094\angle -30° \text{ A}$$

$$I_B = \frac{V_{BN}}{Z_B} = \frac{230.94\angle +90°}{20\angle 60°} = 11.547\angle 30° \text{ A}$$

$$I_Y = \frac{V_{YN}}{Z_Y} = \frac{230.94\angle -150°}{10\angle 30°} = 23.094\angle -180° \text{ A}$$

$$W_1 = V_{RY} I_R \cos \angle V_{RY} \text{ and } I_R$$

$$= 400\times 23.094\times \cos\angle 0+30° = 8{,}000 \text{ watts}$$

$$W_2 = V_{BY} I_B \cos \angle V_{BY} \text{ and } I_B$$

$$= 400 \times 11.547 \times \cos\angle 60°-30° = 4{,}000 \text{ watts}$$

$$W_1 + W_2 = 8,000 + 4,000 = 12,000 \text{ watts}$$

The actual power consumed by the load = $I_R^2 R_R + I_Y^2 R_Y + I_B^2 R_B$

$= 23.094^2 \times 10 + 23.094^2 \times 10 \cos 30° + 11.547^2 \times 20 \cos 60°$

$= 5333.33 + 4618.8 + 1333.33 = 11,285.462 \text{ watts}$

$(W_1 + W_2)$ is not equal to the total power drawn by the load. The wattmeters read more, because, they also measure the power loss that occurs in the neutral wire.

9.16 An unbalanced star connected load is fed from a symmetrical 440 V, 3 phase system. The supply voltage of phase R is 254 $\angle -30°$ V, and the voltage across the load R is 206 $\angle -25°$ V. Draw the vector diagram. Calculate (a) the voltage between the star point of the load and the supply neutral and (b) the voltages across the loads Y and B.
(Kuvempu University)

Solution: The vector diagram of voltages is as shown in Fig. 9.29.

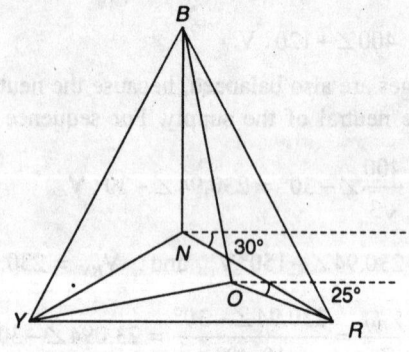

Fig. 9.29

Note: V_{RN} is given as 254 $\angle -30°$ V

Hence, the sequence is RYB

$$V_{RN} = 254 \angle -30° \text{ V}$$

$$V_{YN} = 254 \angle -150° \text{ V}$$

and $$V_{BN} = 254 \angle +90° \text{ V}$$

The voltage between the star point O and the neutral N is given by,

(a) $V_{ON} = V_{OR} + V_{RN} = V_{RN} - V_{RO}$

$$= 254 \angle -30° - 206 \angle -25° = 51.98 \angle -50.21° \text{ V}$$

(b) $V_{YO} = V_{YN} + V_{NO} = V_{YN} - V_{ON}$

$= 254\angle -150° - 51.98\angle -50.21° = 267.78\angle -161.03°\,\text{V}$

(c) $V_{BO} = V_{BN} - V_{ON}$

$= 254\angle 90° - 51.98\angle -50.21° = 295.82\angle 96.46°\,\text{V}$

9.17 Find the currents in each of the three lines shown in Fig. 9.30 and the neutral conductor. The phase sequence is *RYB*.

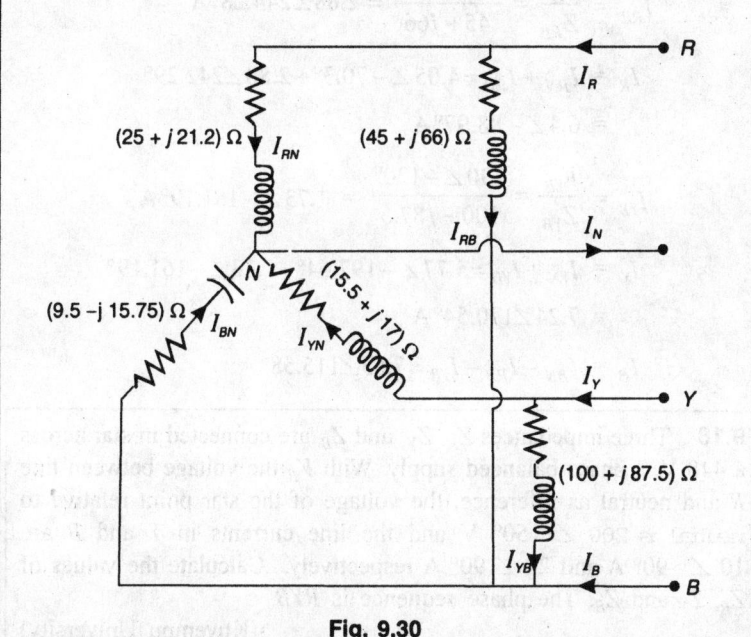

Fig. 9.30

Solution: The phase sequence is *RYB*.

$$V_{RN} = \frac{230}{\sqrt{3}}\angle -30° = 132.79\angle -30°\,\text{V}$$

$$V_{YN} = 132.79\angle -150°\,\text{V}$$

and $\quad V_{BN} = 132.79\angle +90°\,\text{V}$

$\therefore \quad I_{RN} = \dfrac{V_{RN}}{Z_R} = \dfrac{132.79\angle -30°}{25 + j21.2} = 4.05\angle -70.3°\,\text{A}$

$\quad I_{YN} = \dfrac{V_{YN}}{Z_Y} = \dfrac{132.79\angle -150°}{15.5 + j17} = 5.77\angle -197.64°\,\text{A}$

$$I_{BN} = \frac{V_{BN}}{Z_B} = \frac{132.79\angle +90°}{9.5 - j15.75} = 7.22\angle 148.9° \text{A}$$

$$\therefore \quad I_N = I_{RN} + I_{YN} + I_{BN} = 10.45\angle 170.83° \text{A}$$

The three line voltages are:

$$V_{RY} = 230\angle 0° \text{V}, \qquad V_{YB} = 230\angle -120° \text{V}$$

and $\quad V_{BR} = 230\angle +120° \text{V}$

$$\therefore \quad I_{RB} = \frac{V_{RB}}{Z_{RB}} = \frac{230\angle 300°}{45 + j66} = 2.88\angle 244.28° \text{A}$$

$$I_R = I_{RN} + I_{RB} = 4.05\angle -70.3° + 2.88\angle 242.29°$$

$$= 6.4\angle -88.97° \text{A}$$

$$I_{YB} = \frac{V_{YB}}{Z_{YB}} = \frac{230\angle -120°}{100 + j87.5} = 1.73\angle -161.19° \text{A}$$

$$\therefore \quad I_Y = I_{YN} + I_{YB} = 5.77\angle -197.64° + 1.73\angle -161.19°$$

$$= 7.24\angle 170.54° \text{A}$$

$$I_B = I_{BN} - I_{RB} - I_{YB} = 7.63\angle 115.58° \text{A}$$

9.18 Three impedances Z_R, Z_Y and Z_B are connected in star across a 440 V, 3 phase balanced supply. With V_R the voltage between line R and neutral as reference, the voltage of the star point relative to neutral is 200 ∠ 150° V and the line currents in Y and B are 10 ∠ –90° A and 20 ∠ 90° A respectively. Calculate the values of Z_R, Z_Y and Z_B. The phase sequence is RYB.

(Kuvempu University)

Solution

$$V_{RN} = \frac{440}{\sqrt{3}}\angle 0° = 254\angle 0° \text{ V}$$

$$V_{YN} = 254\angle -120° \text{ V}$$

and $\qquad V_{BN} = 254\angle +120° \text{ V}$

As there is no neutral line

$$I_R + I_Y + I_B = 0$$

$$\therefore \quad I_R = -(I_Y + I_B) = -(10\angle -90° + 20\angle 90°)$$

$$= 10\angle -90° \text{A}$$

$$V_{RO} = V_{RN} - V_{ON} = 254\angle 0° - 200\angle 150°$$
$$= 438.75\angle -13.2°\,\text{V}$$

$$\therefore \quad Z_R = \frac{V_{RO}}{I_R} = \frac{438.75\angle -13.2°}{10\angle -90°} = 43.875\angle 76.8°\ \Omega$$

$$= (10 + j\,42.72)\,\Omega$$

$$V_{YO} = V_{YN} - V_{ON} = 254\angle -120° - 200\angle 150°$$
$$= 323.32\angle -81.79°\,\text{V}$$

$$Z_Y = \frac{V_{YO}}{I_Y} = \frac{323.32\angle -81.79°}{10\angle -90°} = 32.332\angle 8.21°$$

$$= (32 + j\,4.62)\,\Omega$$

$$V_{BO} = V_{BN} - V_{ON} = 254\angle +120° - 200\angle 150°$$
$$= 128.59\angle 68.94°\,\text{V}$$

$$\therefore \quad Z_B = \frac{V_{BO}}{I_B} = \frac{128.59\angle 68.94°}{20\angle 90°} = 6.43\angle -21.06°$$

$$= (6 - j\,2.31)\,\Omega$$

9.19 Find the voltage between the neutrals of the load and the supply in the Fig. 9.31 and hence, determine the phase voltages of the star connected unbalanced load.　　(Kuvempu University)

Given:　　$V_{ao} = 250\angle 0°\,\text{V}$,　　　　$V_{bo} = 250\angle -120°\,\text{V}$

and　　　$V_{co} = 250\angle +120°\,\text{V}$

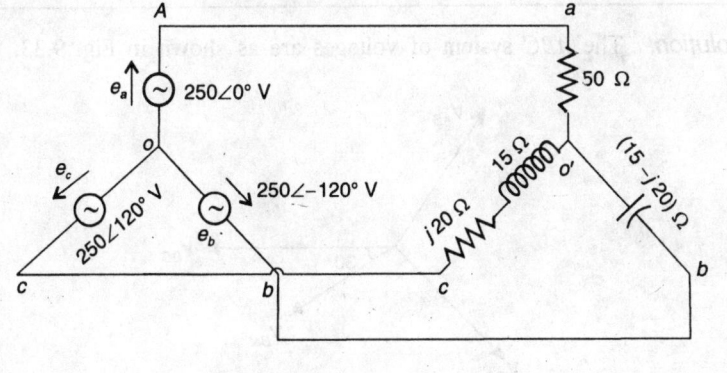

Fig. 9.31

Solution

$$V_{o'o} = \frac{V_{ao}Y_a + V_{bo}Y_b + V_{co}Y_c}{Y_a + Y_b + Y_c}$$

$$\frac{\dfrac{250\angle 0°}{50} + \dfrac{250\angle -120°}{15 - j20} + \dfrac{250\angle +120°}{15 + j20}}{\dfrac{1}{50} + \dfrac{1}{15 - j20} + \dfrac{1}{15 + j20}} = 189 \,\text{V}$$

$$V_{ao'} = V_{ao} + V_{oo'} = V_{ao} - V_{o'o} = 250 - 189 = 61 \,\text{V}$$

$$V_{bo'} = V_{bo} - V_{o'o} = 250\angle -120° - 189 = 381.4\angle -145° \,\text{V}$$

$$V_{co'} = V_{co} - V_{o'o} = 250\angle +120° - 189 = 381.4\angle 145° \,\text{V}$$

9.20 A balanced delta connected load with impedances $10\angle -36.9°\,\Omega$ and a balanced Y connected load are supplied by the same 3 phase, *ABC* system having $V_{CA} = 141.4\angle 240°\,\text{V}$. If $I_B = 40.44\angle 13.41°\,\text{A}$. Find the impedances of the Y connected load.

(Bangalore University)

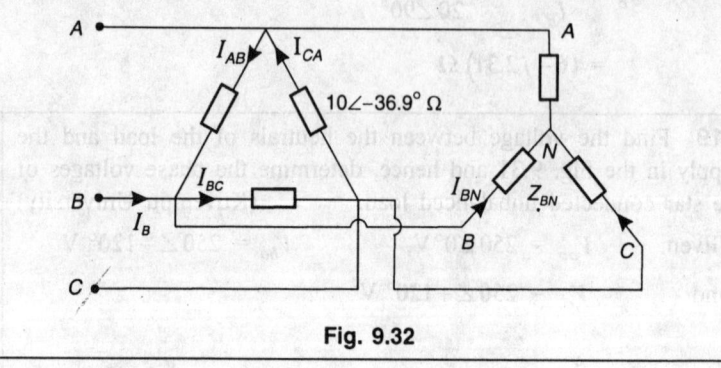

Fig. 9.32

Solution: The *ABC* system of voltages are as shown in Fig. 9.33.

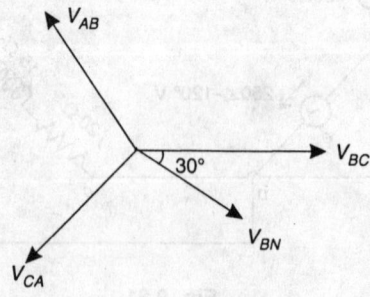

Fig. 9.33

$$V_{AB} = 141.4\angle+120° \text{ V}, \qquad V_{BC} = 141.4\angle0° \text{ V}$$

and $\qquad V_{CA} = 141.4\angle240° \text{ V}$

For *ABC* sequence, the phase voltages lag the line voltages by 30° V.

$\therefore \qquad V_{BN} = \dfrac{V_{BC}}{\sqrt{3}}\angle-30° = \dfrac{141.4}{\sqrt{3}}\angle-30° \text{ V}$

$$I_{AB} = \frac{V_{AB}}{Z_{AB}} = \frac{141.4\angle120°}{10\angle-36.9°} = 14.14\angle156.9° \text{A}$$

$$I_{BC} = \frac{V_{BC}}{Z_{BC}} = \frac{141.4\angle0°}{10\angle-36.9°} = 14.14\angle36.9° \text{A}$$

$$I_{BN} = I_B + I_{AB} - I_{BC}$$

$$= 40.44\angle13.41° + 14.14\angle156.9° - 14.14\angle36.9°$$

$$= 16.35\angle23.19° \text{A}$$

$$Z_{BN} = \frac{V_{BN}}{I_{BN}} = \frac{\dfrac{141.4}{\sqrt{3}}\angle-30°}{16.35\angle23.19°} = 5\angle-53.19° \,\Omega = (3 - j4)\,\Omega$$

9.21 A 3 phase, 3 wire *ABC* system has a line voltage of 440 V. The line currents are $I_A = 19.73\ \angle90°$ A, $I_B = 57.28\ \angle-9.9°$ A and $I_C = 57.28\ \angle189.9°$ A. Obtain the readings of the wattmeters in lines (a) *A* and *B* and (b) *B* and *C*. Take V_{BC} as the reference vector.

(Gulbarga University)

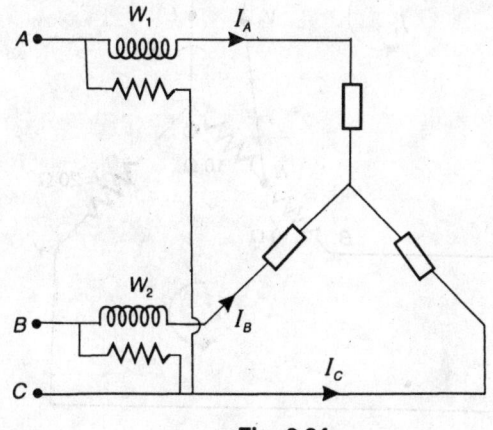

Fig. 9.34

For *ABC* system

If $\qquad V_{BC} = 440 \angle 0° \text{ V}$

then $\qquad V_{AB} = 440 \angle 120° \text{ V}$

and $\qquad V_{CA} = 440 \angle -120° \text{ V}$

(a) When the wattmeters are in lines *A* and *B* as shown in Fig. 9.34.

$$W_1 = V_{AC} I_A \cos \angle V_{AC} \text{ and } I_A$$
$$= 440 \times 19.73 \cos(60° - 90°) = 7,518 \text{ W}$$

$$W_2 = V_{BC} I_B \cos \angle V_{BC} \text{ and } I_B$$
$$= 440 \times 57.28 \cos(0° + 9.9°) = 24,828 \text{ W}$$

(b) When the wattmeters are in lines *B* and *C*,

$$W_1 = V_{BA} I_B \cos \angle V_{BA} \text{ and } I_B$$
$$= 440 \times 57.28 \cos(-60° + 9.9°) = 16,166.58 \text{ W}$$

$$W_2 = V_{CA} \times I_C \times \cos \angle V_{CA} \text{ and } I_C$$
$$= 440 \times 57.28 \cos(-120° - 189.9°) = 16,166.58 \text{ W}$$

9.22 Find the reading of the wattmeter, when the network shown in Fig. 9.35 is connected to a balanced 400 V, 3 phase, 3 wire supply. Neglect instrument losses. Phase sequence is *RYB*.

(Mysore University)

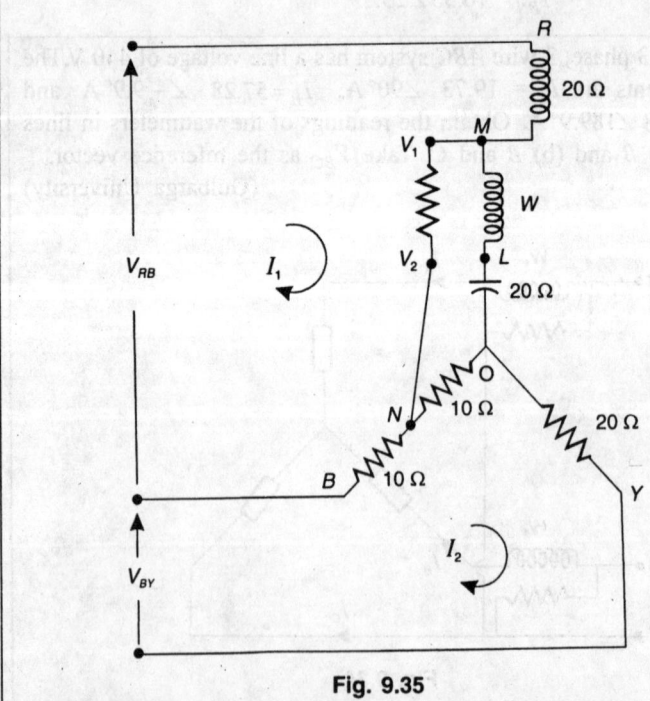

Fig. 9.35

Solution: As the instrument losses are neglected, the current through V_1V_2 and the voltage drop across *ML* are neglected.

The line voltages are:

$$V_{RY} = 440\angle 0°\,\text{V}, \qquad\qquad V_{YB} = 440\angle -120°\,\text{V}$$
and $\quad V_{BR} = 440\angle +120°\,\text{V}$

Assume loop currents I_1 and I_2 as shown.

The loop equations are:

$$20I_1 - 20I_2 = V_{RB} = 400\angle 300° \tag{1}$$
$$-20I_1 + 40I_2 = V_{BY} = 400\angle 60° \tag{2}$$

Solving (1) and (2), we get

$$I_1 = 34.641\angle -30°\,\text{A} \quad \text{and} \quad I_2 = 20\,\text{A}$$
$$\begin{aligned}
V_{v_1 v_2} = V_{MON} &= (-j20)\,I_1 + 10\,(I_1 - I_2)\\
&= -j20 \times 34.641\angle -30° + 10\,(34.641\angle -30° - 20)\\
&= 811.52\angle -107.68°\,\text{V}
\end{aligned}$$

$$\therefore \quad W = V_{v_1 v_2}\,I_1 \cos\angle V_{v_1 v_2} \text{ and } I_1$$
$$= 811.52 \times 34.641 \times \cos(-107.68° + 30°) = 5{,}998.12 \text{ watts.}$$

9.23 Find the reading on the wattmeter, when the network shown in the Fig. 9.36 is connected to a balanced, 400 V, 3 phase supply. The phase sequence is *RYB*. Neglect instrument losses.

(Karnataka University)

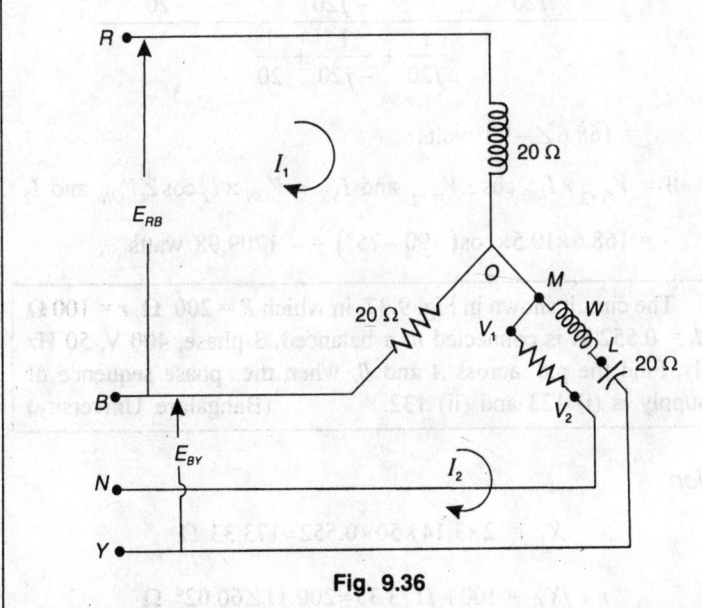

Fig. 9.36

For phase sequence *RYB*

$$E_{RY} = 400\angle 0° \text{ V}$$

$$E_{YB} = 400\angle -120° \text{ V}$$

and

$$E_{BR} = 400\angle +120° \text{ V}$$

$$E_{RN} = \frac{400}{\sqrt{3}}\angle -30° = 230.94\angle -30° \text{ V}$$

$$E_{YN} = 230.94\angle -150° \text{ V}$$

$$E_{BN} = 230.94\angle +90° \text{ V}$$

Assume loop currents I_1 and I_2 as shown.
The loop equations are:

$$(20 + j20)I_1 - 20I_2 = E_{RB} = 400\angle 300° \tag{1}$$

$$-20I_1 + (20 - j20)I_2 = E_{BY} = 400\angle 60° \tag{2}$$

Solving (1) and (2), we get, $I_2 = 10.5\angle 75° \text{ A}$

$$V_{ON} = \frac{E_{RN}Y_R + E_{YN}Y_Y + E_{BN}Y_B}{Y_R + Y_Y + Y_B}$$

$$= \frac{\dfrac{230.94\angle -30°}{j20} + \dfrac{230.94\angle -150°}{-j20} + \dfrac{230.94\angle 90°}{20}}{\dfrac{1}{j20} + \dfrac{1}{-j20} + \dfrac{1}{20}}$$

$$\simeq 168.6\angle -90° \text{ volts}$$

$$\therefore \quad W = V_{v_1 v_2} \times I_{ML} \cos\angle V_{v_1 v_2} \text{ and } I_{ML} = V_{ON} \times I_2 \cos\angle V_{ON} \text{ and } I_2$$

$$= 168.6 \times 10.5 \times \cos(-90 - 75°) = -1709.98 \text{ watts}$$

9.24 The circuit shown in Fig. 9.37, in which $R = 200 \ \Omega$, $r = 100 \ \Omega$ and $L = 0.552$ H is connected to a balanced, 3 phase, 400 V, 50 Hz supply. Find the p.d. across A and B, when the phase sequence of the supply is (i) 123 and (ii) 132. (Bangalore University)

Solution

$$X_L = 2 \times 3.14 \times 50 \times 0.552 = 173.33 \ \Omega$$

$$r + jX_L = 100 + j173.33 = 200.11\angle 60.02° \ \Omega$$

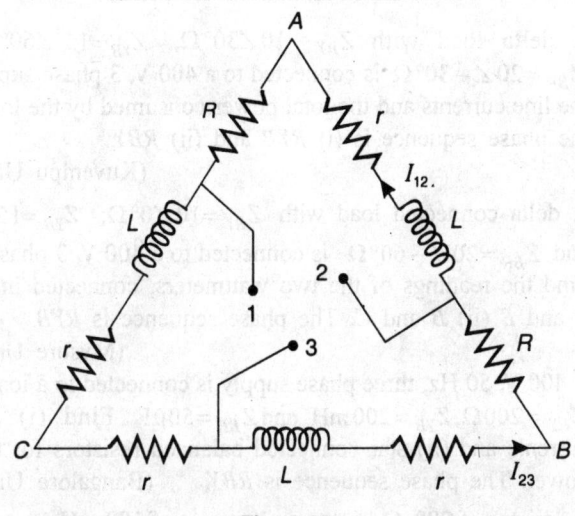

Fig. 9.37

$$(R+r)+jX_L = 300+j173.33 = 346.47\angle30.02° \ \Omega$$

For phase sequence 123

$$V_{12} = 400\angle0° \text{ V}, \qquad\qquad V_{23} = 400\angle-120° \text{ V}$$

and $\quad V_{31} = 400\angle+120° \text{ V}$

$$V_{AB} = (r+jX_L)I_{12} + RI_{23}$$

$$= 200.11\angle60.02° \times \frac{V_{12}}{(R+r)+jX_L} + 200 \times \frac{V_{23}}{(R+r)+jX_L}$$

$$= 200.11\angle60.02° \times \frac{400\angle0°}{346.47\angle30.02°} + 200 \times \frac{400\angle-120°}{346.47\angle30.02°}$$

$$= 0 \text{ volts}$$

(b) For phase sequence 132

$$V_{12} = 400\angle0° \text{ V}$$

$$V_{23} = 400\angle+120° \text{ V}$$

and $\qquad\qquad V_{31} = 400\angle-120° \text{ V}$

$$\therefore V_{AB} = 200.11\angle60.02° \times \frac{400\angle0°}{346.47\angle30.02°} + 200 \times \frac{400\angle120°}{346.47\angle30.02°}$$

$$= 400\angle60° \text{ volts}$$

NUMERICAL PROBLEMS

9.1 A delta load with $Z_{RY} = 10\angle 30°\Omega$, $Z_{YB} = 15\angle 60°\Omega$, and $Z_{BR} = 20\angle - 30°\Omega$ is connected to a 400 V, 3 phase supply. Find the line currents and the total power consumed by the load, when the phase sequence is (i) *RYB* and (ii) *RBY*.

(Kuvempu University)

9.2 A delta connected load with $Z_{RY} = 10\angle 0°\Omega$, $Z_{YB} = 15\angle 30°\Omega$, and $Z_{BR} = 20\angle - 60°\Omega$ is connected to a 400 V, 3 phase supply. Find the readings of the two wattmeters, connected in lines (i) *A* and *B* (ii) *B* and *C*. The phase sequence is *RYB*.

(Mysore University)

9.3 A 400 V, 50 Hz, three phase supply is connected to a load having $Z_{RY} = 200\Omega, Z_{YB} = 200\,\text{mH}$ and $Z_{BR} = 50\mu F$ Find (i) the line currents and (ii) star connected balanced resistors for the same power. The phase sequence is *RBY*. (Bangalore University)

9.4 A resistor of 200 Ω and an inductance of 100 mH are connected in series across lines *R* and *Y* of a 400 V, 50 Hz, 3 phase supply. Find the reading of a voltmeter connected between the junction of the resistor and inductance and the line *B*. The phase sequence is *RYB*. (Mangalore University)

9.5 Three loads are delta connected to a balanced 400 V, 3 phase supply. load *RY* takes 10 kW at 0.8 p.f. lag, Load *YB* takes 15 KVA at 0.6 p.f. lead and load *BR* takes 20 kW at u.p.f. Find the readings of the wattmeters connected to lines *R* and *Y*, to measure the input. The phase sequence is *RBY*.

(Gulbarga University)

9.6 A 3 phase voltage is applied to a load consisting of two equal impedances in series, phase *Y* being connected to the junction. Find the ratio of the currents in the three lines and their relative phase positions. The phase sequence is *RBY*.

(Karnataka University)

9.7 To a balanced, 3 phase supply, an inductive reactance of 10 Ω is connected across *YB* and an impedance *Z* across *RY*. Find *Z* such that $I_Y = 0$. The phase sequence is *RYB*.

(Bangalore University)

9.8 A three phase, 3 wire, 400 V supply is connected to a delta load as shown in Fig. 9.38. Find the reading of the wattmeter, when (i) its potential coil is connected to *A* and (ii) it is connected to *B*. Find also the actual power consumed by the load and compare it with the sum of the two wattmeter readings. The phase sequence is *RBY*. (Kuvempu University)

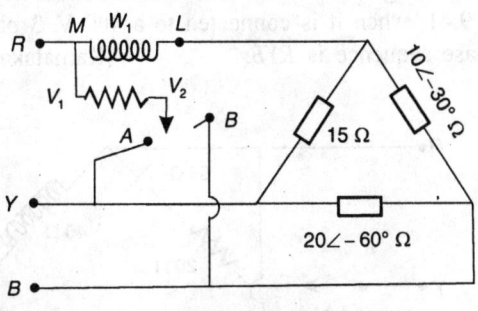

Fig. 9.38

9.9 Find the readings of the two wattmeters connected to a network as shown in Fig. 9.39, when it is connected to a 400 V, 3 phase balanced supply. The phase sequence is *RYB*.

(Mangalore University)

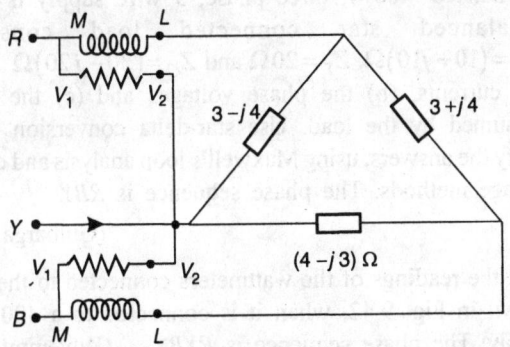

Fig. 9.39

9.10 Find the reading of the wattmeter, when the network shown in Fig. 9.40 is connected to a 400 V, 3 phase supply of phase sequence *RYB*. Neglect instrument losses. (Mysore University)

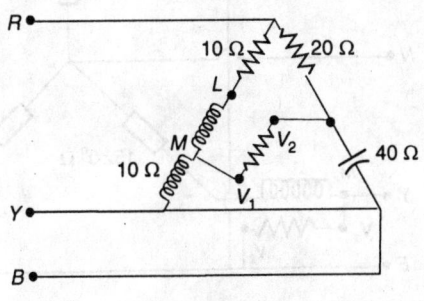

Fig. 9.40

9.11 Find the reading of the wattmeter connected to the network shown in Fig. 9.41, when it is connected to a 400 V, 3 phase, supply. The phase sequence is *RYB*.   (Karnataka University)

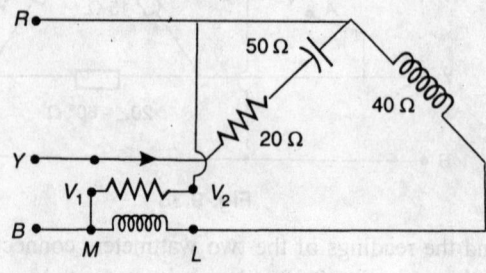

Fig. 9.41

9.12 A balanced 400 V, three phase, 3 wire supply is given to an unbalanced star connected load consisting of $Z_R = (10 + j10)\Omega$, $Z_Y = 20\Omega$ and $Z_C = (20 - j20)\Omega$ Find (a) the line currents, (b) the phase voltages and (c) the total power consumed by the load. Use star-delta conversion method and verify the answers, using Maxwell's loop analysis and displacement voltage methods. The phase sequence is *RBY.*

(Gulbarga University)

9.13 Find the readings of the wattmeters connected to the network as shown in Fig. 9.42, when it is connected to a 400 V, 3 phase supply. The phase sequence is *RYB*.   (Kuvempu University)

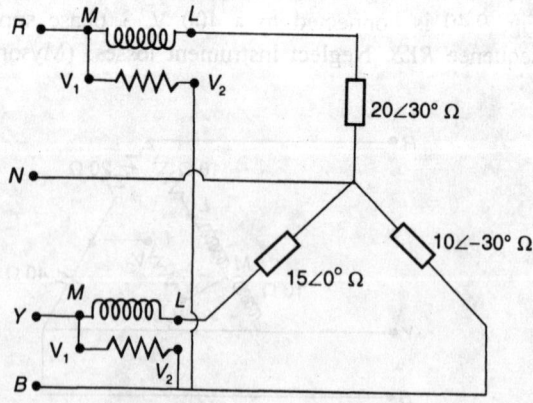

Fig. 9.42

9.14 An unbalanced star connected load is fed from a balanced 400 V, 3 phase supply. The supply voltage $V_{BN} = 230.94 \angle 90° \text{V}$ and the voltage across the load B is $200 \angle 110° \text{V}$. Draw the vector diagram. Calculate (a) the voltage between the star point of the load and neutral of the supply and (b) the voltages across the loads R and Y.

(Mysore University)

9.15 Find the line currents in the network shown in Fig. 9.43 and neutral conductor. The phase sequence is *RBY*. The voltage of the supply is 400 *V*.

(Bangalore University)

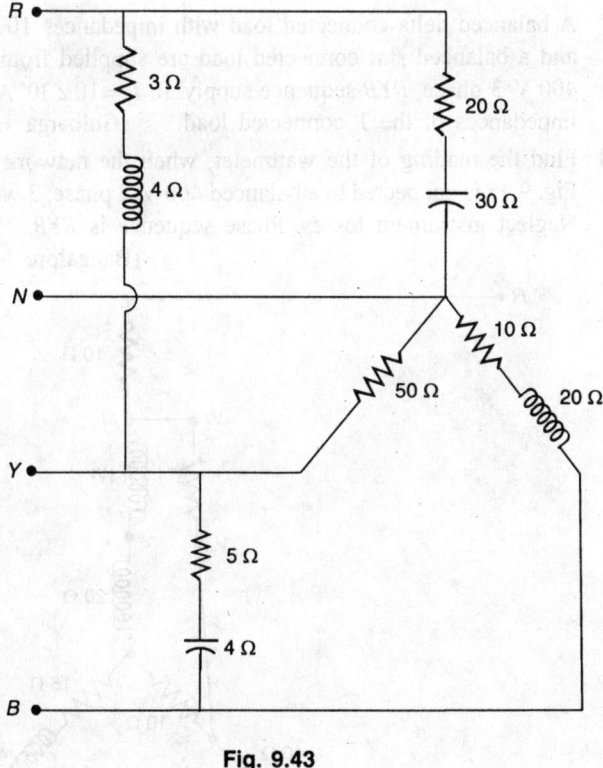

Fig. 9.43

9.16 Find the floating neutral voltage in the circuit shown in Fig. 9.44, and determine the phase voltages across the load and line currents.

(Karnataka University)

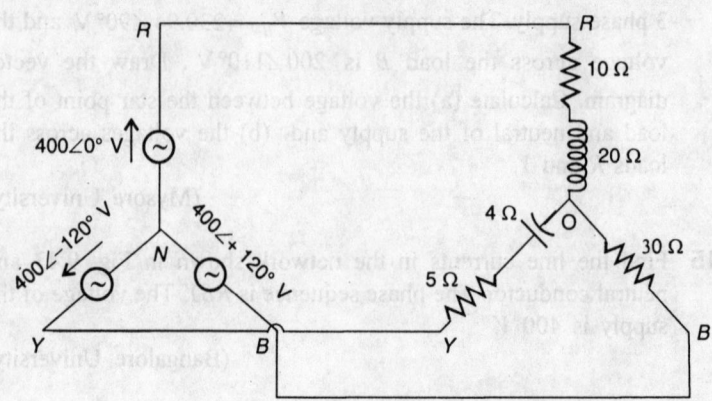

Fig. 9.44

9.17 A balanced delta connected load with impedances $10\angle-30°\,\Omega$ and a balanced star connected load are supplied from the same 400 V, 3 phase, *RYB* sequence supply. If $I_Y = 10\angle30°\,\text{A}$, find the impedances of the *Y* connected load.   (Gulbarga University)

9.18 Find the reading of the wattmeter, when the network shown in Fig. 9.45 is connected to a balanced 400 V, 3 phase, 3 wire supply. Neglect instrument losses. Phase sequence is *RYB*.

(Bangalore University)

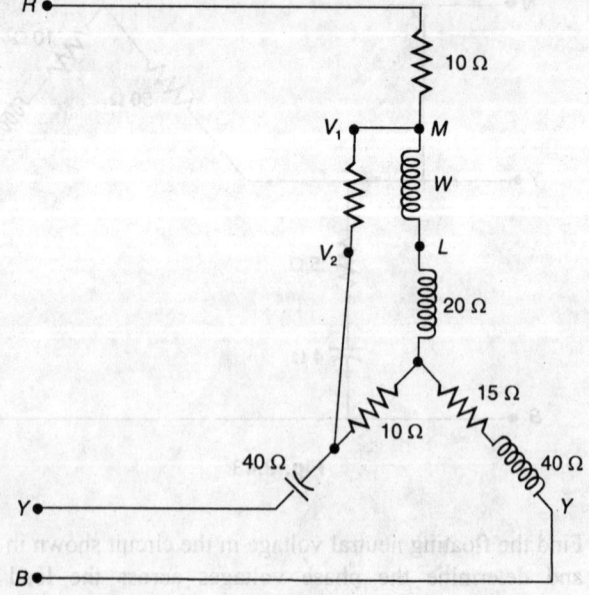

Fig. 9.45

9.19 Find the reading of the wattmeter, when the network shown in Fig. 9.46 is connected to a balanced 400 V, 3 phase, 4 wire supply. Neglect instrument losses. The phase sequence is *RBY*.

(Mangalore University)

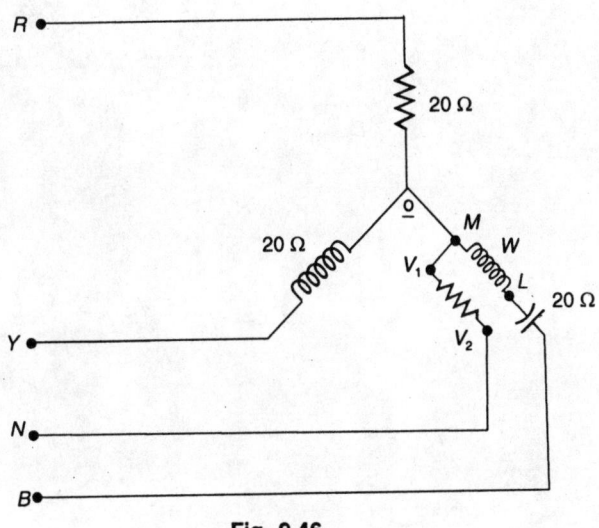

Fig. 9.46

9.20 An unbalanced star connected load comprising of a pure inductor and two resistors is connected to a balanced, 3 phase supply. If the numerical impedance of all the branches is the same, find the voltage across each branch as a percentage of line voltage.

(Karnataka University)

NETWORK ANALYSIS

10

Non-Sinusoidal Waveforms

10.1 INTRODUCTION

In the preliminary studies on electric circuits, we have learnt the analysis of electric circuits, when the sources are either d.c. or sinusoidal. But, frequently we encounter sources having non-sinusoidal waveforms such as square, rectangular, saw-tooth, etc. in which cases, we must be able to write the Eqs for such waveforms and thereafter, try to obtain the responses due to such non-sinusoidal excitations. An excitation having a non-sinusoidal waveform is assumed to consist of an infinite numer of sinusoidal waveforms of different magnitudes and frequencies, whose algebraic sum at every instant is equal to that of the original wave. In practice, it is not necessary to find the Eqs for a large number of components, as the magnitudes of higher frequency components are very small and can be neglected. It is, therefore, required to find the eqs only for the first few components. The component having the same frequency as the original non-sinusoidal wave is called the *first harmonic* or *fundamental component*. The component having the frequency two times that of the original wave is called *second harmonic*. In a similar way, higher order harmonics, such as *third harmonic, fourth harmonic,* etc. are the components having frequencies three times, four times, etc. of that of the original wave. This method of wave analysis is due to a great French mathematician J.B.J. Fourier (1758–1830), and hence called *Fourier Analysis*. The Eq. obtained for the non-sinusoidal wave as a sum of the various harmonics, including a constant component, if any, which is known as the *d.c. component,* is called as the *Fourier Series*.

Before applying this analysis to non-sinusoidal waveforms, the waveform should satisfy certain conditions known as *Dirichlet conditions*. A periodic waveform is one, which satisfies the eq.

$$f(t) = f(t + T) \tag{10.1}$$

Where, T is the period of the waveform.

Any periodic waveform which satisfies the following Dirichlet conditions can be expressed by Fourier Series.

(i) The function describing the non-sinusoidal waveform must be single valued and continuous, except for a finite number of discontinuities in the period T.

(ii) The function must have a finite average value over the period T.

(iii) The function must have a finite number of maxima and minima in the neighbourhood of any point.

Any non-sinusoidal waveform which satisfies the above conditions can be expressed as a Fourier Series as follows:

$$y = f(\theta) = A_0 + A_1 \cos\theta + B_1 \sin\theta + A_2 \cos 2\theta + B_2 \sin 2\theta$$
$$+ \ldots + A_n \cos n\theta + B_n \sin n\theta$$

or $\quad y = f(t) = A_0 + A_1 \cos\omega t + B_1 \sin\omega t + A_2 \cos 2\omega t + B_2 \sin 2\omega t$
$$+ \ldots + A_n \cos n\omega t + B_n \sin n\omega t \qquad (10.2)$$

Where, A_0, A_1 to A_n and B_1 to B_n are constants.

A_0 is the d.c. component. A_1 to A_n are the coefficients of cosine terms. B_1 to B_n are the coefficients of sine terms.

10.2 WAVE ANALYSIS

Wave analysis is nothing but the determination of the above coefficients and writing the eqs for the wave as given in Eq. (10.2).

To find A_0: Multiplying Eq. 10.2 by $d\theta$ and integrating between 0 to 2π, we get

$$\int_0^{2\pi} y \cdot d\theta = \int_0^{2\pi} A_0 \, d\theta + \int_0^{2\pi} A_1 \cos\theta \, d\theta + \int_0^{2\pi} B_1 \sin\theta \, d\theta$$

$$+ \int_0^{2\pi} A_2 \cos 2\theta \, d\theta + \int_0^{2\pi} B_2 \sin 2\theta \, d\theta + \ldots + \int_0^{2\pi} A_n \cos n\theta \, d\theta$$

$$+ \int_0^{2\pi} B_n \sin n\theta \, d\theta \qquad (10.3)$$

In Eq. (10.3), all the terms, except the first term, represent the area under a sine wave or cosine wave over a cycle and hence are zero.

$$\therefore \int_0^{2\pi} y \cdot d\theta = \int_0^{2\pi} A_0 \, d\theta = 2\pi A_0 \quad \therefore \quad A_0 = \frac{1}{2\pi} \int_0^{2\pi} y \cdot d\theta \qquad (10.4)$$

To find B_1 to B_n: Multiplying Eq. (10.2) by $\sin\theta \, d\theta$ and integrating between 0 to 2π, we get

$$\int_0^{2\pi} y \cdot \sin\theta \, d\theta = \int_0^{2\pi} A_0 \sin\theta \, d\theta + \int_0^{2\pi} A_1 \sin\theta \cos\theta \, d\theta + \int_0^{2\pi} B_1 \sin^2\theta \, d\theta$$

$$+ \int_0^{2\pi} A_2 \sin\theta \cos 2\theta \, d\theta + \int_0^{2\pi} B_2 \sin\theta \sin 2\theta \, d\theta \quad + \ldots +$$

$$\int_0^{2\pi} A_n \sin\theta \cos n\theta \, d\theta + \int_0^{2\pi} B_n \sin\theta \sin n\theta \, d\theta \qquad (10.5)$$

In Eq. (10.5), the first term $\int_0^{2\pi} A_0 \sin\theta \, d\theta$ is 0, as it represents the area under a sine wave over a cycle. There are four other types of terms.

(i) $\displaystyle\int_0^{2\pi} \sin^2\theta \, d\theta = \int_0^{2\pi} \frac{1 - \cos 2\theta}{2} \cdot d\theta = \frac{1}{2}\left[\theta - \frac{\sin 2\theta}{2}\right]_0^{2\pi}$

$$= \frac{1}{2} \times 2\pi = \pi$$

(ii) $\displaystyle\int_0^{2\pi} \sin m\theta \cdot \sin n\theta \, d\theta = \int_0^{2\pi} \frac{1}{2}\left[\cos(m-n)\theta - \cos(m+n)\theta\right] d\theta$

$$= \frac{1}{2}\left[\frac{\sin(m-n)\theta}{m-n} - \frac{\sin(m+n)\theta}{m+n}\right]_0^{2\pi} = 0$$

(iii) $\displaystyle\int_0^{2\pi} \sin m\theta \cdot \cos n\theta \, d\theta = \int_0^{2\pi} \frac{1}{2}\left[\sin(m+n)\theta + \sin(m-n)\theta\right] d\theta$

$$= \frac{1}{2}\left[-\frac{\cos(m+n)\theta}{m+n} - \frac{\cos(m-n)\theta}{m-n}\right]_0^{2\pi} = 0$$

(iv) $\displaystyle\int_0^{2\pi} \sin\theta \cos\theta \, d\theta = \int_0^{2\pi} \frac{\sin 2\theta}{2} \, d\theta = \frac{1}{2}\left[-\frac{\cos 2\theta}{2}\right]_0^{2\pi} = 0$

$$\therefore \qquad \int_0^{2\pi} y \cdot \sin\theta \, d\theta = B_1 \pi$$

i.e. $$B_1 = \frac{1}{\pi}\int_0^{2\pi} y \cdot \sin\theta \, d\theta$$

$$\text{III}^{\text{ly}} \ B_2 = \frac{1}{\pi}\int_0^{2\pi} y \cdot \sin 2\theta \, d\theta \qquad (10.6)$$

$$\vdots$$

$$B_n = \frac{1}{\pi}\int_0^{2\pi} y \cdot \sin n\theta \, d\theta$$

To find A_1 to A_n: Multiplying Eq. (10.2) by $\cos\theta \, d\theta$ and integrating between 0 and 2π, we get

$$\int_0^{2\pi} y \cdot \cos\theta \, d\theta = \int_0^{2\pi} A_0 \cos\theta \, d\theta + \int_0^{2\pi} A_1 \cos^2\theta \, d\theta + \int_0^{2\pi} B_1 \sin\theta \cos\theta \, d\theta$$

$$+ \int_0^{2\pi} A_2 \cos\theta \cos 2\theta \, d\theta + \int_0^{2\pi} B_2 \cos\theta \sin 2\theta \, d\theta$$

$$+ \, ... + \int_0^{2\pi} A_2 \cos\theta \cos n\theta \, d\theta + \int_0^{2\pi} B_2 \cos\theta \sin n\theta \, d\theta \qquad (10.7)$$

In Eq. (10.7), all the terms are zero, except $\int_0^{2\pi} A_1 \cos^2\theta \, d\theta$

$$\therefore \quad \int_0^{2\pi} y \cdot \cos\theta \, d\theta = \int_0^{2\pi} A_1 \cos^2\theta \, d\theta = \int_0^{2\pi} A_1 \frac{\cos 2\theta + 1}{2} \, d\theta$$

$$= \frac{A_1}{2} \left[\frac{\sin 2\theta}{2} + \theta \right]_0^{2\pi} = A_1 \pi$$

i.e. $$A_1 = \frac{1}{\pi} \int_0^{2\pi} y \cdot \cos\theta \, d\theta$$

$$\text{III}^{ly} \; A_2 = \frac{1}{\pi} \int_0^{2\pi} y \cdot \cos 2\theta \, d\theta$$

$$\vdots$$

$$A_n = \frac{1}{\pi} \int_0^{2\pi} y \cdot \cos n\theta \, d\theta \qquad (10.8)$$

Using Eqs (10.4), (10.6) and (10.8), all the coefficients of Eq. (10.2) can be found and the eq. for the function $y = f(\theta)$, which describes the non-sinusoidal waveform can be written.

The expressions for A_0, A_n and B_n can also be written in terms of the time period T.

$$A_0 = \frac{1}{T} \int_0^T y \cdot dt \qquad (10.9)$$

$$A_n = \frac{2}{T} \int_0^T y \cdot \cos n\omega t \, dt \qquad (10.10)$$

and $$B_n = \frac{2}{T} \int_0^T y \cdot \sin n\omega t \, dt \qquad (10.11)$$

10.3   WAVE SYMMETRY

Many a times, when we are required to write the Fourier Series, it is not necessary to calculate all the coefficients. By observing the symmetry of the non-sinusoidal waveform, it is possible for us to predict the absence of certain coefficients. The following wave symmetries are helpful in such prediction.

(i) If the area of the +ve loop in a period is equal to the area of the –ve loop, then $A_0 = 0$.

(ii) If the function describing the non-sinusoidal wave is *even*, then $B_n = 0$. The Fourier Series consists of only *cosine* terms.

A function $y = f(\theta)$ is said to be even, when $f(\theta) = f(-\theta)$. The waveforms shown in Figs 10.1 (a), (b), and (c) are described by even functions.

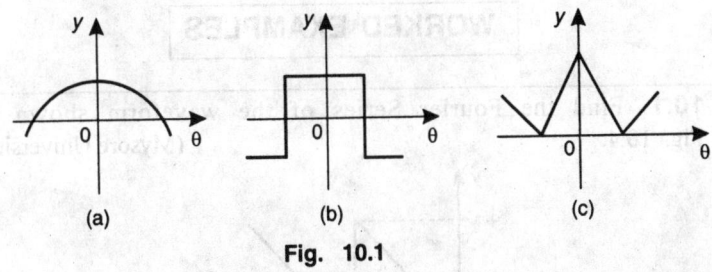

(a)   (b)   (c)

Fig. 10.1

(iii) If the function describing the non-sinusoidal wave is *odd*, then $A_n = 0$. The Fourier Series consists of only *sine* terms.

A function $y = f(\theta)$ is said to be odd when $f(\theta) = -f(-\theta)$. The waveforms shown in Figs 10.2 (a), (b) and (c) are described by odd functions.

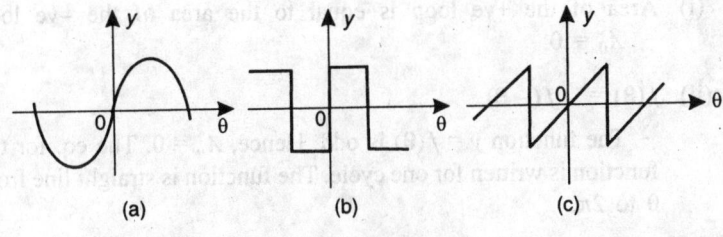

(a)   (b)   (c)

Fig. 10.2

(iv) If the function describing the non-sinusoidal waveform is having *half wave symmetry*, then the series contains only odd harmonics. All even harmonics are absent, i.e. A_2, A_4, B_2, B_4, etc. will be zero. However, the function may contain both cosine and sine terms, unless the function is odd or even.

A function $y = f(\theta)$ is said to be having half wave symmetry, if $f(\theta) = -f(\theta \pm \pi)$ or $f(t) = -f(\theta \pm T/2)$. The waveform shown in Fig. 10.3 is described by a function having half wave symmetry.

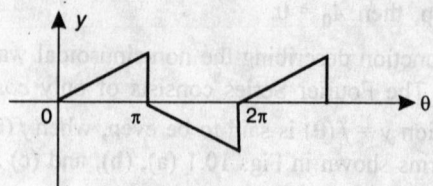

Fig. 10.3

WORKED EXAMPLES

10.1 Find the Fourier Series of the waveform shown in Fig. 10.4. (Mysore University)

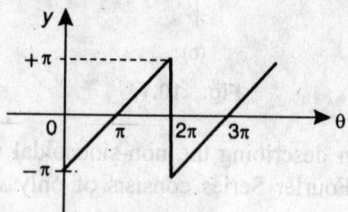

Fig. 10.4

Solution

(i) Area of the +ve loop is equal to the area of the –ve loop

∴ $A_0 = 0$

(ii) $f(\theta) = -f(-\theta)$

∴ The function $y = f(\theta)$ is odd. Hence, $A_n = 0$. The eq. for the function is written for one cycle. The function is straight line from 0 to 2π.

∴ $y = m\,\theta + C = \dfrac{\pi}{\pi}\,\theta - \pi = \theta - \pi$

$B_n = \dfrac{1}{\pi}\displaystyle\int_0^{2\pi} y \cdot \sin n\theta\, d\theta = \dfrac{1}{\pi}\displaystyle\int_0^{2\pi} (\theta - \pi)\sin n\theta\, d\theta$

$= \dfrac{1}{\pi}\displaystyle\int_0^{2\pi} \theta \sin n\theta\, d\theta - \displaystyle\int_0^{2\pi} \sin n\theta\, d\theta$

$= \dfrac{1}{\pi}\left[\left\{-\theta\,\dfrac{\cos n\theta}{n}\right\}_0^{2\pi} - \displaystyle\int_0^{2\pi} -\dfrac{\cos n\theta}{n}\cdot 1\cdot d\theta\right] - 0$

$$= \frac{1}{\pi}\left[\left\{-\theta\frac{\cos n\theta}{n}\right\}_0^{2\pi} + \left\{\frac{\sin n\theta}{n^2}\right\}_0^{2\pi}\right] = -\frac{2}{n}$$

$$\therefore \quad y = f(\theta) = -2\sin\theta - \frac{2}{2}\sin 2\theta - \frac{2}{3}\sin 3\theta \, ... \, -\frac{2}{n}\sin n\theta$$

$$= -2\left(\sin\theta + \frac{1}{2}\sin 2\theta + \frac{1}{3}\sin 3\theta + \, ... \, + \frac{1}{n}\sin n\theta\right)$$

10.2 Write the first four terms of the Fourier Series which will represent half-rectified waveform shown in Fig. 10.5.

(Bangalore University)

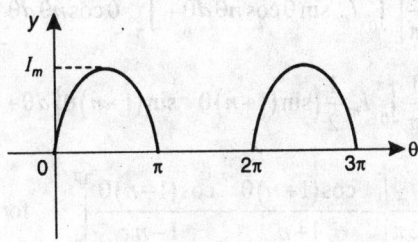

Fig. 10.5

Solution

$$y = I_m \sin\theta \quad \text{from} \quad 0 \text{ to } \pi$$
$$= 0 \qquad\quad \text{from } \pi \text{ to } 2\pi$$

The function does not have any wave symmetry.

$$A_0 = \frac{1}{2\pi}\left[\int_0^\pi I_m \sin\theta\, d\theta + \int_\pi^{2\pi} 0\, d\theta\right] = -\frac{1}{2\pi}(I_m \cos\theta)_0^\pi + 0$$

$$= -\frac{I_m}{2\pi}(-1-1) = \frac{I_m}{\pi} = 0.318\, I_m$$

$$B_n = \frac{1}{\pi}\left[\int_0^\pi I_m \sin\theta \sin n\theta\, d\theta + \int_\pi^{2\pi} 0\sin n\theta\, d\theta\right]$$

$$= \frac{1}{\pi}\left[\int_0^\pi I_m \frac{1}{2}\{\cos(n-1)\theta - \cos(n+1)\theta\}d\theta\right] + 0$$

$$= \frac{I_m}{2\pi}\left[\frac{\sin(n-1)\theta}{n-1} - \frac{\sin(n+1)\theta}{n+1}\right]_0^\pi = 0 \quad \text{for } n \neq 1,$$

$$\therefore \quad B_2 = B_3 = B_4, \text{etc.} = 0$$

$$\therefore \quad B_1 = \frac{1}{\pi}\left[\int_0^\pi I_m \sin\theta \sin\theta\, d\theta + \int_\pi^{2\pi} 0 \sin\theta\, d\theta\right]$$

$$= \frac{I_m}{\pi}\left[\int_0^\pi \sin^2\theta\, d\theta + 0\right] = \frac{I_m}{\pi}\left[\int_0^\pi \frac{1-\cos 2\theta}{2}\, d\theta\right]$$

$$= \frac{I_m}{2\pi}\left[\theta - \frac{\sin 2\theta}{2}\right]_0^\pi = \frac{I_m}{2\pi}\left[\pi - 0 - (0-0)\right] = 0.5\, I_m$$

$$A_n = \frac{1}{\pi}\left[\int_0^\pi I_m \sin\theta \cos n\theta\, d\theta + \int_\pi^{2\pi} 0\cos n\theta\, d\theta\right]$$

$$= \frac{1}{\pi}\int_0^\pi I_m \frac{1}{2}\{\sin(1+n)\theta + \sin(1-n)\theta\}\, d\theta + 0$$

$$= \frac{I_m}{2\pi}\left[-\frac{\cos(1+n)\theta}{1+n} - \frac{\cos(1-n)\theta}{1-n}\right]_0^\pi \quad \text{for } n \neq 1,$$

For $n = 2$

$$A_2 = \frac{I_m}{2\pi}\left[\frac{1}{3} - 1 + \frac{1}{3} - 1\right] = -\frac{2}{3}\frac{I_m}{\pi} = -0.212\, I_m$$

For $n = 3$

$$A_3 = \frac{I_m}{2\pi}\left[-\frac{1}{4} + \frac{1}{2} + \frac{1}{4} - \frac{1}{2}\right] = 0$$

For $n = 4$

$$A_4 = \frac{I_m}{2\pi}\left[\frac{1}{5} - \frac{1}{3} + \frac{1}{5} - \frac{1}{3}\right] = -0.0424\, I_m$$

$$\therefore \quad y = 0.318\, I_m + 0.5\, I_m \sin\theta - 0.212\, I_m \cos 2\theta - 0.0424\, I_m \cos 4\theta$$

10.3 Write down the Fourier Series which represents the waveform shown in Fig. 10.6 upto the third harmonic.

(Kuvempu University)

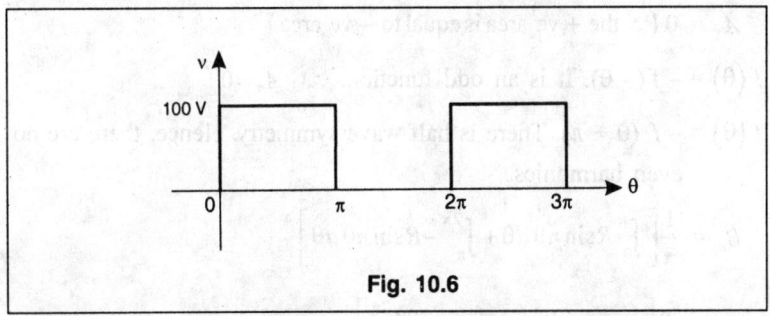

Fig. 10.6

Solution

$$v = 100 \text{ V} \quad \text{for } \theta = 0 \text{ to } \pi, \qquad v = 0 \text{ V} \quad \text{for } \theta = \pi \text{ to } 2\pi$$

$$A_0 = \frac{1}{2\pi}\left[\int_0^\pi 100\,d\theta + \int_\pi^{2\pi} 0\,d\theta\right] = \frac{1}{2\pi}(100\,\theta)_0^\pi = 50$$

$$B_n = \frac{1}{\pi}\int_0^\pi 100\sin n\theta\,d\theta + 0 = \frac{100}{\pi}\left[-\frac{\cos n\theta}{n}\right]_0^\pi = \frac{100}{\pi n}(1 - \cos n\pi)$$

$$\therefore \quad B_1 = \frac{100}{\pi}(1+1) = 63.7,\ B_2 = \frac{100}{2\pi}(1-1) = 0,\ B_3 = \frac{100}{3\pi}(1+1) = 21.2$$

$$A_n = \frac{1}{\pi}\int_0^\pi 100\cos n\theta\,d\theta + 0 = \frac{100}{\pi}\left\{\frac{\sin n\theta}{n}\right\}_0^\pi = 0$$

$$\therefore \quad v = 50 + 63.7\sin\theta + 21.2\sin 3\theta + \dots \text{ volts}$$

10.4 Write down the Fourier Series for the waveform shown in Fig. 10.7. (Karnataka University).

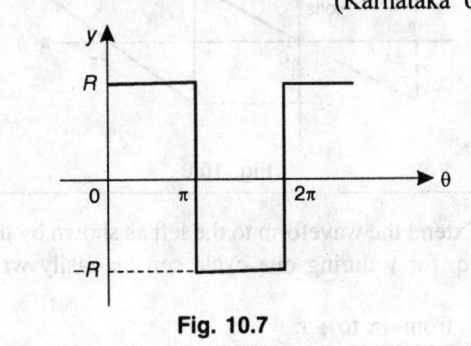

Fig. 10.7

Solution

$$y = R \qquad \text{for } \theta = 0 \text{ to } \pi$$
$$= -R \qquad \text{for } \theta = \pi \text{ to } 2\pi$$

$A_0 = 0$ $(\because$ the $+$ ve area is equal to $-$ ve area$)$

$f(\theta) = -f(-\theta)$. It is an odd function. $\therefore$ $A_n = 0$

$f(\theta) = -f(\theta + \pi)$. There is half wave symmetry. Hence, there are no even harmonics.

$$B_n = \frac{1}{\pi}\left[\int_0^\pi R\sin n\theta\,d\theta + \int_\pi^{2\pi} -R\sin n\theta\,d\theta\right]$$

$$= \frac{R}{\pi}\left[\left\{\frac{-\cos n\theta}{n}\right\}_0^\pi + \left\{\frac{\cos n\theta}{n}\right\}_\pi^{2\pi}\right]$$

$$= \frac{R}{n\pi}\{1 - 2\cos n\pi + \cos 2n\pi\}$$

$\therefore$ $B_1 = \dfrac{R}{\pi}(1+2+1) = \dfrac{4}{\pi}R$, $B_2 = 0$, $B_3 = \dfrac{R}{3\pi}(4) = \dfrac{1}{3}\dfrac{4}{\pi}R$,

$B_4 = 0$, $B_5 = \dfrac{1}{5}\dfrac{4}{\pi}R$, ... $B_n = \dfrac{1}{n}\dfrac{4}{\pi}R$

$\therefore$ $y = \dfrac{4}{\pi}R\left(\sin\theta + \dfrac{1}{3}\sin 3\theta + \dfrac{1}{5}\sin 5\theta + ... + \dfrac{1}{n}\sin n\theta\right)$ volts.

10.5 Find the Fourier Series for the waveform shown in Fig. 10.8.
(Gulbarga University)

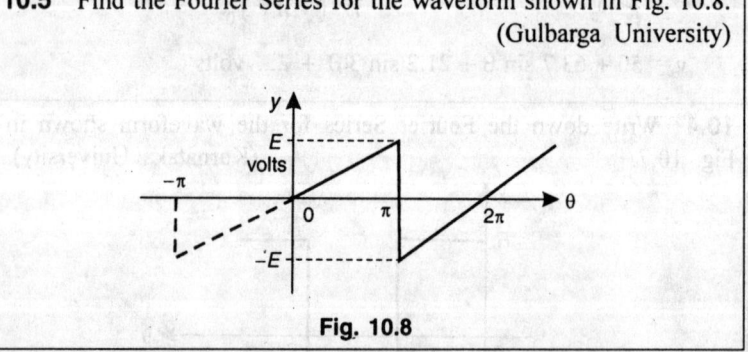

Fig. 10.8

Solution: Extend the waveform to the left as shown by the dotted lines, so that the eq. for y during one cycle can be easily written.

$y = \dfrac{E}{\pi}\theta$ from $-\pi$ to $+\pi$

$A_0 = 0\cdot[\because$ area of the $+$ ve loop is equal to the area of the $-$ ve loop$]$

$A_n = 0\cdot[\because$ it is an odd function $f(\theta) = -f(-\theta)]$

$$B_n = \frac{1}{\pi}\int_{-\pi}^{\pi} \frac{E}{\pi}\theta\cdot\sin n\theta\cdot d\theta = \frac{E}{\pi^2}\left[-\theta\frac{\cos n\theta}{n} + \frac{\sin n\theta}{n^2}\right]_{-\pi}^{\pi}$$

$$= -\frac{E}{n\pi^2}\left[\pi\cos n\pi + \pi\cos n\pi\right]$$

$$= -\frac{2E}{n\pi}\cos n\pi = -\frac{2E}{n\pi}(-1)^n$$

$$\therefore\ B_1 = \frac{2E}{\pi},\ B_2 = \frac{2E}{\pi}\left(-\frac{1}{2}\right),\ B_3 = \frac{2E}{3\pi}\left(\frac{1}{3}\right), B_4 = \frac{2E}{\pi}\left(-\frac{1}{4}\right),\ \text{etc.}$$

$$\therefore\ y = \frac{2E}{\pi}\left(\sin\theta - \frac{1}{2}\sin 2\theta + \frac{1}{3}\sin 3\theta - \frac{1}{4}\sin 4\theta + \ .....\right)\text{volts.}$$

10.6 Write down the Fourier Series for the waveform shown in Fig. 10.9. (Mangalore University)

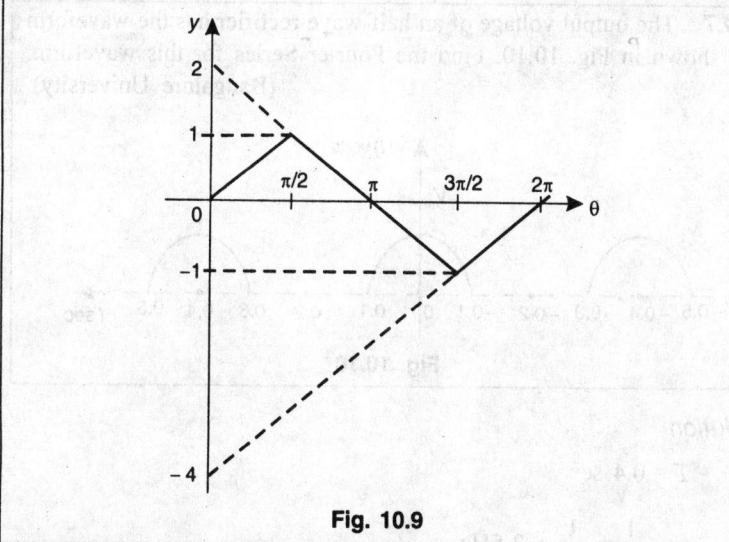

Fig. 10.9

Solution: The intercepts of the lines on y-axix are obtained by the extended dotted lines as shown in the Fig. 10.9.

$A_0 = 0$ [$\because$ Area of +ve loop = Area of – ve loop]

$A_n = 0$ [$\because f(\theta) = -f(-\theta)$, odd function]

Even harmonics are absent, because of halfwave symmetry, i.e.

$f(\theta) = -f(\theta + \pi)$

$$y = \frac{2}{\pi}\theta \qquad \text{from 0 to } \frac{\pi}{2}$$

$$y = \frac{-2}{\pi}\theta + 2 \quad \text{from } \frac{\pi}{2} \text{ to } \frac{3\pi}{2}$$

$$= \frac{2}{\pi}\theta - 4 \quad \text{from } \frac{3\pi}{2} \text{ to } 2\pi$$

$$B_n = \frac{1}{\pi}\left[\int_0^{\pi/2} \frac{2}{\pi}\theta \cdot \sin n\theta \cdot d\theta + \int_{\pi/2}^{3\pi/2}\left(\frac{-2}{\pi}\theta + 2\right)\sin n\theta \cdot d\theta\right.$$

$$\left. + \int_{3\pi/2}^{2\pi}\left(\frac{2}{\pi}\theta - 4\right)\sin n\theta \cdot d\theta\right] = \frac{4}{\pi^2 n^2}\left[\sin\frac{n\pi}{2} - \sin\frac{3n\pi}{2}\right]$$

$$\therefore \quad B_1 = \frac{8}{\pi^2}, B_2 = 0, B_3 = -\frac{8}{\pi^2}\frac{1}{3^2}, B_4 = 0, B_5 = \frac{8}{\pi^2}\frac{1}{5^2}, B_6 = 0, \text{ etc.}$$

$$\therefore \quad y(\theta) = \frac{8}{\pi^2}\left[\sin\theta - \frac{1}{3^2}\sin 3\theta + \frac{1}{5^2}\sin 5\theta - \frac{1}{7^2}\sin 7\theta + \dots\right]$$

10.7 The output voltage of an half-wave rectifier has the waveform as shown in Fig. 10.10. Find the Fourier Series for this waveform.

(Bangalore University)

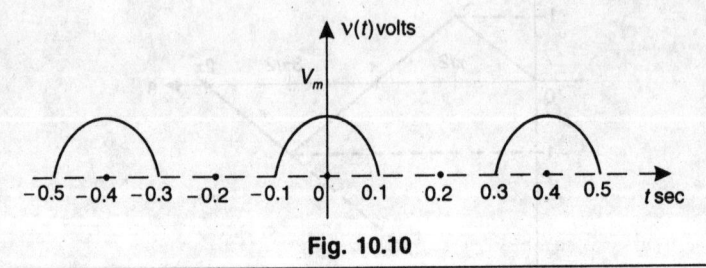

Fig. 10.10

Solution

$$T = 0.4 \text{ sec}$$

$$\therefore \quad f_0 = \frac{1}{T} = \frac{1}{0.4} = 2.5 \text{ Hz}$$

$$\omega_0 = 2\pi f_0 = 2\pi \times 2.5 = 5\pi \text{ rad}$$

The equation for the waveform over one period is :

$$v(t) = V_m \cos 5\pi t \quad \text{for } t = 0 \text{ to } 0.1 \text{ sec}$$

$$= 0 \quad \text{for } t = 0.1 \text{ to } 0.3 \text{ sec}$$

$$= V_m \cos 5\pi t \quad \text{for } t = 0.3 \text{ to } 0.4 \text{ sec}$$

While evaluating the coefficients of the Fourier Series, the above eq.s involve three integrals. But, if the eq. is written for the period starting from – 0.1 sec to 0.3 sec, it results in only two terms for $v(t)$ as shown below.

$$v(t) = V_m \cos 5\pi t \qquad \text{for } t = -0.1 \text{ to } 0.1 \text{sec}$$
$$= 0 \qquad \text{for } t = 0.1 \text{ to } 0.3 \text{sec}$$

The above Eq. is preferable, as it involves only two terms, although both Eqs give the same result.

$$A_0 = \frac{1}{0.4} \left[\int_{-0.1}^{0.1} V_m \cos 5\pi t \cdot dt + \int_{0.1}^{3} 0 \cdot dt \right] = \frac{V_m}{\pi}$$

$$B_n = 0 \quad \left[\because f(\theta) = f(-\theta), \text{ it is an even function} \right]$$

$$A_n = \frac{2}{0.4} \left[\int_{-0.1}^{0.1} V_m \cos 5\pi t \cdot \cos 5n\pi t \cdot dt + 0 \right]$$

$$= 5V_m \int_{-0.1}^{0.1} \frac{1}{2} \left[\cos 5\pi (1+n)t + \cos 5\pi (1-n)t \right] dt$$

$$= \frac{2V_m}{\pi} \frac{\cos n\pi/2}{1-n^2} \quad \text{for } n \neq 1$$

$$\therefore \quad A_2 = \frac{2V_m}{3\pi}, \ A_3 = 0, \ A_4 = -\frac{2V_m}{15\pi}, \ A_5 = 0, \ A_6 = \frac{2V_m}{35\pi}$$

A_1 has to be separately found:

$$A_1 = 5V_m \int_{-0.1}^{0.1} \cos 5\pi t \cdot \cos 5\pi t \cdot dt = 5V_m \int_{-0.1}^{0.1} \cos^2 5\pi t \cdot dt = \frac{V_m}{2}$$

$$\therefore v(t) = \frac{V_m}{\pi} + \frac{V_m}{2} \cos 5\pi t + \frac{2V_m}{3\pi} \cos 10\pi t - \frac{2V_m}{15\pi} \cos 20\pi t$$

$$+ \frac{2V_m}{35\pi} \cos 30\pi t \ \dots\dots$$

10.8 For the square-wave voltage signal shown in Fig. 10.11, write the Fourier Series. (Mangalore University)

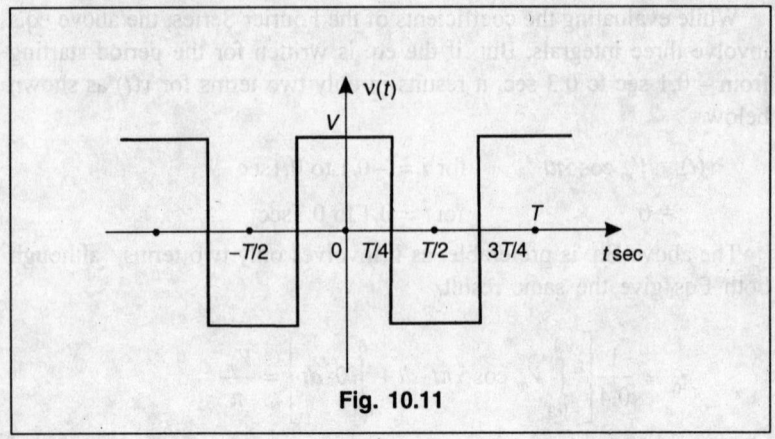

Fig. 10.11

Solution

$$v(t) = V \qquad \text{for } t = 0 \text{ to } t = \frac{\pi}{4} \qquad = -V \text{ for } t = \frac{T}{4} \text{ to } t = \frac{3T}{4}$$

$$= V \qquad \text{for } t = \frac{3T}{4} \text{ to } t = T$$

$A_0 = 0$ ($\because$ the average area over one cycle is zero)

$B_n = 0$ [$\because v(t) = v(-t)$, an even function]

$$v(t) = -v\left(t \pm \frac{T}{2}\right).$$

The waveform has half-wave symmetry. Hence, all even harmonics are absent.

$$A_n = \frac{2}{T}\left[V \int_0^{\frac{T}{4}} \cos n\omega t \cdot dt - V \int_{\frac{T}{4}}^{\frac{3T}{4}} \cos n\omega t \cdot dt + V \int_{\frac{3T}{4}}^{T} \cos n\omega t \cdot dt \right]$$

$$= \frac{2V}{n\omega T}\left[(\sin n\omega t)_0^{\frac{T}{4}} - (\sin n\omega t)_{\frac{T}{4}}^{\frac{T}{4}} + (\sin n\omega t)_{\frac{3T}{4}}^{T} \right]$$

$$= \frac{2V}{n\omega T}\left[2\sin\frac{n\omega T}{4} - 2\sin\frac{3n\omega T}{4} + \sin n\omega T \right]$$

$$= \frac{V}{n\pi}\left[2\sin\frac{n\pi}{2} - 2\sin\frac{3n\pi}{2} + \sin 2n\pi \right] (\because \omega T = 2\pi)$$

$$\therefore \quad A_1 = \frac{4V}{\pi}, \ A_3 = -\frac{4V}{3\pi}, \ A_5 = \frac{4V}{5\pi}, \ A_7 = -\frac{4V}{7\pi}, \text{ etc.}$$

$$\therefore \ v(\theta) \ = \ \frac{4V}{\pi} \left[\cos \theta - \frac{1}{3} \cos 3\theta + \frac{1}{5} \cos 5\theta - \frac{1}{7} \cos 7\theta + \ ..... \right]$$

$$\text{or} \ \ v(t) \ = \ \frac{4V}{\pi} \left[\cos \omega t - \frac{1}{3} \cos 3\omega t + \frac{1}{5} \cos 5\omega t - \frac{1}{7} \cos 7\omega t + ... \right] \text{volts}$$

10.9 For the given sawtooth waveform shown in Fig. 10.12, obtain the Fourier Series. (Karnataka University)

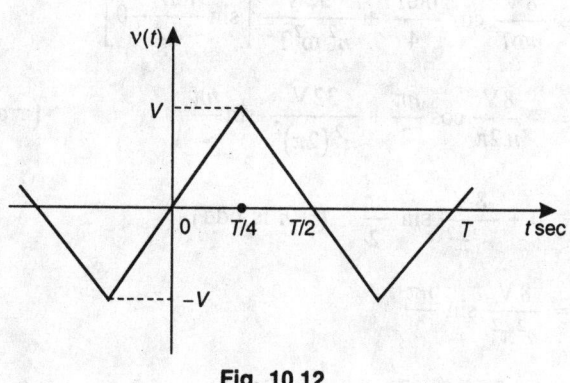

Fig. 10.12

Solution

$$A_0 = 0$$

($\because$ the average area over one cycle is zero)

$$v(t) = -v \ (-t)$$

The function is an odd function.

Hence, $A_n = 0$. Fourier Series contains only sine terms.

$$v(t) = v\left(t \pm \frac{T}{2} \right)$$

The waveform has half-wave symmetry.

Hence, all even harmonics are absent.

Since, the given waveform satisfies two symmetry conditions, it is sufficient to integrate over only one quarter period of the waveform to determine the Fourier coefficients and then multiplying the result by 4.

$$v(t) = \frac{V}{T/4} t = \frac{4V}{T} t \quad \text{from} \quad t = 0 \text{ to } t = \frac{T}{4}$$

i.e. over one quarter cycle

$$\therefore \qquad B_n = 4 \times \frac{2}{T} \int_0^{\frac{T}{4}} v(t) \cdot \sin n\omega t \ dt = \frac{8}{T} \int_0^{\frac{T}{4}} \frac{4V}{T} t \sin n\omega t \ dt$$

where n is odd only

$$= \frac{32\,V}{T^2} \left[\frac{-t \cos n\omega t}{n\omega} \right]_0^{\frac{T}{4}} - \frac{32\,V}{n\omega T^2} \int_0^{\frac{T}{4}} \left[\frac{-\cos n\omega t}{n\omega} \right] dt$$

$$= -\frac{32\,V}{n\omega T^2} \left[\frac{T}{4} \cos \frac{n\omega T}{4} - 0 \right] + \frac{32\,V}{n\omega T^2} \left[\frac{\sin n\omega t}{n\omega} \right]_0^{\frac{T}{4}}$$

$$= \frac{8\,V}{n\omega T} \cos \frac{n\omega T}{4} + \frac{32\,V}{n^2\,\omega^2 T^2} \left[\sin \frac{n\omega T}{\cdot 4} - 0 \right]$$

$$= -\frac{8\,V}{n.2\pi} \cos \frac{n\pi}{2} + \frac{32\,V}{n^2 (2\pi)^2} \sin \frac{n\pi}{2} \qquad (\because \omega T = 2\pi)$$

$$= 0 + \frac{8\,V}{n^2 \pi^2} \sin \frac{n\pi}{2} \qquad (\because n \text{ is odd})$$

$$= \frac{8\,V}{n^2 \pi^2} \sin \frac{n\pi}{2}$$

$$\therefore B_1 = 1, \quad B_2 = 0, \quad B_3 = -\frac{1}{3^2}, \quad B_4 = 0$$

$$B_5 = \frac{1}{5^2}, \quad B_6 = 0, \quad B_7 = -\frac{1}{7^2}$$

$$v(\theta) = 4 \times \frac{8\,V}{\pi^2} \left(\sin\theta - \frac{1}{3^2} \sin 3\theta + \frac{1}{5^2} \sin 5\theta - \frac{1}{7^2} \sin 7\theta + \cdots \right) \text{ volts}$$

or $$v(t) = \frac{32\,V}{\pi^2} \left(\sin\omega t - \frac{1}{3^2} \sin 3\omega t + \frac{1}{5^2} \sin 5\omega t - \frac{1}{7^2} \sin 7\omega t + \cdots \right) \text{ volts}$$

10.4 EFFECTIVE VALUE OF A NON-SINUSOIDALLY VARYING QUANTITY

Let the Eq. for current representing a non-sinusoidal wave be

$$i = I_0 + I_{m1} \sin(\theta + \alpha_1) + I_{m2} \sin(2\theta + \alpha_2) + \cdots + I_{mn} \sin(n\theta + \alpha_n)$$

$$(10.12)$$

Then, the effective value of the current is given by

$$I = \sqrt{\frac{1}{2\pi} \int_0^{2\pi} i^2 \, d\theta}$$

$$= \sqrt{\frac{1}{2\pi} \int_0^{2\pi} \left[I_0 + I_{m1} \sin(\theta + \alpha_1) + I_{m2} \sin(2\theta + \alpha_2) + \cdots + I_{mn} \sin(n\theta + \alpha_n) \right]^2 d\theta}$$

The expansion of i^2 consists of 4 types of terms.

(i) $\quad \int_0^{2\pi} I_0^2 \, d\theta = 2\pi I_0^2$

(ii) $\quad \int_0^{2\pi} 2I_0 I_{mk} \sin(k\theta + \alpha_k) = 0$

(iii) $\quad \int_0^{2\pi} 2I_{mk} I_{ml} \sin(k\theta + \alpha_k) \sin(l\theta + \alpha_l) \quad = \quad \pi \cos(\alpha_k - \alpha_l)$

$$\text{if } k = l \text{ and } \alpha_k \neq \alpha_l$$
$$= \quad \pi \text{ if } k = l \text{ and } \alpha_k = \alpha_l$$
$$= \quad 0 \text{ if } k \neq l$$

(iv) $\quad \int_0^{2\pi} I_{mk}^2 \sin^2(k\theta + \alpha_k) = I_{mk}^2 \int_0^{2\pi} \frac{1 - \cos 2(k\theta + \alpha_k)}{2} \, d\theta = I_{mk}^2 \, \pi$

$$\therefore \quad I = \sqrt{\frac{1}{2\pi} \left\{ I_0^2 \, 2\pi + \pi \left[I_{m1}^2 + I_{m2}^2 + I_{m3}^2 + \cdots + I_{mn}^2 \right] \right\}}$$

$$= \sqrt{\left[I_0^2 + \frac{I_{m1}^2 + I_{m2}^2 + I_{m3}^2 + \cdots + I_{mn}^2}{2} \right]}$$

$$= \sqrt{I_0^2 + I_1^2 + I_2^2 + I_3^2 + \cdots + I_n^2} \tag{10.13}$$

Similarly, if the voltage describing a non-sinusoidal wave is

$$e = E_0 + E_{m1} \sin(\theta + \phi_1) + E_{m2} \sin(2\theta + \phi_2) + \cdots + E_{mn} \sin(n\theta + \phi_n) \tag{10.14}$$

Then, its effective value is given by

$$E = \sqrt{E_0^2 + E_1^2 + E_2^2 + E_3^2 + \cdots + E_n^2} \tag{10.15}$$

10.5 POWER DUE TO A NON-SINUSOIDAL WAVE

When a voltage due to a non-sinusoidal wave given by Eq. (10.14) is applied to an electrical circuit, then the resulting current is given by the Eq. (10.12).

Power is a scalar quantity and hence, only its average value has to be considered.

$$\therefore \quad P = \frac{1}{2\pi} \int_0^{2\pi} ei \, d\theta$$

$$= \frac{1}{2\pi} \int_0^{2\pi} \left[\{E_0 + E_{m1} \sin(\theta + \phi_1) + E_{m2} \sin(2\theta + \phi_2) \right.$$

$$+ \cdots + E_{mn} \sin(n\theta + \phi_n) \}$$

$$\left. \{I_0 + I_{m1} \sin(\theta + \alpha_1) + I_{m2} \sin(2\theta + \alpha_2) + \cdots + I_{mn} \sin(n\theta + \alpha_n) \} \right] d\theta$$

$$(10.16)$$

In the above expression, we have the following four types of terms.

(i) $\displaystyle \int_0^{2\pi} E_0 I_0 \, d\theta = 2\pi E_0 I_0$

(ii) $\displaystyle \int_0^{2\pi} E_0 I_{mk} \sin(k\theta + \alpha_k) = 0$

(iii) $\displaystyle \int_0^{2\pi} I_0 E_{mk} \sin(k\theta + \phi_k) = 0$

(iv) $\displaystyle E_{mk} I_{ml} \int_0^{2\pi} \sin(k\theta + \phi_k) \sin(l\theta + \alpha_l) d\theta$

$$\begin{aligned}
&= 0 && \text{if } K \neq l \\
&= \pi && \text{if } K = l \text{ and } \phi_k = \alpha_l \\
&= \pi \cos(\phi_k - \alpha_l) && \text{if } K = l \text{ and } \phi_k \neq \alpha_l \\
&= \pi \cos(\phi_k - \alpha_k)
\end{aligned}$$

Substituting the above results in Eq. (10.16), we get

$$P = E_0 I_0 + \frac{E_{m1} I_{m1}}{2} \cos(\phi_1 - \alpha_1) + \frac{E_{m2} I_{m2}}{2} \cos(\phi_2 - \alpha_2)$$

$$+ \cdots + \frac{E_{mn} I_{mn}}{2} \cos(\phi_n - \alpha_n)$$

$$= E_0 I_0 + E_1 I_1 \cos(\phi_1 - \alpha_1) + E_2 I_2 \cos(\phi_2 - \alpha_2)$$

$$+ \cdots + E_n I_n \cos(\phi_n - \alpha_n) \quad (10.17)$$

10.6 CIRCUIT ANALYSIS WHEN THE WAVEFORMS ARE NON-SINUSOIDAL

When a voltage due to a non-sinusoidal wave is applied to an electric circuit, the impedances offered by the circuit to the different harmonic components are different. This is because, eventhough the resistance is same for all components, the reactances X_L and X_C are different, as they are frequency dependent. Hence, currents due to various harmonics of the voltage, have to be separately calculated, both in magnitude and phase. If the voltage applied is as given in Eq. (10.14), results in a current as given in Eq. (10.12), then, the effective values of voltage, current and power can be calculated using Eqs (10.15), (10.13), and (10.17) respectively. Then, the p.f. of the circuit is given by

$$\text{p.f.} = \frac{P}{EI} \tag{10.18}$$

WORKED EXAMPLES

10.10 For the circuit given in Fig. 10.13, find the current, power, voltage drop across inductance and eq. of the current wave, when a voltage $e = 141.4 \sin \omega t + 70.7 \sin (3\omega t + 30°) - 28.28 \sin (5\omega t - 20°)$ volts is applied across it. Take $\omega = 377$ rad/sec.

(Mangalore University)

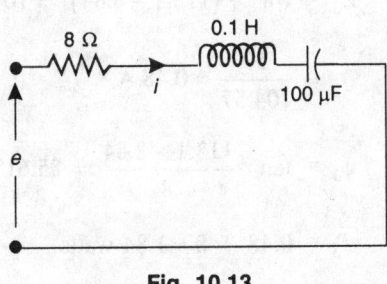

Fig. 10.13

Solution

Fundamental component

$$E_1 = \frac{141.4}{\sqrt{2}} = 100 \text{ Volts,}$$

$$X_{L1} = 377 \times 0.1 = 37.7 \ \Omega$$

$$X_{C1} = \frac{1}{377 \times 100 \times 10^{-6}} = 26.53 \ \Omega$$

$$Z_1 = \sqrt{8^2 + (37.7 - 26.53)^2} = 13.74 \ \Omega$$

$$I_1 = \frac{E_1}{Z_1} = \frac{100}{13.74} = 7.29 \ A,$$

$$\alpha_1 = \tan^{-1} \frac{37.7 - 26.53}{8} = 54.39^\circ \ (\text{lagging as } X_{L1} > X_{C1})$$

$$P_1 = I_1^2 \ R = 7.29^2 \times 8 = 425.15 \ \text{watts}$$

$$V_{L1} = I_1 \ X_{L1} = 7.29 \times 37.7 = 274.83 \ V$$

$$\therefore \quad i_1 = 7.29 \sqrt{2} \sin(\omega t - 54.39^\circ) \ A = 10.31 \sin(\omega t - 54.39^\circ) \ A$$

3rd harmonic

$$E_3 = \frac{70.7}{\sqrt{2}} = 50 \ V$$

$$X_{L3} = 3 \ X_{L1} = 3 \times 37.7 = 113.1 \ \Omega,$$

$$X_{C3} = \frac{X_{C1}}{3} = \frac{26.53}{3} = 8.84 \ \Omega$$

$$Z_3 = \sqrt{8^2 + (113.1 - 8.84)^2} = 104.57 \ \Omega,$$

$$I_3 = \frac{50}{104.57} = 0.48 \ A$$

$$\phi_3 = \tan^{-1} \frac{113.1 - 8.84}{8} = 85.61^\circ \ (\text{lagging}),$$

$$P_3 = 0.48^2 \times 8 = 1.84 \ \text{watts}$$

$$V_{L3} = I X_{L3} = 0.48 \times 113.1 = 54.29 \ V$$

$$\therefore \quad i_3 = 0.48 \sqrt{2} \sin(3\omega t + 30^\circ - 85.61^\circ)$$

$$= 0.68 \sin(3\omega t - 55.61^\circ) \ A$$

5th harmonic

$$E_5 = \frac{28.28}{\sqrt{2}} = 20 \ V,$$

$$X_{L5} = 5 \, X_{L1} = 5 \times 37.7 = 188.5 \, \Omega,$$

$$X_{C5} = \frac{X_{C1}}{5} = \frac{26.53}{5} = 5.31 \, \Omega$$

$$Z_5 = \sqrt{8^2 + (188.5 - 5.31)^2} = 183.37 \, \Omega,$$

$$I_5 = \frac{20}{183.37} = 0.109 \, A$$

$$\phi_5 = \tan^{-1} \frac{188.5 - 5.31}{8} = 87.5° \, (\text{lagging}),$$

$$P_5 = 0.109^2 \times 8 = 0.095 \, \text{watts}$$

$$V_{L5} = 0.109 \times 188.5 = 20.55 \, \text{Volts}$$

$$\therefore \quad i_5 = -0.109\sqrt{2} \sin (5\omega t - 20° - 87.5°) \, A$$

$$= -0.154 \sin (5\omega t - 107.5°) \, A$$

$$i = i_1 + i_2 + i_3 = 10.31 \sin (\omega t - 54.39°)$$

$$+ 0.68 \sin (3\omega t - 55.61°) - 0.154 \sin (5\omega t - 107.5°) \, A$$

$$I = \sqrt{I_1^2 + I_3^2 + I_5^2}$$

$$= \sqrt{7.29^2 + 0.48^2 + 0.109^2} = 7.306 \, A$$

$$P = P_1 + P_3 + P_5 = 425.15 + 1.84 + 0.095$$

$$= 427.09 \, \text{watts}$$

$$V_L = \sqrt{V_{L1}^2 + V_{L3}^2 + V_{L5}^2}$$

$$= \sqrt{274.83^2 + 54.29^2 + 20.55^2} = 280.895 \, V$$

10.11 To the circuit shown in Fig. 10.14, a voltage
$$v = 141.4 \sin \omega t + 70.7 \sin (3\omega t + 30°)$$
$$- 28.28 \sin (5\omega t - 20°) \, \text{volts is applied.}$$
Find (i) the total current, (ii) current in each branch, (iii) total power dissipated and power dissipated by each branch and (iv) equation of the total current. (Bangalore University)

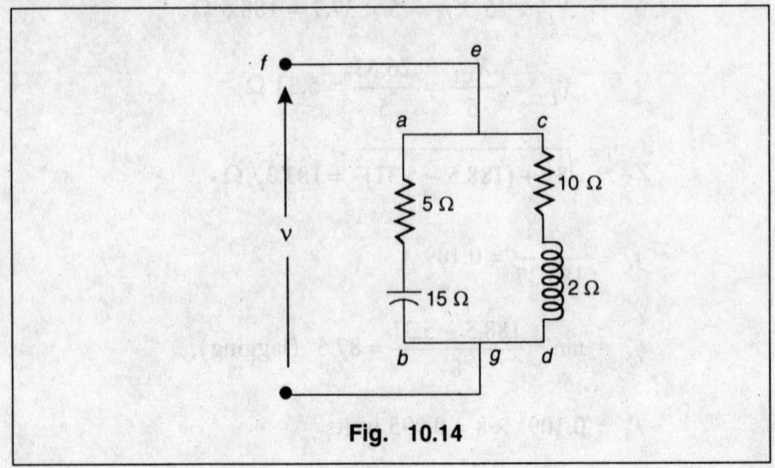

Fig. 10.14

Solution

Fundamental component

$$V_1 = \frac{141.4}{\sqrt{2}} = 100 \text{ V},$$

$$I_{ab1} = \frac{100}{5 - j15} = 6.33 \underline{|71.57°} \text{ A} = (2 + j6) \text{ A}$$

$$I_{cd1} = \frac{100}{10 + j2} = 9.81 \underline{|-11.34°} \text{ A} = (9.62 - j1.93) \text{ A}$$

$$I_{fe1} = I_{ab1} + I_{cd1} = (11.62 + j4.07)\text{A} = 12.31 \underline{|19.3°} \text{ A}$$

$$\therefore \quad i_{fe1} = 12.31 \sqrt{2} \sin(\omega t + 19.3°) \text{ A}$$

$$= 17.41 \sin(\omega t + 19.3°) \text{ A}$$

$$P_{ab1} = (I_{ab1})^2 \times 5 = 6.33^2 \times 5 = 200 \text{ watts},$$

$$P_{cd1} = (I_{cd1})^2 \times 10 = (9.81)^2 \times 10 = 962.36 \text{ watts}$$

3rd harmonic

$$V_3 = \frac{70.7}{\sqrt{2}} = 50 \text{ V},$$

$$I_{ab3} = \frac{50}{5 - j5} = 7.07 \underline{|45°} \text{ A} = (5 + j5) \text{ A}, \quad \left[\because X_{C3} = \frac{X_{C1}}{3} \right]$$

$$I_{cd3} = \frac{50}{10 + j6} = 4.3 \lfloor -30.99° \text{ A} = (3.68 - j\, 2.21) \text{ A}$$

$$[\because X_{L3} = 3X_{L1}]$$

$$I_{fe3} = I_{ab3} + I_{cd3} = (8.68 + j\, 2.21) \text{ A} = 9.12 \lfloor 17.82° \text{ A}$$

$$\therefore \quad i_{fe3} = 9.12\sqrt{2} \sin (3\omega t + 30° + 17.82°) \text{ A}$$

$$= 12.9 \sin (3\omega t + 47.82°) \text{ A}$$

$$P_{ab3} = 7.07^2 \times 5 = 250 \text{ watts,}$$

$$P_{cd3} = 4.3^2 \times 10 = 184 \text{ watts}$$

5th harmonic

$$V_5 = \frac{28.28}{\sqrt{2}} = 20 \text{ V,}$$

$$I_{ab5} = \frac{20}{5 - j3} = 3.43 \lfloor 30.91° \text{ A} = (2.94 + j\, 1.76) \text{ A,}$$

$$\left[\because X_{C5} = \frac{X_{C1}}{5} \right]$$

$$I_{cd5} = \frac{20}{10 + j\, 10} = 1.414 \lfloor -45° = (1 - j\, 1) \text{ A} \qquad [\because X_{L5} = 5X_{L1}]$$

$$I_{fe5} = I_{ab5} + I_{cd5} = 3.94 + j\, 0.76 = 4.01 \lfloor 10.92° \text{ A}$$

$$\therefore \quad i_{fe5} = 4.01\sqrt{2} \sin (5\omega t - 20° + 10.92°) \text{ A}$$

$$= 5.67 \sin (5\omega t - 9.08°) \text{ A}$$

$$P_{ab5} = 3.43^2 \times 5 = 58.8 \text{ watts,}$$

$$P_{cd5} = 1.414^2 \times 10 = 20 \text{ watts}$$

$$\therefore \quad I = \sqrt{I_{fe1}^2 + I_{fe3}^2 + I_{fe5}^2} = \sqrt{12.31^2 + 9.12^2 + 4.01^2} = 15.84 \text{ A}$$

$$I_{ab} = \sqrt{I_{ab1}^2 + I_{ab3}^2 + I_{ab5}^2} = \sqrt{6.33^2 + 7.07^2 + 3.43^2} = 10.1 \text{ A}$$

$$I_{cd} = \sqrt{I_{cd1}^2 + I_{cd3}^2 + I_{cd5}^2} = \sqrt{9.81^2 + 4.3^2 + 1.414^2} = 10.81 \text{ A}$$

$$P_{ab} = P_{ab1} + P_{ab3} + P_{ab5} = 200 + 250 + 58.8 = 508.8 \text{ W}$$

$$P_{cd} = P_{cd1} + P_{cd3} + P_{cd5} = 962.36 + 184 + 20 = 1166.36 \text{ W}$$

$$P_{Total} = 508.8 + 1166.36 = 1675.16 \text{ watts}$$

The Eq. for the current is

$$i = i_{fe1} + i_{fe3} + i_{fe5}$$

$$= 17.41 \sin\left(\omega t + 19.3°\right) + 12.9 \sin\left(3\omega t + 47.82°\right)$$
$$-5.67 \sin\left(5\omega t + 9.08°\right) \text{ A}$$

10.12 Find the power and p.f. of the circuit to which the voltage applied and the resulting current are given by,

$$e = 100\sin\left(\omega t + 30°\right) - 50\sin\left(3\omega t + 60°\right) + 25\sin 5\omega t \text{ volts}$$
and $i = 20\sin\left(\omega t - 30°\right) + 15\sin\left(3\omega t + 30°\right) + 10\cos\left(5\omega t - 60°\right)$ amps

(Kuvempu University)

Solution: Equation for the current is re-written as

$$i = 20\sin\left(\omega t - 30°\right) + 15\sin\left(3\omega t + 30°\right) + 10\sin\left(5\omega t + 30°\right) \text{ amps}$$

$$E = \sqrt{E_1^2 + E_3^2 + E_5^2}$$

$$= \sqrt{\left(\frac{100}{\sqrt{2}}\right)^2 + \left(\frac{50}{\sqrt{2}}\right)^2 + \left(\frac{25}{\sqrt{2}}\right)^2} = 81 \text{ V}$$

$$I = \sqrt{I_1^2 + I_3^2 + I_5^2}$$

$$= \sqrt{\left(\frac{20}{\sqrt{2}}\right)^2 + \left(\frac{15}{\sqrt{2}}\right)^2 + \left(\frac{10}{\sqrt{2}}\right)^2} = 19.04 \text{ A}$$

$$P = E_1 I_1 \cos\left(\phi_1 - \alpha_1\right) + E_3 I_3 \cos\left(\phi_3 - \alpha_3\right) + E_5 I_5 \cos\left(\phi_5 - \alpha_5\right)$$

$$= \frac{100}{\sqrt{2}} \times \frac{20}{\sqrt{2}} \cos\left(30° + 30°\right) + \frac{-50}{\sqrt{2}} \times \frac{15}{\sqrt{2}} \cos\left(60° - 30°\right)$$

$$+ \frac{25}{\sqrt{2}} \times \frac{10}{\sqrt{2}} \cos\left(0° - 30°\right) = 283.5 \text{ watts}$$

$$\text{p.f.} = \frac{P}{EI} = \frac{283.5}{1541} = 0.1837$$

10.13 Find the effective voltage, effective current, average power and p.f. of the network, if the applied voltage and the resulting current are given by

$$v = 200 + 100 \cos\left(500t + 30°\right) + 75 \cos\left(1500t + 60°\right) \text{ volts}$$
and $i = 3.53 \cos\left(500t + 75°\right) + 3.55 \cos\left(1500t + 78.45°\right)$ amps

(Karnataka University)

Solution: As all the functions are cosine, there is no necessity to convert them into sine functions.

$$V = \sqrt{V_0^2 + V_1^2 + V_3^2} = \sqrt{200^2 + \left(\frac{100}{\sqrt{2}}\right)^2 + \left(\frac{75}{\sqrt{2}}\right)^2} = 218.66 \text{ Volts}$$

$$I = \sqrt{I_1^2 + I_3^2} = \sqrt{\left(\frac{3.53}{\sqrt{2}}\right)^2 + \left(\frac{3.55}{\sqrt{2}}\right)^2} = 3.54 \text{ A}$$

$$P = E_0 I_0 + E_1 I_1 \cos\left(\phi_1 - \alpha_1\right) + E_3 I_3 \cos\left(\phi_3 - \alpha_3\right)$$

$$= 0 + \frac{100}{\sqrt{2}} \times \frac{3.53}{\sqrt{2}} \cos\left(30° - 75°\right) + \frac{75}{\sqrt{2}} \times \frac{3.55}{\sqrt{2}} \cos\left(60° - 78.45°\right)$$

$$= 124.8 + 126.28 = 251.08 \text{ W}$$

$$\text{p.f.} = \frac{P}{EI} = \frac{251.08}{218.66 \times 3.54} = 0.325$$

10.14 A voltage $v = (50 + 25 \sin 500t + 10 \sin 1500t + 5 \sin 2500t)$ volts is applied to the terminals of a network and the resulting current is

$$i = \{5 + 2.23 \sin\left(500\,t - 26.6°\right) + 0.566 \sin\left(1500\,t - 56.3°\right)$$
$$+ 0.186 \sin\left(2500t - 68.2°\right)\} \text{ amps.}$$

Find the effective voltage, current, power, and p.f. of the circuit.

(Mangalore University)

Solution

$$V = \sqrt{V_0^2 + V_1^2 + V_3^2 + V_5^2}$$

$$= \sqrt{50^2 + \left(\frac{25}{\sqrt{2}}\right)^2 + \left(\frac{10}{\sqrt{2}}\right)^2 + \left(\frac{5}{\sqrt{2}}\right)^2} = 53.6 \text{ Volts}$$

$$I = \sqrt{I_0^2 + I_1^2 + I_3^2 + I_5^2}$$

$$= \sqrt{5^2 + \left(\frac{2.23}{\sqrt{2}}\right)^2 + \left(\frac{0.566}{\sqrt{2}}\right)^2 + \left(\frac{0.186}{\sqrt{2}}\right)^2} = 5.25 \text{ A}$$

$$P = E_0 I_0 + E_1 I_1 \cos\left(\phi_1 - \alpha_1\right) + E_3 I_3 \cos\left(\phi_3 - \alpha_3\right) + E_5 I_5 \cos\left(\phi_5 - \alpha_5\right)$$

$$= \left(50 \times 5\right) + \frac{25}{\sqrt{2}} \times \frac{2.23}{\sqrt{2}} \cos\left(0 + 26.6°\right) + \frac{10}{\sqrt{2}} \times \frac{0.566}{\sqrt{2}} \cos\left(0 + 56.3°\right)$$

$$+ \frac{5}{\sqrt{2}} \times \frac{0.186}{\sqrt{2}} \cos(0 + 68.2°) = 276.5 \text{ watts}$$

$$\text{p.f.} = \frac{P}{EI} = \frac{276.5}{53.6 \times 5.25} = 0.983$$

NUMERICAL PROBLEMS

10.1 Write down the Fourier Series for the half-rectified sine-wave shown in Fig. 10.15. (Bangalore University)

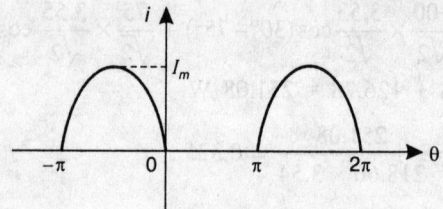

Fig. 10.15

10.2 Write down the Fourier Series for the half-rectified sinewave shown in Fig. 10.16. (Kuvempu University)

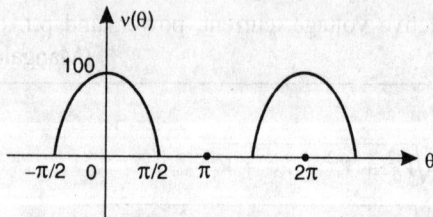

Fig. 10.16

10.3 Express the Eq. for the waveform shown in Fig. 10.17 by Fourier Series. (Mysore University)

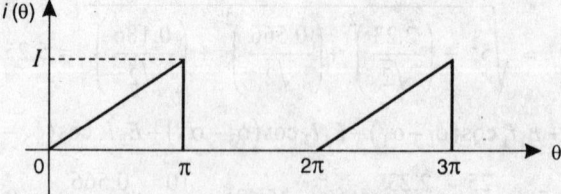

Fig. 10.17

10.4 Write the Fourier Series for the waveform shown in Fig. 10.18.
(Mangalore University)

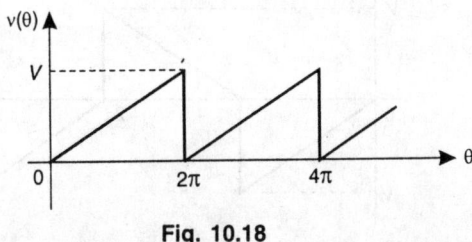

Fig. 10.18

10.5 Write the Fourier Series for the waveform shown in Fig. 10.19.
(Karnataka University)

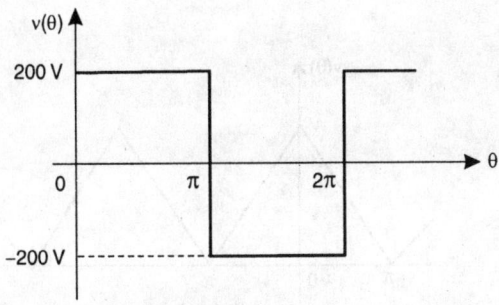

Fig. 10.19

10.6 Write the Fourier Series for the waveform shown in Fig. 10.20.
(Gulbarga University)

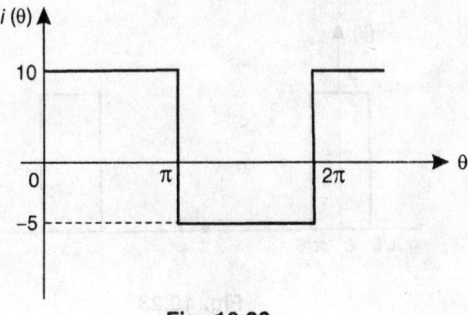

Fig. 10.20

10.7 Write the Fourier Series for the waveform shown in Fig. 10.21.
(Mangalore University)

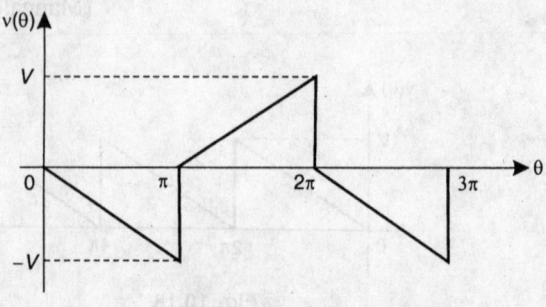

Fig. 10.21

10.8 Write the Fourier Series for the waveform shown in Fig. 10.22.
(Karnataka University)

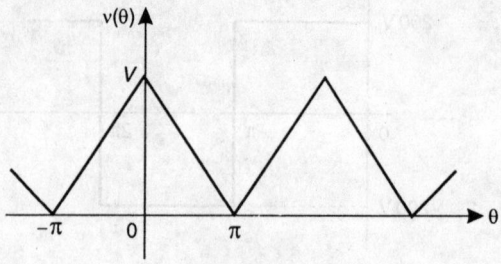

Fig. 10.22

10.9 Write the Fourier Series for the rectangular pulse show in
Fig. 10.23. (Bangalore University)

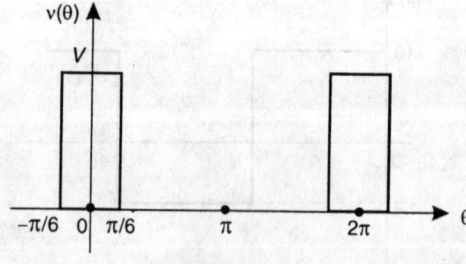

Fig. 10.23

10.10 Write the Fourier Series for the waveform shown in Fig. 10.24.
(Mysore University)

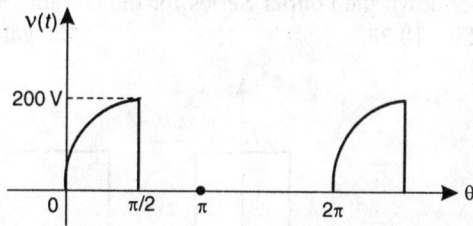

Fig. 10.24

10.11 For the full-rectified sine wave shown in Fig. 10.25, find the Fourier Series expression. (Kuvempu University)

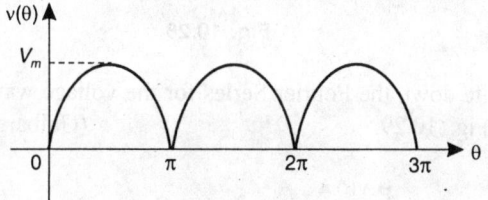

Fig. 10.25

10.12 For the full-rectified sine wave shown in Fig. 10.26, find the Fourier Series expression. (Gulbarga University)

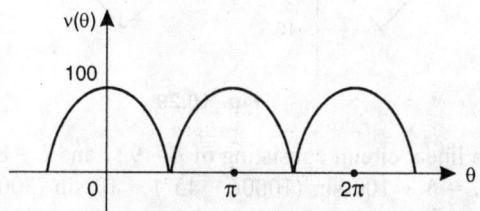

Fig. 10.26

10.13 Write down the Fourier Series for the voltage waveform shown in Fig. 10.27. (Karnataka University)

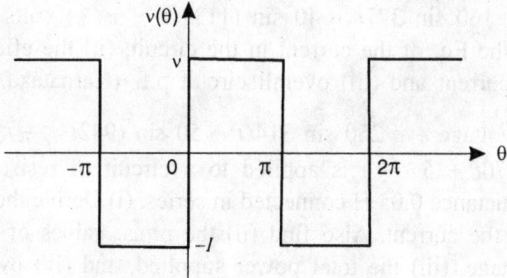

Fig. 10.27

10.14 Write down the Fourier Series for the current waveform shown in Fig. 10.28. (Bangalore University)

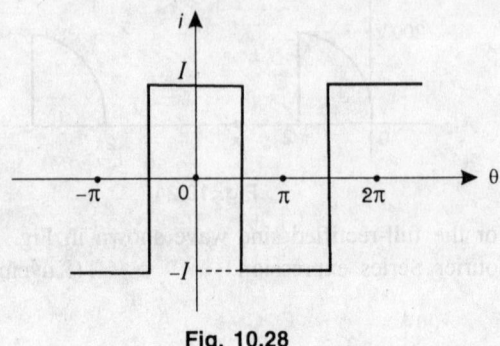

Fig. 10.28

10.15 Write down the Fourier Series for the voltage waveform shown in Fig. 10.29. (Gulbarga University)

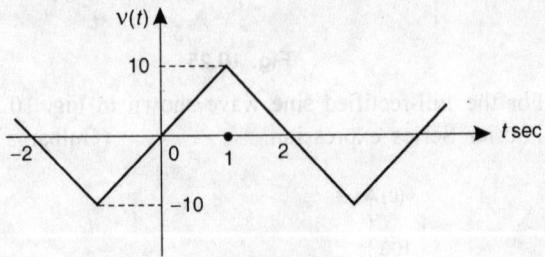

Fig. 10.29

10.16 In a linear circuit consisting of $R = 9\ \Omega$ and $L = 8$ mH, a current of $i = 5 + 100 \sin (1000t + 45°) + 10 \sin (3000t + 60°)$ A is flowing. Find (i) the Eq. of the applied voltage and (ii) the average power. (Mangalore University)

10.17 A series circuit consisting of 8 Ω resistor, 0.0053 H inductor and a 332 μF capacitor is connected to a source of e.m.f. $e = 160 \sin 377t + 40 \sin (1131\ t + \pi/3)$ volts. Determine (i) the Eq. of the current in the circuit, (ii) the effective value of current and (iii) overall circuit p.f. (Karnataka University)

10.18 A voltage $e = 250 \sin 314\ t + 50 \sin (942t + \pi/3) + 20 \sin (1570t + 5\pi/6)$ is applied to a circuit of resistance 20 Ω, inductance 0.05 H connected in series. (i) Derive the expression for the current. Also find (ii) the r.m.s. values of current and voltage, (iii) the total power supplied, and (iv) overall p.f. (Kuvempu University)

10.19 An alternating voltage represented by expression $v = 200 \sin 314t + 20 \sin 942t$ volts is applied to a coil of resistance 25 Ω, inductance 0.02 H, connected in series with a condenser of 40 μF. Derive the expression for the current. Find also the p.f. and the power input.
(Mysore University)

10.20 A voltage of $v = 141.4 \sin \omega t + 70.7 \sin (3\omega t + 60°) - 28.28 \sin (5\omega t - 30°)$ is applied to a parallel circuit consisting of a resistance of 8 Ω, inductive reactance of 6 Ω in one branch and a resistance of 4 Ω, a capacitive reactance of 3 Ω in another branch. Find (i) equation for the total current, (ii) total power dissipated and (iii) the p.f. of the circuit.

(Gulbarga University)

10.21 Find the effective values of voltage and current, power and p.f. if the voltage applied to a circuit and the resulting currents are given by:

(a) $e = 100 \sin (\omega t + 30°) - 50 \cos 2\omega t + 25 \sin (5\omega t + 150°)$ V
and $i = 20 \sin (\omega t + 40°) + 10 \sin (2\omega t + 30°) - 5 \sin (5\omega t - 50°)$ A
(Mangalore University)

(b) $e = 100 \sin (\omega t + 70°) - 60 \sin (2\omega t + 30°) + 30 \sin (3\omega t + 60°)$ V and $i = 50 \cos (\omega t - 60°) + 30 \sin (2\omega t + 70°) - 15 \cos (5\omega t - 30°)$ A (Karnataka University)

(c) $e = 100 \sin (\omega t - 20°) + 50 \sin (3\omega t + 60°) - 25 \cos (5\omega t - 30°)$ V and $i = 20 \cos (\omega t - 60°) - 10 \sin (3\omega t + 15°) + 5 \sin (5\omega t - 70°)$ A (Mysore University)

(d) $e = 200 + 100 \cos (500t + 30°) + 75 \cos (1500t + 60°)$ V
and $i = 3.53 \cos (500t + 75°) + 3.55 \cos (1500t + 78.45°)$ A
(Gulbarga University)

10.22 A two element series circuit with $R = 10 \Omega$ amd $L = 0.02$ H carries a current $i = (5 \sin 100t + 3 \sin 300t + 2 \sin 500t)$ A. Find the effective value of the applied voltage and the average power.
(Kuvempu University)

10.23 A three element circuit of $R = 5 \Omega$ in series with a parallel combination of L and C. At $\omega = 500$ rad/sec., the corresponding reactances are $j2 \Omega$ and $- j8 \Omega$ respectively. Find the total current, if the applied voltage is $e = 50 + 20 \sin 500t + 10 \sin 1000t$ volts.
(Bangalore University)

Initial Conditions

11.1 INTRODUCTION

Any electrical network consists of voltage sources, current sources, resistances, inductances and capacitances. When such networks are to be analysed, the integro-differential Eqs are written and solved. The general solution of such an Eq. consists of two parts, viz. (i) *Complementary function* and (ii) *Particular integral*.

The complementary function is the solution of the *homogeneous equation*, which also represents the *transient response* of the system. The transient response entirely depends on the type, value and arrangement of the elements in the network. The particular integral represents the *steady-state response* of the system, which depends not only on the system, but also on the excitation. The complementary function is the general solution of the homogeneous Eq. and the particular integral is the particular solution of the non-homogeneous Eq.

While solving a differential Eq. of the nth order, we come across n number of constants in the complementary function, which are to be evaluated to get the exact solution. To evaluate these constants, n number of initial conditions are required.

The *initial conditions* of a network are the conditions prevailing in the elements of the network at the instant of closing the switch at $t = 0$.

In a switching operation, $t = 0$ is taken as reference. The initial conditions in a network may be the voltages across the various elements, the currents through them or charges existing on them at the time of switching operation, i.e. at $t = 0$. Immediately before a switching operation, these quantities are referred as v (0–), i (0–), q (0–), at $t = 0-$. Immediately after the switching operation, these quantities are referred as v (0+), i (0+), q (0+), at $t = 0+$. Knowing the values of voltages, currents and charges on the various elements at $t = 0-$, and the changes introduced immediately after the switching operation, i.e.

at $t = 0+$, additional Eqs can be written, which can be solved simultaneously with the general differential Eq., to evaluate the constants. The conditions existing on the various elements of the network at $t = \infty$, are called the *final conditions*.

The initial conditions in a network depends on the past history of the network prior to $t = 0-$ and the network structure at $t = 0+$, after switching. They also depend on the nature of the elements in the network. It is assumed that the switching time is zero.

The knowledge of the initial conditions is useful in the following ways:

1. They are necessary to evaluate the constants during the solution of a differential Eq.
2. The knowledge of the behaviour of the elements at the instant of switching is indispensable in the understanding of nonlinear switching circuits.
3. The knowledge of the initial values of one or more derivatives of a response, are helpful in anticipating the form of response, thus serving as the check on the solution.
4. Useful in getting to know the elements individually and in combination, which is essential in the analysis of networks.

11.2 INITIAL CONDITIONS IN ELEMENTS

While solving problems on networks, using integro-differential Eqs in mathematics, initial conditions are always given and the job of a mathematician is to find the solution using the given initial conditions. But, an engineer has to correctly determine the initial conditions first, before finding the general solution, which is more difficult.

Knowing the values of voltages and currents of the elements at $t = 0-$, finding these values at $t = 0+$, constitutes the evaluation of initial conditions.

(i) The resistor

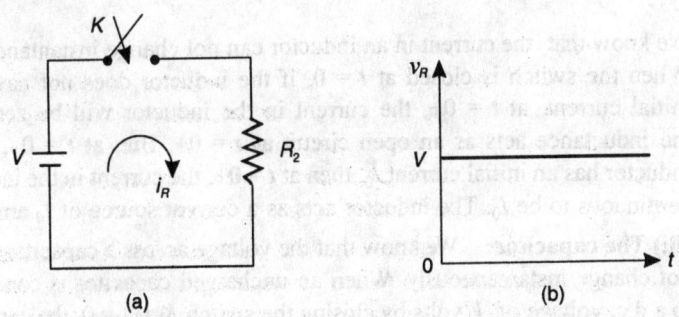

(a)   (b)

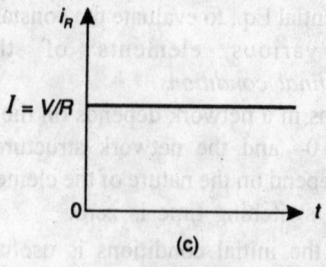

Fig. 11.1

In a resistor, the current and the voltage are related by $v = iR$. When a step voltage of V volts as shown in Fig. 11.1 (b) is applied to a resistor of R Ω, by closing the switch K at $t = 0$, as shown in Fig. 11.1 (a), the current will also be a step function as shown in Fig. 11.1 (c) and is given by, $I = V/R$. The waveform of the current is the same as the waveform of the voltage, i.e. the current through a resistor changes instantaneously, if the voltage changes instantaneously. Similarly, the voltage across the resistor also changes instantaneously, when the current through it changes instantaneously.

(ii) The inductor

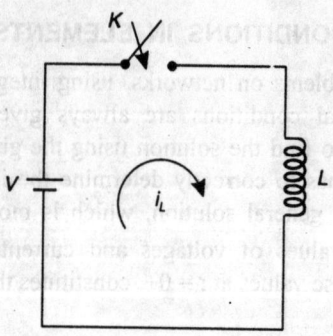

Fig. 11.2

We know that the current in an inductor can not change instantaneously. When the switch is closed at $t = 0$, if the inductor does not have any initial current, at $t = 0+$, the current in the inductor will be zero, i.e. the inductance acts as an open circuit at $t = 0+$. But, at $t = 0-$, if the inductor has an initial current I_0, then at $t = 0+$, the current in the inductor continuous to be I_0. The inductor acts as a current source of I_0 amperes.

(iii) The capacitor: We know that the voltage across a capacitance can not change instantaneously. When an uncharged capacitor is connected to a d.c. voltage of V volts by closing the switch K at $t = 0$, the capacitor

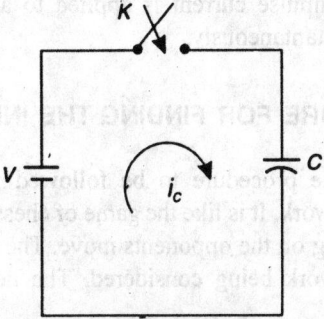

Fig. 11.3

acts as a short circuit and hence, the current flows instantaneously. This is because, when there is no charge on the capacitor, the voltage across it is also zero and amounts to short circuit. If the capacitor has an initial charge of q_0 coulombs at $t = 0-$, then, at $t = 0+$, the capacitor is equivalent to a voltage source $V_0 = q_0/C$.

Table 11.1 gives the conditions of the elements, before and after closing the switch.

Table 11.1: Initial conditions

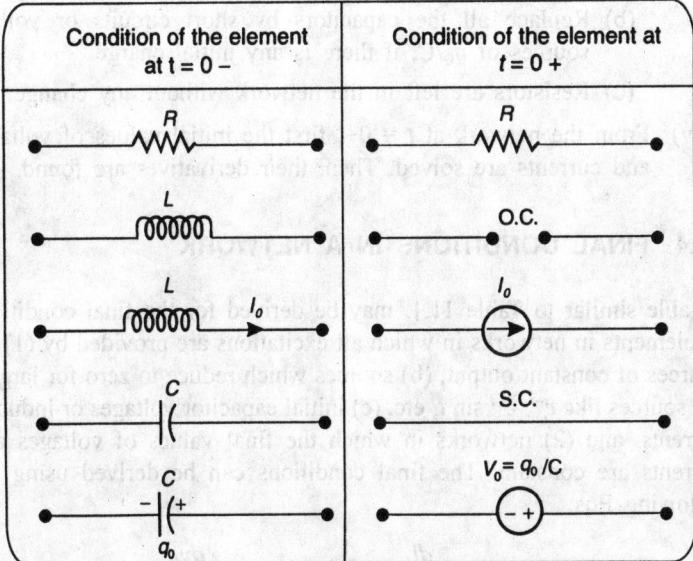

There are some exceptions to the initial conditions of the elements as given in Table 11.1. They are:

(i) When an impulse voltage is applied to an inductance, its current changes instantaneously.

(ii) When an impulse current is applied to a capacitor, its voltage changes instantaneously.

11.3 PROCEDURE FOR FINDING THE INITIAL CONDITIONS

There is no unique procedure to be followed for finding the initial conditions of a network. It is like the game of chess in which, the strategy is chosen depending on the opponents move. The procedure depends on the particular network being considered. The normal procedure is as follows.

(i) The initial values of voltages and currents, i.e. before closing the switch at $t = 0-$, can be found directly from the schematic diagram of the given network.

(ii) For each element of the network, we must find out, what happens to the element at $t = 0+$, i.e. after closing the switch.

(iii) A new equivalent network $t = 0+$ is constructed as per the following rules:

 (a) Replace all the inductors by open circuits or current sources having values of current flowing at $t = 0-$.

 (b) Replace all the capacitors by short circuits or voltage sources of q_0/C, if there is any initial charge.

 (c) Resistors are left in the network without any change.

(iv) From the network at $t = 0+$, first the initial values of voltages and currents are solved. Then, their derivatives are found.

11.4 FINAL CONDITIONS IN A NETWORK

A table similar to Table 11.1, may be derived for the final conditions of elements in networks in which all excitations are provided by (1) (a) sources of constant output, (b) sources which reduce to zero for large t, i.e. sources like e^{-t}, $e^{-t} \sin t$, etc. (c) initial capacitor voltages or inductor currents, and (2) networks in which the final values of voltages and currents are constant. The final conditions can be derived using the following Eqs.

$$v_L = L\frac{di_L}{dt} \quad \text{and} \quad i_C = C\frac{dv_C}{dt}$$

The derivatives have zero value, when the quantity in the steady state is constant. Table 11.2 gives the initial and final conditions of the elements.

Table 11.2: Initial and final conditions of elements

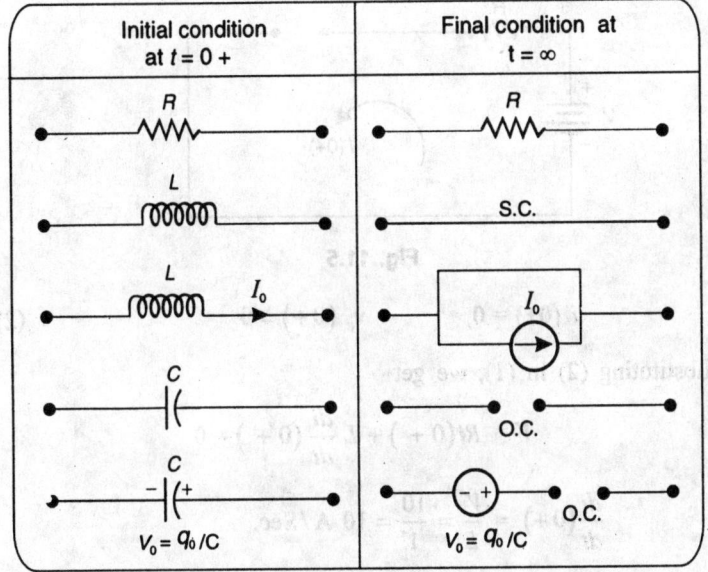

Initial condition at $t = 0+$	Final condition at $t = \infty$
R	R
L	S.C.
L I_0	I_0
C	O.C.
C $V_0 = q_0/C$	$V_0 = q_0/C$ O.C.

WORKED EXAMPLES

11.1 In the circuit shown in Fig. 11.4, $V = 10$ V, $R = 10\ \Omega$, $L = 1$ H, $C = 10\ \mu$F, and $v_c(0) = 0$, find $i(0+)$, $\dfrac{di}{dt}(0+)$ and $\dfrac{d^2i}{dt^2}(0+)$.

(Kuvempu University)

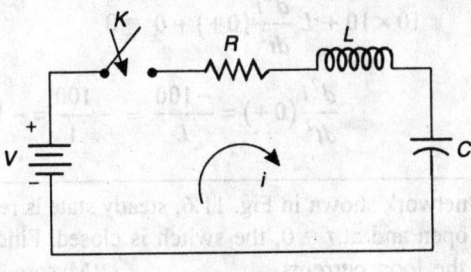

Fig. 11.4

Solution: When the switch K is closed at $t = 0$,

$$V = Ri + L\frac{di}{dt} + \frac{1}{C}\int i\,dt \qquad (1)$$

At $t = 0+$, the network is as shown in Fig. 11.5.

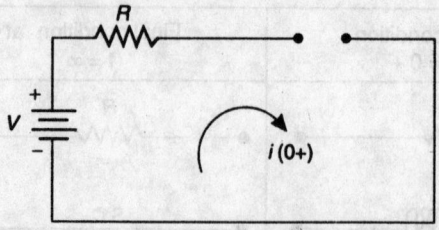

Fig. 11.5

∴ $\qquad i\,(0+) = 0, \qquad v_c\,(0+) = 0$ $\qquad\qquad$ (2)

Substituting (2) in (1), we get

$$V = Ri(0+) + L\frac{di}{dt}(0+) + 0$$

∴ $\qquad \dfrac{di}{dt}(0+) = \dfrac{V}{L} = \dfrac{10}{1} = 10 \text{ A/Sec.}$

Differentiating (1), we get

$$R\frac{di}{dt} + L\frac{d^2i}{dt^2} + \frac{i}{C} = 0$$

Substituting initial conditions, we get

$$R\frac{di}{dt}(0+) + L\frac{d^2i}{dt^2}(0+) + \frac{i(0+)}{C} = 0$$

i.e. $\qquad 10 \times 10 + L\dfrac{d^2i}{dt^2}(0+) + 0 = 0$

∴ $\qquad \dfrac{d^2i}{dt^2}(0+) = \dfrac{-100}{L} = -\dfrac{100}{1} = -100 \text{ A/Sec}^2$

11.2 In the network shown in Fig. 11.6, steady state is reached with the switch K open and at $t = 0$, the switch is closed. Find the initial values of all the loop currents. (Mysore University)

Solution: We don't know the currents and voltages across the inductor and the capacitors before closing the switch K and they have to be found. At $t = 0-$, i.e. under steady state conditions, K is open. L acts a s.c. and C_1 and C_2 act as open circuits.

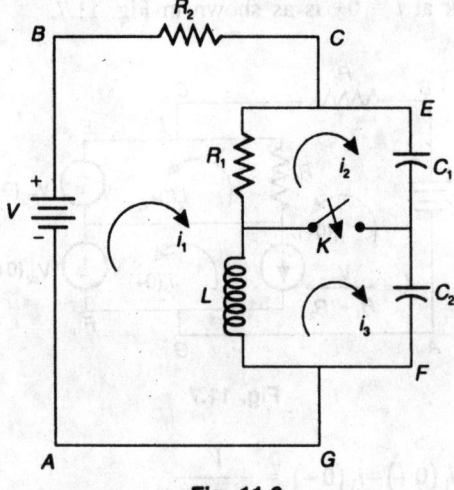

Fig. 11.6

$$i_1(0-) = i_L(0-) = i_L(0+) = \frac{V}{R_1 + R_2} \qquad (1)$$

The total voltage across the capacitors is equal to the voltage across R_1.

$$V_{C_1}(0-) + V_{C_2}(0-) = i_1(0-) R_1 = \frac{R_1}{R_1 + R_2} V \qquad (2)$$

The charges on the capacitors are same as they are in series.

$$q_1(0-) = q_2(0-)$$

i.e.

$$C_1 V_{C_1}(0-) = C_2 V_{C_2}(0-)$$

∴

$$\frac{V_{C_1}(0-)}{V_{C_2}(0-)} = \frac{C_2}{C_1}$$

$$\frac{V_{C_1}(0-) + V_{C_2}(0-)}{V_{C_2}(0-)} = \frac{C_2 + C_1}{C_1} = \frac{V\left(\dfrac{R_1}{R_1 + R_2}\right)}{V_{C_2}(0-)}$$

∴

$$V_{C_2}(0-) = V\left(\frac{C_1}{C_1 + C_2}\right)\left(\frac{R_1}{R_1 + R_2}\right) = V_{C_2}(0+) \qquad (3)$$

III^{ly}

$$V_{C_1}(0-) = V\left(\frac{C_2}{C_1 + C_2}\right)\left(\frac{R_1}{R_1 + R_2}\right) = V_{C_1}(0+) \qquad (4)$$

The network at $t = 0+$ is as shown in Fig. 11.7.

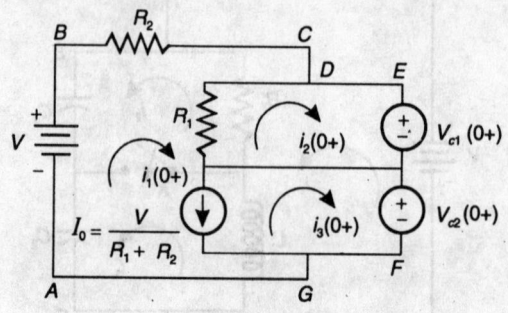

Fig. 11.7

$$i_1(0+) - i_3(0+) = \frac{V}{R_1 + R_2}$$

But
$$i_1(0+) = \frac{V}{R_1 + R_2}$$

$$\therefore \quad i_3(0+) = 0$$

For the closed loop of $i_2(0+)$

$$-\left[i_1(0+) - i_2(0+)\right]R_1 + V_{C_1}(0+) = 0$$

i.e.
$$\left[i_1(0+) - i_2(0+)\right]R_1 = V_{C_1}(0+)$$

i.e.
$$i_1(0+)R_1 - i_2(0+)R_1 = V \frac{C_2}{C_1 + C_2} \frac{R_1}{R_1 + R_2}$$

$$\therefore \quad i_2(0+) = \frac{\dfrac{VR_1}{R_1 + R_2} - V \dfrac{C_2}{C_1 + C_2} \dfrac{R_1}{R_1 + R_2}}{R_1}$$

$$= \frac{V}{R_1 + R_2} \frac{C_1}{C_1 + C_2}$$

11.3 In the network of the Fig. 11.8 (a), the switch K is closed at $t = 0$, with the capacitor uncharged. Find the values for $i, \dfrac{di}{dt}, \dfrac{d^2 i}{dt^2}$ at $t = 0+$, for element values as follows. $V = 100$ V, $R = 1,000 \ \Omega$ and $C = 1 \ \mu F$. (Karnataka University)

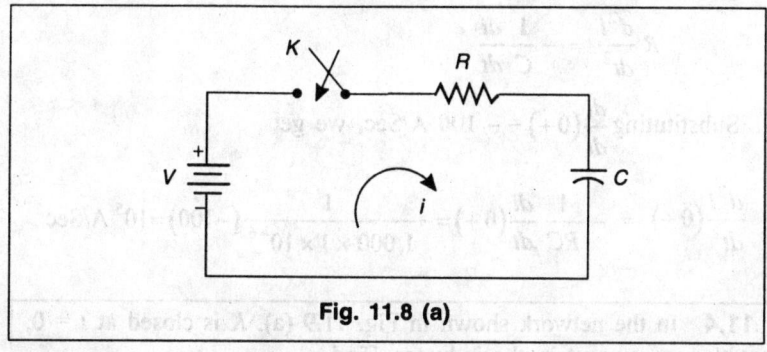

Fig. 11.8 (a)

Solution: The network at $t = 0+$ is as shown in Fig. 11.8 (b)

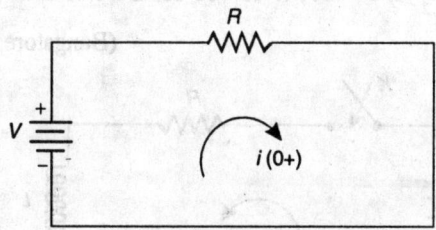

Fig. 11.8 (b)

$$\therefore \qquad i(0+) = \frac{V}{R} = \frac{100}{1,000} = 0.1 \text{ A} \qquad (1)$$

When K is closed at $t = 0$

$$V = Ri + \frac{1}{C} \int i \, dt \quad \text{on differentiation, we get}$$

$$0 = R \frac{di}{dt} + \frac{i}{C} \qquad (2)$$

or $$\qquad 0 = R \frac{di}{dt}(0+) + \frac{i(0+)}{C}$$

Substituting $i(0+) = 0.1$ A, we get

$$\frac{di}{dt}(0+) = -\frac{i(0+)}{RC} = -\frac{0.1}{1,000 \times 1 \times 10^{-6}} = -100 \text{ A/Sec} \qquad (3)$$

From (2), we can write

$$R \frac{di}{dt} = -\frac{i}{C} \quad \text{on differentiation, we get}$$

$$R \frac{d^2i}{dt^2} = -\frac{1}{C} \frac{di}{dt}$$

Substituting $\frac{di}{dt}(0+) = -100$ A/Sec, we get

$$\frac{d^2i}{dt^2}(0+) = -\frac{1}{RC} \frac{di}{dt}(0+) = -\frac{1}{1,000 \times 1 \times 10^{-6}}(-100) = 10^5 \text{ A/Sec}^2$$

11.4 In the network shown in Fig. 11.9 (a), K is closed at $t = 0$, with zero current in the inductor. Find

i, $\dfrac{di}{dt}$, and $\dfrac{d^2i}{dt^2}$ at $t = 0+$, if $R = 10\ \Omega$, $L = 1$ H and $V = 100$ V.

(Bangalore University)

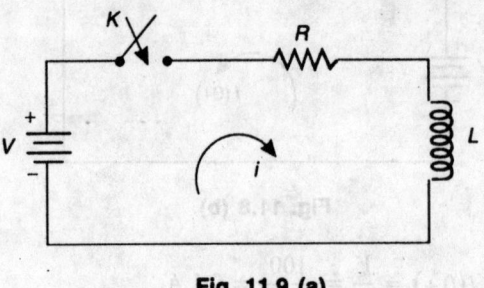

Fig. 11.9 (a)

Solution: Network at $t = 0+$ is as shown in Fig. 11.9 (b)

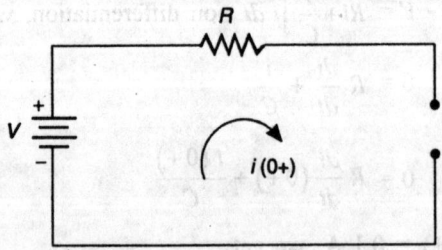

Fig. 11.9 (b)

$$i(0+) = 0 \qquad\qquad (1)$$

When K is closed at $t = 0$

$$V = Ri + L\frac{di}{dt} \qquad\qquad (2)$$

$$\therefore \qquad \frac{di}{dt} = \frac{V - Ri}{L} \qquad\qquad (3)$$

i.e. $\qquad \dfrac{di}{dt}(0+) = \dfrac{V - Ri\,(0+)}{L} = \dfrac{V}{L} = \dfrac{100}{1} = 100 \text{ A/Sec}$

Differentiating (3), we get

$$\frac{d^2 i}{dt^2} = -\frac{R}{L}\frac{di}{dt}$$

i.e. $\qquad \dfrac{d^2 i}{dt^2}(0+) = -\dfrac{R}{L}\dfrac{di}{dt}(0+) = -\dfrac{10}{1} \times 100 = -1{,}000 \text{ A/Sec}^2$

11.5 In the network shown in Fig. 11.10, K is changed from position a to b at $t = 0$. Solve for i, $\dfrac{di}{dt}$, $\dfrac{d^2 i}{dt^2}$ at $t = 0 +$, if $R = 1{,}000\ \Omega$, $L = 1\,H$, $C = 0.1\,\mu F$ and $V = 100\,V$. Assume that the cpacitor is initially uncharged. (Mangalore University)

Solution: When the switch K is in position a, L acts as s.c.

$$\therefore \qquad i(0-) = \frac{V}{R} = 100/1{,}000 = 0.1 \text{ A}$$

$$\therefore \qquad i(0+) = 0.1 \text{ A}$$

i.e. the inductor has an initial current of 0.1 A.

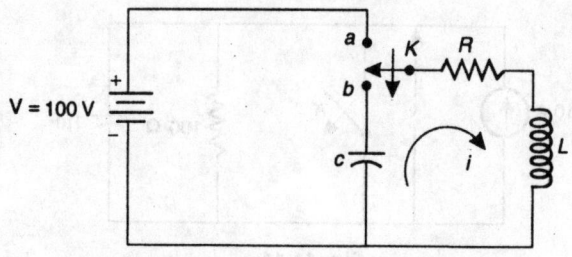

Fig. 11.10

When K is changed from a to b, $i(0+) = 0.1$ A

When K is at b.

$$Ri + L\frac{di}{dt} + \frac{1}{C}\int i\ dt = 0 \qquad\qquad (1)$$

i.e. $$Ri(0+) + L\frac{di}{dt}(0+) + \frac{1}{C}\int i(0+)dt = 0$$

It is given that, $$\frac{1}{C}\int i(0+)dt = v_C(0+) = 0$$

$$\therefore \quad 1,000 \times 0.1 + 1\frac{di}{dt}(0+) + 0 = 0$$

i.e. $$\frac{di}{dt}(0+) = -100 \text{ A/Sec}$$

Differentiating (1) $$R\frac{di}{dt} + L\frac{d^2i}{dt^2} + \frac{i}{C} = 0$$

i.e. $$R\frac{di}{dt}(0+) + L\frac{d^2i}{dt^2}(0+) + \frac{i(0+)}{C} = 0$$

$$1,000(-100) + 1\frac{d^2i}{dt^2}(0+) + \frac{0.1}{0.1 \times 10^{-6}} = 0$$

$$\therefore \quad \frac{d^2i}{dt^2}(0+) = -9 \times 10^5 \text{ A/Sec}^2$$

11.6 In the circuit shown in Fig. 11.11, switch K is opened at $t = 0$, find the values of v, $\dfrac{dv}{dt}$, and $\dfrac{d^2v}{dt^2}$ at $t = 0+$.

(Karnataka University)

Solution: When the switch K is closed, all the current flows through the s.c. The capacitor is not charged as it acts as an o.c.

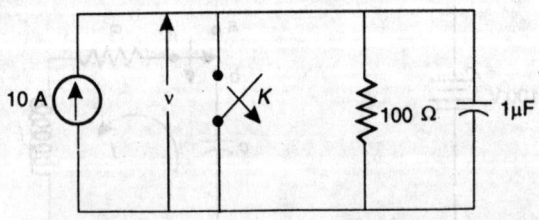

Fig. 11.11

$$\therefore \quad v_C(0-) = 0 = v(0+)$$

When the switch K is opened

$$10 = \frac{v}{100} + C\frac{dv}{dt}$$

i.e.
$$\frac{dv}{dt} = \left(10 - \frac{v}{100}\right)\frac{1}{C} \qquad (1)$$

i.e.
$$\frac{dv}{dt}(0+) = \left(10 - \frac{v(0+)}{100}\right)\frac{1}{C}$$

$$= (10 - 0)\frac{1}{1 \times 10^{-6}} = 10^7 \text{ V/Sec}$$

Differentiating Eq. (1), we get

$$\frac{d^2v}{dt^2} = 0 - \left(\frac{dv}{dt}\frac{1}{100}\right)\frac{1}{C}$$

i.e.
$$\frac{d^2v}{dt^2}(0+) = -\frac{1}{1 \times 10^{-6}}\left(\frac{1}{100} \times 10^7\right) = 10^{11} \text{ V/Sec}^2$$

11.7 In the network shown in Fig. 11.12, $v_C(0-) = 1$ V, $i_2(0-) = 0$ A. The switches Sw_1 and Sw_2 are closed at $t = 0$. Find i_1, i_2 and their first and second order derivatives at $t = 0+$.

(Bangalore University)

Solution

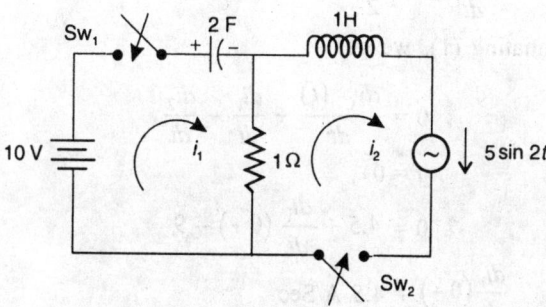

Fig. 11.12

$$v_C(0-) = v_C(0+) = 1 \text{ V}$$

$$i_2(0-) = i_2(0+) = 0 \text{ A}$$

$$i_1(0+) = \frac{(10-1)}{1} = 9 \text{ A}$$

When Sw_1 and Sw_2 are closed at $t = 0$

$$10 = v_C(t) + (i_1 - i_2)1 \qquad (1)$$

$$5 \sin 2t = (i_2 - i_1)1 + L\frac{di_2}{dt} \qquad (2)$$

(1) + (2), gives

$$10 + 5 \sin 2t = v_C(t) + \frac{di_2}{dt}$$

At $\qquad t = 0+,$

$$10 + 0 = 1 + \frac{di_2}{dt}(0+)$$

$\therefore \qquad \dfrac{di_2}{dt}(0+) = 9 \text{ A/Sec}$

From (1) $\qquad 10 = v_C(0+) + i_1(0+) - i_2(0+)$

$$10 = 1 + i_1(0+) - 0$$

$\therefore \qquad i_1(0+) = 9 \text{ A}$

$$v_C(t) = \frac{1}{C}\int i_1 \, dt \quad \text{On differentiation, we get}$$

$$\frac{dv_C(t)}{dt} = \frac{1}{C}i_1$$

i.e. $\qquad \dfrac{dv_C(0+)}{dt} = \dfrac{9}{2} = 4.5 \text{ V/Sec}$

Differentiating (1), we get

$$0 = \frac{dv_C(t)}{dt} + \frac{di_1}{dt} - \frac{di_2}{dt} \qquad (3)$$

At $\qquad t = 0+,$

$$0 = 4.5 + \frac{di_1}{dt}(0+) - 9$$

$\therefore \qquad \dfrac{di_1}{dt}(0+) = 4.5 \text{ A/Sec}$

Differentiating (2), we get

$$5 \times 2 \cos 2t = \left(\frac{di_2}{dt} - \frac{di_1}{dt}\right) + \frac{d^2 i_2}{dt^2}$$

At $\qquad t = 0+,$

$$10 = (9 - 4.5) + \frac{d^2 i_2}{dt^2}(0+)$$

$\therefore \qquad \dfrac{d^2 i_2}{dt^2}(0+) = 5.5 \text{ A/Sec}^2$

$$i_1 = C \frac{dv_C}{dt} = 2 \frac{dv_C}{dt}$$

$$\frac{di_1}{dt} = 2 \frac{d^2 v_C}{dt^2}$$

$$\therefore \quad \frac{d^2 v_C}{dt^2}(0+) = \frac{1}{2} \frac{di_1}{dt}(0+) = \frac{4.5}{2} = 2.25 \text{ V/Sec}^2$$

Differentiating (3), we get

$$0 = \frac{d^2 v_C}{dt^2} + \frac{d^2 i_1}{dt^2} - \frac{d^2 i_2}{dt^2}$$

At $t = 0+$,

$$0 = 2.25 + \frac{d^2 i_1}{dt^2}(0+) - 5.5$$

$$\therefore \quad \frac{d^2 i_1}{dt^2}(0+) = 3.25 \text{ A/Sec}^2$$

11.8 In the circuit shown in Fig. 11.13, the switch was in position a for sufficiently long time to have achieved steady state. At $t = 0$, the switch was changed from a to b. Determine I_L and v_C, their first and second order derivatives at $t = 0+$.

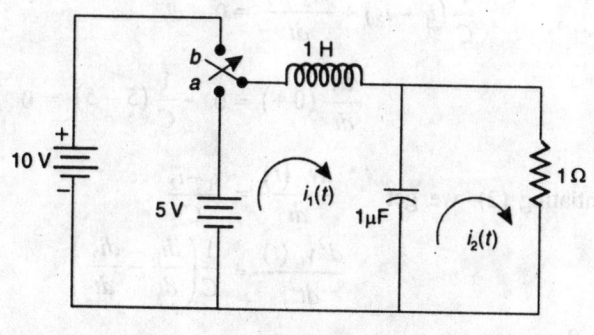

Fig. 11.13

Solution: When K is on a, L acts as s.c. and C acts as an o.c.

$$\therefore \quad i_1(0-) = \frac{5}{1} = 5 \text{ A} = i_1(0+)$$

$$i_2(0-) = 5 \text{ A} = i_2(0+)$$

$$v_C(0-) = 5 \text{ V} = v_C(0+)$$

When K is switched on to b

$$10 = 1\frac{di_1}{dt} + \frac{1}{C}\int(i_1 - i_2)\,dt \qquad\qquad (1)$$

Substituting the initial conditions, we get

$$10 = \frac{di_1}{dt}(0+) + 5, \qquad \frac{di_1}{dt}(0+) = 5 \text{ A/Sec}$$

$$v_C(t) = \frac{1}{C}\int(i_1 - i_2)\,dt \qquad\qquad (2)$$

i.e. $\qquad \dfrac{dv_C(t)}{dt} = \dfrac{1}{C}(i_1 - i_2)$

i.e. $\qquad \dfrac{dv_C}{dt}(0+) = \dfrac{1}{C}(5-5) = 0$

Differentiating (1), we get

$$0 = \frac{d^2 i_1}{dt^2} + \frac{i_1 - i_2}{C}, \qquad \therefore \frac{d^2 i_1}{dt^2}(0+) = 0$$

$$\frac{1}{C}\int(i_1 - i_2)\,dt + i_2(t)\times 1 = 0$$

i.e. $\qquad \dfrac{1}{C}(i_1 - i_2) + \dfrac{di_2(t)}{dt} = 0$

i.e. $\qquad \dfrac{di_2}{dt}(0+) = 0 - \dfrac{1}{C}(5-5) = 0$

Differentiating (2), we get $\qquad \dfrac{dv_C(t)}{dt} = \dfrac{i_1 - i_2}{C}$

$$\therefore \qquad \frac{d^2 v_C(t)}{dt^2} = \frac{1}{C}\left(\frac{di_1}{dt} - \frac{di_2}{dt}\right)$$

At $t = 0+$

$$\frac{d^2 v_C}{dt^2}(0+) = \frac{1}{C}\left(\frac{di_1}{dt}(0+) - \frac{di_2}{dt}(0+)\right)$$

$$= \frac{1}{C}(5 - 0) = \frac{1}{1\times 10^{-6}} \qquad (5)$$

$$= 5\times 10^6 \text{ V/Sec}^2$$

11.9 In the network shown in Fig. 11.14, determine the initial values of i_1, i_2 and v_C and their derivatives, if the switch is closed at $t = 0$. Given $v_C (0-) = 1$ V $= v_C (0+)$, $i_1 (0-) = 0 = i_1 (0+)$, and $\omega = 300$ rad/Sec. (Mangalore University)

Solution

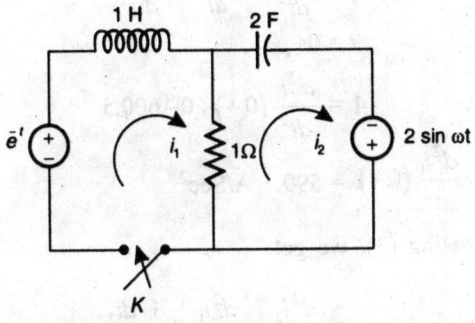

Fig. 11.14

When K is closed at $t = 0$

$$e^{-t} = 1\frac{di_1}{dt} + 1(i_1 - i_2) \tag{1}$$

$$2 \sin \omega t = (i_2 - i_1)1 + \frac{1}{2}\int i_2\, dt \tag{2}$$

Substituting initial conditions in (2), we get

$$0 = i_2(0+) - i_1(0+) + v_C(0+)$$

$$i_2(0+) = i_1(0+) - v_C(0+) = 0 - 1 = -1 \text{ A}$$

Substituting initial conditions in Eq. (1), we get

$$e^{-0} = 1\frac{di_1}{dt}(0+) + 1\left[i_1(0+) - i_2(0+)\right]$$

i.e. $$1 = 1\frac{di_1}{dt}(0+) + 1[0+1]$$

$\therefore$ $$\frac{di_1}{dt}(0+) = 0$$

Differentiating Eq. (2), we get

$$2\,\omega \cos \omega t = \frac{di_2}{dt} - \frac{di_1}{dt} + \frac{i_2}{2} \tag{3}$$

At $t = 0+$

$$2\,\omega = \frac{di_2}{dt}(0+) - 0 + \frac{-1}{2}$$

$$\therefore \quad \frac{di_2}{dt}(0+) = (2 \times 300) + 0.5 = 600.5 \text{ A/Sec}$$

Differentiating (1), we get

$$e^{-t}(-1) = \frac{d^2 i_1}{dt^2} + \frac{di_1}{dt} - \frac{di_2}{dt}$$

At $\quad t = 0+,$

$$-1 = \frac{d^2 i_1}{dt^2}(0+) + 0 - 600.5$$

$$\therefore \quad \frac{d^2 i_1}{dt^2}(0+) = 599.5 \text{ A/Sec}^2$$

Differentiating (3), we get

$$-2\omega^2 \sin \omega t = \frac{d^2 i_2}{dt^2} - \frac{d^2 i_1}{dt^2} + \frac{1}{2}\frac{di_2}{dt}$$

At $\quad t = 0+,$

$$\frac{d^2 i_2}{dt^2}(0+) - 599.5 + \frac{600.5}{2} = 0$$

$$\therefore \quad \frac{d^2 i_2}{dt^2}(0+) = 599.5 - 300.25 = 299.25 \text{ A/Sec}^2$$

$$v_C = \frac{1}{C} \int i_2 \, dt$$

i.e.

$$\frac{dv_C}{dt} = \frac{i_2}{C}$$

At $\quad t = 0+,$

$$\frac{dv_C}{dt}(0+) = \frac{i_2(0+)}{C} = \frac{1}{2}(-1) = -\frac{1}{2} \text{ V/Sec}$$

$$\frac{d^2 v_C}{dt^2} = \frac{1}{C}\frac{di_2}{dt}$$

i.e. $\quad \dfrac{d^2 v_C}{dt^2}(0+) = \dfrac{1}{C}(600.5) = \dfrac{600.5}{2} = 300.25 \text{ V/Sec}^2$

11.10 For the network shown in Fig. 11.15, determine the initial values of $i_1, i_2, i_3, v_{C1}, v_{C2}$ and their first and second derivatives given, $i_2(0-) = 0$, $i_a(0-) = 1$ A, $v_{C1}(0-) = 2$ V, $V_{C2}(0-) = 0$ V. At $t = 0$, all switches are closed. (Gulbarga University)

Solution:

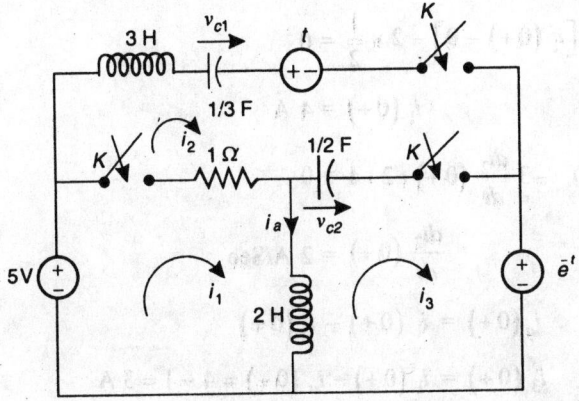

Fig. 11.15

Writing differential eqs for all the loops.

$$5 - 1\left(i_1 - i_2\right) - 2\frac{di_a}{dt} = 0 \tag{1}$$

$$-3\frac{di_2}{dt} + v_{C_1} - t - v_{C_2} - \left(i_2 - i_1\right)1 = 0 \tag{2}$$

$$2\frac{di_a}{dt} + v_{C_2} + e^{-t} = 0 \tag{3}$$

From (2), at $t = 0+$,

$$-3\frac{di_2}{dt}\left(0+\right) + v_{C_1}\left(0+\right) - 0 - v_{C_2}\left(0+\right) - \left[i_2\left(0+\right) - i_1\left(0+\right)\right] = 0$$

$$-3\frac{di_2}{dt}\left(0+\right) + 2 - 0 - 0 - \left[0 - i_1\left(0+\right)\right] = 0$$

i.e. $-3\dfrac{di_2}{dt}\left(0+\right) + 2 + i_1\left(0+\right) = 0$ \hfill (4)

From (3), at $t = 0+$,

$$2\frac{di_a}{dt}\left(0+\right) + v_{C_2}\left(0+\right) - 1 = 0$$

i.e. $\qquad 2\dfrac{di_a}{dt}\left(0+\right) + 0 - 1 = 0$

i.e. $\qquad \dfrac{di_a}{dt}\left(0+\right) = \dfrac{1}{2}$ A/Sec

From (1), at $t = 0+$,

$$5 - \left[i_1(0+) - 0\right] - 2 \times \frac{1}{2} = 0$$

$\therefore$ $\qquad\qquad\qquad i_1(0+) = 4$ A

From (4), $-3 \dfrac{di_2}{dt}(0+) + 2 + 4 = 0$

$\therefore$ $\qquad\qquad\qquad \dfrac{di_2}{dt}(0+) = 2$ A/Sec

Also $\qquad i_a(0+) = i_1(0+) - i_3(0+)$

$\therefore$ $\qquad i_3(0+) = i_1(0+) - i_a(0+) = 4 - 1 = 3$ A

Differentiating (1), (2), and (3), we get

$$-\frac{di_1}{dt} + \frac{di_2}{dt} - 2\frac{d^2 i_a}{dt^2} = 0 \qquad\qquad (4)$$

$$-3\frac{d^2 i_2}{dt^2} + \frac{dv_{C_1}}{dt} - 1 - \frac{dv_{C2}}{dt} - \frac{d_{i_2}}{dt} + \frac{di_1}{dt} = 0 \qquad\qquad (5)$$

$$2\frac{d^2 i_a}{dt^2} + \frac{dv_{C2}}{dt} + e^{-t} = 0 \qquad\qquad (6)$$

We know $\qquad v_{C_1} = \dfrac{1}{C} \displaystyle\int i_2\, dt = 3 \int i_2\, dt$

i.e. $\qquad\qquad \dfrac{dv_{C_1}}{dt} = 3\, i_2$

i.e. $\qquad \dfrac{dv_{C_1}}{dt}(0+) = 3\, i_2(0+) = 0$ V/Sec

$$v_{C_2} = 2 \int (i_2 - i_3)\, dt$$

i.e. $\qquad\qquad \dfrac{dv_{C_2}}{dt} = 2(i_2 - i_3)$

i.e. $\qquad \dfrac{dv_{C_2}}{dt}(0+) = 2(0 - 3) = -6$ V/Sec

From (6), $2\dfrac{d^2 i_a}{dt^2}(0+) - 6 + 1 = 0$

$$\therefore \qquad \frac{d^2 i_a}{dt^2}(0+) = \frac{5}{2} \text{ A/Sec}^2$$

From (4), $\quad -\dfrac{di_1}{dt}(0+) + 2 - 2 \times \dfrac{5}{2} = 0$

$$\therefore \qquad \frac{di_1}{dt}(0+) = -3 \text{ A/Sec}$$

We know $\qquad i_a = i_1 - i_3$

$$\frac{di_a}{dt} = \frac{di_1}{dt} - \frac{di_3}{dt}$$

i.e. $\qquad \dfrac{di_3}{dt}(0+) = \dfrac{di_1}{dt}(0+) - \dfrac{di_a}{dt}(0+)$

$$= -3 - 1/2 = -7/2 \text{ A/Sec}$$

From (5), $\quad -3 \dfrac{d^2 i_2}{dt^2}(0+) + 0 - 1 + 6 - 2 - 3 = 0$

$$\therefore \qquad \frac{d^2 i_a}{dt^2}(0+) = 0$$

Now $\qquad \dfrac{d^2 i_a}{dt^2} = \dfrac{d^2 i_1}{dt^2} - \dfrac{d^2 i_3}{dt^2}$

Differentiating (4) and (6), we get

$$-\frac{d^2 i_1}{dt^2} = \frac{d^2 i_2}{dt^2} - 2 \frac{d^3 i_a}{dt^3} = 0 \tag{7}$$

$$2\frac{d^3 i_a}{dt^3} + \frac{d^2 v_{C2}}{dt^2} - e^{-t} = 0 \tag{8}$$

$$2\frac{d^3 i_a}{dt^3} + \frac{d}{dt}\left(\frac{dv_{C2}}{dt}\right) - e^{-t} = 0$$

i.e. $\quad 2\dfrac{d^3 i_a}{dt^3} + \dfrac{d}{dt}\left[\dfrac{d}{dt}\{2\int (i_2 - i_3)dt\}\right] - e^{-t} = 0$

$$2\frac{d^3 i_a}{dt^3} + 2\left(\frac{di_2}{dt} - \frac{di_3}{dt}\right) - e^{-t} = 0$$

i.e. $\quad 2\dfrac{d^3 i_a}{dt^3}(0+) + 2\left[2 + \dfrac{7}{2}\right] - 1 = 0$

$$\therefore \qquad \frac{d^3 i_a}{dt^3}(0+) = -5 \text{ A/Sec}^3$$

From (7), $\qquad -\frac{d^2 i_1}{dt^2}(0+) + 0 + 10 = 0$

$$\therefore \qquad \frac{d^2 i_1}{dt^2}(0+) = 10 \text{ A /Sec}^2$$

We know

$$\frac{d^2 i_3}{dt^2}(0+) = \frac{d^2 i_1}{dt^2}(0+) - \frac{d^2 i_a}{dt^2}(0+)$$

$$= 10 - \frac{5}{2} = 7.5 \text{ A/Sec}^2$$

11.11 For the network shown in Fig. 11.16. Find the values of nodal voltages, their first and second derivatives, all evaluated at $t = 0+$. For $t < 0$, all switches are closed. At $t = 0$, they are opened. $i_a(0-) = 1$, $v_2(0-) = 0$. (Bangalore University)

Solution

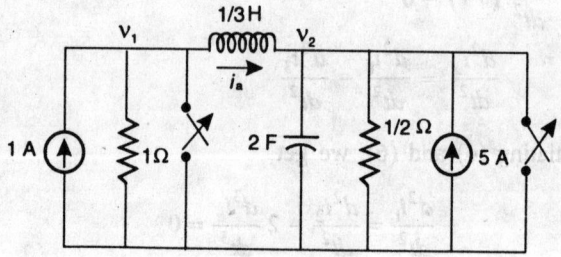

Fig. 11.16

$$i_a(0+) = 1 \text{ A}, \qquad v_2(0+) = 0 \text{ V}$$

At $t = 0$, the switches are opened, then

For node 1, $\quad 1 - \dfrac{v_1}{1} - i_a = 0$ \hfill (1)

$$i_a = \frac{1}{L} \int (v_1 - v_2)\, dt = 3 \int (v_1 - v_2)\, dt \qquad (2)$$

For node v_2

$$i_a - 2 \frac{dv_2}{dt} - \frac{v_2}{1/2} + 5 = 0$$

i.e. $\qquad i_a - 2 \dfrac{dv_2}{dt} - 2 v_2 + 5 = 0$ \hfill (3)

From (1), $\quad 1 - v_1(0+) - 1_a(0+) = 0$

i.e. $\qquad\qquad\qquad v_1(0+) = 1 - 1 = 0$ volts

From (3), $\quad i_a(0+) - 2\dfrac{dv_2}{dt}(0+) - 2v_2(0+) + 5 = 0$

i. e. $\qquad\qquad 1 - 2\dfrac{dv_2}{dt}(0+) - 0 + 5 = 0$

$\therefore \qquad\qquad\quad \dfrac{dv_2}{dt}(0+) = 3$ V/Sec

From (2), $\qquad\qquad \dfrac{di_a}{dt} = 3(v_1 - v_2) \qquad\qquad\qquad (4)$

i.e. $\qquad\qquad \dfrac{di_a}{dt}(0+) = 3(0 - 0) = 0$ A/Sec

Differentiating (1), we get

$$-\dfrac{dv_1}{dt} - \dfrac{di_a}{dt} = 0$$

$\therefore \qquad\qquad \dfrac{dv_1}{dt}(0+) = -\dfrac{di_a}{dt}(0+) = 0$ V/Sec $\qquad\qquad (5)$

Differentiating (3), we get

$$\dfrac{di_a}{dt} - 2\dfrac{d^2v_2}{dt^2} - 2\dfrac{dv_2}{dt} = 0$$

At $t = 0+$,

$$0 - 2\dfrac{d^2v_2}{dt^2}(0+) - 2 \times 3 = 0$$

$\therefore \qquad\qquad \dfrac{d^2v_2}{dt^2}(0+) = -3$ V/Sec^2

Differentiating (4), we get

$$\dfrac{d^2i_a}{dt^2} = 3\left(\dfrac{dv_1}{dt} - \dfrac{dv_2}{dt}\right)$$

$\therefore \qquad\qquad \dfrac{d^2i_a}{dt^2}(0+) = 3(0 - 3) = -9$ A/Sec^2

Differentiating (5), we get

$$\dfrac{d^2v_1}{dt^2}(0+) = -\dfrac{d^2i_a}{dt^2}(0+) = 9$$ V/Sec^2

11.12 For the circuit shown in Fig. 11.17, the switch K is closed at $t = 0$, connecting the battery to an unenergised network. Determine (i) v_1 and v_2 at $t = 0+$ and at $t = \infty$, (ii) 1st and 2nd order derivatives of v_1 and v_2 at $t = 0+$.   (Mangalore University)

Solution

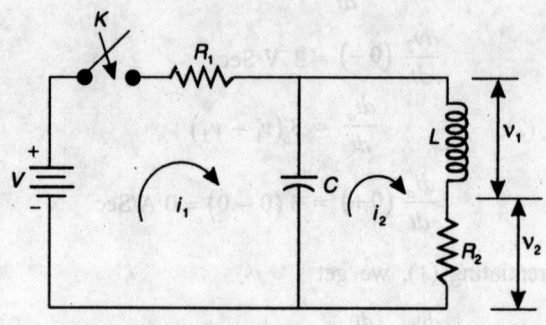

Fig. 11.17

At $t = \infty$, L acts as a s.c. and C acts as an o.c.

$$v_1(\infty) = 0$$

$$v_2(\infty) = \frac{V}{R_1 + R_2} R_2$$

At $t = 0+$, L acts as an o.c. and C acts as a s.c.

$$i_2(0-) = 0 \qquad\qquad i_1(0-) = 0$$

$\therefore \qquad\qquad i_2(0+) = 0 \qquad\qquad i_1(0+) = \dfrac{V}{R_1}$

$$v_1(0+) + v_2(0+) = 0$$

but $\qquad\qquad v_2(0+) = 0 \qquad \therefore\; i_2(0+) = 0$

$\therefore \qquad\qquad v_1(0+) = 0$

When K is closed at $t = 0$

$$V - i_1 R_1 - \frac{1}{C} \int (i_1 - i_2)\, dt = 0 \tag{1}$$

Differentiating (1), we get

$$0 - \frac{di_1}{dt} R_1 - \frac{1}{C}(i_1 - i_2) = 0 \tag{2}$$

i.e. $\quad -R_1 \dfrac{di_1}{dt}(0+) = \dfrac{1}{C}\left[\dfrac{V}{R_1} - 0\right]$

$\therefore \qquad \dfrac{di_1}{dt}(0+) = -\dfrac{V}{CR_1^2}$

$$L\dfrac{di_2}{dt} = V - i_1 R_1 - i_2 R_2 \qquad (3)$$

i.e. $\quad L\dfrac{di_2}{dt}(0+) = V - \dfrac{V}{R_1}R_1 - 0 = 0$

$\therefore \qquad \dfrac{di_2}{dt}(0+) = 0$

$$v_1 = V - i_1 R_1 - i_2 R_2$$

$$\dfrac{dv_1}{dt} = 0 - R_1\dfrac{di_1}{dt} - R_2\dfrac{di_2}{dt} \qquad (4)$$

$\therefore \qquad \dfrac{dv_1}{dt}(0+) = -R_1 \times \dfrac{-V}{CR_1^2} - 0 = \dfrac{V}{CR_1}$

$$v_2 = i_2 R_2$$

$$\dfrac{dv_2}{dt} = R_2\dfrac{di_2}{dt} \qquad (5)$$

$\therefore \qquad \dfrac{dv_2}{dt}(0+) = 0$

Differentiating (2), we get

$$-R_1\dfrac{d^2 i_1}{dt^2} - \dfrac{1}{C}\left(\dfrac{di_1}{dt} - \dfrac{di_2}{dt}\right) = 0$$

i.e. $\quad -R_1\dfrac{d^2 i_1}{dt^2}(0+) = \dfrac{1}{C}\left[-\dfrac{V}{CR_1^2} - 0\right]$

$\therefore \qquad \dfrac{d^2 i_1}{dt^2}(0+) = +\dfrac{V}{C^2 R_1^3}$

Differentiating (3), we get

$$L\dfrac{d^2 i_2}{dt^2} = 0 - R_1\dfrac{di_1}{dt} - R_2\dfrac{di_2}{dt}$$

i.e. $\quad \dfrac{d^2 i_2}{dt^2}(0+) = \dfrac{1}{L}\left[-R_1\dfrac{-V}{CR_1^2} - 0\right] = +\dfrac{V}{CLR_1}$

Differentiating (4), we get

$$\dfrac{d^2 v_1}{dt^2} = -R_1\dfrac{d^2 i_1}{dt^2} - R_2\dfrac{d^2 i_2}{dt^2}$$

i.e. $$\frac{d^2v_1}{dt^2}(0+) = -R_1\left(+\frac{V}{C^2R_1^3}\right) - R_2\left(+\frac{V}{CLR_1}\right)$$

$$= -\frac{V}{C^2R_1^2} - \frac{VR_2}{CLR_1}$$

Differentiating (5), we get

$$\frac{d^2v_2}{dt^2} = R_2\frac{d^2i_2}{dt^2}$$

i.e. $$\frac{d^2v_2}{dt^2}(0+) = R_2\left(+\frac{V}{CLR_1}\right) = +\frac{VR_2}{CLR_1}$$

11.13 In the circuit shown in Fig. 11.18, the capacitor is initially uncharged. Switch K is closed at time $t = 0$. The initial value of the current is found to be 25 mA through the CRO. The transient disappears (reduces to 2% of its initial value) after a time 0.1 Sec. By the classical method, determine the (i) value of R, (ii) value of C, and (iii) expression for the current $i(t)$ for $t > 0$.

(Mysore University)

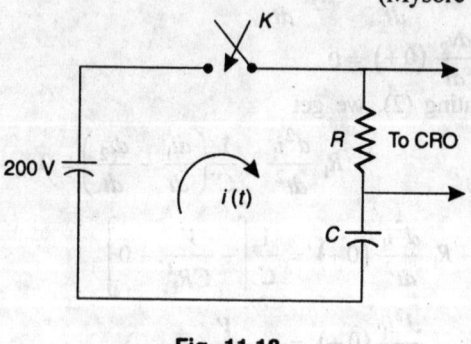

Fig. 11.18

Solution: When K is closed at $t = 0$

$$200 = \frac{1}{C}\int i\, dt + i R$$

i.e. $$R\frac{di}{dt} + \frac{i}{C} = 0$$

i.e. $$\left(RD + \frac{1}{C}\right) i = 0$$

$$(RCD + 1) = 0$$

i.e. $$D = -\frac{1}{RC}$$

$\therefore$ $$i = Ke^{-t/RC} \tag{1}$$

Where, K is a constant

At $t = 0+$,

$$i(0+) = 25 \times 10^{-3} \text{ A}$$

Substituting this in Eq. (1), we get
$$25 \times 10^{-3} = K$$

$\therefore$ $$i = 25 \times 10^{-3} e^{-t/RC} \tag{2}$$

At $t = 0+$, C acts as a short circuit

$\therefore$ $$i(0+) = \frac{200}{R} = 25 \times 10^{-3}$$

$\therefore$ $$R = \frac{200 \times 10^3}{25} = 8000 \ \Omega$$

After 0.1 Sec, $$i = 25 \times 10^{-3} \times \frac{2}{100} \text{ amps}$$

$\therefore$ $$25 \times 10^{-3} \times \frac{2}{100} = 25 \times 10^{-3} e^{-0.1/8000 \, C}$$

i.e. $$e^{-0.1/8000 \, C} = \frac{2}{100}$$

i.e. $$e^{0.1/8000 \, C} = 50$$

i.e. $$\frac{0.1}{8000 \, C} \log e = \log 50 \qquad \text{on taking logarithm,}$$
$$\log_{10} e = \log_{10} 2.71828 = 0.434$$

$$\frac{0.1}{8000 \, C} \times 0.434 = 1.699$$

$\therefore$ $$C = \frac{0.1 \times 0.434}{8000 \times 1.699} = 3.19 \times 10^{-6} \text{ F} = 3.19 \ \mu F$$

11.14 In the network shown in Fig. 11.19, the switch is opened at $t = 0$, after the network has attained the steady state with the switch closed. (a) Find an expression for the voltage across the switch at $t = 0+$, (b) If the parameters are adjusted such that $i(0+) = 1$ A and $\frac{di}{dt}(0+) = -1$ A/Sec. What is the value of the derivative of the voltage across the switch? (Kuvempu University)

Solution

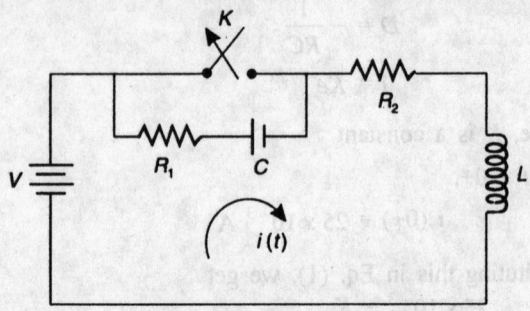

Fig. 11.19

When K is closed, L acts as a S.C. and C acts as an O.C.

$$\therefore \qquad i(0-) = \frac{V}{R_2} = i\,(0+) \text{ and } v_C\,(0-) = 0 = v_C\,(0+)$$

When K is opened at $t = 0$,

$$v_K = R_1\, i(t) + \frac{1}{C}\, \int i\, dt$$

i.e. $\qquad v_K\,(0+) = R_1\, i(0+) + 0 = R_1\, \dfrac{V}{R_2} = V\, \dfrac{R_1}{R_2}$

$$\frac{dv_K}{dt} = R_1\, \frac{di}{dt} + \frac{i}{C}$$

i.e. $\qquad \dfrac{dv_K}{dt}\,(0+) = R_1\,(-1) + \dfrac{1}{C} = \dfrac{1}{C} - R_1$

11.15 The network shown in the Fig. 11.20, has two independent node pairs. If the switch K is opened at $t = 0$, find the following quantities at $t = 0+$, (i) v_1, (ii) v_2, (iii) $\dfrac{dv_1}{dt}$ and (iv) $\dfrac{dv_2}{dt}$.

(Gulbarga University)

Solution

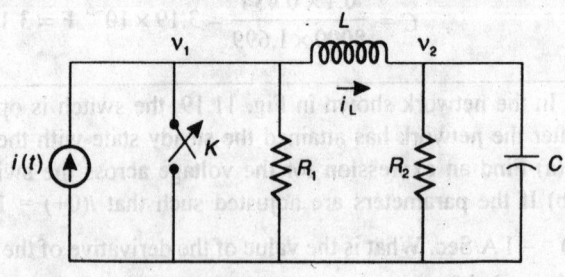

Fig. 11.20

When K is closed, all the current flows through the S.C. and hence

$$i_L(0-) = 0, \qquad \therefore \quad i_L(0+) = 0$$

$$v_C(0-) = 0 \qquad \therefore \quad v_C(0+) = 0$$

$$\therefore \qquad v_2(0+) = 0$$

When K is opened at $t = 0$

For node v_1, $\qquad i(t) = \dfrac{v_1}{R_1} + i_L \qquad\qquad (1)$

At $t = 0+$,

$$i(0+) = \frac{v_1(0+)}{R_1} + 0$$

$$\therefore \qquad v_1(0+) = R_1 \, i(0+) \qquad\qquad (2)$$

For node v_2, $\quad i_L - \dfrac{v_2}{R_2} - C\dfrac{dv_2}{dt} = 0 \qquad\qquad (3)$

At $t = 0+$, $\quad -C\dfrac{dv_2}{dt}(0+) = 0$

$$\therefore \qquad \frac{dv_2}{dt}(0+) = 0$$

Differentiating Eq. (1), we get

$$\frac{di}{dt} = \frac{1}{R_1}\frac{dv_1}{dt} + \frac{di_L}{dt}$$

$$\frac{di}{dt}(0+) = \frac{1}{R_1}\frac{dv_1}{dt}(0+) + \frac{1}{L}R_1\,i(0+)$$

$$\left[\because \qquad i_L = \frac{1}{L}\int(v_1 - v_2)\,dt \right.$$

$$\therefore \qquad \frac{di_L}{dt} = \frac{v_1 - v_2}{L}$$

$$\frac{di_L}{dt}(0+) = \frac{1}{L}\left[R_1\,i(0+) - 0\right] = \frac{R_1\,i(0+)}{L} \left. \right]$$

$$\therefore \quad \frac{dv_1}{dt}(0+) = R_1\left[\frac{di}{dt}(0+) - \frac{1}{L}R_1\,i(0+)\right] = R_1\left[-1 - \frac{R_1}{L}\right]$$

$$= -R_1\left[1| + \frac{R_1}{L}\right]$$

11.16 In the circuit shown in Fig. 11.21 (a), the switch K is closed at $t = 0$. Show that at $t = 0$, $\dfrac{di_1}{dt} = \dfrac{V_0}{R}\left[\omega\cos\omega t - \dfrac{\sin\omega t}{RC}\right]$ and $\dfrac{di_2}{dt} = \dfrac{V_0\sin\omega t}{L}$.

(Bangalore University)

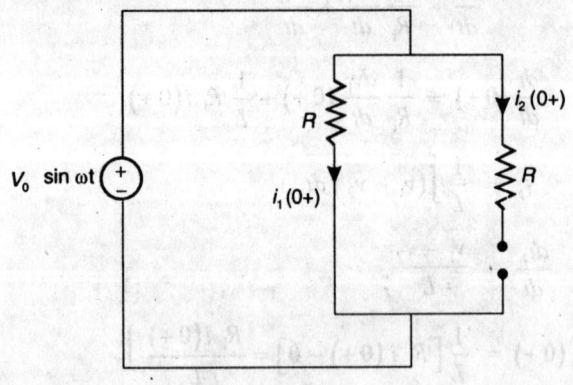

Fig. 11.21 (a)

Solution: Network at $t = 0+$ is as shown in Fig. 11.21 (b).

Fig. 11.21 (b)

$$i_1(0+) = \frac{V_0\sin\omega t}{R}$$

and

$$i_2(0+) = 0$$

Writing loop Eq. for i_1, after closing K

$$V_0 \sin \omega t = R i_1 + \frac{1}{C} \int i_1 \, dt \quad \text{On differentiation, we get}$$

$$V_0 \, \omega \cos \omega t = R \frac{di_1}{dt} + \frac{i_1}{C}$$

At $t = 0+$, we get

$$V_0 \, \omega \cos \omega t = R \frac{di_1}{dt}(0+) + \frac{1}{C} \frac{V_0 \sin \omega t}{R}$$

$$\therefore \quad \frac{di_1}{dt}(0+) = \frac{V_0}{R}\left[\omega \cos \omega t - \frac{\sin \omega t}{RC}\right]$$

Writing loop Eq. for i_2, $V_0 \sin \omega t = R i_2 + L \dfrac{di_2}{dt}$

At $t = 0+$

$$V_0 \sin \omega t = 0 + L \frac{di_2}{dt}(0+)$$

$$\therefore \quad \frac{di_2}{dt}(0+) = \frac{V_0}{L} \sin \omega t$$

11.17 In the network shown in the Fig. 11.22 (a), the switch K is closed at $t = 0$. The network being initially unenergised, find $i_1(0+)$, $i_2(0+)$, $\dfrac{di_1}{dt}(0+)$, $\dfrac{di_2}{dt}(0+)$, $\dfrac{d^2 i_1}{dt^2}(0+)$, and $\dfrac{d^2 i_2}{dt^2}(0+)$.

(Karnataka University)

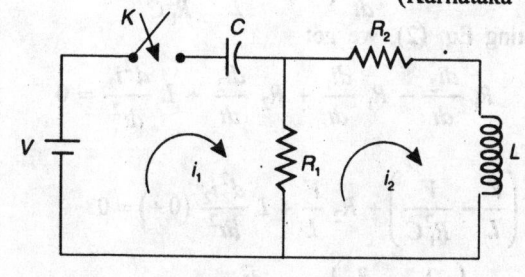

Fig. 11.22 (a)

Solution

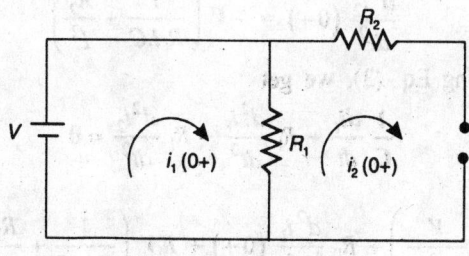

Fig. 11.22 (b)

Network at $t = 0+$, is as shown in Fig. 11.22 (b)

$$i_1(0+) = \frac{V}{R_1} \quad \text{and} \quad i_2(0+) = 0$$

When K is closed at $t = 0$, the loop Eqs are:

$$\frac{1}{C}\int i_1 \, dt + R_1(i_1 - i_2) = V \tag{1}$$

$$R_1(i_2 - i_1) + R_2 i_2 + L\frac{di_2}{dt} = 0 \tag{2}$$

Substituting the initial conditions in (2), we get

$$-R_1\frac{V}{R_1} + L\frac{di_2}{dt}(0+) = 0$$

$$\therefore \qquad \frac{di_2}{dt}(0+) = \frac{V}{L}$$

Differentiating Eq. (1), we get

$$\frac{i_1}{C} + R_1\frac{di_1}{dt} - R_1\frac{di_2}{dt} = 0 \tag{3}$$

At $t = 0+$,

$$\frac{V}{R_1 C} + R_1\frac{di_1}{dt}(0+) - R_1\frac{V}{L} = 0$$

$$\therefore \qquad \frac{di_1}{dt}(0+) = \frac{V}{L} - \frac{V}{R_1^2 C}$$

Differentiating Eq. (2), we get

$$R_1\frac{di_2}{dt} - R_1\frac{di_1}{dt} + R_2\frac{di_2}{dt} + L\frac{d^2 i_2}{dt^2} = 0$$

At $t = 0+$,

$$R_1\frac{V}{L} - R_1\left(\frac{V}{L} - \frac{V}{R_1^2 C}\right) + R_2\frac{V}{L} + L\frac{d^2 i_2}{dt^2}(0+) = 0$$

$$-V\left(\frac{1}{R_1 C} + \frac{R_2}{L}\right) = L\frac{d^2 i_2}{dt^2}(0+)$$

$$\therefore \qquad \frac{d^2 i_2}{dt^2}(0+) = -V\left(\frac{1}{R_1 LC} + \frac{R_2}{L^2}\right)$$

Differentiating Eq. (3), we get

$$\frac{1}{C}\frac{di_1}{dt} + R_1\frac{d^2 i_1}{dt^2} - R_1\frac{d^2 i_2}{dt^2} = 0$$

At $t = 0+$,

$$\frac{1}{C}\left(\frac{V}{L} - \frac{V}{R_1^2 C}\right) + R_1\frac{d^2 i_1}{dt^2}(0+) + R_1 V\left(\frac{1}{R_1 LC} + \frac{R_2}{L^2}\right) = 0$$

i.e. $\dfrac{d^2 i_1}{dt^2}(0+) = \dfrac{1}{R_1}\left[\dfrac{1}{C}\left(\dfrac{V}{R_1^2 C} - \dfrac{V}{L}\right) - VR_1\left(\dfrac{1}{R_1 LC} + \dfrac{R_2}{L^2}\right)\right]$

11.18 The network given in Fig. 11.23 consists of two coupled coils and a capacitor. At $t = 0$, the switch K is closed. Find the initial values of v_a, $\dfrac{dv_a}{dt}$ and $\dfrac{d^2 v_a}{dt^2}$. (Mysore University)

Solution

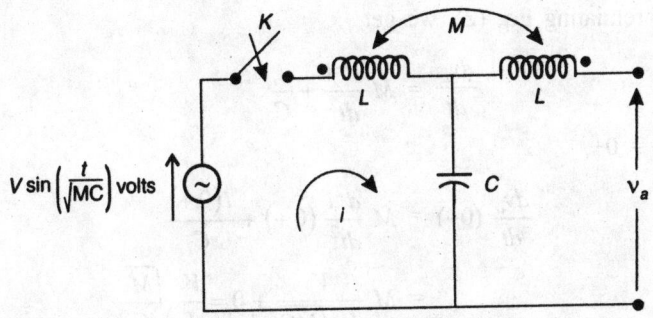

Fig. 11.23

When the switch K is closed at $t = 0$, the inductance acts as an open circuit

∴ $i(0+) = i(0-) = 0,$

$v_c(0+) = 0$ and $v_a(0+) = 0$

As there is no current through the second coil, there is no mutually induced e.m.f. in the first coil.

For loop 1

∴ $V \sin\left(\dfrac{t}{\sqrt{MC}}\right) = L\dfrac{di}{dt} + \dfrac{1}{C}\displaystyle\int i\,dt$ (1)

At $t = 0+$,

$$0 = L\dfrac{di}{dt}(0+) + 0$$

∴ $\dfrac{di}{dt}(0+) = 0$

For loop 2, $v_a = M\dfrac{di}{dt} + \dfrac{1}{C}\displaystyle\int i\,dt$ (2)

Differentiating Eq. (1), we get

$$\dfrac{V}{\sqrt{MC}}\cos\left(\dfrac{t}{\sqrt{MC}}\right) = L\dfrac{d^2 i}{dt^2} + \dfrac{i}{C}$$ (3)

At $t = 0+$,

$$\frac{V}{\sqrt{MC}} = L \frac{d^2 i}{dt^2}(0+) + \frac{i(0+)}{C}$$

i.e.

$$\frac{V}{\sqrt{MC}} = L \frac{d^2 i}{dt^2}(0+) + 0$$

$\therefore$

$$\frac{d^2 i}{dt^2}(0+) = \frac{V}{L\sqrt{MC}}$$

Differentiating Eq. (2), we get

$$\frac{dv_a}{dt} = M \frac{d^2 i}{dt^2} + \frac{i}{C} \tag{4}$$

At $t = 0+$,

$$\frac{dv_a}{dt}(0+) = M \frac{d^2 i}{dt^2}(0+) + \frac{i(0+)}{C}$$

$$= M \frac{V}{L\sqrt{MC}} + 0 = \frac{V}{L}\sqrt{\frac{M}{C}}$$

Differentiating Eq. (3), we get

$$-\frac{V}{MC} \sin \frac{t}{\sqrt{MC}} = L \frac{d^3 i}{dt^3} + \frac{1}{C}\frac{di}{dt}$$

At $t = 0+$,

$$0 = L \frac{d^3 i}{dt^3}(0+) + \frac{1}{C}\frac{di}{dt}(0+) = L \frac{d^3 i}{dt^3}(0+) + 0$$

$\therefore$

$$\frac{d^3 i}{dt^3}(0+) = 0$$

Differentiating Eq. (4), we get

$$\frac{d^2 v_a}{dt^2} = M \frac{d^3 i}{dt^3} + \frac{1}{C}\frac{di}{dt}$$

At $t = 0+$,

$$\frac{d^2 v_a}{dt^2}(0+) = M (0) + \frac{1}{C}(0) = 0$$

11.19   In the network shown in Fig. 11.24, the switch K is changed from position a to b at $t = 0$, steady state being established at position a. Find i_1, i_2, and i_3 at $t = 0+$.   (Bangalore University)

Solution

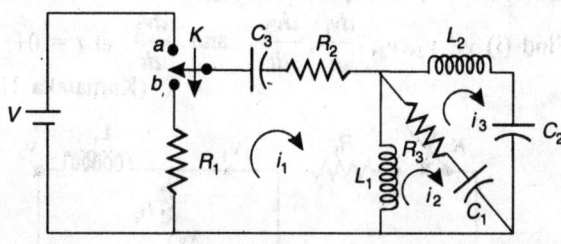

Fig. 11.24

When the switch K is on a, the circuit is under steady state condition. C_3 acts as an O.C. Hence

$$i_{L1}(0-) = 0 = i_{L1}(0+)$$

But $\quad\quad i_{L1}(0+) = i_1(0+) - i_2(0+) = 0$

$\therefore\quad\quad\quad i_1(0+) = i_2(0+)$

C_2 also acts as an O.C. Hence

$$i_{L2}(0-) = 0 = i_{L2}(0+) = i_3(0+)$$

At the open circuit terminals of C_3, a voltage V exists. Therefore, $V_{C3}(0-) = V_{C3}(0+) = V$ volts. When the switch K is changed from a to b at $t = 0$, the $t = 0+$ network is as shown in Fig. 11.25.

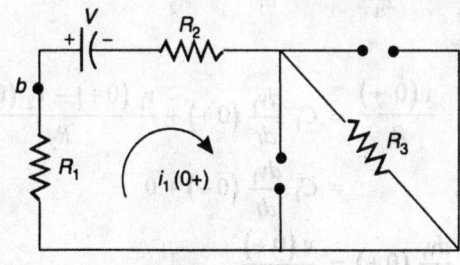

Fig. 11.25

From the circuit in Fig. 11.27, we get

$$V = -i_1(0+)(R_1 + R_2 + R_3)$$

$\therefore\quad\quad i_1(0+) = i_2(0+) = -\dfrac{V}{R_1 + R_2 + R_3}$

11.20 In the network shown in Fig. 11.26, the switch K is closed at $t = 0$. At $t = 0-$, all capacitor voltages and inductor currents are zero. Find (i) v_1, v_2, v_3, $\dfrac{dv_1}{dt}$, $\dfrac{dv_2}{dt}$, and $\dfrac{dv_3}{dt}$ at $t = 0+$.

(Karnataka University)

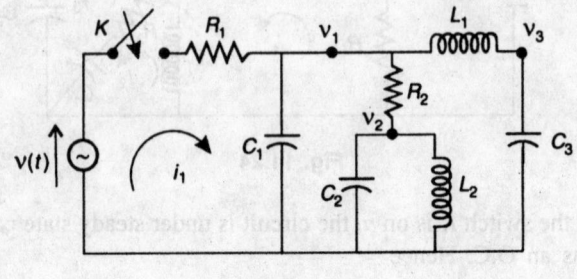

Fig. 11.26

Solution: When K is closed at $t = 0$, the inductances L_1 and L_2 act as open circuits. The capacitances C_1, C_2, and C_3 act as short circuits.

$$\therefore \quad i_{L_1}(0+) = i_{L_2}(0+) = i_{R_2}(0+) = 0$$

and

$$v_1(0+) = v_1(0-) = 0,$$
$$v_2(0+) = v_2(0-) = 0,$$
$$v_3(0+) = v_3(0-) = 0$$

At node v_1, $\dfrac{v(t)}{R_1} = C_1 \dfrac{dv_1}{dt} + \dfrac{v_1 - v_2}{R_2}$ \hfill (1)

At $t = 0+$,

$$\frac{v(0+)}{R_1} = C_1 \frac{dv_1}{dt}(0+) + \frac{v_1(0+) - v_2(0+)}{R_2}$$

$$= C_1 \frac{dv_1}{dt}(0+) + 0$$

$$\frac{dv_1}{dt}(0+) = \frac{v(0+)}{R_1 C_1}$$

At node v_2, $\dfrac{v_1 - v_2}{R_2} = C_2 \dfrac{dv_2}{dt} + \dfrac{1}{L_2} \displaystyle\int v_2 \, dt$ \hfill (2)

At $t = 0+$,

$$\frac{v_1(0+) - v_2(0+)}{R_2} = C_2 \frac{dv_2}{dt}(0+) + 0$$

$$\therefore \qquad \frac{dv_2}{dt}(0+) = 0$$

$$\because \qquad v_1(0+) = 0$$

and $\qquad v_2(0+) = 0$

At node 3, $\quad \dfrac{1}{L_1} \displaystyle\int (v_1 - v_3)\, dt = C_3 \dfrac{dv_3}{dt}$ \qquad (3)

At $t = 0+$,

$$0 = C_3 \frac{dv_3}{dt}(0+)$$

$$\therefore \qquad v_1(0+) = 0$$

and $\qquad v_3(0+) = 0$

$$\therefore \qquad \frac{dv_3}{dt}(0+) = 0$$

NUMERICAL PROBLEMS

11.1 The circuit shown in Fig. 11.27, is under steady-state, when the switch K is open. The switch is closed at $t = 0$. Find the initial and final values of the voltages and currents in each element.

(Kuvempu University)

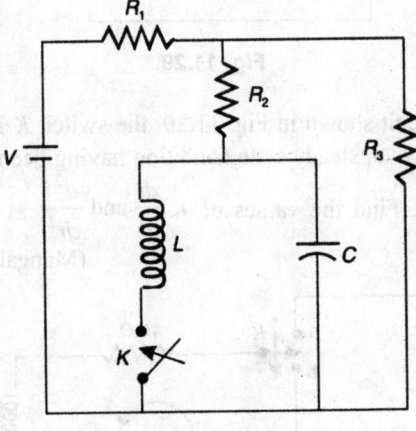

Fig. 11.27

11.2 Find the first derivatives of voltages and currents for the circuit shown in Fig. 11.28. (Mysore University)

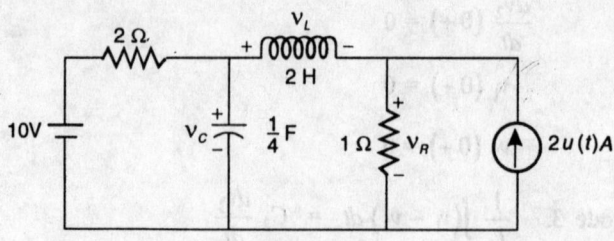

Fig. 11.28

11.3 In the network shown in Fig. 11.29, the switch K is closed at $t = 0$. The current waveform is observed in a CRO. The initial value of the current is measured to be 0.01 A. The time constant of the circuit is 0.025 Sec. Find (i) the value of R, (ii) the value of C and (iii) the Eq. for $i(t)$.　　　(Bangalore University)

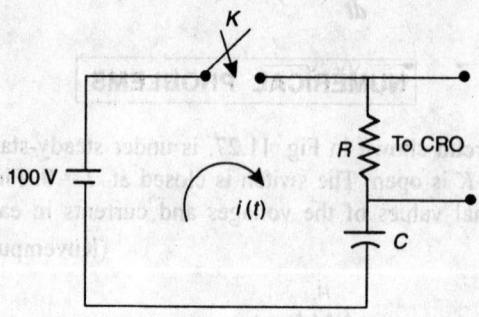

Fig. 11.29

11.4 In the circuit shown in Fig. 11.30, the switch K is changed from a to b at $t = 0$, steady state condition having been reached before switching. Find the values of i, $\dfrac{di}{dt}$ and $\dfrac{d^2i}{dt^2}$ at $t = 0+$.

　　　(Mangalore University)

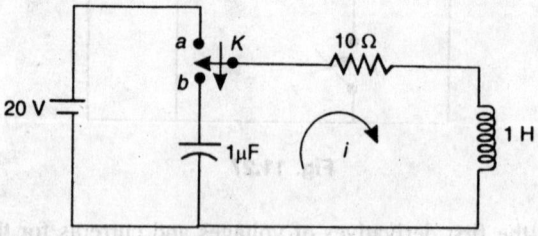

Fig. 11.30

11.5 For the network shown in Fig. 11.31, determine i, $\dfrac{di}{dt}$ and $\dfrac{d^2i}{dt^2}$ at $t = 0+$. The switch K is closed at $t = 0$.

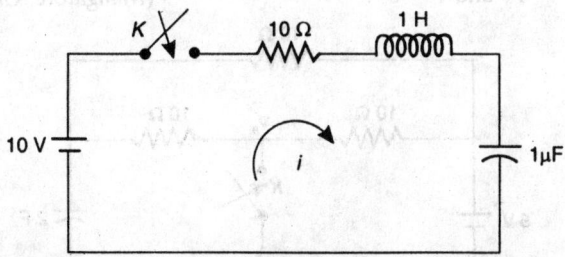

Fig. 11.31

11.6 For the network shown in Fig. 11.32, the switch is closed at $t = 0$. The initial capacitor voltage and inductor current are zero. Find v_1, v_2, $\dfrac{dv_1}{dt}$ and $\dfrac{dv_2}{dt}$ at $t = 0+$. (Bangalore University)

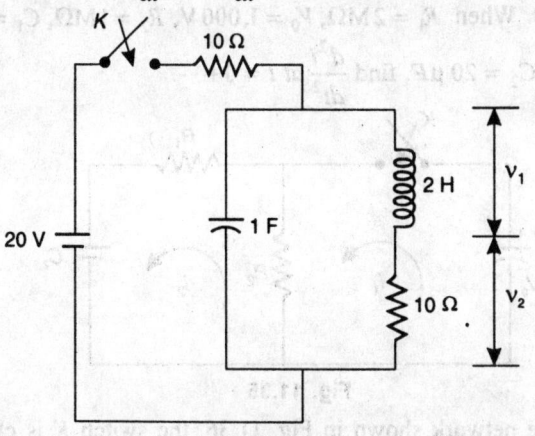

Fig. 11.32

11.7 In the network shown in Fig. 11.33, the switch K is opened at $t = 0$. Obtain the values of v, $\dfrac{dv}{dt}$ and $\dfrac{d^2v}{dt^2}$ at $t = 0+$.

(Gulbarga University)

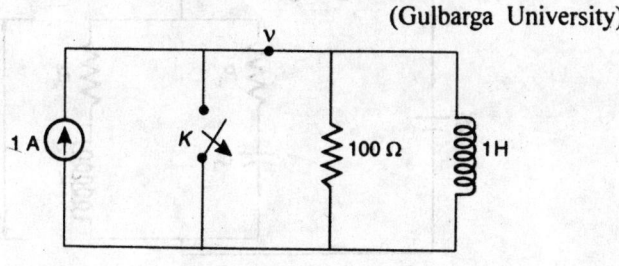

Fig. 11.33

11.8 In the network shown in Fig. 11.34, steady state is reached when K is open. At $t = 0$, the switch is closed. Determine V_a at $t = 0-$ and $t = 0+$. (Mangalore University)

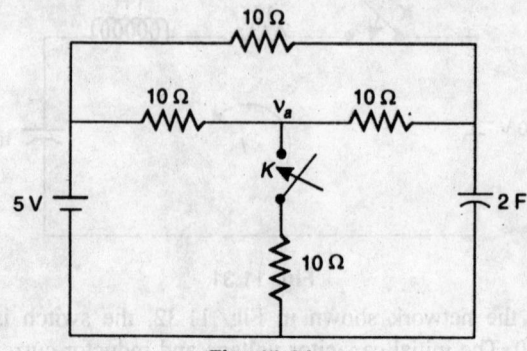

Fig. 11.34

11.9 In the network shown in Fig. 11.35, the switch K is closed at $t = 0$. When $R_1 = 2\,M\Omega$, $V_0 = 1,000\,V$, $R_2 = 1\,M\Omega$, $C_1 = 10\,\mu F$ and $C_2 = 20\,\mu F$, find $\dfrac{d^2 i}{dt^2}$ at $t = 0+$.

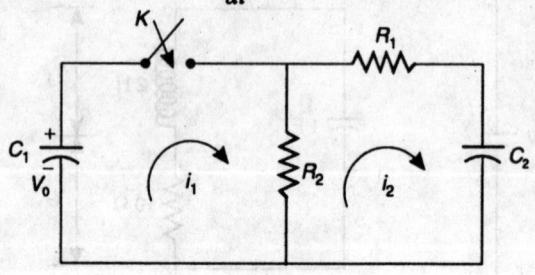

Fig. 11.35

11.10 In the network shown in Fig. 11.36, the switch K is closed at $t = 0$. Determine v_1, $\dfrac{dv_1}{dt}$ and $\dfrac{d^2 v_1}{dt^2}$ at $t = 0+$. (Bangalore University)

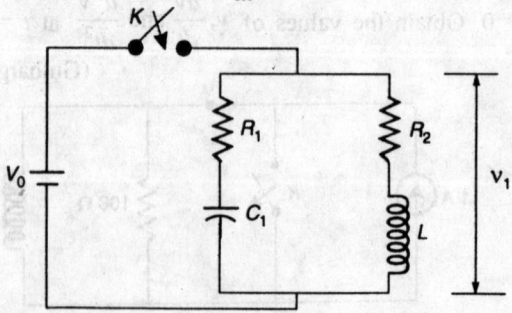

Fig. 11.36

11.11 In the circuit shown in Fig. 11.37, the switch K is thrown from a to b at $t = 0$, steady state conditions having been reached before switching. Find i, $\dfrac{di}{dt}$ and $\dfrac{d^2i}{dt^2}$ at $t = 0+$.

(Karnataka University)

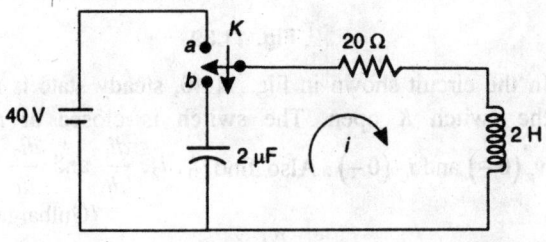

Fig. 11.37

11.12 In the circuit shown in Fig. 11.38, the switch K is closed at $t = 0$. Solve for (i) v_1 and v_2 at $t = 0+$, (ii) v_1 and v_2 at $t = \infty$, (iii) $\dfrac{dv_1}{dt}$ and $\dfrac{dv_2}{dt}$ at $t = 0+$, and (iv) $\dfrac{d^2v_2}{dt^2}$ at $t = 0+$.

(Mysore University)

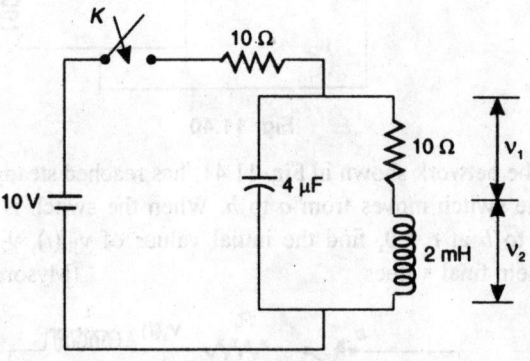

Fig. 11.38

11.13 In the circuit shown in Fig. 11.39, the switch K is opened at $t = 0$. Find the values of v_1, v_2, $\dfrac{dv_1}{dt}$ and $\dfrac{dv_2}{dt}$ at $t = 0+$.

(Mysore University)

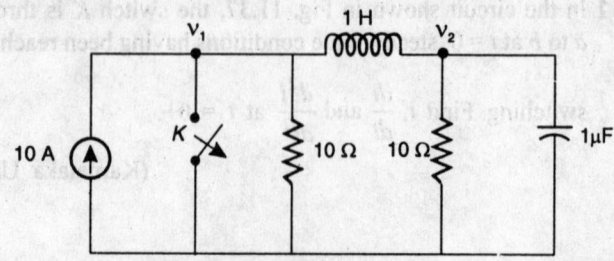

Fig. 11.39

11.14 In the circuit shown in Fig. 11.40, steady state is reached with the switch K open. The switch is closed at $t = 0$. Find $v_c(0-)$ and $i_C(0-)$. Also find i_1, i_2, $\dfrac{di_1}{dt}$ and $\dfrac{di_2}{dt}$ at $t = 0+$.

(Gulbarga University)

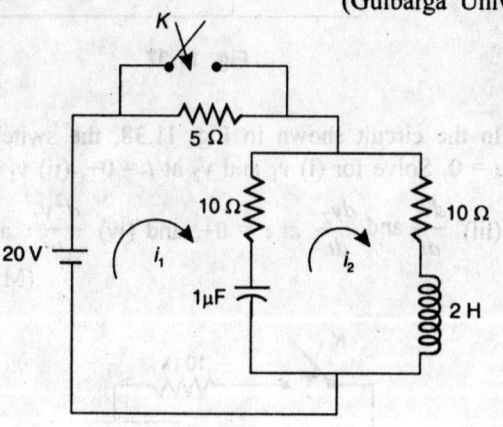

Fig. 11.40

11.15 The network shown in Fig. 11.41, has reached steady state before the switch moves from a to b. When the switch is moved from a to b at $t = 0$, find the initial values of $v_1(t)$, $v_2(t)$ and also their final values.

(Mysore University)

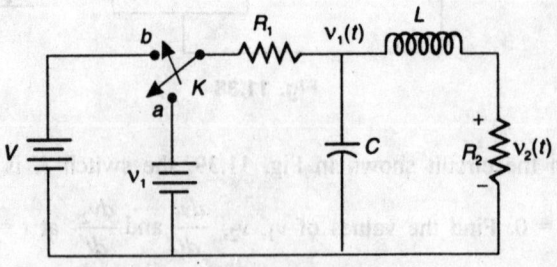

Fig. 11.41

11.16 In the network shown in Fig. 11.42, steady state is reached with the switch K open. $V = 100$ V, $R_1 = 10\,\Omega$, $R_2 = 20\,\Omega$, $R_3 = 20\,\Omega$, $L = 1$ H and $C = 1\mu F$. At $t = 0$, the switch is closed. Find $i_1, i_2, \dfrac{di_1}{dt}$ and $\dfrac{di_2}{dt}$ at $t = 0+$. Also find $\dfrac{di_1}{dt}$ at $t = \infty$.

(Mangalore University)

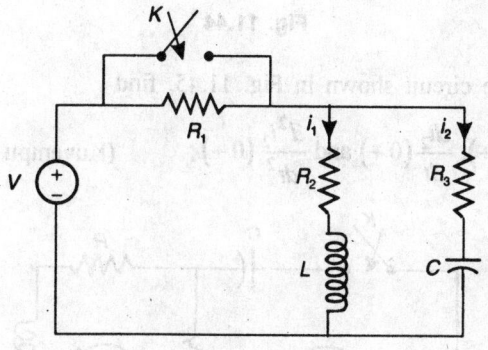

Fig. 11.42

11.17 In the network shown in Fig. 11.43, the switch K is closed at $t = 0$. Find v_a at $t = 0+$ and the voltage across the capacitor C_1 at $t = \infty$. (Bangalore University)

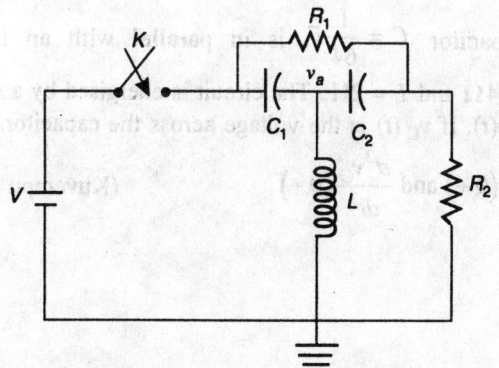

Fig. 11.43

11.18 For the circuit shown in Fig. 11.44, the switch K is changed from a to b at $t = 0$. Solve for i, $\dfrac{di}{dt}, \dfrac{d^2i}{dt^2}$ at $t = 0+$.

(Mysore University)

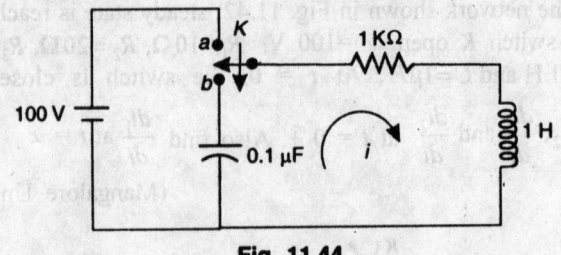

Fig. 11.44

11.19 For the circuit shown in Fig. 11.45, find

$$\frac{di_1}{dt}(0+), \frac{di_2}{dt}(0+) \text{ and } \frac{d^2i_2}{dt^2}(0+). \qquad \text{(Kuvempu University)}$$

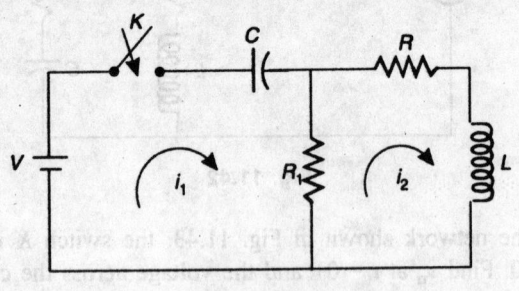

Fig. 11.45

11.20 A capacitor $C = \frac{1}{64}$ F is in parallel with an inductor of $R = 24\Omega$ and $L = 2H$. The circuit is energised by a step current of $2\,u(t)$. If $v_C(t)$ is the voltage across the capacitor, determine,

$$\frac{d^2 v_C}{dt^2}(0+) \text{ and } \frac{d^3 v_C}{dt^3}(0+). \qquad \text{(Kuvempu University)}$$

CHAPTER

12

Laplace Transformation

12.1 INTRODUCTION

The Laplace transform method of solving differential Eqs, i.e. the operational calculus was invented by Oliver Heaviside (1850–1985), a British engineer. The basis for substantiating the work of Heaviside was found in the writings of the French mathematician, Pierre Simon Laplace (1749–1825). The Laplace transform methods of solving differential eqs offer a number of advantages over the classical method. They are:

(i) This method gives the total solution, i.e. the complementary function and the particular integral in one operation.

(ii) The initial conditions are incorporated into the problem as one of the first steps rather than as the last step.

(iii) The initial conditions are automatically specified in the transformed Eqs.

(iv) The solution of differential Eqs is routine and progresses systematically.

(v) Tables of Laplace transform and inverse Laplace transform are readily available for us, while solving differential Eqs.

The procedure for solving a network using Laplace transform method is as follows:

(i) The integro-differential Eq. is written for the given network.

(ii) The Laplace transform of the Eq. is written.

(iii) The transformed Eq. is revised after inserting initial conditions to obtain the revised Laplace transform.

(iv) The inverse Laplace transform of the revised transformed Eq. gives the total solution.

12.2 THE LAPLACE TRANSFORMATION

If $f(t)$ is a function of time, then its Laplace transform $F(s)$ is written as:

$$\mathcal{L}\left[f(t)\right] = F(s) = \int_{0-}^{\infty} f(t)\, e^{-st}\, dt \tag{12.1}$$

$$\text{Provided, } \int_{0-}^{\infty} |f(t)|\, e^{-\sigma t}\, dt < \infty \tag{12.2}$$

for real positive σ.

The symbol "$\mathcal{L}$" stands for "Laplace transform of".

Here, $s = \sigma + j\omega$ is a complex number.

The condition (12.2) is satisfied by most of the functions $f(t)$, encountered in engineering, since $e^{-\sigma t}$ is a powerful reducing factor. Functions of the form e^{-at^n} does not satisfy (12.2), but such functions are seldom required to describe the driving functions in engineering problems.

The lower limit 0– used in Eq. (12.2), needs some discussion. In electric circuits, the time $t = 0$ is taken as the reference time, i.e. the time at which the switch is closed. The time $t = 0-$, indicates the time just before the switching operation. The time $t = 0+$, indicates the time just after the switching operation. When $f(0-) = f(0+)$, the choice of 0– or 0 has no significance. We have learnt in the previous chapter on "Initial Conditions" that, $i_L\left(0-\right) = i_L\left(0+\right)$ and $v_C\left(0-\right) = v_C\left(0+\right)$. Hence, when we are dealing with electric circuits, which are excited by sources other than impulse sources, there is no distinction between 0–, 0 or 0+. However, if there is an impulse source $\delta\left(t\right)$ at $t = 0$, $t = 0-$ must be used, so that the impulse function is included. Instead of making an exception for networks containing impulse sources, $t = 0-$, is used throughout the discussion on Laplace transforms.

12.3 LAPLACE TRANSFORM OF STANDARD FUNCTIONS

(i) Unit step function $f(t) = u(t)$

The unit step function is described by the Eq. (12.3) and is as shown in Fig. 12.1.

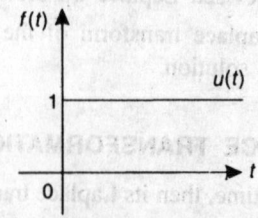

Fig. 12.1

$$u(t) = 1 \text{ for } t \geq 0$$

$$= 0 \text{ for } t < 0 \tag{12.3}$$

The unit step function $u(t)$ represents a switching operation without the presence of a switch in the circuit. If the driving voltage is represented by $Vu(t)$, it means that, a voltage V is applied to the circuit at $t = 0$ eventhough, a switch is not shown in the circuit to perform this operation.

$$\mathcal{L}\left[u(t)\right] = \int\limits_{0-}^{\infty} 1.e^{-st}\, dt = \left[-\frac{1}{s}e^{-st}\right]_{0-}^{\infty} = \frac{1}{s} \tag{12.4}$$

(ii) $f(t) = e^{at}$, where a is a constant

$$\mathcal{L}\left[e^{at}\right] = \int\limits_{0-}^{\infty} e^{at}\, e^{-st}\, dt = \int\limits_{0-}^{\infty} e^{-(s-a)t}\, dt$$

$$= \left[-\frac{1}{s-a}e^{-(s-a)t}\right]_{0-}^{\infty} = \frac{1}{s-a} \tag{12.5}$$

(iii) $f(t) = \sin \omega t$

$$\mathcal{L}(\sin \omega t) = \mathcal{L}\left[\frac{1}{2j}\left(e^{j\omega t} - e^{-j\omega t}\right)\right]$$

$$= \frac{1}{2j}\left[\frac{1}{s-j\omega} - \frac{1}{s+j\omega}\right]$$

$$= \frac{1}{2j}\frac{2j\omega}{s^2 + \omega^2} = \frac{\omega}{s^2 + \omega^2} \tag{12.6}$$

(iv) $f(t) = \cos \omega t$

$$\mathcal{L}\left[\cos \omega t\right] = \mathcal{L}\left[\frac{1}{2j}\left(e^{j\omega t} - e^{-j\omega t}\right)\right]$$

$$= \frac{1}{2}\left[\frac{1}{s-j\omega} + \frac{1}{s+j\omega}\right]$$

$$= \frac{1}{2}\frac{2s}{s^2 + \omega^2} = \frac{s}{s^2 + \omega^2} \tag{12.7}$$

(v) $f(t) = t^n$

$$\mathcal{L}\left[t^n\right] = \int\limits_{0-}^{\infty} t^n . e^{-st} . dt = \left[t^n \frac{e^{-st}}{-s}\right]_{0-}^{\infty} - \int\limits_{0-}^{\infty} \frac{e^{-st}}{-s} n t^{n-1} . dt$$

$$= 0 + \frac{n}{s} \int\limits_{0-}^{\infty} t^{n-1} . e^{-st} dt = \frac{n}{s} \mathcal{L}\left[t^{n-1}\right]$$

$$= \frac{n}{s} \frac{n-1}{s} \mathcal{L}[t^{n-2}]$$

$$= \frac{n}{s} \frac{n-1}{s} \frac{n-2}{s} \cdots \frac{2}{s} \frac{1}{s} \mathcal{L}\left[t^{n-n}\right]$$

$$= \frac{n}{s} \frac{n-1}{s} \frac{n-2}{s} \cdots \frac{2}{s} \frac{1}{s} \frac{1}{s} = \frac{n!}{s^{n+1}} \quad\quad (3.8)$$

III^{ly} $\quad \mathcal{L}\left[t^{n-1}\right] = \frac{(n-1)!}{s^n} \quad\quad\quad \mathcal{L}\left[t^3\right] = \frac{3!}{s^4}$

$$\mathcal{L}\left[t^2\right] = \frac{2!}{s^3} \quad \text{and} \quad \mathcal{L}[t] = \frac{1}{s^2}$$

(vi) $t^n f(t)$

$$\mathcal{L}[f(t)] = F(s) = \int\limits_{0-}^{\infty} f(t) e^{-st} dt$$

$$\mathcal{L}[f'(t)] = F'(s) = \int\limits_{0-}^{\infty} -t f(t) e^{-st} dt = \mathcal{L}[-t f(t)]$$

$$\mathcal{L}[f''(t)] = F''(s) = \int\limits_{0-}^{\infty} t^2 f(t) e^{-st} dt = \mathcal{L}\left[t^2 f(t)\right]$$

$$\vdots$$

$$\mathcal{L}\left[f^n(t)\right] = F^n(s)(-1)^n = \int\limits_{0-}^{\infty} t^n f(t) e^{-st} dt$$

$$\therefore \quad \mathcal{L}\left[t^n f(t)\right] = (-1)^n F^n(s) \quad\quad (12.9)$$

(vii) $f(t) = \sinh \omega t$

$$L\left[\sinh \omega t\right] = L\left[\frac{e^{\omega t} - e^{-\omega t}}{2}\right]$$

$$= \frac{1}{2}\left[\frac{1}{s-\omega} - \frac{1}{s+\omega}\right] = \frac{\omega}{s^2 - \omega^2} \qquad (12.10)$$

(viii) $f(t) = \cosh \omega t$

$$L\left[\cosh \omega t\right] = L\left[\frac{e^{\omega t} + e^{-\omega t}}{2}\right]$$

$$= \frac{1}{2}\left[\frac{1}{s-\omega} + \frac{1}{s+\omega}\right] = \frac{s}{s^2 - \omega^2} \qquad (12.11)$$

(ix) Laplace transform of derivatives

$$L\left[f'(t)\right] = \int_{0-}^{\infty} f'(t) e^{-st} \, dt$$

$$= \left[e^{-st} f(t)\right]_{0-}^{\infty} - \int_{0-}^{\infty} f(t) e^{-st} (-s).dt$$

$$= -f(0-) + s \int_{0-}^{\infty} f(t) e^{-st} \, dt = s F(s) - f(0-)$$

III^{ly} $L\left[f''(t)\right] = s^2 F(s) - sf(0-) - f'(0-)$

In general

$$L\left[f^n(t)\right] = s^n F(s) - s^{n-1} f(0-) - s^{n-2} f'(0-) \cdots f^{n-1}(0-)$$

$$(12.12)$$

(x) Laplace transform of integrals

$$L\left[\int_{0-}^{t} f(t) \, dt\right] = \int_{0-}^{\infty}\left[\int_{0-}^{t} f(t) \, dt\right] e^{-st} \, dt$$

$$= \left[-\frac{e^{-st}}{s} \int_{0-}^{t} f(t) \, dt\right]_{0-}^{\infty} + \frac{1}{s} \int_{0-}^{t} f(t) e^{-st} \, dt$$

$$= 0 + \frac{F(s)}{s} = \frac{F(s)}{s} \qquad (12.13)$$

If the integral has the limits $-\infty$ to t instead of $0-$ to t, then

$$\int_{-\infty}^{t} f(t)\, dt = \int_{-\infty}^{0-} f(t)\, dt + \int_{0-}^{t} f(t)\, dt$$

The first term on the right hand side of the above eq. is constant and can be represented as $f(-\infty)$ or $f(0-)$.

$$\therefore \; L\left[\int_{-\infty}^{t} f(t)\, dt\right] = L\left[f(0-) + \int_{0-}^{t} f(t)\, dt\right] = \frac{f(0-)}{s} + \frac{F(s)}{s}$$

$$(12.14)$$

If $f(t)$ is a current, then $f(0-)$ represents the initial charge $q(0-)$. If $f(t)$ is a voltage, then $f(0-)$ represents the flux linkages $\psi(0-) = L\, i\, (0-)$.

(xi) Property of linearity

If $F(s)$ is the Laplace transform of $f(t)$, then, $L\left[K f(t)\right] = K\, L\left[f(t)\right] = K\, F(s)$, where K is a constant.

(xii) Property of superposition

If $F_1(s), F_2(s) \cdots F_n(s)$, are the Laplace transforms of $f_1(t), f_2(t) \cdots f_n(t)$, then

$$L\left[f_1(t) + f_2(t) + \cdots + f_n(t)\right] = F_1(s) + F_2(s) + \cdots + F_n(s)$$

12.4 THE INVERSE LAPLACE TRANSFORM

The function $f(t)$ can be found from the given $F(s)$, by finding the inverse Laplace transform of $F(s)$.

$$\therefore \qquad f(t) = L^{-1}\left[F(s)\right] \qquad (12.15)$$

The symbol L^{-1} is read as "Laplace inverse of". The inverse Laplace transformation is given by the complex inversion integral given in 12.16.

$$f(t) = \frac{1}{2\pi j} \int_{\sigma_1 - j\infty}^{\sigma_1 + j\infty} F(s)\, e^{st} \cdot dt \qquad (12.16)$$

However, Eq. (12.16) is seldom used to find $f(t)$, because, ready-made formulae are available for finding inverse Laplace transformation for given standard form of $F(s)$. If $F(s)$ is not in the standard form for which $f(t)$ can be readily found, it must be converted into the standard form and then its inverse is found. The property of the "uniqueness" of the Laplace transformation, i.e. no two different functions have the same Laplace transformation, helps us to find $f(t)$ for a given $F(s)$, using standard transformation, without resorting to the complex integration given in (12.16).

12.5 PROCEDURE TO USE LAPLACE TRANSFORMATION

When Laplace transformation method is used to solve an electrical network, the procedure is as follows:

1. The integro-differential Eqs are written for the given network.
2. On applying Laplace transformation, the transformed Eqs are written, inserting the initial conditions into them.
3. The transformed equations are manipulated algebraically, such that, they are in standard forms, for which inverse Laplace transform can be found.
4. The inverse Laplace transforms of these Eqs give the required solutions.

12.6 PARTIAL FRACTION EXPANSION

Consider a differential Eq. of the general form as shown in 12.17.

$$a_0 \frac{d^n i}{dt^n} + a_1 \frac{d^{n-1} i}{dt^{n-1}} + \cdots + a_{n-1} \frac{di}{dt} + a_n i = v(t) \qquad (12.17)$$

Where $a_0, a_1, \ldots a_{n-1}, a_n$ are constant coefficients. On taking Laplace transformation of (12.17) and manipulation, the Eq. for the transformed unknown $I(s)$ may be written as in (12.18).

$$I(s) = \frac{\mathcal{L}\left[v(t) + \text{initial condition terms}\right]}{a_0 s^n + a_1 s^{n-1} + \cdots + a_{n-1} s + a_n} = \frac{P(s)}{Q(s)} \qquad (12.18)$$

If $P(s) / Q(s)$ is in standard form, its inverse Laplace transformation gives $i(t)$. However, $I(s)$ must be written as a sum of simple terms, before applying inverse Laplace. For writing $I(s)$ as a sum of simple terms, the condition to be satisfied by (12.18) is that the order of $P(s)$ must be less than the order of $Q(s)$. If this condition is not satisfied, $P(s)$ is divided by $Q(s)$ and the expansion is written as in (12.19).

$$I(s) = \frac{P(s)}{Q(s)} = b_0 + b_1 s + b_2 s^2 + \cdots + b_{m-n} s^{m-n} + \frac{P_1(s)}{Q(s)} \qquad (12.19)$$

Where, b_0, b_1 ... b_{m-n} are constant coefficients. m is the order of $P(s)$ and n is the order of $Q(s)$ and $m > n$. The new function $P_1(s)/Q(s)$, now satisfies the condition that the order of $P_1(s)$ is less than the order of $Q(s)$. After splitting $P_1(s)/Q(s)$ into a sum of simple terms, $\mathcal{L}^{-1}$ $I(s)$ gives $i(t)$.

The procedure for writing $P_1(s)/Q(s)$ as sum of several simple terms is as follows:

1. $Q(s) = 0$ is known as the characteristic Eq.

2. $Q(s)$ is factorised and is written as in (12.20).

$$Q(s) = a_0 s^n + a_1 s^{n-1} + ... + a_{n-1} s + a_n$$

$$= a_0 (s - s_1)(s - s_2)...(s - s_n) \tag{12.20}$$

Where s_1, s_2 ... s_n are the n roots of $Q(s)$.

3. The roots of $Q(s)$ may be of three types:

    (i) real and unequal, (ii) complex conjugates and (iii) multiple roots.

4. If the roots of $Q(s) = 0$ are real and unequal, then

$$\frac{P_1(s)}{Q(s)} = \frac{P_1(s)}{(s-s_1)(s-s_2)\cdots(s-s_n)}$$

$$= \frac{A_1}{s-s_1} + \frac{A_2}{s-s_2} + \cdots + \frac{A_n}{s-s_n} \tag{12.21}$$

Where A_1, A_2 ... A_n are real constants called as *residues*.

5. If the roots are of $Q(s) = 0$ are complex conjugates, then

$$\frac{P_1(s)}{Q(s)} = \frac{P_1(s)}{Q_1(s)(s+\alpha+j\omega)(s+\alpha-j\omega)}$$

$$= \frac{A_1}{(s+\alpha+j\omega)} + \frac{A_1^*}{(s+\alpha-j\omega)} \tag{12.22}$$

Where, A_1^* is the complex conjugate of A_1. When the roots are conjugates, the partial fraction expansion coefficients are also conjugates. An expansion of the type given in (12.22) is necessary for each pair of complex conjugate roots.

6. If the roots of $Q(s) = 0$ are of multiplicity r, then

$$\frac{P_1(s)}{Q(s)} = \frac{P_1(s)}{(s-s_1)^r} = \frac{A_{11}}{s-s_1} + \frac{A_{12}}{(s-s_1)^2} + ... + \frac{A_{1r}}{(s-s_1)^r} \tag{12.23}$$

There will be similar terms for every other repeated root.

In the expansion of a quotient of polynomials by partial fraction, it may be necessary to use a combination of the three rules given above depending on the nature of the roots of $Q(s) = 0$.

12.7 HEAVISIDE'S PARTIAL FRACTION EXPANSION THEOREM

The coefficients of the partial fraction expansion of $P_1(s)/Q(s)$ are given by this theorem.

(a) When the roots of $Q(s) = 0$ are real and unequal, then

$$\frac{P_1(s)}{Q(s)} = \frac{A_1}{s-s_1} + \frac{A_2}{s-s_2} + \cdots + \frac{A_n}{s-s_n}$$

The coefficients $A_1, A_2, \ldots A_n$ are given by

$$A_j = \left(s-s_j\right) \frac{P_1(s)}{Q(s)}\bigg|_{s=s_j}$$

Where $\qquad j = 1, 2, 3 \ldots n$ \hfill (12.24)

(b) When the roots of $Q(s) = 0$ are complex conjugates

$$\frac{P_1(s)}{Q(s)} = \frac{A_1}{\left(s+\alpha+j\omega\right)} + \frac{\overset{*}{A_1}}{\left(s+\alpha-j\omega\right)}$$

The coefficient A_1 is given by

$$A_1 = \left(s+\alpha+j\omega\right) \frac{P_1(s)}{Q(s)}\bigg|_{s=-\alpha-j\omega} \hfill (12.25)$$

Once A_1 is found, A_1^* can be easily known.

(c) When the roots of $Q(s) = 0$ are multiple roots

$$\frac{P_1(s)}{Q(s)} = \frac{P_1(s)}{\left(s-s_j\right)^r} = \frac{A_{j1}}{s-s_j} + \frac{A_{j2}}{\left(s-s_j\right)^2} + \cdots + \frac{A_{jn}}{\left(s-s_j\right)^n} + \cdots + \frac{A_{jr}}{\left(s-s_j\right)^r}$$

Then, the coefficients are given by

$$A_{jn} = \frac{1}{(r-n)!} \frac{d^{r-n}}{ds^{r-n}} \left[\left(s-s_j\right)^n \frac{P_1(s)}{Q(s)}\right]_{s=s_j} \hfill (12.26)$$

The general Eq. (12.26), used to find A_{jn} appears to be quite complicated, but when it is actually used to solve the problems, it is much easier, as will be found in the problems solved later in this chapter.

12.8 FIRST SHIFTING THEOREM

Statement: If $F(s)$ is the Laplace transform of $f(t)$, then

$$L\left[e^{-at} f(t)\right] = F(s+a)$$

For example:

$$L\left[\sin \omega t\right] = \frac{\omega}{s^2 + \omega^2}$$

$$\therefore \quad L\left[e^{-at} \sin \omega t\right] = \frac{\omega}{(s+a)^2 + \omega^2}$$

Proof:

$$L\left[e^{-at} f(t)\right] = \int_{0-}^{\infty} e^{-at} f(t)e^{-st}\ dt = \int_{0-}^{\infty} f(t)e^{-(s+a)t}\ dt \qquad (12.27)$$

Equation (12.27) is exactly similar to Eq. (12.28), except that s is replaced by $(s + a)$

$$L[f(t)] = \int_{0-}^{\infty} f(t)e^{-st}\ dt = F(s) \qquad (12.28)$$

$$\therefore \quad L\left[e^{-at} f(t)\right] = L\left[f(t)\right]_{s=s+a} = F(s+a) \qquad (12.29)$$

12.9 SECOND SHIFTING THEOREM

Statement: If $L[f(t)] = F(s)$, then $L\left[f(t-a)u(t-a)\right] = e^{-as} F(s)$

Example: $L[tu(t)] = \dfrac{1}{s^2}$

$$\therefore \quad L\left[(t-a)u(t-a)\right] = e^{-as} L[tu(t)] = e^{-as} \frac{1}{s^2}$$

Proof:

$$f(t-a)u(t-a) = 0 \text{ for } t < a$$

$$= f(t-a) \text{ for } t \ge a$$

$$L\left[f(t-a)u(t-a)\right] = \int_{0-}^{\infty}\left[f(t-a)u(t-a)\right]e^{-st}.dt$$

$$= \int_{0-}^{a}\left[f(t-a)u(t-a)\right]e^{-st}.dt + \int_{a}^{\infty}\left[f(t-a)u(t-a)\right]e^{-st}.dt$$

$$= 0 + \int_{a}^{\infty} f(t-a)e^{-st}.dt = \int_{a}^{\infty} f(u)e^{-s(u+a)}.du \quad \left| \begin{array}{l} \text{Putting } t - a = u \\ \qquad\qquad dt = du \end{array}\right.$$

When $t = a, \ u = 0$; $\qquad t = \infty, \ u = \infty$

$$\mathcal{L}\left[f\left(t-a\right)u\left(t-a\right)\right] = e^{-as}\int_0^\infty f\left(u\right)e^{-su}\,du = e^{-as}\,F\left(s\right) \qquad (12.30)$$

12.10 CONVOLUTION THEOREM

Statement: If $F_1(s)$ and $F_2(s)$ are the Laplace transforms of $f_1(t)$ and $f_2(t)$ respectively, then

$$\mathcal{L}\left[\int_0^t f_1\left(T\right)f_2\left(t-T\right)dT\right] = \mathcal{L}\left[\int_0^t f_1\left(t-T\right)f_2\left(T\right)dT\right]$$

$$= \mathcal{L}\left[f_1\left(t\right)*f_2\left(t\right)\right]$$

$$= \mathcal{L}\left[f_2\left(t\right)*f_1\left(t\right)\right] = F_1\left(s\right)F_2\left(s\right)$$

Where, T is a dummy variable for t.

$f(t) = f_1\left(t\right)*f_2\left(t\right)$ is read as "the convolution of $f_1(t)$ and $f_2(t)$". The integrals in the above statement are called as convolution integrals.

With the help of the convolution theorem, it is possible to convert the Laplace transform of the convolution of two functions into a product of two transformed functions. This theorem helps us to find the inverse Laplace transform of complicated transformed functions, when the application of the standard formulae fail.

Proof:

$$\mathcal{L}\left[\int_0^t f_1\left(t-T\right)f_2\left(T\right)dT\right] = \int_0^\infty\left[\int_0^t f_1\left(t-T\right)f_2\left(T\right)dT\right]e^{-st}\cdot dt$$

$$(12.31)$$

To prove the theorem, it is necessary to have two integrals to vary from 0 to ∞. Here, we introduce a shifted unit step function $u\left(t-T\right)$ as described in Fig. 12.2.

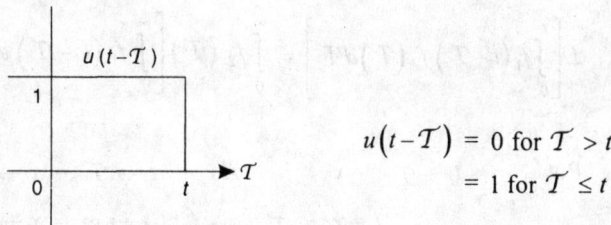

$$u\left(t-T\right) = 0 \text{ for } T > t$$
$$= 1 \text{ for } T \leq t$$

Fig. 12.2

Here, t is a constant and $\mathcal{T}$ is a variable.

$$\therefore \quad \int_0^t f_1(t-\mathcal{T})d\mathcal{T} = \int_0^\infty f_1(t-\mathcal{T})u(t-\mathcal{T})d\mathcal{T} \tag{12.32}$$

Substituting (12.32) in (12.31), we get

$$\mathcal{L}\left[\int_0^t f_1(t-\mathcal{T})f_2(\mathcal{T})d\mathcal{T}\right] = \int_0^\infty \left[\int_0^\infty f_1(t-\mathcal{T})u(t-\mathcal{T})f_2(\mathcal{T})d\mathcal{T}\right]e^{-st}.dt$$

As $f_1(t)$ and $f_2(t)$ are Laplace transformable, the variables of the two integrations may be interchanged.

$$\therefore \mathcal{L}\left[\int_0^t f_1(t-\mathcal{T})f_2(\mathcal{T})d\mathcal{T}\right] = \int_0^\infty f_2(\mathcal{T})\left[\int_0^\infty f_1(t-\mathcal{T})u(t-\mathcal{T})e^{-st}.dt\right]d\mathcal{T}$$

$$\tag{12.33}$$

When t is the variable, the function $u(t-\mathcal{T})$ exists only for values of $t \geq \mathcal{T}$, as explained in Fig. 12.3.

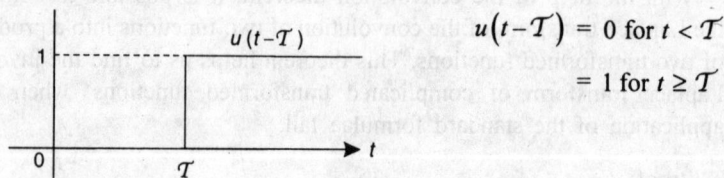

$$u(t-\mathcal{T}) = 0 \text{ for } t < \mathcal{T}$$
$$= 1 \text{ for } t \geq \mathcal{T}$$

Fig. 12.3

Here, $\mathcal{T}$ is a constant and t is a variable.

$$\therefore \mathcal{L}\int_0^\infty f_1(t-\mathcal{T})u(t-\mathcal{T})dt = \int_{\mathcal{T}}^\infty f_1(t-\mathcal{T})dt \tag{12.34}$$

In view of Eq. (12.34), Eq. (12.33) may be written as

$$\mathcal{L}\left[\int_0^t f_1(t-\mathcal{T})f_2(\mathcal{T})d\mathcal{T}\right] = \int_0^\infty f_2(\mathcal{T})\left[\int_{\mathcal{T}}^\infty f_1(t-\mathcal{T})e^{-st}.dt\right]d\mathcal{T}$$

$$\tag{12.35}$$

Put $\quad t - \mathcal{T}^- = x \qquad$ As $\quad t \to \mathcal{T} = x \to 0$

$$t = x + \mathcal{T}^- \qquad\qquad t \to \infty = x \to \infty$$

$$\therefore \qquad\qquad dt = dx$$

Equation (12.35) may be written as

$$\mathcal{L}\left[\int_0^t f_1(t-T)f_2(T)dt\right]=\int_0^\infty f_2(T)\left[\int_0^\infty f_1(x)e^{-s(x+T)}\cdot dx\right]dT$$

$$=\int_0^\infty f_2(T)e^{-sT}\cdot dT\left[\int_0^\infty f_1(x)e^{-sx}\cdot dx\right]=F_2(s)F_1(s)\qquad(12.36)$$

12.11 INITIAL VALUE THEOREM

This theorem helps us to find the initial value of the function $f(t)$, directly from the transformed function $F(s)$.

Statement: If $f(t)$ and $f'(t)$ are Laplace transformable, then, the behaviour of $f(t)$ in the neighbourhood of $t = 0-$, corresponds to the behaviour of $sF(s)$, in the neighbourhood of $s = \infty$.

i.e. $$f(0-)=\underset{t\to 0}{\text{Lt}}\ f(t)=\underset{s\to\infty}{\text{Lt}}\ sF(s)\qquad(12.37)$$

Proof:

We know that

$$\mathcal{L}[f'(t)]=sF(s)-f(0-)$$

i.e. $$\int_{0-}^\infty f'(t)e^{-st}\cdot dt = sF(s)-f(0-)$$

i.e. $$\underset{s\to\infty}{\text{Lt}}\int_0^\infty f'(t)e^{-st}\cdot dt = \underset{s\to\infty}{\text{Lt}}\left[sF(s)-f(0-)\right]$$

i.e. $$0=\underset{s\to\infty}{\text{Lt}}\ [sF(s)-f(0-)]$$

$\therefore$ $$f(0-)=\underset{s\to\infty}{\text{Lt}}\ sF(s)$$

12.12 FINAL VALUE THEOREM

This theorem helps us to find the value of the function $f(t)$, directly from the transformed function $F(s)$.

Statement: If $f(t)$ and $f'(t)$ are Laplace transformable, then the behaviour of $f(t)$ in the neighbourhood of $t = \infty$, corresponds to the behaviour of $SF(s)$ in the neighbourhood of $s = 0$.

i.e. $$f(\infty)=\underset{t\to\infty}{\text{Lt}}\ f(t)=\underset{s\to 0}{\text{Lt}}\ sF(s)\qquad(12.38)$$

Proof:

We know that $\quad \mathcal{L}\left[f'(t)\right] = sF(s) - f(0-)$

i.e. $\quad \displaystyle\int_{0-}^{\infty} f'(t) e^{-st} \cdot dt = sF(s) - f(0-)$

$$\underset{s\to 0}{\text{Lt}} \int_{0-}^{\infty} f'(t) e^{-st} \cdot dt = \underset{s\to 0}{\text{Lt}} \left[sF(s) - f(0-)\right].$$

i.e. $\quad \displaystyle\int_{0-}^{\infty} f'(t) \, dt = \underset{s\to 0}{\text{Lt}} \left[sF(s) - f(0-)\right]$

i.e. $\quad \underset{t\to\infty}{\text{Lt}} \displaystyle\int_{0-}^{t} f'(t) \, dt = \underset{s\to 0}{\text{Lt}} \left[sF(s) - f(0-)\right]$

i.e. $\quad \underset{t\to\infty}{\text{Lt}} \left[f(t)\right]_{0-}^{t} = \underset{s\to 0}{\text{Lt}} \left[sF(s) - f(0-)\right]$

i.e. $\quad \underset{t\to\infty}{\text{Lt}} \left[f(t) - f(0-)\right] = \underset{s\to 0}{\text{Lt}} \left[sF(s) - f(0-)\right]$

$\therefore \quad f(\infty) - f(0-) = \underset{s\to 0}{\text{Lt}} sF(s) - f(0-)$

$[\because f(0-) \text{ is independent of both } t \text{ and } s]$

$\therefore \quad f(\infty) = \underset{s\to 0}{\text{Lt}} sF(s)$

12.13 LAPLACE TRANSFORM OF UNIT IMPULSE FUNCTION $\delta(\tau)$

A function having an extremely large magnitude and an extremely short duration is known as an *impulse function*. Impulse functions have caused a great deal of controversy among mathematicians and engineers, because, its existence is not conventionally justifiable. French mathematician Laurent Schewartz has introduced a new mathematical entity, which provides justification for some of the properties of the impulse function on a rigorous basis.

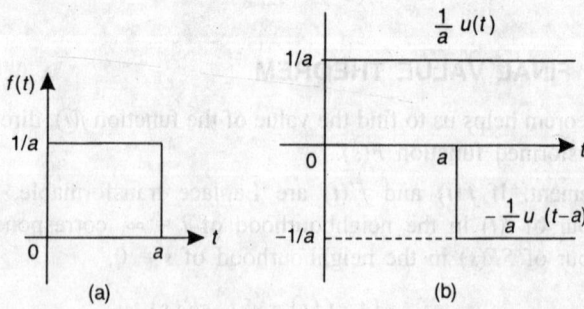

Fig. 12.4

Consider a pulse $f(t)$ as shown in Fig. 12.4 (a), which is equivalent to the algebraic sum of two functions as shown in Fig. 12.4 (b). The Eq. for the function $f(t)$ is as given in (12.39).

$$f(t) = \frac{1}{a}\left[u(t) - u(t-a)\right] \tag{12.39}$$

For any value of a, the area of the pulse is 1. As a approaches zero, the pulse approaches an extremely large height and an extremely short duration, the area still remaining 1. Such a function is known as a unit impulse function, whose Eq. is as given in (12.40)

$$\therefore \qquad \delta(t) = \underset{a\to 0}{\text{Lt}} \frac{1}{a}\left[u(t) - u(t-a)\right] \tag{12.40}$$

The graphical representation of $\delta(t)$ is as shown in Fig. 12.5.

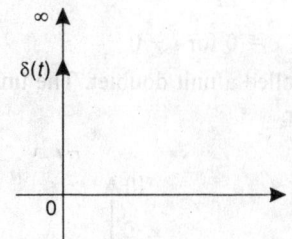

Fig. 12.5

$$\mathcal{L}\left[\delta(t)\right] = \underset{a\to 0}{\text{Lt}} \frac{1}{a}\left[\frac{1}{s} - \frac{e^{-as}}{s}\right] = \underset{a\to 0}{\text{Lt}} \frac{1}{a}\left[\frac{1-e^{-as}}{s}\right]$$

$$= 1 \text{ by } L \text{ Hospital's Rule}$$

$$\therefore \qquad \mathcal{L}\left[\delta(t)\right] = 1$$

A $\delta(t)$ is an impulse function having a magnitude A.

$$\therefore \qquad \mathcal{L}\left[A\,\delta(t)\right] = A$$

Delayed unit impulse function

Figure 12.5 (a) shows a delayed unit impulse function which is defined as:

$$\delta(t-a) = 0 \qquad \text{for } t \neq a$$

$$\text{and}$$

$$\int_{-\infty}^{\infty} \delta(t-a)\,dt = 1$$

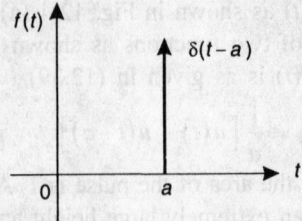

Fig. 12.5 (a)

Unit doublet function

When the unit impulse function $\delta(t)$ is differentiated w.r.t. t, we get

$$\delta'(t) = \frac{d}{dt}\left[\delta(t)\right] = +\infty \text{ and } -\infty \text{ for } t = 0$$

$$= 0 \text{ for } t \neq 0$$

This function is called a unit doublet. The unit doublet is graphically represented as under.

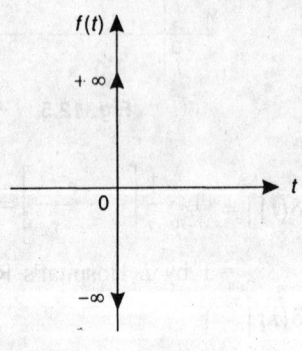

Fig. 12.5 (b)

The Laplace transform of a unit doublet is given by

$$\mathcal{L}\left[\delta'(t)\right] = \mathcal{L}\left[\frac{d}{dt}\delta(t)\right]$$

$$= s\left[\mathcal{L}\{\delta(t)\}\right] = s \times 1 = s$$

12.14 UNIT STEP FUNCTION

A unit step function and its Laplace transformation is already described in section 12.3 (i) of this chapter. The various types of shifted unit step functions are shown in Figs 12.6, 12.7 and 12.8 respectively.

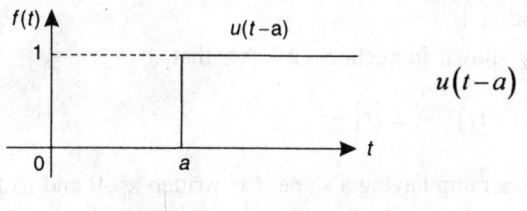

$$u(t-a) = 1 \text{ for } t \geq a$$
$$= 0 \text{ for } t < a$$

Fig. 12.6

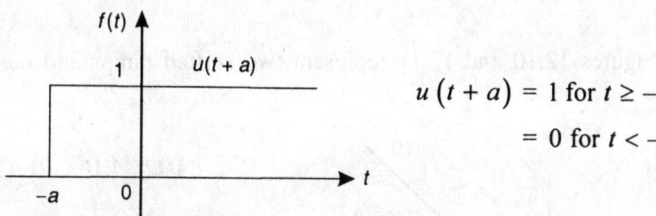

$$u(t+a) = 1 \text{ for } t \geq -a$$
$$= 0 \text{ for } t < -a$$

Fig. 12.7

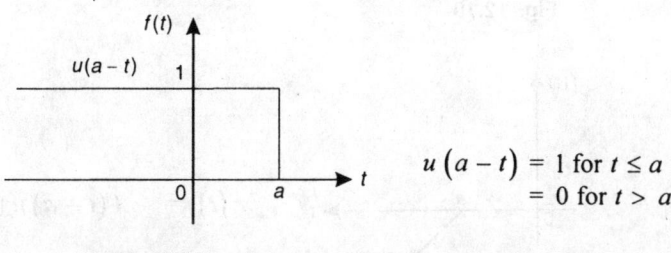

$$u(a-t) = 1 \text{ for } t \leq a$$
$$= 0 \text{ for } t > a$$

Fig. 12.8

12.15 UNIT RAMP FUNCTIONS

A unit ramp function is as shown in Fig. 12.9. $r(t) = t$ is the unit ramp function.

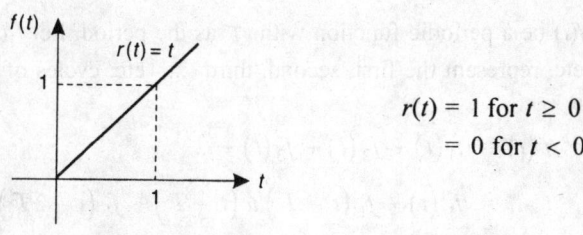

$$r(t) = 1 \text{ for } t \geq 0$$
$$= 0 \text{ for } t < 0$$

Fig. 12.9

The slope of the ramp always represents its magnitude. For a unit ramp, the slope is 1.

It is already shown in section 12.3 (v), that

$$\mathcal{L}\left[r\left(t\right)\right] = \mathcal{L}\left[t\right] = \frac{1}{s^2}$$

Equation for a ramp having a slope A is written as At and its Laplace transform is given by

$$\mathcal{L}\left[At\right] = \frac{A}{s^2}$$

Figures 12.10 and 12.11 represent two shifted ramps and their Eqs.

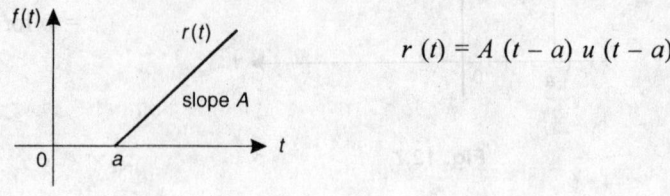

$$r\left(t\right) = A\left(t - a\right)u\left(t - a\right)$$

Fig. 12.10

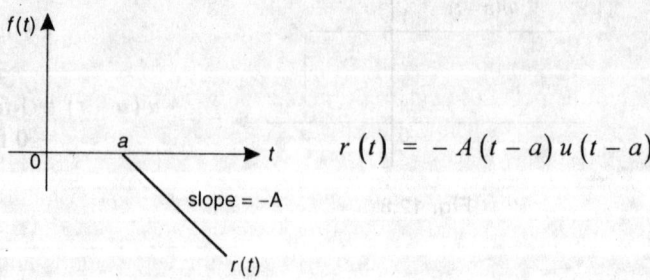

$$r\left(t\right) = -A\left(t - a\right)u\left(t - a\right)$$

Fig. 12.11

12.16 LAPLACE TRANSFORM OF PERIODIC FUNCTIONS

Let $f(t)$ be a periodic function with $\mathcal{T}$ as the period. Let $f_1(t), f_2(t), f_3(t)$ ..., etc. represent the first, second, third ..., etc. cycles of the periodic wave.

Then $f\left(t\right) = f_1\left(t\right) + f_2\left(t\right) + f_3\left(t\right) + ...$

$$= f_1\left(t\right) + f_1\left(t - T\right)u\left(t - T\right) + f_1\left(t - 2T\right)u\left(t - T\right)$$
$$+ f_3\left(t - 3T\right)u\left(t - 3T\right) + ...$$

Then $F(s) = F_1(s) + e^{-Ts} F_1(s) + e^{-2Ts} F_1(s) + e^{-3Ts} F_1(s) + \cdots$

$$= F_1(s)\left[1 + e^{-Ts} + e^{-2Ts} + e^{-3Ts} + \ldots\right]$$

$$= F_1(s)\left[1 - e^{-Ts}\right]^{-1} = \frac{F_1(s)}{1 - e^{-Ts}} \tag{12.41}$$

12.17 IMPULSE RESPONSE

Consider a network of impedance $Z(t)$, as shown in Fig. 12.12, to which a unit impulse voltage $\delta(t)$ is applied, resulting in an impulse current $I_\delta(t)$.

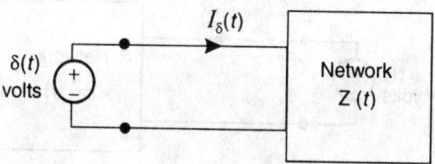

Fig. 12.12

$$I_\delta(t) = \frac{\delta(t)}{Z(t)}$$

or $I_\delta(s)$ $= \dfrac{1}{Z(s)}$ or $Z(s) = \dfrac{1}{I_\delta(s)}$

If we know the response for any other excitation, then the excitation causing this response may be found from the knowledge of the response to an impulse function.

Let $i(t)$ be the new response for an excitation $v(t)$

then, $v(t) = i(t) Z(t)$

or $V(s)$ $= I(s) Z(s) = \dfrac{I(s)}{I_\delta(s)}$ $\left[\because Z(s) = \dfrac{1}{I_\delta(s)}\right]$

Thus, the response of a system to any arbitrary function can be obtained in terms of the impulse response.

12.18 STEP RESPONSE (DUMAHEL'S SUPERPOSITION INTEGRAL)

Consider a network of impedance $Z(t)$, as shown in Fig. 12.13, to which a unit step voltage $u(t)$ is applied, resulting in a current response of $\alpha(t)$.

$$\alpha(t) = \frac{u(t)}{Z(t)}$$

$$\therefore \quad \alpha(s) = \frac{1}{sZ(s)}$$

Put $\quad \dfrac{1}{Z(s)} = H(s)$

$$\therefore \quad \alpha(s) = \frac{H(s)}{s}$$

$$\text{or } \alpha(t) = \mathcal{L}^{-1}\left[\frac{H(s)}{s}\right] \tag{12.42}$$

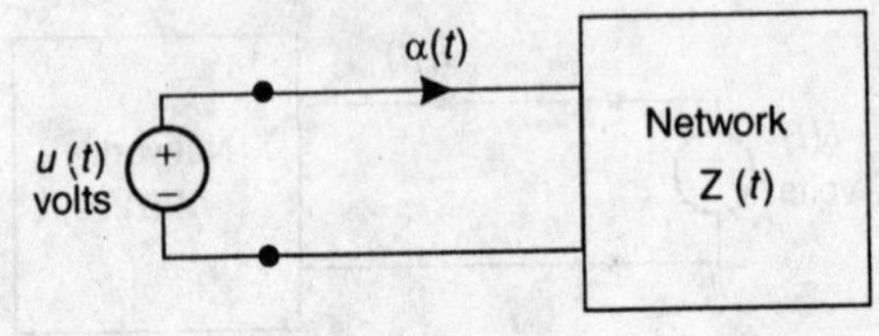

Fig. 12.13

Let an excitation $v(t)$ be applied to the same network of $Z(t)$ and $i(t)$ be the response.

Then $\quad i(t) = \dfrac{v(t)}{Z(t)}$

$$\text{or } I(s) = \frac{V(s)}{Z(s)} = V(s)H(s) = \frac{H(s)}{s} \times s\,V(s)$$

$$\therefore \quad i(t) = \mathcal{L}^{-1}\frac{H(s)}{s} \times s\,V(s) = \mathcal{L}^{-1}\,F_1(s)\,F_2(s) \tag{12.43}$$

Where $F_1(s) = \dfrac{H(s)}{s}$

$$\therefore \quad f_1(t) = \alpha(t) \qquad \text{[from (12.42)]}$$

$$F_2(s) = s\,V(s)$$

$$\therefore \quad f_2(t) = v'(t) + v(0-) = v'(t) + v(0+)$$

$$\therefore \quad i(t) = \int_0^t \alpha(\mathcal{T})\{v'(t-\mathcal{T}) + v(0+)\}\,d\mathcal{T}$$

$$\text{(from convolution theorem)}$$

$$= \int_0^t v(0+)\alpha(\mathcal{T}) + \int_0^t \alpha(\mathcal{T}) \times v'(t-\mathcal{T})\,d\mathcal{T}$$

$$= v(0+)\,\alpha(t) + \int_0^t \alpha(\mathcal{T})v'(t-\mathcal{T})\,d\mathcal{T} \tag{12.44}$$

The integral in Eq. (12.44) is known as *Dumahel's Superposition Integral*. Thus, the response of a system to any arbitrary function can be obtained in terms of the unit response.

12.19 TRANSFORMED NETWORKS

For solving electrical networks using Laplace transformation, it is necessary for us to know the transformed equivalents of all the elements present in the network, taking into consideration the initial values on them. The elements to be considered are the resistance, the inductance and the capacitance.

1. The resistance

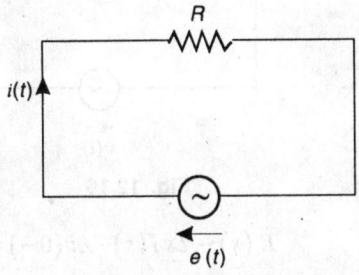

Fig. 12.14

Consider a circuit as shown in Fig. 12.14, in which a resistance R is supplied by a voltage source $e(t)$, due to which, a current $i(t)$ flows through it. Then

$$i(t) = \frac{e(t)}{R} \tag{12.45}$$

or $I(s)$
$$= \frac{E(s)}{R} \tag{12.46}$$

The transformed circuit is written as shown in Fig. 12.15, satisfying Eq. (12.46).

From Eqs (12.45) and (12.46), we observe that the resistance remains unchanged in the transformed network.

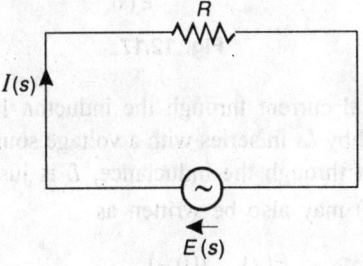

Fig. 12.15

2. The inductance

Consider a circuit as shown in Fig. 12.16, in which an inductance L is supplied by a voltage source $e(t)$, due to which a current $i(t)$, flows through it.

Then $\qquad e(t) = L\dfrac{di(t)}{dt} \qquad$ i.e. $E(s) = L\left[SI(s) - i(0-)\right]$

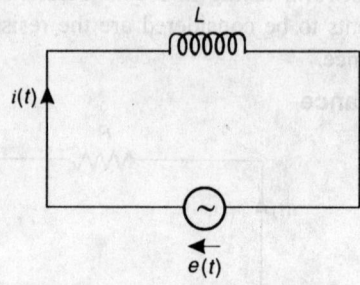

Fig. 12.16

i.e. $\qquad\qquad E(s) - Ls\,I(s) + Li\,(0-) = 0$

$$I(s) = \frac{E(s) + Li(0-)}{Ls} \qquad\qquad (12.47)$$

The transformed circuit satisfying Eq. (12.47) is as shown in Fig. 12.17.

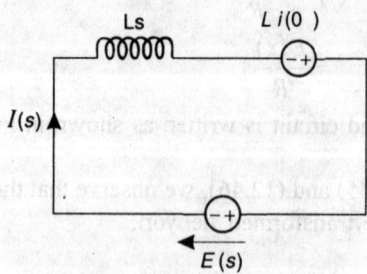

Fig. 12.17

$i(0-)$ is the initial current through the inductor. In the transformed circuit L is replaced by Ls in series with a voltage source $Li(0-)$. If there is no initial current through the inductance, L is just replaced by Ls.

Equation (12.47) may also be written as

$$I(s) = \frac{E(s)}{Ls} + \frac{i(0-)}{s} \qquad\qquad (12.48)$$

The transformed circuit satisfying Eq. (12.48) is as shown in Fig. 12.18. In the transformed circuit. L is replaced by Ls in parallel with a current source $i(0-)/s$ with polarity as shown. If there is no initial current through the inductance, it is just replaced by Ls.

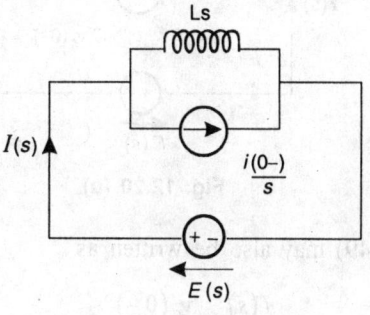

Fig. 12.19

The transformed circuit for the circuit shown in Fig. 12.17, may be either the one shown in Fig. 12.17 or Fig. 12.18.

3. The capacitance

Consider a circuit as shown in Fig. 12.19, in which a capacitance C is supplied by a voltage source $e(t)$, due to which a current $i(t)$, flows through it.

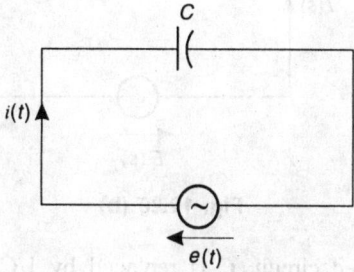

Fig. 12.19

Then
$$i(t) = C\,\frac{de(t)}{dt}$$

i.e.
$$I(s) = C\left[sE(s) - v_c(0-)\right]$$
$$= C\,s\,E(s) - C v_c(0-) \tag{12.49}$$

Where, $v_c(0-)$ is the initial voltage on the capacitor. The transformed circuit satisfying Eq. (12.49) is as shown in Fig. 12.20 (a).

In the transformed circuit, C is replaced by $1/Cs$ in parallel with a current source $C v_c(0-)$, with the polarity as shown. If there is no initial voltage on the capacitor, it is just replaced by $1/Cs$.

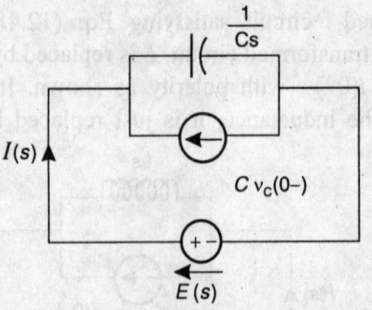

Fig. 12.20 (a)

Equation (12.49) may also be written as

$$E(s) = \frac{I(s)}{Cs} + \frac{v_c(0-)}{s} \qquad (12.50)$$

The transformed circuit satisfying Eq. (12.50) is as shown in Fig. 12.20 (b).

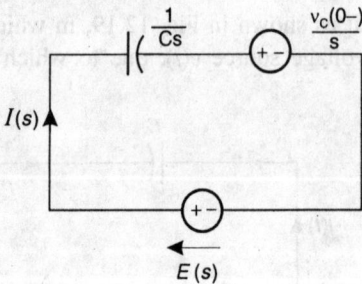

Fig. 12.20 (b)

In the transformed circuit, C is replaced by $1/Cs$ in series with a voltage source $v_c(0-)/s$. If there is no initial charge on the capacitor, it is just replaced by $1/Cs$.

The transformed circuit, for the circuit shown in Fig. 12.19, may be either the one shown in Fig. 12.20 (a) or (b).

12.20 GATE FUNCTION g_{t0} (*t*)

The gate function helps to determine the Laplace transform of discrete periodic functions. The gate function has a height of 1 and a period of $\mathcal{T}$. If starts at any point $t = t_0$ and ends at $t = t_0 + \mathcal{T}$, where $\mathcal{T}$ is the period of the gate function. Fig. 12.21, shows a gate function which starts at $t = t_0$, whose Eq. can be written as

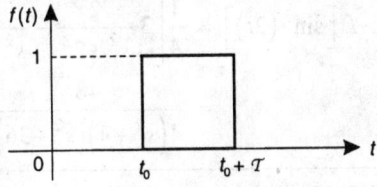

Fig. 12.21

$$g_{t_0}(t) = u(t-t_0) - u\{t-(t_0+T)\} \tag{12.51}$$

$$\therefore \quad G_{t_0}(s) = e^{-t_0 s}\frac{1}{s} e^{-(t_0+T)}\frac{1}{s} = \frac{e^{-t_0 s}}{s}\left[1-e^{-Ts}\right] \tag{12.52}$$

When $t_0 = 0$, i.e. for a gate function which starts from the origin

$$G_0(s) = \frac{1}{s}\left[1-e^{-Ts}\right] \tag{12.53}$$

Worked examples 12.19 and 12.20, shows how gate function can be used to find the Laplace transform of discrete periodic functions.

WORKED EXAMPLES

12.1 Find the Laplace transforms of the following functions.
    (i) sin 2*t*.sin 3*t*

$$\sin A \cdot \sin B = \frac{1}{2}\left[\cos(A-B) - \cos(A+B)\right]$$

$$\therefore \quad \sin 2t \cdot \sin 3t = \frac{1}{2}\left[\cos t - \cos 5t\right]$$

$$\mathcal{L}\left[\sin 2t \cdot \sin 3t\right] = \mathcal{L}\left[\frac{1}{2}\left[\cos t - \cos 5t\right]\right]$$

$$= \frac{1}{2}\left[\frac{s}{s^2+1^2} - \frac{s}{s^2+5^2}\right] = \frac{12s}{\left(s^2+1\right)\left(s^2+25\right)}$$

    (ii) $\sin^3 2t$

$$\sin 3\theta = 3\sin\theta - 4\sin^3\theta$$

$$\therefore \quad \sin 3(2t) = 3\sin(2t) - 4\sin^3(2t)$$

i.e. $\quad \sin^3(2t) = \dfrac{3\sin(2t) - \sin 6t}{4}$

$$\mathcal{L}\left[\sin^3(2t)\right] = \frac{1}{4}\left[3\,\frac{2}{s^2+2^2} - \frac{6}{s^2+6^2}\right]$$

$$= \frac{48}{4\left(s^2+4\right)\left(s^2+36\right)}$$

(iii) $\cos^3 3t$

$$\cos 3\theta = 4\cos^3\theta - 3\cos\theta$$

$\therefore$ $$\cos 3(3t) = 4\cos^3(3t) - 3\cos(3t)$$

i.e. $$\cos^3(3t) = \frac{\cos 9t + 3\cos 3t}{4}$$

$$\mathcal{L}\left[\cos^3(3t)\right] = \frac{1}{4}\left[\frac{s}{s^2+9^2} + \frac{3s}{s^2+3^2}\right]$$

$$= \frac{s\left(s^2+63\right)}{\left(s^2+81\right)\left(s^2+9\right)}$$

(iv) $\sin(\omega t + \theta)$

$$\sin(\omega t + \theta) = \sin\omega t\cos\theta + \cos\omega t\sin\theta$$

$$\mathcal{L}\left[\sin(\omega t + \theta)\right] = \cos\theta\,\frac{\omega}{s^2+\omega^2} + \sin\theta\,\frac{s}{s^2+\omega^2}$$

$$= \frac{\omega\cos\theta + s\sin\theta}{s^2+\omega^2}$$

(v) $t\sin at$ and $t\cos at$

From the first shifting theorem, we know that

$$\mathcal{L}\left[e^{-at}f(t)\right] = F(s+a)$$

$\therefore$ $$\mathcal{L}\left[e^{iat}t\right] = \frac{1}{(s-ia)^2} = \frac{(s+ia)^2}{(s^2+a^2)}$$

$$\frac{s^2-a^2+2ias}{(s^2+a^2)^2} = \mathcal{L}\left[t(\cos at + i\sin at)\right]$$

$\therefore$ $$\mathcal{L}(t\cos at) = \frac{s^2-a^2}{(s^2+a^2)^2}$$

$$\text{and} \quad L(t \sin at) = \frac{2as}{(s^2+a^2)^2}$$

(vi) $e^{-3t}(2\cos 5t - 3\sin 5t)$

$$L\left[e^{-3t}(2\cos 5t - 3\sin 5t)\right]$$

$$= L\left[(2\cos 5t - 3\sin 5t)\right]_{s=s+3}$$

$$= \left[\frac{2s}{s^2+5^2} - 3\frac{5}{s^2+5^2}\right]_{s=s+3}$$

$$= \frac{2(s+3)}{(s+3)^2+25} - \frac{15}{(s+3)^2+25} = \frac{2s-9}{s^2+6s+34}$$

(vii) $\dfrac{1}{2a^2}\sin h\, at . \sin at$

$$L\left[\frac{1}{2a^2}\sin h\, at . \sin at\right] = \frac{1}{2a^2}L\left[\frac{e^{at}-e^{-at}}{2}\sin at\right]$$

$$= \frac{1}{4a^2}\left[\frac{a}{(s-a)^2+a^2} - \frac{a}{(s+a)^2+a^2}\right]$$

$$= \left[\frac{s}{\left[(s-a)^2+a^2\right]\left[(s+a)^2+a^2\right]}\right]$$

12.2 Find the Inverse Laplace transforms of the following functions.

(i) $F(s) = \dfrac{1}{(s+a)^5}$ $\quad : L^{-1}\dfrac{1}{(s+a)^5} = e^{-at}\, L^{-1}\dfrac{1}{s^5}\dfrac{4!}{4!} = e^{-at}\dfrac{t^4}{24}$

(ii) $F(s) = \dfrac{s-3}{s^2+4s+13}$ $\quad : L^{-1}\dfrac{s-3}{s^2+4s+13} = L^{-1}\dfrac{s+2-5}{(s+2)^2+3^3}$

$$= e^{-2t}\, L^{-1}\dfrac{s-5}{s^2+3^2} = e^{-2t}\left(\cos 3t - \frac{5}{3}\sin 3t\right)$$

(iii) $F(s) = \dfrac{s^2-3s+4}{s^3}$ $\quad : L^{-1}\dfrac{s^2-3s+4}{s^3} = L^{-1}\left[\dfrac{1}{s} - \dfrac{3}{s^2} + \dfrac{4}{s^3}\right]$

$$= 1 - 3t + 2t^2$$

(iv) $F(s) = \dfrac{s+2}{s^2-4s+13}$ $\quad : L^{-1}\dfrac{s+2}{s^2-4s+13} = L^{-1}\dfrac{s+2}{(s-2)^2+3^2}$

$$= L^{-1}\dfrac{(s-2)+4}{(s-2)^2+3^2}$$

$$= \mathcal{L}^{-1} \frac{s-2}{(s-2)^2 + 3^2} + \mathcal{L}^{-1} \frac{4}{(s-2)^2 + 3^2}$$

$$= e^{2t} \left(\cos 3t + \frac{4}{3} \cdot \sin 3t \right)$$

(v) $F(s) = \dfrac{2s^2 - 6s + 5}{s^3 - 6s^2 + 11s - 6}$

$$= \frac{2s^2 - 6s + 5}{s^3 - s^2 - 5s^2 + 5s + 6s - 6}$$

$$\frac{2s^2 - 6s + 5}{s(s-1) - 5s(s-1) + 6(s-1)} = \frac{2s^2 - 6s + 5}{(s-1)(s - 5s + 6)}$$

$$= \frac{2s^2 - 6s + 5}{(s-1)(s - 2s - 3s + 6)} = \frac{2s^2 - 6s + 5}{(s-1)\{s(s-2) - 3(s-2)\}}$$

$$= \frac{2s^2 - 6s + 5}{(s-1)(s-2)(s-3)} = \frac{A}{s-1} + \frac{B}{s-2} + \frac{C}{s-3}$$

$$A = (s-1) \frac{2s^2 - 6s + 5}{(s-1)(s-2)(s-3)} \bigg|_{s=1} = \frac{1}{2}$$

$$B = (s-2) \frac{2s^2 - 6s + 5}{(s-1)(s-2)(s-3)} \bigg|_{s=2} = -1$$

and $C = (s-3) \dfrac{2s^2 - 6s + 5}{(s-1)(s-2)(s-3)} \bigg|_{s=3} = \dfrac{5}{2}$

$$\mathcal{L}^{-1} \frac{2s^2 - 6s + 5}{s^3 - 6s^2 + 11s - 6} = \mathcal{L}^{-1} \left[\frac{1}{2(s-1)} - \frac{1}{s-2} + \frac{5}{2(s-3)} \right]$$

$$f(t) = \frac{e^t}{2} - e^{2t} + \frac{5}{2} e^{3t}$$

(vi) $F(s) = \dfrac{4s+5}{(s-1)^2 (s+2)}$

Let $\dfrac{4s+5}{(s-1)^2 (s+2)} = \dfrac{A_{11}}{s-1} + \dfrac{A_{12}}{(s-1)^2} + \dfrac{B}{s+2}$

$$A_{12} = (s-1)^2 \frac{4s+5}{(s-1)^2(s+2)}\bigg|_{s=1} = 3$$

$$A_{11} = \frac{1}{(2-1)!}\frac{d}{ds}\left[(s-1)^2\frac{4s+5}{(s-1)^2(s+2)}\right]_{s=1} = \frac{1}{3}$$

$$B = (s+2)\frac{4s+5}{(s-1)^2(s+2)}\bigg|_{s=-2} = -\frac{1}{3}$$

$$\therefore\ \mathcal{L}^{-1}\frac{4s+5}{(s-1)^2(s+2)} = \mathcal{L}^{-1}\left[\frac{1}{3(s-1)} + \frac{3}{(s-1)^2} - \frac{1}{3(s+2)}\right]$$

$$= \frac{1}{3}e^t + 3e^t\,t - \frac{1}{3}e^{-2t}$$

(vii) $F(s) = \dfrac{5s+3}{(s-1)(s^2+2s+5)}$

Let $\dfrac{5s+3}{(s-1)(s^2+2s+5)} = \dfrac{A}{s-1} + \dfrac{Bs+C}{s^2+2s+5}$

$\therefore \qquad 5s+3 = A(s^2+2s+5) + (Bs+C)(s-1)$

Equating the coefficients on both sides, we get

$$A + B = 0$$
$$2A - B + C = 5$$
and $$5A - C = 3$$

$$A = (s-1)\frac{5s+3}{(s-1)(s^2+2s+5)}\bigg|_{s=1} = 1$$

$$B = -1 \quad \text{and} \quad C = 2$$

$$\therefore \mathcal{L}^{-1}\frac{5s+3}{(s-1)(s^2+2s+5)} = \mathcal{L}^{-1}\left[\frac{1}{s-1} + \frac{-s+2}{s^2+2s+5}\right]$$

$$= e^t + \mathcal{L}^{-1}\frac{-s-1+1+2}{(s+1)^2+4} = e^t + \mathcal{L}^{-1}\frac{-(s+1)+3}{(s+1)^2+2^2}$$

$$= e^t - e^{-t}\cos 2t + \frac{3}{2}e^{-t}\sin 2t$$

(viii) $F(s) = \dfrac{10(s^2 + 2s + 2)}{s^2 + 9s + 20}$

$$\dfrac{10(s^2 + 2s + 2)}{s^2 + 9s + 20} = \dfrac{10(s^2 + 9s + 20 - 7s - 18)}{s^2 + 9s + 20}$$

$$= 10\left[1 - \dfrac{7s + 18}{s^2 + 9s + 20}\right] = 10\left[1 - \dfrac{7s + 18}{(s+4)(s+5)}\right]$$

$$= 10\left[1 - \left\{\dfrac{A_1}{s+4} + \dfrac{A_2}{s+5}\right\}\right]$$

$$A_1 = (s+4)\dfrac{7s+18}{(s+4)(s+5)}\bigg|_{s=-4} = -10$$

$$A_2 = (s+5)\dfrac{7s+18}{(s+4)(s+5)}\bigg|_{s=-5} = 17$$

$$\therefore \quad \mathcal{L}^{-1}\dfrac{10(s^2+2s+2)}{s^2+9s+20} = \mathcal{L}^{-1}10\left[1 - \left\{\dfrac{-10}{s+4} + \dfrac{17}{s+5}\right\}\right]$$

$$\therefore \quad f(t) = 10\delta(t) + 100e^{-4t} - 170e^{-5t}$$

(ix) $F(s) = \dfrac{10}{(s+3)^2(s+4)^3}$

Let $\dfrac{10}{(s+3)^2(s+4)^3}$

$$= \dfrac{A_{11}}{s+3} + \dfrac{A_{12}}{(s+3)^2} + \dfrac{B_{11}}{(s+4)} + \dfrac{B_{12}}{(s+4)^2} + \dfrac{B_{13}}{(s+4)^3}$$

$$A_{12} = (s+3)^2\dfrac{10}{(s+3)^2(s+4)^3}\bigg|_{s=-3} = 10$$

$$A_{11} = \dfrac{1}{(2-1)!}\dfrac{d}{ds}\left[(s+3)^2\dfrac{10}{(s+3)^2(s+4)^3}\right]_{s=-3} = -30$$

$$B_{13} = (s+4)^3\dfrac{10}{(s+3)^2(s+4)^3}\bigg|_{s=-4} = 10$$

$$B_{12} = \frac{1}{(3-2)!} \frac{d}{ds} \left[(s+4)^3 \frac{10}{(s+3)^2 (s+4)^3} \right]_{s=-4} = 20$$

$$B_{11} = \frac{1}{(3-1)!} \frac{d^2}{ds^2} \left[(s+4)^3 \frac{10}{(s+3)^2 (s+4)^3} \right]_{s=-4} = 30$$

$$\therefore \quad \mathcal{L}^{-1} \frac{10}{(s+3)^2 (s+4)^3}$$

$$= \frac{-30}{s+3} + \frac{10}{(s+3)^2} + \frac{30}{(s+4)} + \frac{20}{(s+4)^2} + \frac{10}{(s+4)^3}$$

$$\therefore \quad f(t) = 10te^{-3t} - 30e^{-3t} + 30e^{-4t} + 20te^{-4t} + 5t^2 e^{-4t}$$

(x) $F(s) = \dfrac{5}{s^2 + 6s + 10}$

$$\frac{5}{s^2 + 6s + 10} = \frac{5}{s^2 + 6s + 9 + 1} = \frac{5}{(s+3)^2 - j^2}$$

$$= \frac{5}{(s+3+j)(s+3-j)}$$

$$= \frac{A_1}{(s+3+j)} + \frac{A_1^*}{(s+3-j)}$$

$$A_1 = (s+3+j) \frac{5}{(s+3+j)(s+3-j)} \Bigg|_{s=-3-j}$$

$$= -\frac{5}{2} j = \frac{5}{2} e^{-j\pi/2}$$

$$\therefore \quad A_1^* = \frac{5}{2} e^{j\pi/2}$$

$$F(s) = \frac{5}{s^2 + 6s + 10} = \frac{\frac{5}{2} e^{-j\pi/2}}{(s+3+j)} + \frac{\frac{5}{2} e^{j\pi/2}}{(s+3-j)}$$

$$f(t) = \frac{5}{2} e^{-j\pi/2} e^{-(3+j)t} + \frac{5}{2} e^{j\pi/2} e^{-(3-j)t}$$

$$= \frac{5}{2} e^{-3t} \left[e^{-j(\pi/2+t)} + e^{j(\pi/2+t)} \right]$$

$$= \frac{5}{2} e^{-3t} [\cos(\pi/2 + t) - j\sin(\pi/2 + t)$$
$$+ \cos(\pi/2 + t) + j\sin(\pi/2 + t)]$$

$$= \frac{5}{2} e^{-3t} \, 2 \cos\left(\pi/2 + t\right) = -\frac{5}{2} e^{-3t} \, 2 \sin t$$

$$= -5 e^{-3t} \sin t$$

12.3 Solve the following differential Eqs using Laplace transformation.

(i) $\dfrac{d^3 y}{dt^3} + 2 \dfrac{d^2 y}{dt^2} - \dfrac{dy}{dt} - 2y = 0$. Given $y\,(0) = \, y'\,(0) = 0$

and $y''\,(0) = 6$.

Taking L.T., the above Eq. can be written as

$$\left[s^3 \, Y(s) - s^2 \, y(0) - sy'(0) - y''(0) \right]$$

$$+ 2 \left[s^2 Y(s) - sy(0) - y'(0) \right] - \left[sY(s) - y(0) \right] - 2Y(s) = 0$$

i.e. $\qquad \left[s^3 \, Y(s) - 0 - 0 - 6 \right] + 2 \left[s^2 \, Y(s) - 0 - 0 \right]$

$$- \left[sY(s) - 0 \right] - 2Y(s) = 0$$

i.e. $\qquad \left[s^3 + 2s^2 - s - 2 \right] Y(s) = 6$

or $\qquad Y(s) = \dfrac{6}{s^3 + 2s^2 - s - 2} = \dfrac{6}{s\left(s^2 - 1\right) + 2\left(s^2 - 1\right)}$

$$= \dfrac{6}{(s+1)(s-1)(s+2)} = \dfrac{-3}{s+1} + \dfrac{1}{s-1} + \dfrac{2}{s+2}$$

$$y(t) = -3e^{-t} + e^{t} + 2e^{-2t}$$

(ii) $\dfrac{d^2 x}{dt^2} - \dfrac{2\,dx}{dt} + x = e^{t}$, given $x = 2$ and $\dfrac{dx}{dt} = -1$ at $t = 0$

Taking L.T., the above Eq. becomes

$$\left[s^2 \, X(s) - s\,x(0) - x'(0) \right] - 2 \left[s\,X(s) - x\,(0) \right] + X(s) = \dfrac{1}{s-1}$$

i.e. $[s^2 X(s) - 2s + 1] - 2\,[sX(s) - 2] + X(s) = \dfrac{1}{s-1}$

i.e. $X(s)\left[s^2 - 2s + 1 \right] - 2s + 1 + 4 = \dfrac{1}{s-1}$

$$X\,(s) = \dfrac{\dfrac{1}{s-1} + 2s - 5}{s^2 - 2s + 1} = \dfrac{2s^2 - 7s + 6}{(s-1)^3}$$

$$= \dfrac{A_{11}}{s-1} + \dfrac{A_{12}}{(s-1)^2} + \dfrac{A_{13}}{(s-1)^3}$$

$$A_{13} = (s-1)^3 \frac{2s^2-7s+6}{(s-1)^3}\bigg|_{s=1} = 1$$

$$A_{12} = \frac{1}{(3-2)!}\frac{d}{ds}\left[(s-1)^3 \frac{2s^2-7s+6}{(s-1)^3}\right]_{s=1} = -3$$

$$A_{11} = \frac{1}{(3-1)!}\frac{d^2}{ds^2}\left[(s-1)^3 \frac{2s^2-7s+6}{(s-1)^3}\right]_{s=1} = 2$$

$$\therefore X(s) = \frac{2}{s-1} - \frac{3}{(s-1)^2} + \frac{1}{(s-1)^3}$$

On taking inverse L.T., we get

$$x(t) = 2e^t - 3\,e^t\,.\,t + e^t\,\frac{t^2}{2}$$

(iii) $\dfrac{d^4 y}{dt^4} - k^4 y = 0$, given $y(0) = 1$, $y'(0)=y''(0)=y'''(0) = 0$

On taking L.T., the above Eq. can be written as

$$\left[s^4 Y(s)-s^3 y(0)-s^2 y'(0)-sy''(0)-y'''(0)\right]-k^4 Y(s)= 0$$

i.e. $\left[s^4 Y(s)-s^3 -0-0-0\right]-k^4 Y(s) = 0$

i.e. $Y(s)\left[s^4 -k^4\right] = s^3$

or $\qquad Y(s) = \dfrac{s^3}{s^4 -k^4}$

$$\frac{s^3}{s^4 -k^4} = \frac{s^3}{(s^2 +k^2)(s^2 -k^2)} = \frac{As+B}{s^2 +k^2} + \frac{Cs+D}{s^2 -k^2}$$

i.e. $(As+B)(s^2 -k^2) + (Cs+D)(s^2 +k^2) = s^3$

Equating the coefficients on both sides, we get

$$A + C = 1$$
$$C - A = 0$$
$$B = 0$$

and $\qquad D = 0$

$\therefore \qquad A = 1/2$

and $\qquad C = 1/2$

$$\therefore \qquad Y(s) = \frac{s}{2\left(s^2 + k^2\right)} + \frac{s}{2\left(s^2 - k^2\right)}$$

$$\therefore \qquad y(t) = 1/2\left[\cos kt + \cos h\, kt\right]$$

(iv) $y + \int\limits_0^t y\, dt = 1 - e^{-t}$

On taking L.T.

$$Y(s) + \frac{Y(s)}{s} = \frac{1}{s} - \frac{1}{s+1} = \frac{1}{s(s+1)}$$

$$\therefore \qquad Y(s) = \frac{1}{(s+1)^2}$$

$$\therefore \qquad y(t) = e^{-t}\, t$$

12.4 Find the initial and final values of the following functions.

(i) $i(t) = 3\, e^{-t} - e^{-2t}$

$$I(s) = \frac{3}{s+1} - \frac{1}{s+2} = \frac{2s+5}{s^2 + 3s + 2}$$

$$i(0) = \mathop{\text{Lt}}_{t \to 0} i(t) = \mathop{\text{Lt}}_{s \to \infty} sI(s) = \mathop{\text{Lt}}_{s \to \infty} \frac{2s^2 + 5s}{s^2 + 3s + 2}$$

$$= \mathop{\text{Lt}}_{s \to \infty} \frac{2 + 5/s}{1 + 3/s + 2/s^2} = 2$$

$$i(\infty) = \mathop{\text{Lt}}_{t \to \infty} f(t) = \mathop{\text{Lt}}_{s \to 0} sI(s) = \mathop{\text{Lt}}_{s \to 0} \frac{2s^2 + 5s}{s^2 + 3s + 2} = 0$$

(ii) $i(t) = 5u(t) - 3e^{-2t}$

$$I(s) = \frac{5}{s} - \frac{3}{s+2} = \frac{2s + 10}{s^2 + 2s}$$

$$i(0) = \mathop{\text{Lt}}_{t \to 0} i(t) = \mathop{\text{Lt}}_{s \to \infty} sI(s) = \mathop{\text{Lt}}_{s \to \infty} \frac{2s^2 + 10s}{s^2 + 2s}$$

$$= \mathop{\text{Lt}}_{s \to \infty} \frac{2 + 10/s}{1 + 2/s} = 2$$

$$i(\infty) = \mathop{\text{Lt}}_{t \to \infty} i(t) = \mathop{\text{Lt}}_{s \to 0} sI(s) = \mathop{\text{Lt}}_{s \to 0} \frac{2s^2 + 10s}{s^2 + 2s}$$

$$= \mathop{\text{Lt}}_{s \to \infty} \frac{2s + 10}{s + 2} = 5$$

(iii) $\qquad f(t) = 1 + e^{-t}\left(\sin t + \cos t\right)$

$$F(s) = \frac{1}{s} + \frac{1}{(s+1)^2 + 1} + \frac{s+1}{(s+1)^2 + 1}$$

$$= \frac{1}{s} + \frac{s+2}{(s+1)^2 + 1} = \frac{1}{s} + \frac{s+2}{s^2 + 2s + 2}$$

$$f(0) = \underset{t \to 0}{\text{Lt}}\ f(t) = \underset{s \to \infty}{\text{Lt}}\ sF(s) = \underset{s \to \infty}{\text{Lt}}\ 1 + \frac{s^2 + 2s}{s^2 + 2s + 2}$$

$$= \underset{s \to \infty}{\text{Lt}}\ 1 + \frac{\dfrac{1}{s} + 2}{\dfrac{1}{s} + 2 + \dfrac{2}{s}} = 1 + 1 = 2$$

$$f(\infty) = \underset{t \to \infty}{\text{Lt}}\ f(t) = \underset{s \to 0}{\text{Lt}}\ sF(s)$$

$$= \underset{s \to \infty}{\text{Lt}}\ 1 + \frac{s^2 + 2s}{s^2 + 2s + 2} = 1$$

12.5 Using Convolution theorem, find the Laplace inverse of the following functions.

(i) $\qquad F(s) = \dfrac{1}{s(s+1)}$

Let $\qquad F_1(s) = \dfrac{1}{s}, \qquad \therefore\ f_1(t) = 1$

$$F_2(s) = \frac{1}{s+1}, \qquad \therefore\ f_2(t) = e^{-t}$$

$$\mathcal{L}^{-1}\ F(s) = \mathcal{L}^{-1}\ F_1(s)\ F_2(s) = \int_0^t f_1(t-T) f_2(T)\, dT$$

$$= \int_0^t 1\, e^{-T}\, dT = \left[-e^{-T}\right]_0^t = 1 - e^{-t}$$

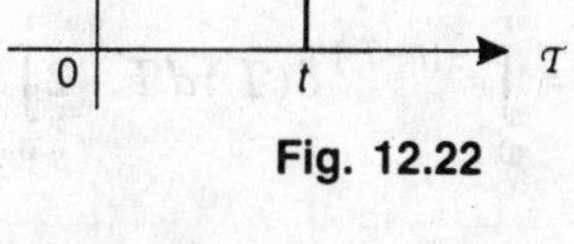

Since, $f_1(t) = 1$
$f_1(t-T) = 1$

Fig. 12.22

(ii) $F(s) = \dfrac{1}{(s-a)^2}$

Let $\quad F_1(s) = \dfrac{1}{s-a}$

$\therefore \quad f_1(t) = e^{at}$

$\quad\quad F_2(s) = \dfrac{1}{s-a}$

$\therefore \quad f_2(t) = e^{at}$

$\quad \mathcal{L}^{-1}\, F(s) = \mathcal{L}^{-1}\, F_1(s)\, F_2(s)$

$$= \int_0^t f_1(t-T)\, f_2(T)\, dT$$

$$= \int_0^t e^{a(t-T)}\, e^{at}\, dT = \int_0^t e^{at}\, dT$$

$$= \Big[e^{at}\, T\Big]_0^t = t\, e^{at}$$

(iii) $F(s) = \dfrac{s}{(s+1)(s+2)}$

Let $\quad F_1(s) = \dfrac{1}{s+1}$

$\therefore \quad f_1(t) = e^{-t}$

$\quad\quad F_2(s) = \dfrac{s}{s+2} = \dfrac{s+2-2}{s+2} = 1 - \dfrac{2}{s+2}$

$\therefore \quad f_2(t) = \delta(t) - 2e^{-2t}$

$\therefore \mathcal{L}^{-1}\, F(s) = \mathcal{L}^{-1}\, F_1(s)\, F_2(s)$

$$= \int_0^t f_1(t-T)\, f_2(T)\, dT$$

$$= \int_0^t e^{-(t-T)}\Big[\delta(T) - 2e^{-2T}\Big]\, dT$$

$$= \int_0^t e^{-(t-T)}\delta(T)\, dT - 2\int_0^t e^{-(t-T)}e^{-2T}\, dT \quad (1)$$

The first integral is not continuous and exists only when $T = 0$, because, the unit impulse function $\delta(t)$ exists only between the limits $t = 0-$ to $t = 0+$. No other function exists between these limits.

At $T = 0$, $\delta(T) = 1$ and $\delta(T) = 0$ elsewhere

$$\therefore \quad \int_0^t e^{-(t-T)} \delta(T) dT = \int_0^t e^{-t} \, dT = \left(e^{-t} T \right)_0^t = t \, e^{-t}$$

The second integral:

$$-2 \int_0^t e^{-(t-T)} e^{-2T} \, dT = -2 \int_0^t e^{-(t+T)} dT$$

$$= -2 \left[e^{-(t+T)} (-1) \right]_0^t$$

$$= 2 e^{-(t+T)} \Big|_0^t = 2e^{-2t} - 2e^{-t}$$

$$\therefore \quad \mathcal{L}^{-1} F(s) = \left(te^{-t} + 2e^{-2t} - 2e^{-t} \right)$$

(iv) $F(s) = \dfrac{1}{(s^2 + 1)^2}$

Let $F_1(s) = \dfrac{1}{s^2 + 1}$

$\therefore \quad f_1(t) = \sin t$

$F_2(s) = \dfrac{1}{s^2 + 1}$

$\therefore \quad f_2(t) = \sin t$

$\therefore \mathcal{L}^{-1} F(s) = \mathcal{L}^{-1} F_1(s) F_2(s)$

$$= \int_0^t f_1(t-T) f_2(T) dT$$

$$= \int_0^t \sin(t-T) \sin T \, dT$$

$$= \frac{1}{2} \left[\int_0^t \{ \cos(t - 2T) - \cos t \} dT \right]$$

$$= \frac{1}{2}\left[\left\{-\frac{1}{2}\sin\left(t-\mathcal{T}\right)\right\}_0^t - \left\{\cos t \cdot \mathcal{T}\right\}_0^t\right]$$

$$= \frac{1}{2}\left[\left\{-\frac{1}{2}\left\{\sin(-t)-\sin t\right\} - \left\{t\cos t - 0\right\}\right]\right]$$

$$= \frac{1}{2}\left(\sin t - t\cos t\right)$$

(v) $F(s) = \dfrac{s}{s(s+1)(s+2)}$

Let $F_1(s) = \dfrac{1}{s(s+1)} = \dfrac{A}{s} + \dfrac{B}{s+1} = \dfrac{1}{s} - \dfrac{1}{s+1}$

$\therefore \quad f_1(t) = 1-e^{-t}, \qquad F_2(s) = \dfrac{1}{s+2}$

$\therefore \quad f_2(t) = e^{-2t}$

$$\mathcal{L}^{-1} F(s) = \int_0^t f_1(t-\mathcal{T})\, f_2(\mathcal{T})\, d\mathcal{T}$$

$$= \int_0^t \left\{1-e^{-(t-\mathcal{T})}\right\} e^{-2\mathcal{T}}\, d\mathcal{T}$$

$$= \int_0^t e^{-2\mathcal{T}}\, d\mathcal{T} - \int_0^t e^{-(t+\mathcal{T})} d\mathcal{T}$$

$$= \left[-\frac{1}{2}e^{-2\mathcal{T}}\right]_0^t + \left[e^{-(t+\mathcal{T})}\right]_0^t$$

$$= -\frac{1}{2}e^{-2t} + \frac{1}{2} + e^{-2t} - e^{-t} = \frac{1}{2} + \frac{1}{2}e^{-2t} - e^{-t}$$

12.6 Find the Laplace Eq. of the function which has a saw tooth wave form as shown in Fig. 12.23.

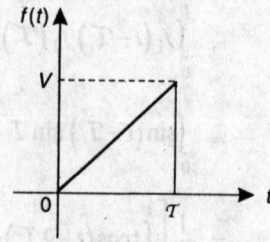

Fig. 12.23

Solution: The waveform in Fig. 12.23 may be split into ramp and step functions as shown in Fig. 12.24.

$$\therefore \quad f(t) = \frac{V}{T} tu(t) - Vu(t - T) - \frac{V}{T}(t - T)u(t - T)$$

$$F(s) = \frac{V}{T}\frac{1}{s^2} - V e^{-Ts}\frac{1}{s} - \frac{V}{T}e^{-Ts}\frac{1}{s^2}$$

$$= \frac{V}{Ts^2}\left[1 - (Ts+1)e^{-Ts}\right]$$

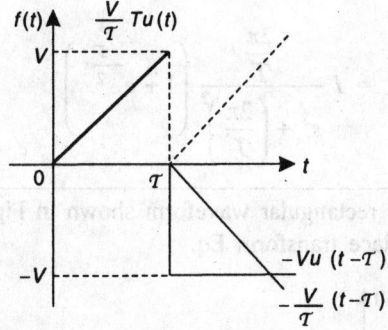

Fig. 12.24

12.7 Write the Eq. for the sinusoidal waveform shown in Fig. 12.25, and find its Laplace transform.

Solution: The sinusoidal waveform shown in Fig. 12.25 may be written as a sum of two sine waves as shown in Fig. 12.26.

Fig. 12.25

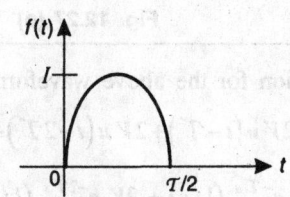

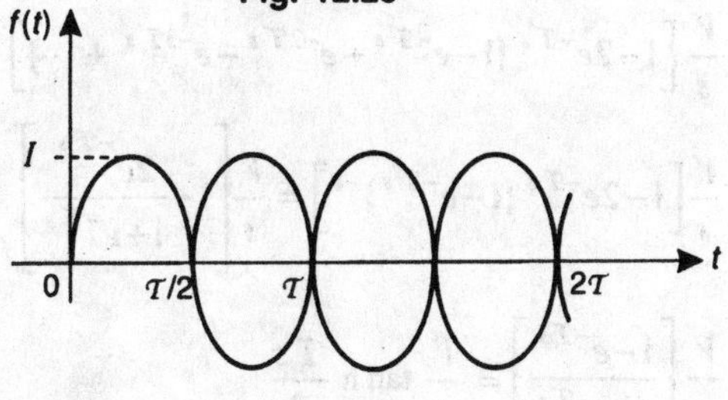

Fig. 12.26

$$\omega = 2\pi f = \frac{2\pi}{T}$$

$$\therefore \quad f(t) = I\sin\frac{2\pi}{T}t\, u(t) + I\sin\frac{2\pi}{T}(t-T/2)\,u\,(t-T/2)$$

$$F(s) = I\frac{\dfrac{2\pi}{T}}{s^2+\left(\dfrac{2\pi}{T}\right)^2} + I\,e^{-\frac{T}{2}s}\,\frac{\dfrac{2\pi}{T}}{s^2+\left(\dfrac{2\pi}{T}\right)^2}$$

$$= I\,\frac{\dfrac{2\pi}{T}}{s^2+\left(\dfrac{2\pi}{T}\right)^2}\left(1+e^{-\frac{T}{2}s}\right)$$

12.8 For the rectangular waveform shown in Fig. 12.27 (a) write down the Laplace transform Eq.

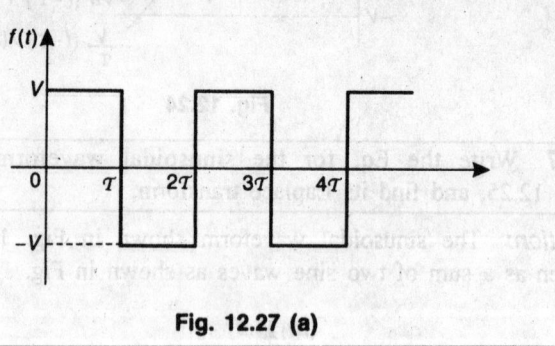

Fig. 12.27 (a)

Solution: Equation for the above waveform may be written as

$$f(t) = V\,u(t) - 2V\,u(t-T) + 2V\,u(t-2T) - 2V\,u(t-3T) + \cdots$$

$$F(s) = \frac{V}{s} - 2V\,e^{-Ts}\,(1/s) + 2V\,e^{-2Ts}\,(1/s) - 2V\,e^{-3Ts}\,(1/s) + \cdots$$

$$= \frac{V}{s}\left[1 - 2e^{-Ts}\{1 - e^{-Ts} + e^{-2Ts} - e^{-3Ts} + \cdots\}\right]$$

$$= \frac{V}{s}\left[1 - 2e^{-Ts}\{1 - e^{-Ts}\}^{-1}\right] = \frac{V}{s}\left[1 - \frac{2e^{-Ts}}{1+e^{-Ts}}\right]$$

$$= \frac{V}{s}\left[\frac{1-e^{-Ts}}{1+e^{-Ts}}\right] = \frac{V}{s}\tanh\frac{Ts}{2}$$

Alternate solution: Consider only one cycle of the wave as shown in Fig. 12.27 (b).

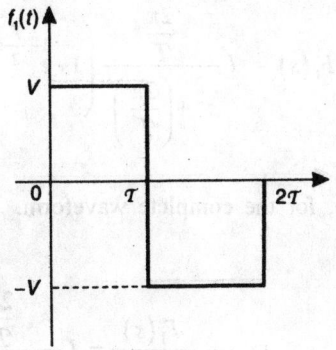

Fig. 12.27 (b)

$$f_1(t) = V u(t) - 2V u(t - 2T) + u(t - 2T)$$

$$F_1(s) = \frac{V}{s} - 2V e^{-Ts} \frac{1}{s} + \frac{e^{-2Ts}}{s}$$

$$= \frac{V}{s}\left[1 - 2e^{-Ts} + e^{-2Ts}\right] = \frac{V}{s}\left(1 - e^{-Ts}\right)^2$$

∴ For the complete periodic cycle as shown in Fig. 12.24 (a).

$$F(s) = \frac{F_1(s)}{1 - e^{-2Ts}} = \frac{V}{s}\frac{(1 - e^{-Ts})^2}{1 - e^{-2Ts}}$$

$$= \frac{V}{s}\frac{1 - e^{-Ts}}{1 + e^{-Ts}} = \frac{V}{s}\tan h \frac{Ts}{2}$$

12.9 Find the Laplace transform Eq. for the half rectified sine wave shown in Fig. 12.28.

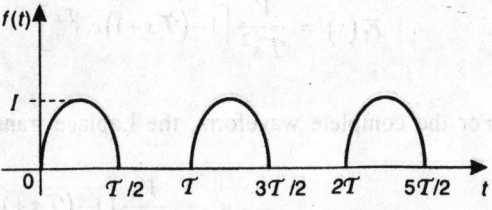

Fig. 12.28

Solution: From Prob. 12.7, we know that the Laplace transform Eq. of the first cycle of waveform given in Fig. 12.28 is

$$F_1(s) = I \frac{\frac{2\pi}{T}}{s^2 + \left(\frac{2\pi}{T}\right)^2} \left(1 + e^{-\frac{T}{2}s}\right)$$

Therefore, for the complete waveform.

$$F(s) = \frac{F_1(s)}{1 - e^{-Ts}} = I \frac{\frac{2\pi}{T}}{s^2 + \left(\frac{2\pi}{T}\right)^2} \left[\frac{\left(1 - e^{-\frac{T}{2}s}\right)}{1 - e^{-Ts}}\right]$$

12.10 Write the Eq. of the saw tooth waveform given in Fig. 12.29 and find its Laplace transform.

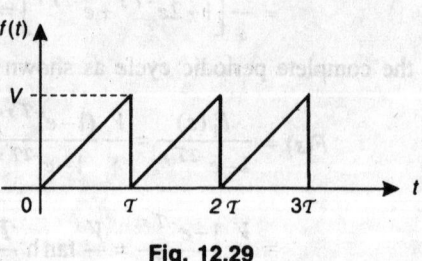

Fig. 12.29

Solution: From Prob. 12.6, we know that the Laplace transform Eq. of the first cycle is given by

$$F_1(s) = \frac{V}{Ts^2}\left[1 - (Ts+1)e^{-Ts}\right]$$

∴ For the complete waveform, the Laplace transform Eq. is

$$F(s) = \frac{F_1(s)}{1 - e^{-Ts}} = \frac{\frac{V}{Ts^2}\left[1 - (Ts+1)e^{-Ts}\right]}{1 - e^{-Ts}}$$

12.11 For the rectangular pulse shown in Fig. 12.30, write the Eq., and find its Laplace transform.

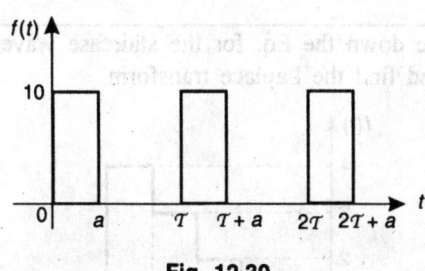

Fig. 12.30

Solution For the first cycle of the waveform, the Eq. is

$$f_1(t) = 10\left[u(t) - u(t-a)\right]$$

$$F_1(s) = 10\left[\frac{1}{s} - e^{-as}\frac{1}{s}\right] = \frac{10}{s}\left[1 - e^{-as}\right]$$

For the complete waveform, the Eq. is

$$F(s) = \frac{F_1(s)}{1 - e^{-Ts}} = \frac{10}{s}\left[\frac{1 - e^{-as}}{1 - e^{-Ts}}\right]$$

12.12 For the waveform shown in Fig. 12.31, write the Laplace transform Eq.

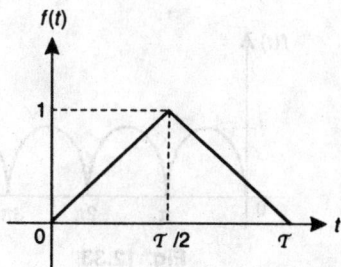

Fig. 12.31

Solution

$$f(t) = \frac{2}{T} t\, u(t) - \frac{4}{T}(t - T/2)\, u(t - T/2) + \frac{2}{T}(t - T)\, u(t - T)$$

$$F(s) = \frac{2}{T}\frac{1}{s^2} - \frac{4}{T} e^{-\frac{T}{2}s}\frac{1}{s^2} + \frac{2}{T} e^{-Ts}\frac{1}{s^2}$$

$$= \frac{2}{T s^2}\left[1 - 2 e^{-\frac{T}{2}s} + e^{-Ts}\right] = \frac{2}{T s^2}\left(1 - e^{-\frac{T}{2}s}\right)^2$$

12.13 Write down the Eq. for the staircase waveform shown in Fig. 12.32 and find the Laplace transform.

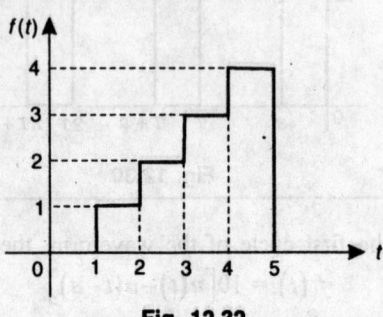

Fig. 12.32

Solution

$$f(t) = u\left(t-1\right) + u(t-2) + u(t-3) + u\left(t-4\right) - 4u(t-5)$$

$$\therefore \quad F(s) = \frac{1}{s}\left[e^{-s} + e^{-2s} + e^{-3s} + e^{-4s} - 4e^{-5s}\right]$$

12.14 For the full-rectified waveform shown in Fig. 12.33, find the Laplace transform Eq.

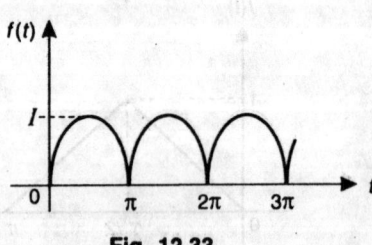

Fig. 12.33

Solution: From Prob. 12.7, we know that the Laplace transform Eq. for the one cycle of the above waveform is given as

$$F_1\left(s\right) = I\frac{\dfrac{2\pi}{T}}{s^2 + \left(\dfrac{2\pi}{T}\right)^2}\left(1 + e^{-\frac{T}{2}s}\right)$$

But, in this case, $\mathcal{T} = 2\pi$

$$\therefore \quad F_1(s) = I \frac{1}{s^2+1}\left(1+e^{-\pi s}\right)$$

For the complete waveform, the Laplace transform Eq. is

$$F(s) = \frac{F_1(s)}{1-e^{-\pi s}} = \frac{1}{s^2+1} \frac{\left(1+e^{-\pi s}\right)}{\left(1-e^{-\pi s}\right)}$$

$$= \frac{1}{s^2+1} \coth \frac{\pi s}{2} \quad (\because \mathcal{T} = \pi \text{ for the}$$
$$\text{complete waveform)}$$

12.15 Write the Eq. for the waveform shown in Fig. 12.34 and find its Laplace transform.

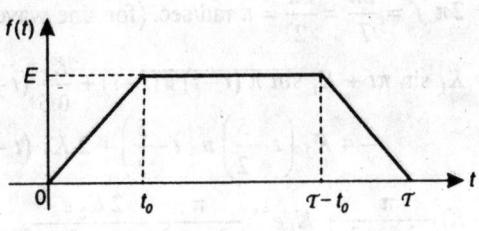

Fig. 12.34

Solution

$$f(t) = \frac{E}{t_0} t \, u(t) - \frac{E}{t_0}\left(t-t_0\right) u\left(t-t_0\right)$$

$$- \frac{E}{t_0}\left[\{t-\left(\mathcal{T}-t_0\right)\} u\{t-\left(\mathcal{T}-t_0\right)\}\right] + \frac{E}{t_0}\left(t-\mathcal{T}\right) u\left(t-\mathcal{T}\right)$$

$$F(s) = \frac{E}{t_0}\frac{1}{s^2} - \frac{E}{t_0}e^{-t_0 s}\frac{1}{s^2} - \frac{E}{t_0}\frac{1}{s^2}e^{-(\mathcal{T}-t_0)s}\frac{1}{s^2} + \frac{E}{t_0}e^{-\mathcal{T}s}\frac{1}{s^2}$$

$$= \frac{E}{t_0 s^2}\left[1-e^{-t_0 s} - e^{-(\mathcal{T}-t_0)s} + e^{-\mathcal{T}s}\right]$$

$$= \frac{E}{t_0 s^2}\left[\left(1-e^{-t_0 s}\right) + e^{-\mathcal{T}s}\left(1-e^{-t_0 s}\right)\right]$$

$$= \frac{E}{t_0 s^2}\left(1-e^{-t_0 s}\right)\left(1+e^{-\mathcal{T}s}\right)$$

12.16 The waveform shown in Fig. 12.35 is sinusoidal in the interval $t = 0$ to $t = 1$, and is an isosceles triangle from $t = 2$ to $t = 3$. For all other t, $v = 0$. Write the expression for $v(t)$, using step, ramp and sine functions and find its Laplace transform.

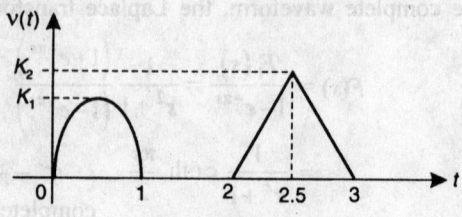

Fig. 12.35

Solution

$$\omega = 2\pi f = \frac{2\pi}{T} = \frac{2\pi}{2} = \pi \text{ rad/sec. (for sine wave)}$$

$$v(t) = K_1 \sin \pi t + K_1 \sin \pi (t-1) u(t-1) + \frac{K_2}{0.5}(t-2) u(t-2)$$

$$- 4 K_2 \left(t - \frac{5}{2}\right) u\left(t - \frac{5}{2}\right) + 2 K_2 (t-3) u(t-3)$$

$$V(s) = K_1 \frac{\pi}{s^2 + \pi^2} + K_1 e^{-s} \frac{\pi}{s^2 + \pi^2} + \frac{2 K_2 e^{-2s}}{s^2}$$

$$- \frac{4 K_2 e^{-\frac{5}{2}s}}{s^2} + \frac{2 K_2 e^{-3s}}{s^2}$$

$$= K_1 \frac{\pi}{s^2 + \pi^2} \left(1 + e^{-s}\right) + \frac{2K_2}{s^2}\left(e^{-2s} - 2e^{-\frac{5}{2}s} + e^{-3s}\right)$$

12.17 For the waveform shown in Fig. 12.36, write down the Laplace transform Eq.

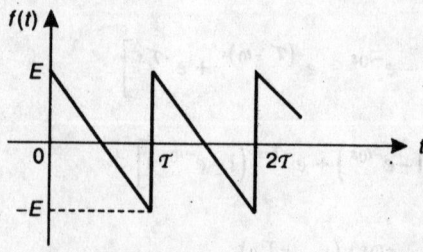

Fig. 12.36

Solution: Equation for the first cycle is

$$f_1(t) = -\frac{2E}{T}\,t u(t) + E u(t) + E u(t-T) + \frac{2E}{T}(t-T)u(t-T)$$

$$F_1(s) = -\frac{2E}{T}\frac{1}{s^2} + E\frac{1}{s} + E e^{-Ts}\frac{1}{s} + \frac{2E}{T}e^{-Ts}\frac{1}{s^2}$$

$$= -\frac{2E}{T}\frac{1}{s^2}\left(1 - e^{-Ts}\right) + \frac{E}{s}\left(1 + e^{-Ts}\right)$$

$$\therefore\quad F(s) = \frac{F_1(s)}{1 - e^{-Ts}} = -\frac{2E}{Ts^2} + \frac{E}{s}\left(\frac{1 + e^{-Ts}}{1 - e^{-Ts}}\right) = -\frac{2E}{Ts^2} + \frac{E}{s}\coth\frac{Ts}{2}$$

12.18 Sketch the following signals.

(i) $\sin\dfrac{2\pi}{T}t$

(ii) $\sin\dfrac{2\pi}{T}t\,u(t)$

(iii) $\sin\dfrac{2\pi}{T}t\,u\left(t - \dfrac{\pi}{4}\right)$

(iv) $\sin\dfrac{2\pi}{T}\left(t - \dfrac{T}{4}\right)u(t)$

(v) $\sin\dfrac{2\pi}{T}\left(t - \dfrac{T}{4}\right)u\left(t - \dfrac{T}{4}\right)$

(vi) $\sin\dfrac{2\pi}{T}\left(t - \dfrac{T}{8}\right)u\left(t - \dfrac{T}{4}\right)$

Solution

(i) $\sin\dfrac{2\pi}{T}t$

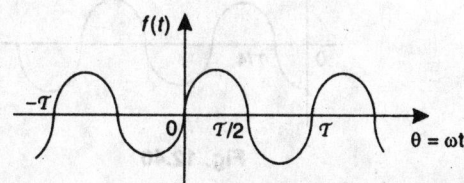

Fig. 12.37

This waveform starts at $t = 0$ and exists for all values of t.

(ii) $\sin\dfrac{2\pi}{T}t\,u(t)$

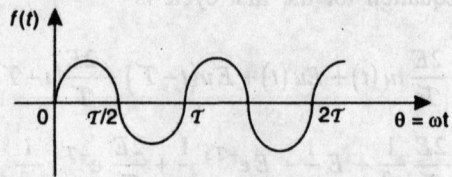

Fig. 12.38

This waveform starts at $t = 0$ and exists only for $t \geq 0$.

(iii) $\sin \dfrac{2\pi}{T} t \cdot u\left(t - \dfrac{\pi}{4}\right)$

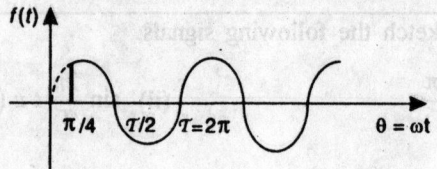

Fig. 12.39

This waveform starts at $t = 0$ and exists for all values of $t \geq \dfrac{\pi}{4}$.

(iv) $\sin \dfrac{2\pi}{T}\left(t - \dfrac{T}{4}\right) u(t)$

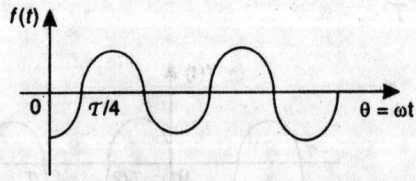

Fig. 12.40

This waveform starts at $t = \dfrac{T}{4}$ and exists for all values of $t \geq 0$.

(v) $\sin \dfrac{2\pi}{T}\left(t - \dfrac{T}{4}\right) u\left(t - \dfrac{T}{4}\right)$

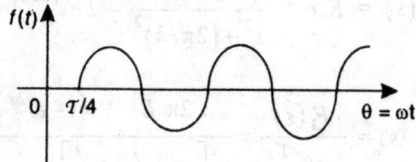

Fig. 12.41

This waveform starts at $t = \dfrac{T}{4}$ and exists for all values of $t \geq \dfrac{T}{4}$.

(vi) $\sin \dfrac{2\pi}{T}\left(t - \dfrac{T}{8}\right) u\left(t - \dfrac{T}{4}\right)$

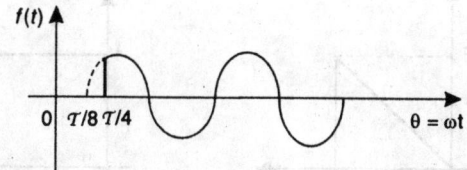

Fig. 12.42

This waveform starts at $t = \dfrac{T}{8}$ and exists for all values of $t \geq \dfrac{T}{4}$

12.19 Find the Laplace transform of the periodic train of one cycle sinusoidal pulses, as shown in Fig. 12.43.

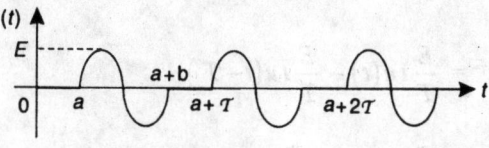

Fig. 12.43

Solution: The period of the waveform is

$$a + T - a = T \text{ or } a + b - a = b, \text{ i.e. } T = b$$

Equation for the first cycle of the train of pulses is given by

$$f_1(t) = E \sin \dfrac{2\pi}{b}(t - a) u(t - a) - E \sin \dfrac{2\pi}{b}$$
$$(t - a - b) u(t - a - b)$$

$$F_1(s) = .E.e^{-as} \frac{2\pi/b}{s^2 + (2\pi/b)^2} - Ee^{-(a+b)s} \frac{2\pi/b}{s^2 + (2\pi/b)^2}$$

$$\therefore \quad F(s) = \frac{F_1(s)}{1-e^{-Ts}} = \frac{2\pi\, Ee^{-as}\left(1-e^{-bs}\right)}{b\left[s^2 + \left(\dfrac{2\pi}{b}\right)^2\right]\left(1-e^{-bs}\right)}$$

12.20 Using gate function, find the Laplace transform of the saw tooth function shown in Fig. 12.44.

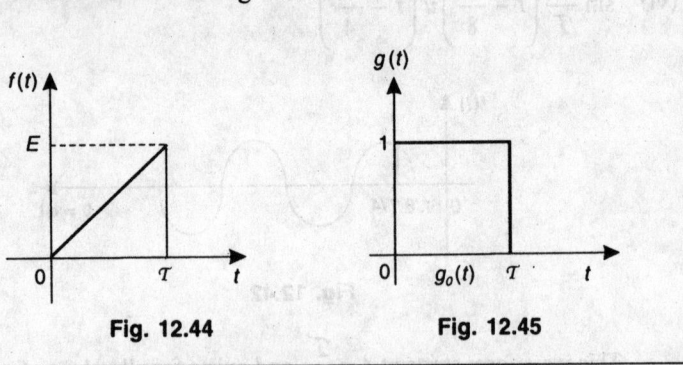

Fig. 12.44   **Fig. 12.45**

Solution: Using the gate function $g_0(t)$ as shown in Fig. 12.45, the Eq. for the saw tooth function may be written as

$$f(t) = \frac{E}{T} t\, g_0(t) = \frac{E}{T} t\left[u(t) - u(t-T)\right]$$

$$= \frac{E}{T} t\, u(t) - \frac{E}{T} t\, u(t-T)$$

$$= \frac{E}{T} t\, u(t) - \frac{E}{T}\left[(t - T + T)\, u(t-T)\right]$$

$$= \frac{E}{T} tu(t) - \frac{E}{T}\left[(t-T)\, u(t-T) - \frac{E}{T} T u(t-T)\right]$$

$$= \frac{E}{Ts^2}\left[1 - e^{-Ts} - Ts\, e^{-Ts}\right] = \frac{E}{Ts^2}\left[1 - (1+Ts)e^{-Ts}\right]$$

This answer tallies with the answer obtained in problem 12.6.

12.21 Using gate function, find the Laplace transform of the discrete sinewave shown in Fig. 12.46.

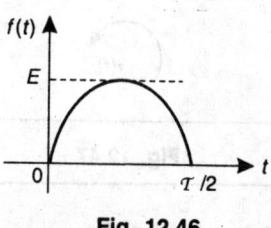

Fig. 12.46

Solution: Using gate function, the eq. for the discrete sine wave shown in Fig. 12.46, may be written as

$$f(t) = \left\{ E\sin\frac{2\pi}{T}t \right\} g_0(t)$$

$$= \left\{ E\sin\frac{2\pi}{T}t \right\} \left\{ u(t) - u\left(t - \frac{T}{2}\right) \right\}$$

$$= E\sin\frac{2\pi}{T}t\, u(t) - E\sin\frac{2\pi}{T}\left(t - \frac{T}{2} + \frac{T}{2}\right) u\left(t - \frac{T}{2}\right)$$

$$= E\sin\frac{2\pi}{T}t\, u(t) + E\sin\frac{2\pi}{T}\left(t - \frac{T}{2}\right) u\left(t - \frac{T}{2}\right)$$

$$\left[\because \sin\left(\frac{T}{2} + \theta\right) = \sin(\pi + \theta) = -\sin\theta \right]$$

$$\therefore F(s) = \frac{E(2\pi/T)}{s^2 + (2\pi/T)^2} + Ee^{-\frac{T}{2}s}\frac{2\pi/T}{s^2 + (2\pi/T)^2}$$

$$= \frac{E(2\pi/T)}{s^2 + (2\pi/T)^2}\left[1 + e^{-\frac{T}{2}s} \right]$$

This answer tallies with the answer obtained in problem 12.7.

12.22 For the circuit shown in Fig. 12.47, find an expression for $i(t)$, when the switch K is closed at $t = 0$.

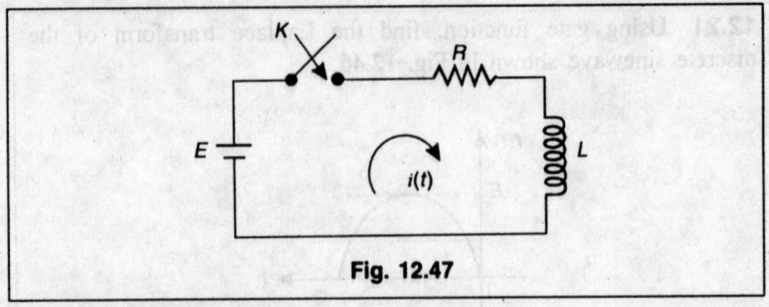

Fig. 12.47

Solution: When K is closed at $t = 0$

$$L \frac{di}{dt} + Ri = E \qquad \qquad \text{On taking L.T.}$$

$$L\big[sI(s) - i(0-)\big] + R\,I(s) = \frac{E}{s}$$

$$\because \qquad i(0-) = i(0+) = 0$$

$$\therefore \quad (R + Ls)I(s) = \frac{E}{s}$$

i.e. $\qquad I(s) = \dfrac{E}{s(R+Ls)} = \dfrac{E}{L}\left[\dfrac{1}{s(s+R/L)}\right]$

$$= \frac{E}{L}\left[\frac{L/R}{s} - \frac{L/R}{s+R/L}\right]$$

$$\therefore \qquad i(t) = \frac{E}{L}\left[\frac{L}{R} - \frac{L}{R}\,e^{-\frac{R}{L}t}\right] = \frac{E}{R}\left(1 - e^{-\frac{R}{L}t}\right)$$

12.23 For the circuit shown in Fig. 12.48, find an expression for $i(t)$, when the switch K is closed at $t = 0$. Assume that there is no initial charge on the capacitor.

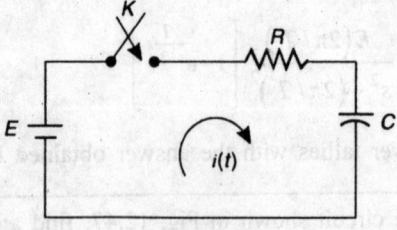

Fig. 12.48

Solution: When the switch K is clsoed at $t = 0$

$$Ri + \frac{1}{C} \int idt = E$$

On taking L.T.

$$R I(s) + \frac{1}{C} \left[\frac{I(s)}{s} + \frac{q(0-)}{s} \right] = \frac{E}{s}$$

But, $\quad q(0-) = 0$

$$\therefore \quad I(s) \left[R + \frac{1}{Cs} \right] = \frac{E}{s}$$

i.e. $\qquad I(s) = \dfrac{E}{s \left(R + \dfrac{1}{Cs} \right)} = \dfrac{E/R}{s + \dfrac{1}{RC}}$

i.e. $\qquad i(t) = \dfrac{E}{R} e^{-t/RC}$

12.24 In the circuit shown in Fig. 12.49, if the capacitor is initially charged to IV, find an expression for $i(t)$, when the switch K is closed at $t = 0$.

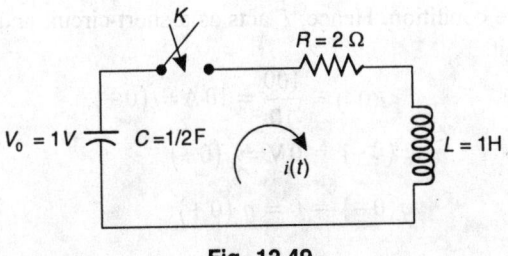

Fig. 12.49

Solution: When the switch K is closed at $t = 0$

$$L \frac{di}{dt} + Ri + \frac{1}{C} \int idt = 0$$

On taking L.T.

$$L \left[sI(s) - i(0-) \right] + RI(s) + \frac{1}{C} \left[\frac{I(s)}{s} + \frac{q(0-)}{s} \right] = 0$$

As $\qquad i(0-) = 0 \text{ A}$

and $\qquad \dfrac{q(0-)}{C} = V_0 = 1 \text{ V}$

$$1 \times s\, I\, (s) + 2I\, (s) + 2\, \frac{I(s)}{s} - \frac{1}{s} = 0$$

or $$I(s) = \frac{1}{s^2 + 2s + 2} = \frac{1}{(s+1)^2 + 1}$$

i.e. $$i(t) = e^{-t} \sin t$$

12.25 In the circuit shown in Fig. 12.50, the switch $\dot{K}$ is closed and the steady state is reached. At $t = 0$, the switch is opened. Find the expression for the current in the inductor using Laplace transform.

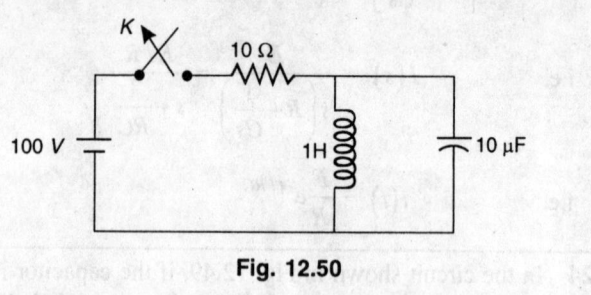

Fig. 12.50

Solution: Initially, the switch K is closed and the circuit is under steady-state condition. Hence, L acts as a short-circuit and C acts as an open-circuit.

∴ $$i(0-) = \frac{100}{10} = 10\,\text{A} = i(0+)$$

$$v_c\,(0-) = 0\,\text{V} = v_c\,(0+)$$

$$q(0-) = 0 = q\,(0+)$$

When K is opened, $L\dfrac{di}{dt} + \dfrac{1}{C}\displaystyle\int i dt = 0$. Taking L.T.

$$L\left[sI\,(s) - i\,(0-)\right] + \frac{1}{C}\left[\frac{I(s)}{s} + \frac{q(0-)}{s}\right] = 0$$

i.e. $$Ls\,I(s) - Li\,(0-) + \frac{I(s)}{Cs} = 0$$

i.e. $$1 \times sI(s) - 1 \times 10 + \frac{I(s)}{10 \times 10^{-6}\,s} = 0$$

i.e. $$I(s)\left(s + \frac{10^5}{s}\right) = 10$$

i.e.
$$I(s)\left(\frac{s^2+10^5}{s}\right) = 10$$

$\therefore$
$$I(s) = \frac{10s}{s^2+10^5} = \frac{10s}{s^2+\left(10^{\frac{5}{2}}\right)^2}$$

$\therefore$
$$i(t) = 10\cos\left(10^{\frac{5}{2}}\right)t$$

12.26 In the network shown in Fig. 12.51, all the initial currents are zero. Obtain $i_1(t)$ and $i_2(t)$ by L.T. method.

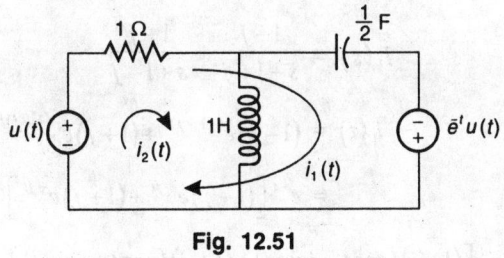

Fig. 12.51

Solution: The transformed circuit is as shown in Fig. 12.52.

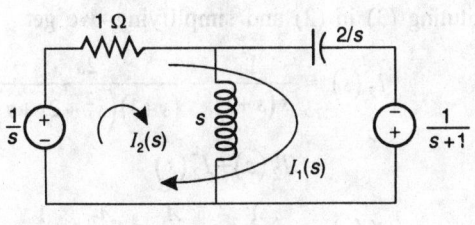

Fig. 12.52

The loop Eqs are:

$$\left(1+\frac{2}{s}\right)I_1(s) + 1\times I_2(s) = \frac{1}{s} + \frac{1}{s+1}$$

i.e.
$$(s+1)(s+2)I_1(s)+s(s+1)I_2(s) = (2s+1) \qquad (1)$$

$$1\times I_1(s) + (1+s)I_2(s) = \frac{1}{s}$$

i.e.
$$sI_1(s)+s(s+1)I_2(s) = 1 \qquad (2)$$

Solving (1) and (2), we get

$$I_1(s) = \frac{2s}{s^2 + 2s + 2} = \frac{2s}{(s+1)^2 + 1}$$

$$= \frac{2s}{(s+1+j)(s+1-j)} \qquad (3)$$

$$= \frac{A}{(s+1+j)} + \frac{\overset{*}{A}}{(s+1-j)}$$

Where

$$A = \frac{2s}{(s+1-j)}\bigg|_{s=-1-j} = 1-j,$$

∴

$$\overset{*}{A} = 1+j$$

∴

$$I_1(s) = \frac{1-j}{s+1+j} + \frac{1+j}{s+1-j}$$

$$i_1(t) = (1-j)e^{(-1-j)t} + (1+j)e^{-(1-j)t}$$

$$= e^{-t}\left[(1-j)e^{-jt} + (1+j)e^{+jt}\right]$$

$$= e^{-t}\left[(1-j)(\cos t - j\sin t) + (1+j)(\cos t + j\sin t)\right]$$

$$= 2e^{-t}(\cos t - \sin t)$$

Substituting (3) in (2) and simplifying, we get

$$I_2(s) = \frac{1}{s(s+1)} - \frac{2s}{(s+1)(s^2 + 2s + s)}$$

$$= I_2'(s) + I_2''(s)$$

$$I_2'(s) = \frac{1}{s(s+1)} = \frac{A_1}{s} + \frac{A_2}{s+1} = \frac{1}{s} - \frac{1}{s+1}$$

∴

$$I_2'(t) = 1 - e^{-t} \qquad (4)$$

$$I_2''(s) = -\frac{2s}{(s+1)(s+1+j)(s+1-j)}$$

$$= \frac{A_1}{s+1} + \frac{A_2}{s+1+j} + \frac{\overset{*}{A_2}}{s+1-j}$$

$$= \frac{2}{s+1} + \frac{-1-j}{s+1+j} + \frac{-1+j}{s+1-j}$$

$\therefore \qquad i_2''(t) = 2e^{-t} + e^{-t}[(-1-j)(\cos t - j\sin t) + (-1+j)$
$$\qquad\qquad\qquad\qquad\qquad (\cos t + j\sin t)]$$
$$\qquad = 2e^{-t}(1-\cos t - \sin t) \qquad\qquad\qquad (5)$$

$\therefore \qquad i_2(t) = i_2'(t) + i_2''(t) = \left(1-e^{-t}\right) + 2e^{-t}\left(1-\cos t - \sin t\right)$

12.27 For the circuit shown in Fig. 12.53, using Thevenin's theorem, find the voltage $v_2(t)$, when the switch is opened at $t = 0$. Assume all the initial conditions to be zero.

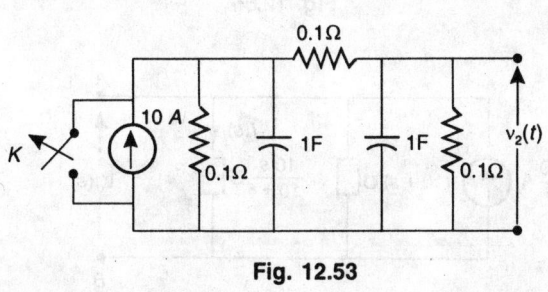

Fig. 12.53

Solution: When the switch K is opened at $t = 0$, the transformed circuit is as shown in Fig. 12.54.

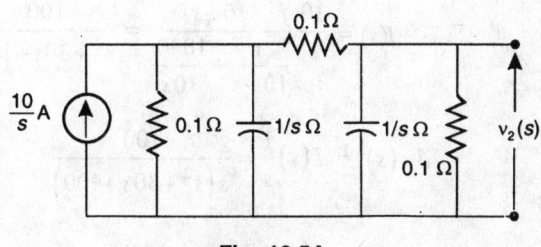

Fig. 12.54

Replacing all the impedances by their admittances, the circuit can be written as shown in Fig. 12.55.

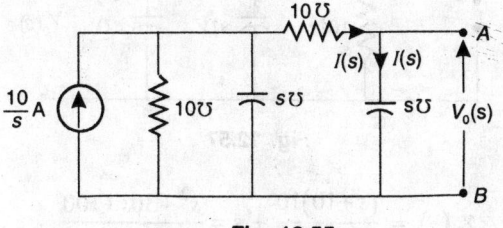

Fig. 12.55

For finding Thevenin's equivalent voltage and impedance, consider $10\,\mho$ across AB as the load. Then the circuit can be written as shown in Fig. 12.56.

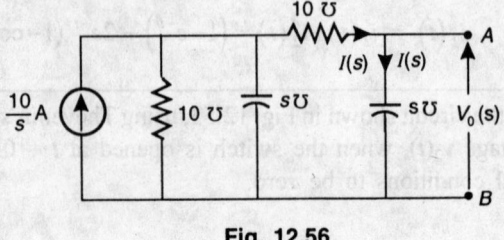

Fig. 12.56

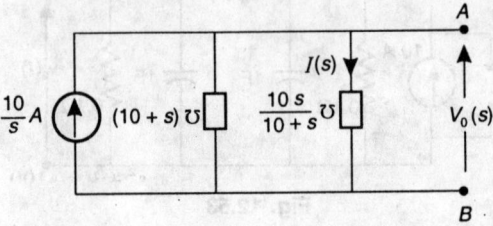

Fig. 12.56 (a)

$$I(s) = \frac{10}{s}\,\frac{\dfrac{1}{10+s}}{\dfrac{1}{10+s}+\dfrac{10+s}{10s}} = \frac{100}{s^2+30s+100}$$

$$V_0(s) = I(s)\frac{1}{s} = \frac{100}{s\left(s^2+30s+100\right)}$$

To find $Z_0(s)$, the circuit is as shown in Fig. 12.54.

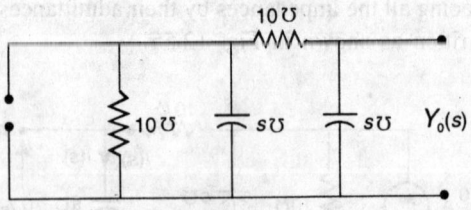

Fig. 12.57

$$Y_0(s) = \frac{(s+10)10}{s+10+10} + s = \frac{s^2+30s+100}{s+20}$$

$$\therefore \qquad Z_0(s) = \frac{s+20}{s^2+30s+100}$$

The Thevenin's equivalent circuit is as shown in Fig. 12.58.

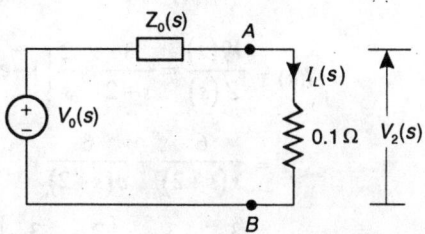

Fig. 12.58

$$I_L(s) = \frac{V_0(s)}{Z_0(s)+0.1} = \frac{\dfrac{100}{s\left(s^2+30s+100\right)}}{\dfrac{s+20}{s^2+30s+100}+0.1}$$

$$= \frac{1000}{s\left(s^2+40s+300\right)} = \frac{1000}{s(s+10)(s+30)}$$

$$= \frac{10/3}{s} - \frac{5}{s+10} + \frac{5/3}{s+30}$$

$$\therefore \qquad i_L(t) = \frac{10}{3} - 5\,e^{-10t} + \frac{5}{3}\,e^{-30t}$$

$$\therefore \qquad v_2(t) = 0.1 i_L(t) = \frac{1}{3} - 0.5 e^{-10t} + \frac{0.5}{3}e^{-30t}$$

12.28 The impulse response of a network is as shown in Fig. 12.59. If the excitation is $3e^{-2t}u(t)$, determine the response.

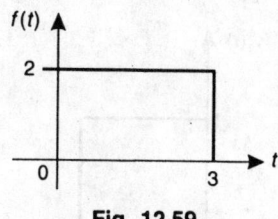

Fig. 12.59

Solution: $\qquad I_\delta(t) = 2u(t) - 2u(t-3)$

$$\therefore \qquad I_\delta(s) = \frac{2}{s}\left[1 - e^{-3s}\right] = \frac{1}{Z(s)}$$

$$\therefore \qquad v_i(t) = 3e^{-2t}u(t) \qquad V_i(s) = \frac{3}{s+2}$$

$$I(s) = \frac{V_i(s)}{Z(s)} = \frac{3}{s+2} \times \frac{2}{s}\left[1 - e^{-3s}\right]$$

$$= \frac{6}{s(s+2)} - \frac{6}{s(s+2)}e^{-3s}$$

$$= \frac{3}{s} - \frac{3}{s+2} - \left\{\frac{3}{s} - \frac{3}{s+2}\right\}e^{-3s}$$

$$\therefore \qquad i(t) = 3u(t) - 3e^{-2t} - \{3u(t-3) - 3e^{-2(t-3)}\}$$

12.29 In the network shown in Fig. 12.60, $L = CR_1^2$ and $R_1 = R_2$. If $v_1(t)$ is a voltage pulse of 1 V amplitude and $\mathcal{T}$ sec. duration, show that $v_2(t)$ is also a pulse and find its amplitude and duration.

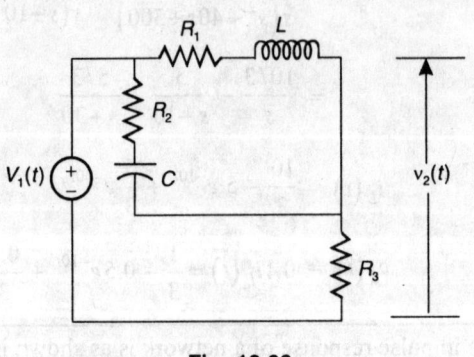

Fig. 12.60

Solution: The pulse $v_1(t)$ is as shown in Fig. 3.61.

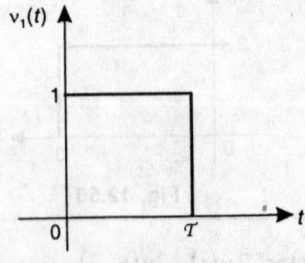

Fig. 12.61

$$v_i(t) = u(t) - u(t - \mathcal{T})$$

$$\therefore \qquad V_1(s) = \frac{1}{s}\left[1 - e^{-\mathcal{T}s}\right]$$

The transformed circuit is a shown in Fig. 12.62.

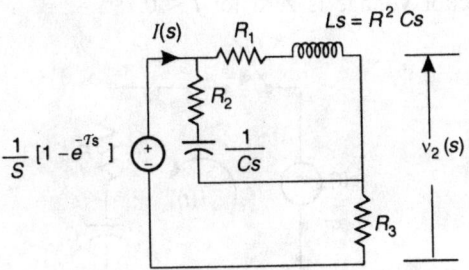

Fig. 12.62

$$Z(s) = \frac{\left(R_1 + \dfrac{1}{Cs}\right)\left(R_1 + R_1^2\, Cs\right)}{\left(R_1 + \dfrac{1}{Cs}\right) + \left(R_1 + R_1^2\, Cs\right)} + R_3 = R_1 + R_3$$

$$\therefore \qquad I(s) = \frac{\dfrac{1}{s}\left[1 - e^{-\mathcal{T}s}\right]}{R_1 + R_3}$$

$$\therefore \qquad V_2(s) = \frac{R_3}{R_1 + R_3}\left[\frac{1}{s} - \frac{1}{s}e^{-\mathcal{T}s}\right]$$

$$v_2(t) = \frac{R_3}{R_1 + R_3}\left[u(t) - u(t - \mathcal{T})\right]$$

The pulse $v_2(t)$ is as shown in Fig. 12.63

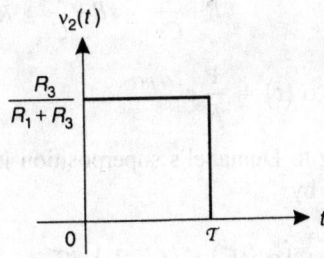

Fig. 12.63

$v_2(t)$ is a pulse of magnitude $\dfrac{R_3}{R_1 + R_3}$ and duration $\mathcal{T}$.

12.30 Find the current in the R–C circuit shown in Fig. 12.64, if the source is $(A + A_2 t)\, u(t)$, using Dumahel's superposition integral. The capacitor voltage is zero for $t < 0$.

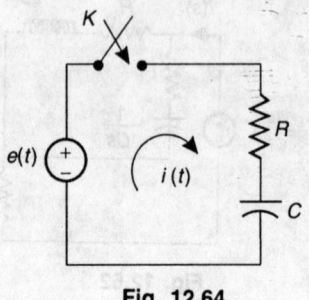

Fig. 12.64

Solution: Let a unit step function $u(t)$ be applied to the circuit and $\alpha(t)$ be the response as shown in Fig. 12.65 (a), and its transformed circuit is as shown in Fig. 12.65 (b).

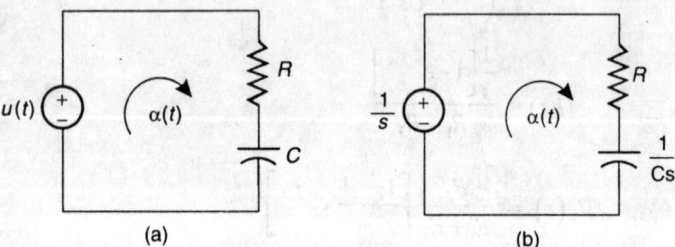

(a)   (b)

Fig. 12.65

$$\alpha(s) = \frac{\dfrac{1}{s}}{R + \dfrac{1}{Cs}} = \frac{1}{sR + \dfrac{1}{C}} = \frac{1}{R\left(s + \dfrac{1}{RC}\right)}$$

$$\therefore \qquad \alpha(t) = \frac{1}{R} e^{-t/RC}$$

Then, according to Dumahel's superposition integral, the response $i(t)$ for $e(t)$ is given by

$$i(t) = e(0+)\, \alpha(t) + \int_0^t \alpha(\mathcal{T})\, e'(t - \mathcal{T})\, d\mathcal{T}$$

$$= 0 + \int_0^t \frac{1}{R} e^{-T/RC} A_2 . dT \qquad \qquad \therefore \qquad e\,(t) = A + A_2'$$

$$= \frac{1}{R}(-RC)\left\{ e^{-T/RC} A_2 \right\}_0^t \qquad \qquad e'\,(t) = A_2$$

$$= A_2 C \left[1 - e^{-t/RC} \right] \qquad \qquad e'\,(t-T) = A_2$$

NUMERICAL PROBLEMS

12.1 Find the Laplace transform of the following functions.

(i) $10\,t^3 - 5\cos 3t + 8\sin t$ (ii) $e^{-3t} \sin^3 3t$

(iii) $e^{3t} \cos^3 2t$ (iv) $t^2 e^{-at} \cos \omega t$

(v) $t \cos(\omega t + \theta)$

12.2 Find the inverse Laplace transforms of the following functions.

(i) $\dfrac{2s+6}{s^2+6s+5}$ (ii) $\dfrac{2s}{(s^2+4)(s^2+5)}$

(iii) $\dfrac{s+5}{s^2+2s+5}$ (iv) $\dfrac{1}{(s+1)(s+1)^2}$

(v) $\dfrac{s^2+2s+1}{(s+2)(s^2+4)}$ (vi) $\dfrac{s+2}{s^2-4s+12}$

(vii) $\dfrac{2s^2-6s+5}{s^3-6s^2+11s-6}$ (viii) $\dfrac{s^3-s^2-3s+9}{s^2(s^2+9)}$

12.3 Solve the following differential Eqs using Laplace transform method.

(i) $\dfrac{d^2 i}{dt^2} + 4\dfrac{di}{dt} + 8i = 8\,u(t)$, given $i(0+) = 3$ and $\dfrac{di}{dt}(0+) = -4$

(ii) $\dfrac{d^2 x}{dt^2} - 2\dfrac{dx}{dt} + x = e^t$, given $x(0+) = 2$ and $x'(0+) = -1$

(iii) $\dfrac{d^2 i}{dt^2} + 2\dfrac{di}{dt} + 4i = -4\sin 2t$, given $i'(0+) = 1$ and $i'(0+) = 1$

(iv) $\dfrac{d^2 i}{dt^2} + 4\dfrac{di}{dt} + 3i = -12e^{-3t}$, given $i(0+) = 0$ and $i'(0+) = 4$

(v) $2\dfrac{d^3 i}{dt^3} + 9\dfrac{d^2 i}{dt^2} + 13\dfrac{di}{dt} + 6i = 0$, given $i(0+) = 0$,

$i'(0+) = 1$ and $i''(0+) = -1$

12.4 Find the initial and final values of the following functions.

(i) $\dfrac{1}{s(s^2 - a^2)}$

(ii) $\dfrac{s^3 + 7s^2 + 5}{s(s^3 + 3s^2 + 4s + 2)}$

(iii) $\dfrac{2s + 3}{(s+1)(s+3)}$

(iv) $\dfrac{e^{-2s}(s+2)}{s^3 + 5s}$

(v) $\dfrac{2(s+1)(s+3)}{(s+2)(s+6)}$

(vi) $\dfrac{(s+a)\sin\theta + b\cos\theta}{(s+a)^2 + b^2}$

(vii) $\dfrac{8(s^2 + 2s + 1)}{(s+2)(s^2 + 4)}$

12.5 Find the inverse Laplace transform of the following functions, using convolution theorem.

(i) $\dfrac{s}{(s^2 + a^2)^2}$

(ii) $\dfrac{s}{(s^2 + a)(s^2 + 25)}$

(iii) $\dfrac{1}{s(s^2 - a^2)}$

(iv) $\dfrac{s+1}{s(s^2 + 4)}$

(v) $\dfrac{5}{s^2(s+2)^2}$

12.6 Using convolution theorem, find $v(t)$, in the circuit shown in Fig. 12.66.

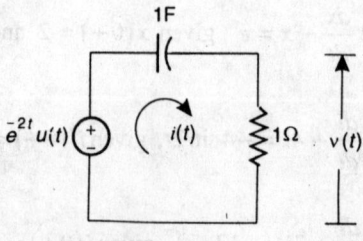

Fig. 12.66

12.7 For the circuit shown in Fig. 12.67, find $v(t)$ using convolution theorem.

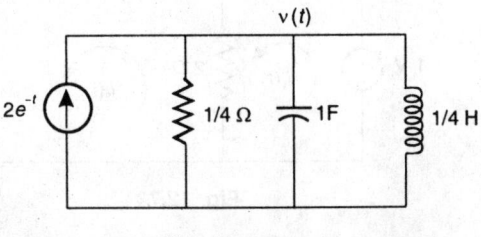

Fig. 12.67

12.8 In the circuit shown in Fig. 12.68, the switch is thrown from position 1 to 2 at $t = 0$. Just before the switch is thrown, the initial conditions are $i(0-) = 2$ A and $v_C(0-) = 2$ V. Find $i(t)$ after the switching action, using Laplace transform method.

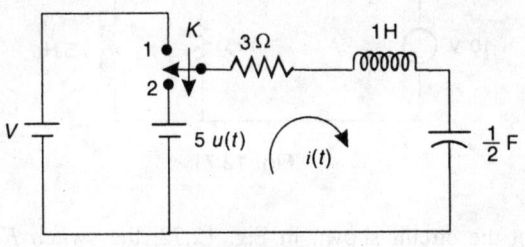

Fig. 12.68

12.9 In the circuit shown in Fig. 12.69, the switch is closed at $t = 0$, with zero initial conditions. Find $i(t)$ using L.T. method.

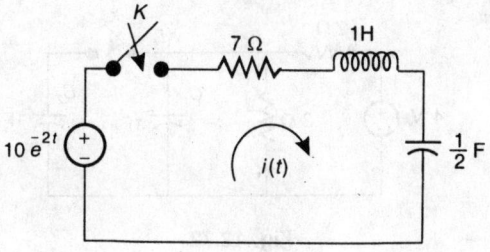

Fig. 12.69

12.10 In the circuit shown in Fig. 12.70, find $i_2(t)$ after the switch is closed at $t = 0$, using transformed circuit.

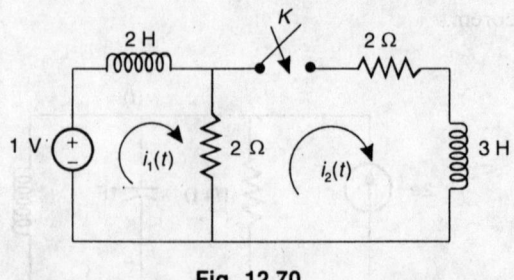

Fig. 12.70

12.11 In the circuit shown in Fig. 12.71, the switch is closed at $t = 0$, find $v_L(t)$ using transformed circuit.

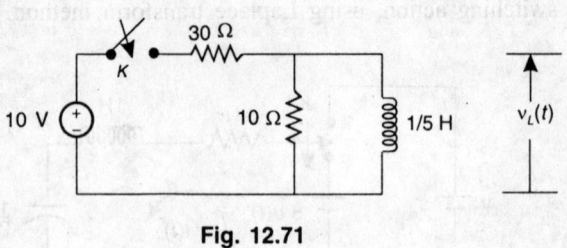

Fig. 12.71

12.12 In the circuit shown in Fig. 12.72, the switch K is closed at $t = 0$, after steady state is reached. Find $v(t)$, given $v_{c_1}(0-) = 2$ V and $v_{c_2}(0-) = 0$.

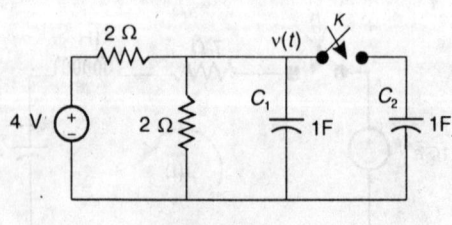

Fig. 12.72

12.13 At $t = 0$,. the switch K is opened in the network shown in Fig. 12.73. Find the values of $v_1(t)$ and $v_2(t)$ for all $t > 0$, using L.T. method.

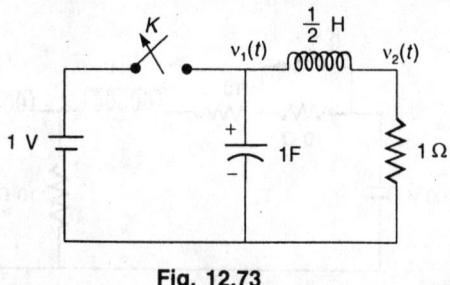

Fig. 12.73

12.14 The circuit in Fig. 12.74 shows an *R–L–C* series circuit. A voltage of $v = 2 \cos 2t$ volts is applied at $t = 0$. If $i(0-) = 1$ A and $v_C(0-) = 1$ V, find $i(t)$ for $t > 0$, using L.T. method.

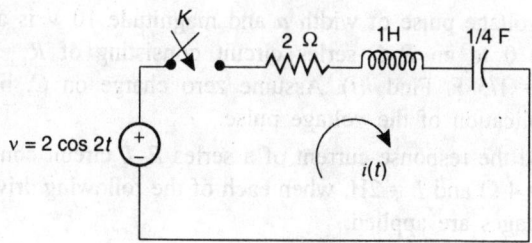

Fig. 12.74

12.15 In the network shown in Fig. 12.75, the network is in steady state with the switch *K* open. The switch is closed at $t = 0$. Find the expression for $i_2(t)$ for $t > 0$.

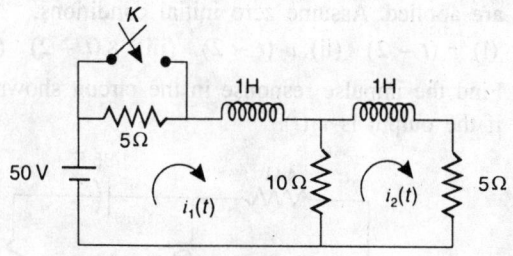

Fig. 12.75

12.16 In the network shown in Fig. 12.76, the switch *K* is closed at $t = 0$, steady state having been reached previously. Find the current $i(t)$ using Thevenin's theorem. Use L.T. method.

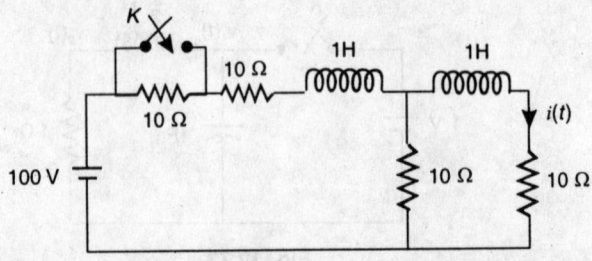

Fig. 12.76

12.17 A pulse voltage of width a and magnitude 20 V is applied at $t = 0$, to an R–L series circuit consisting of $R = 5\ \Omega$ and $L = 3$ H. Find $i(t)$ using L.T. method. Assume zero initial conditions.

12.18 A voltage pulse of width a and magnitude 10 V is applied at $t = 0$ to an R–C series circuit consisting of $R = 1\Omega$ and $C = 1/5$ F. Find $i(t)$. Assume zero charge on C, before the application of the voltage pulse.

12.19 Find the response current of a series R–L circuit consisting of $R = 4\ \Omega$ and $L = 2$H, when each of the following driving force voltages are applied.

  (i) Unit ramp voltage $r\ (t - 5)$

  (ii) Unit impulse voltage $\delta\ (t - 5)$

  (iii) Unit step voltage $u\ (t - 5)$

  (iv) Unit doublet $\delta'\ (t - 5)$

    Assume zero initial conditions.

12.20 Find the current $i(t)$ in a series R–C circuit consisting of $R = 4\ \Omega$ and $C = 1/5$ F, when each of the following voltages are applied. Assume zero initial conditions.

  (i) $r\ (t - 2)$ (ii) $u\ (t - 2)$ (iii) $\delta\ (t - 2)$ (iv) $\delta'\ (t - 2)$

12.21 Find the impulse response in the circuit shown in Fig. 12.77, if the output is $v_L(t)$.

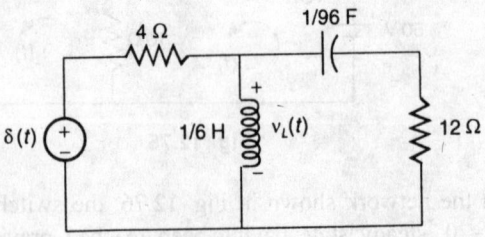

Fig. 12.77

12.22 The network shown in Fig. 12.78 is initially in relaxed state. When the source is 10 $u(t)$ volts, the transform of the input current is $10/(2s + 4)$. The circuit is brought to its initial state once again. Find the impedance and input response $V_s(t)$, when the source is a current generator of $5e^{-2t}$ amperes.

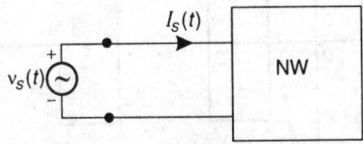

Fig. 12.78

12.23 Given the following sources and the results they produce in a single element circuit. Deduce the type of element and its value in ohms, henrys, farads as the case may be. If the source is a voltage, the response is current and vice versa.

Source	*Response*
(i) $i(t) = 5\delta(t)$	$10\ u(t)$
(ii) $i(t) = 5\ u(t)$	$3\delta(t)$
(iii) $e(t) = 10\ u(t)$	$5\delta(t)$
(iv) $i(t) = 3/2\ \delta(t)$	$\dfrac{9}{4}\delta(t)$
(v) $e(t) = 1/3\ \delta(t)$	$3u(t)$

12.24 The periodic current waveform is as shown in Fig. 12.79. Fin its Laplace transform Eq.

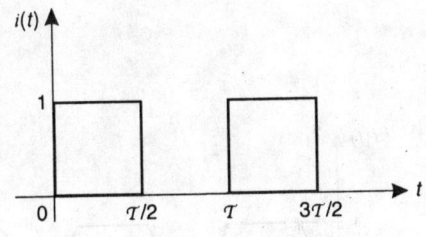

Fig. 12.79

12.25 Write down the Laplace transform Eq. for the waveform shown in Fig. 12.80.

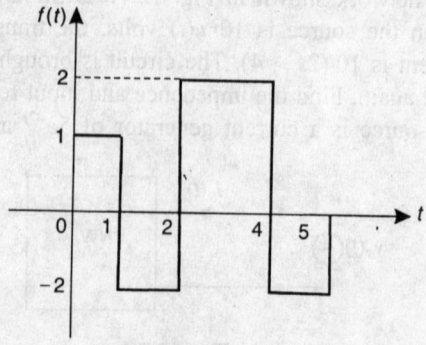

Fig. 12.80

12.26 The waveform shown in Fig. 12.81 is made up of straight line segments. Write an Eq. for this wave and find its Laplace transformation.

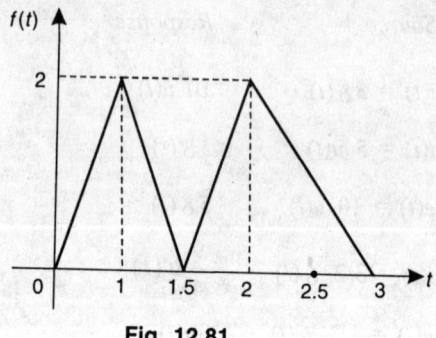

Fig. 12.81

12.27 For the waveform shown in Fig. 12.82, write the transformed Eq.

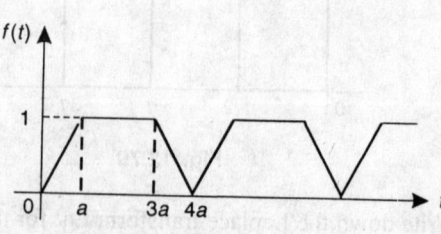

Fig. 12.82

12.28 For the waveform shown in Fig. 12.73, write the transformed Eq.

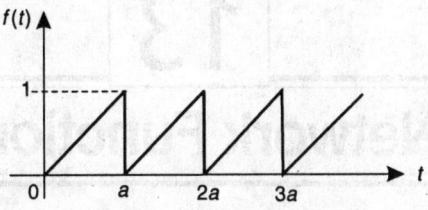

Fig. 12.83

Network Functions

13.1 INTRODUCTION

Network functions give the relation between the transformed excitation to the transformed response. They are either transformed impedance functions or transformed admittance functions. If the transformed excitation and the transformed response are referred to a particular port, the network functions are called *immittance functions*. An immittance function may be an impedance function or an admittance function. If the transformed excitation is referred to one-port and the transformed response is referred to another port, then the network functions are called as *transfer functions*. The transfer functions may be impedance functions, admittance functions or dimensionless ratios of similar quantities, referred to two different ports. The network functions are always defined for a network consisting of only passive elements under zero state, i.e. when the network does not contain any independent sources or the elements of the network does not possess any initial voltages or currents, but may contain dependent sources. By observing the nature of the network function, it is possible to predict the behaviour of the output response and its stability.

13.2 ONE-PORT NETWORK

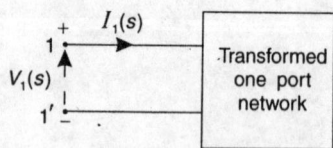

Fig. 13.1

One-port network is one, which has only two input terminals as shown in Fig. 13.1. The reference direction of voltages and currents of the port are always as shown. Only driving point functions are defined w.r.t. one-port network.

The driving point impedance and the driving point admittance of one-port network are given by

$$Z_{11}(s) = \frac{V_1(s)}{I_1(s)} \quad \text{and} \quad Y_{11}(s) = \frac{I_1(s)}{V_1(s)}$$

13.3 TWO-PORT NETWORK

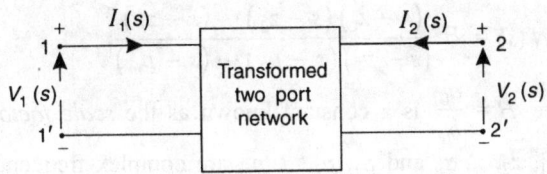

Fig. 13.2

A two-port network is one, which contains two ports, one input port and one output port or two input terminals and two output terminals as shown in Fig. 13.2.

The driving point impedance and admittance at input port are given by

$$Z_{11}(s) = \frac{V_1(s)}{I_1(s)} \quad \text{and} \quad Y_{11}(s) = \frac{I_1(s)}{V_1(s)}$$

The driving point impedance and admitttance at output port are given by

$$Z_{22}(s) = \frac{V_2(s)}{I_2(s)} \quad \text{and} \quad Y_{22}(s) = \frac{I_2(s)}{V_2(s)}$$

The transfer impedance and admittance functions are given by

$$Z_{12}(s) = \frac{V_2(s)}{I_1(s)} \quad \text{and} \quad Y_{12}(s) = \frac{I_2(s)}{V_1(s)}$$

The voltage ratio transfer function is given by

$$G_{12}(s) = \frac{V_2(s)}{V_1(s)}$$

The current ratio transfer function is given by

$$\alpha_{12}(s) = \frac{I_2(s)}{I_1(s)}$$

13.4 POLES AND ZEROS OF NETWORK FUNCTIONS

In general, any transformed network function may be written as

$$N(s) = \frac{P(s)}{Q(s)} = \frac{a_0 s^n + a_1 s^{n-1} + a_2 s^{n-2} + \cdots a_{n-1} s + a_n}{b_0 s^m + b_1 s^{m-1} + b_2 s^{m-2} + \cdots b_{m-1} s + b_m} \quad (13.1)$$

Where, $a_0, a_1, a_2, \ldots a_{n-1}, a_n$ and $b_0, b_1, b_2, \ldots b_{m-1}, b_m$ are constant coefficients, which are real and positive, if the network is in zero state. n is the number of zeros and m is the number of poles.

Equation (13.1) may be written as

$$N(s) = H \frac{(s - z_1)(s - z_2) \cdots (s - z_n)}{(s - p_1)(s - p_2) \cdots (s - p_m)} \quad (13.2)$$

Where $H = \dfrac{a_0}{b_0}$ is a constant known as the *scale factor*

$z_1, z_2, \ldots z_n$ and $p_1, p_2, \ldots p_m$ are complex frequencies.

$N(s) = 0$, when $s = z_1, z_2, \ldots z_n$ and hence, they are called as the *zeros* of $N(s)$.

$N(s) = \infty$ when $s = p_1, p_2, \ldots p_n$ and hence, they are called as the *poles* of $N(s)$.

The salient features of poles and zeros of a network function N(s) are:

1. The network function $N(s)$ can be completely described by its poles, zeros and the scale factor.
2. At poles, $N(s) = \infty$ and at zeros, $N(s) = 0$. Hence, poles and zeros are called *critical complex frequencies*. At any other complex frequency, the network function has a finite value.
3. For a rational network function, the number of poles is always equal to the number of zeros.
4. If the poles and zeros have distinct values, then they are called simple poles and zeros.
5. When r number of poles or zeros have the same value, then the poles or zeros are said to be of multiplicity r.
6. When $n > m$, then the pole at infinity is of multiplicity $(n - m)$.
7. The poles of a network function determine the nature of the waveform of the time variation of the response.
8. The poles and zeros of the network function determine the magnitude of the response.
9. For a capacitor, $Z(s) = 1/Cs$. This network function has a pole at $s = 0$ and a zero at $s = \infty$. Hence, a capacitor behaves as an open circuit at the pole frequency $s = 0$ and a short circuit at $s = \infty$.

10. For an inductor, $Z(s) = sL$. This network function has a zero at $s = 0$ and a pole at $s = \infty$. Hence, an inductor behaves as a short circuit at $s = 0$ and an open circuit at $s = \infty$.

13.5 POLE–ZERO PLOT

Poles and zeros are complex frequencies of the form $s = \sigma \pm j\omega$, where σ is the real part and $j\omega$ is the imaginary part and hence, they are plotted on the s plane. The location of a zero on the s-plane is marked as o and the pole as x. The s-plane in which, all the poles and zeros of the network function are plotted is known as a pole-zero plot. For example, consider the network function as given in (13.3).

$$N(s) = \frac{s(s+3)(s-2)}{(s+4)(s-5)(s+3+j2)(s+3-j2)} \tag{13.3}$$

The zeros are: $s = 0$, $s = -3$ and $s = 2$.
The poles are: $s = -4$, $s = 5$, $s = -3 -j\,2$ and $s = -3 + j\,2$.
The pole–zero plot of the above function is as shown in Fig. 13.3.

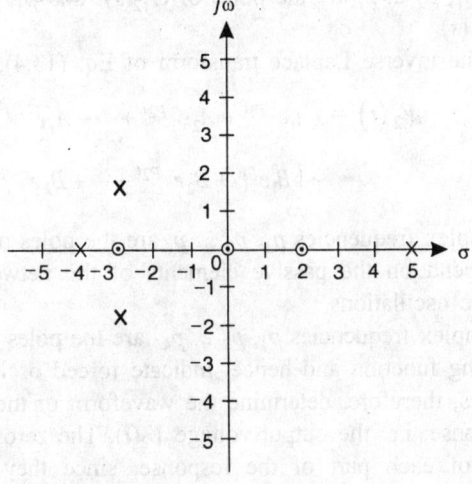

Fig. 13.3

There are four finite poles and three finite zeros. But, for any rational function, the number of poles must be always equal to the number of zeros. The fourth zero is assumed to be at infinity.

13.6 PHYSICAL SIGNIFICANCE OF POLES AND ZEROS

Consider the driving point impedance function given by $Z(s) = V(s)/I(s)$. The zero of this function implies zero voltage for a finite driving current and hence, means a short circuit. The pole of this function implies zero current for a finite driving voltage and hence, means an open circuit.

Consider the driving point admittance function $Y(s) = I(s)/V(s)$. The zero of this function implies zero current for a finite driving voltage and hence, means an open circuit. The pole of this function implies zero voltage for a finite driving current and hence, means a short circuit.

Consider the voltage ratio transfer function.

$$G_{12}(s) = \frac{V_2(s)}{V_1(s)} \quad \text{or} \quad V_2(s) = G_{12}(s)V_1(s)$$

Both $G_{12}(s)$ and $V_1(s)$ are the ratios of polynomials in s. Expanding these two functions using partial fractions, $V_2(s)$ may be written as,

$$V_2(s) = \left(\frac{A_1}{s + p_1} + \frac{A_2}{s + p_2} + \cdots + \frac{A_j}{s + p_j} \right)$$
$$+ \left(\frac{B_1}{s + p_1} + \frac{B_2}{s + p_2} + \cdots + \frac{B_k}{s + p_k} \right) \quad (13.4)$$

Where, $p_1, p_2 \ldots p_j$ are the poles of $G_{12}(s)$ and $p_1, p_2 \ldots p_k$ are the poles of $V_1(s)$.

Taking the inverse Laplace transform of Eq. (13.4), we get

$$V_2(t) = (A_1 e^{-p_1 t} + A_2 e^{-p_2 t} + \cdots + A_j e^{-p_j t})$$
$$+ \left(B_1 e^{p_1 t} + B_2 e^{-p_2 t} + \cdots + B_k e^{-p_k t} \right) \quad (13.5)$$

The complex frequencies $p_1, p_2 \ldots p_j$ are the poles of $G_{12}(s)$, which entirely depend on the passive elements of the network and hence, indicate free oscillations.

The complex frequencies $p_1, p_2 \ldots p_k$ are the poles of $V_1(s)$, which is the forcing function and hence, indicate forced oscillations.

The poles, therefore, determine the waveform of the time variation of the response, i.e. the output voltage $V_2(t)$. The zeros determine the magnitude of each part of the response, since they determine the magnitudes of $A_1, A_2 \ldots A_j$ and $B_1, B_2 \ldots B_k$.

WORKED EXAMPLES

13.1 Consider the Eq., for the transformed current $I(s)$ as given in (13.6).

$$I(s) = \frac{5(s+1)}{s(s+2)(s+3)} \quad (13.6)$$

Solution: Here, 5 is the scale factor

$$z_1 = -1 \text{ is the zero, } p_1 = 0, p_2 = -2$$

and $\qquad p_3 = -3$ are the poles.

After expanding Eq. (13.6) into partial fractions, we get

$$I(s) = \frac{5}{6}\frac{1}{s} + \frac{5}{2}\frac{1}{(s+2)} - \frac{10}{3}\frac{1}{(s+3)}$$

$$\therefore \qquad i(t) = \frac{5}{6} + \frac{5}{2}e^{-2t} - \frac{10}{3}e^{-3t} = i_1(t) + i_2(t) + i_3(t)$$

The pole–zero plot and the components of the response $i(t)$ are marked as shown in Fig. 13.4 (a) and (b) respectively.

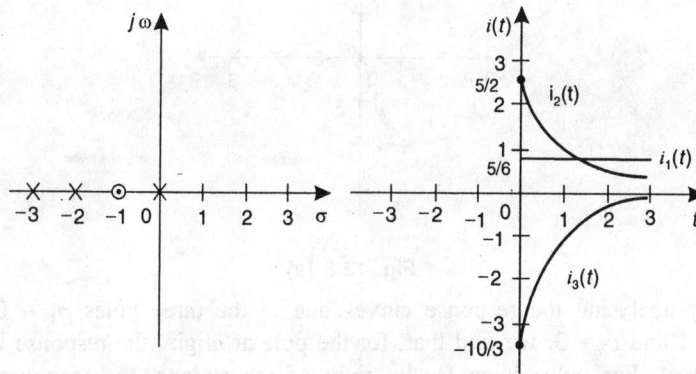

Fig. 13.4

From the study of the above plots the following conclusions can be made.

(i) The component of the response due to the pole at the origin, i.e. $p_1 = 0$ is time invariant as represented by $i_1(t)$.

(ii) The component of the response due to the pole $p_2 = -2$ is having a smaller magnitude and decays at a slower rate.

(iii) The component of the response due to the pole $p_3 = -3$ is having a larger magnitude and decays at a faster rate.

(iv) If the zero is moved, closer to a pole, the magnitude of the component due to that pole gradually decreases and finally vanishes completely, when the zero coincides with it.

13.2 Consider the Eq. for the transformed voltage $V(s)$ given by

$$V(s) = \frac{s-1}{s(s-2)(s-3)}.$$ On partial fraction expansion, we get

$$V(s) = -\frac{1}{2}\frac{1}{s} + \frac{1}{4}\frac{1}{(s-2)} + \frac{2}{3}\frac{1}{s-3}$$

$$\therefore \qquad v(t) = -\frac{1}{2} + \frac{1}{4}e^{2t} + \frac{2}{3}e^{3t} = v_1(t) + v_2(t) + v_3(t)$$

Solution: The pole–zero plot and the response curves of the individual components of $v(t)$ are plotted as shown in Figs 13.5 (a) and (b) respectively.

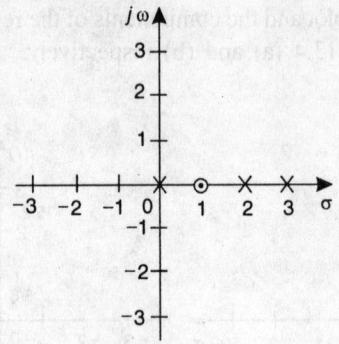

Fig. 13.5 (a)

By analysing the response curves due to the three poles $p_1 = 0$, $p_2 = 2$ and $p_3 = 3$, we find that, for the pole at origin, the response is constant. For poles lying to the right of the s-plane, the responses increase exponentially with time. The rate of increase is more, as the pole position is away from the $j\omega$ axis. Therefore, we conclude that, if any pole of the output response lies on the right-half of the s-plane, the response becomes unstable.

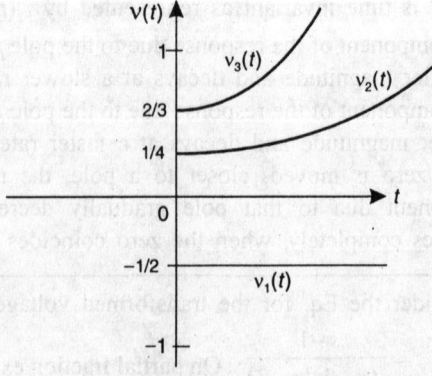

Fig. 13.5 (b)

13.3 Let $I(s) = \dfrac{2}{s^2 + 9} = \dfrac{2}{s^2 + 3^2} = \dfrac{2}{(s + j3)(s - j3)}$

The poles are $p_1 = -j3$ and $p_2 = +j3$

The time response is $i(t) = \dfrac{2}{3} \sin 3t$

Solution: The pole-zero plot and the response curve are as shown in Figs 13.6 (a) and (b) respectively.

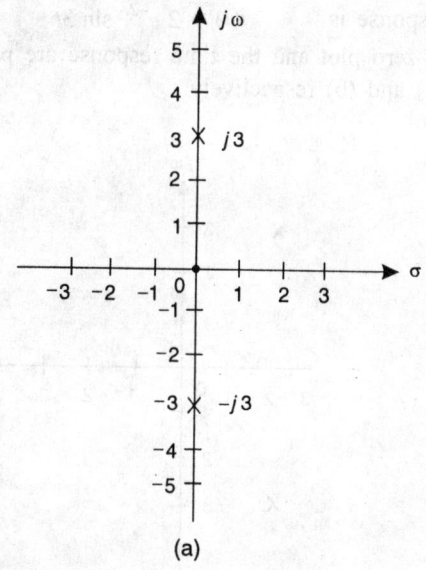

(a)

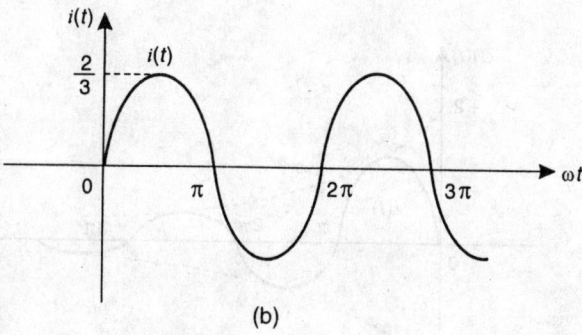

(b)

Fig. 13.6

By analysing the above figures, we find that, if the poles, which are complex conjugates lie on the $j\omega$ axis, the time response is oscillatory in nature and is bounded. The oscillatory response which is bounded is considered to be a stable response.

13.4 Let
$$I(s) = \frac{2}{(s+2)^2 + 9} = \frac{2}{(s+2)^2 + 3^2}$$
$$= \frac{2}{(s+2+j3)(s+2-j3)} \tag{13.7}$$

Solution: The poles are $\quad p_1 = -2 - j\,3$

and $\quad\quad\quad\quad\quad\quad\quad p_2 = -2 + j\,3$

The time response is $\quad\quad i(t) = 2\,e^{-2t}\sin 3t$

The pole–zero plot and the time response are plotted as shown in Figs 13.7 (a) and (b) respectively.

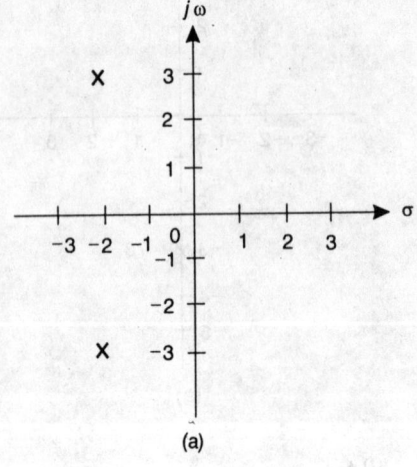

(a)

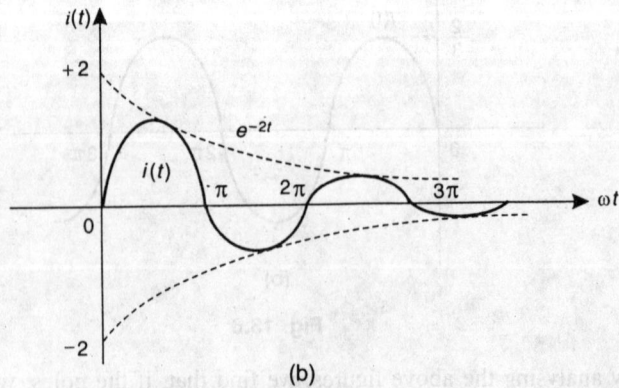

(b)

Fig. 13.7

By analysing the above figure, we find that, if the pole is with two complex conjugates lie on the left ..., the time response is oscillatory

By analysing the above figures, we find that, if the poles, which are complex conjugates lie on the left half of the *s*-plane, then the time response is oscillatory but it decays. The decaying of oscillations is faster as the poles, which are complex conjugates move away from the $j\omega$ axis. The output response is stable.

13.5 Let
$$I(s) = \frac{2}{(s-2)^2 + 9} = \frac{2}{(s-2)^2 + 3^2}$$
$$= \frac{2}{(s-2+j3)\,(s-2-j3)}$$

Solution: The poles are

$$p_1 = 2 - j\,3 \qquad \text{and} \qquad p_2 = 2 + j\,3$$

The time response $i(t) = 2\,e^{2t}\,\sin 3t$

The pole–zero plot and the time response are plotted as shown in Figs 13.8 (a) and (b) respectively.

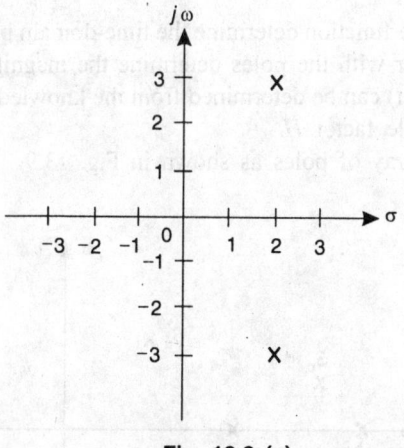

Fig. 13.8 (a)

By analysing the above figures, we find that, if the poles, which are complex conjugates, lie on the right half of the *s*-plane, then the time response is oscillatory and the amplitude of the oscillation increases exponentially and hence, the output response is unstable.

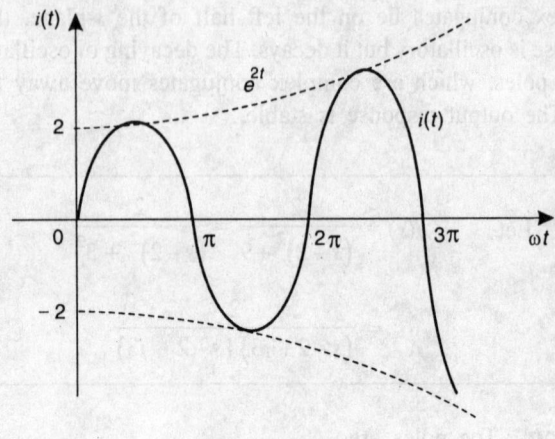

Fig. 13.8 (b)

13.7 TIME DOMAIN RESPONSE FROM POLE-ZERO PLOT

Let the transformed function $I(s)$ be given as,

$$I(s) = \frac{P(s)}{Q(s)} = H \frac{(s - s_1)(s - s_2) \cdots (s - s_n)}{(s - s_a)(s - s_b) \cdots (s - s_m)} \qquad (13.8)$$

The poles of the function determine the time-domain behaviour of $i(t)$. The zeros together with the poles determine the magnitude of each of the terms of $i(t)$. $i(t)$ can be determined from the knowledge of the poles, zeros and the scale factor H.

Consider an array of poles as shown in Fig. 13.9.

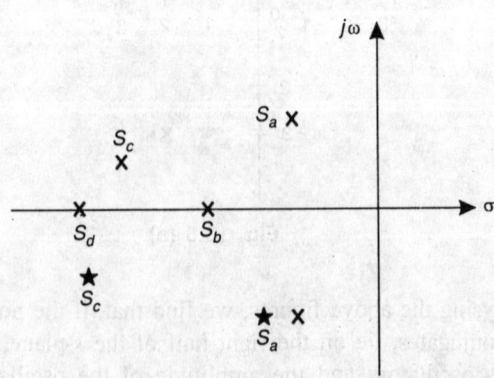

Fig. 13.9

The pair of poles s_a and $\overset{*}{s}_a$ the pair s_c and $\overset{*}{s}_c$ correspond to oscillatory expressions in the time domain. The poles s_b and s_d correspond to exponential decaying expression in the time domain. The total response corresponding to these poles is found by adding each of the individual expressions.

$$\therefore \quad i(t) = K_a e^{s_a t} + \overset{*}{K}_a e^{\overset{*}{s}_a t} + K_b E^{s_b t} + K_c e^{s_c t} + \overset{*}{K}_c e^{\overset{*}{s}_c t} + K_d e^{s_d t}$$

(13.9)

Expanding Eq. (13.8) by partial fractions, we get

$$I(s) = \frac{K_a}{s - s_a} + \frac{K_b}{s - s_b} + \cdots + \frac{K_r}{s - s_r} + \cdots + \frac{K_m}{s - s_m} \quad (13.10)$$

Any of the coefficients (or residues) can be found by Heaviside's partial fraction expansion method.

$$K_r = H \frac{(s-s_1)(s-s_2)\cdots(s-s_n)}{(s-s_a)(s-s_b)\cdots(s-s_r)\cdots(s-s_m)}(s-s_r)\Bigg|_{s=s_r}$$

(13.11)

Substituting $s = s_r$ in Eq. (13.11), we get

$$K_r = H \frac{(s_r-s_1)(s_r-s_2)\cdots(s_r-s_n)}{(s_r-s_a)(s_r-s_b)\cdots(s_r-s_m)}$$

(13.12)

Equation (13.12) is composed of factors of the general forms $s_r - s_n$, where, both s_r and s_n are known complex numbers. The difference of two complex numbers is another complex number, which may be written in the polar form as

$$s_r - s_n = M_{nr} e^{j\phi_{nr}}$$

(13.13)

Where, M_{nr} is the magnitude of the phasor $(s_r - s_n)$ and ϕ_{nr}, is the phase angle of this phasor. The difference of the two complex quantities s_r and s_n is illustrated in Fig. 13.10.

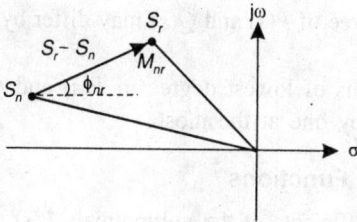

Fig. 13.10

The term $s_r - s_n$ is interpreted as a phasor directed from s_n to s_r. The magnitude M_{nr} is the distance from s_n to s_r. The phase angle ϕ_{nr}, is the angle of the line from s_n to s_r w.r.t. the reference axis. All the terms of this type in Eq. (13.12) can be expressed in terms of their magnitude and phase angle and hence, Eq. (13.12) may be written as

$$K_r = H \frac{M_{1r} M_{2r} M_{3r} \cdots M_{nr}}{M_{ar} M_{br} M_{cr} \cdots M_{mr}} e^{j\left(\phi_{1r} + \phi_{2r} + \cdots - \phi_{ar} - \phi_{br} \cdots\right)} \qquad (13.14)$$

Equation (13.14) gives the magnitude and phase of K_r. The procedure for finding the coefficients K_r in the partial fraction expansion represented by Eq. (13.10) can be summarised as follows.

1. Plot the poles and zeros of $I(s) = P(s)/Q(s)$ to scale on the s-plane.
2. Measure the distance from each of the other finite poles and zeros to the given pole s_r.
3. Measure the angle from each of the other finite poles and zeros to the given pole s_r.
4. Substitute these quantities in Eq. (13.14) to evaluate K_r.

This procedure is best illustrated in worked examples 13.12 and 13.13.

13.8 RESTRICTIONS ON POLE AND ZERO LOCATIONS

(a) For Driving Point Functions (Immittance Functions)

1. The coefficients of the $P(s)$ and $Q(s)$ of $N(s) = P(s)/Q(s)$ must be real and positive.
2. The poles and zeros of $N(s)$ must be conjugates, if they are imaginary or complex.
3. The real parts of all poles and zeros must be negative or zero. If the real part is zero, then the poles and zeros must be simple.
4. The polynomials $P(s)$ and $Q(s)$ should not have missing terms between those of highest and lowest degree, unless all even or odd terms are missing.
5. The degree of $P(s)$ and $Q(s)$ may differ by either zero or one only.
6. The terms of lowest degree in $P(s)$ and $Q(s)$ may differ in degree by one at the most.

(b) For Transfer Functions

1. The coefficients of the polynomials $P(s)$ and $Q(s)$ of $N(s) = P(s)/Q(s)$ must be real and those of $Q(s)$ must be positive.

2. The poles and zeros of $N(s)$ must be conjugates, if they are imaginary or complex.

3. The real parts of all poles must be negative or zero. If the real part of the pole is zero, it must be simple. This includes the origin also.

4. The polynomials of $Q(s)$ should not have any missing terms between those of highest and lowest degree, unless all even or odd terms are missing.

5. The polynomials of $P(s)$ may have terms missing between the terms of highest and lowest degree and some of the co-efficients may be negative.

6. The degree of $P(s)$ may be as small as zero and is independent of the degree of $Q(s)$.

7. (a) For $G_{12}(s)$ and $\alpha_{12}(s)$, the maximum degree of $P(s)$ is the degree of $Q(s)$.

   (b) For $Z_{12}(s)$ and $Y_{12}(s)$, the maximum degree of $P(s)$ is the degree of $Q(s)$ plus one.

WORKED EXAMPLES

13.6 For the network shown in Fig. 13.11, with port 2 open, find (i) input impedance at port 1 and (ii) the voltage ratio transfer function G_{12}. (Mysore University)

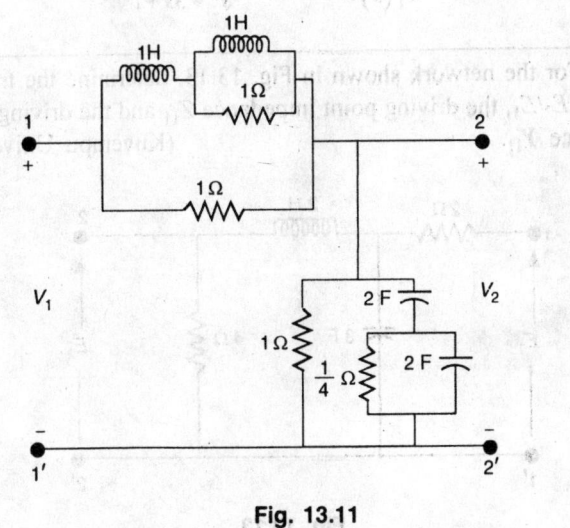

Fig. 13.11

Solution: Replacing all the parameters of the circuit by transformed quantities, i.e. V_1 by $V_1(s)$ V_2 by $V_2(s)$, 1 H by s, 1Ω by 1, 2 F by $1/2s$ and simplifying, the transformed network can be written as shown in Fig. 13.12.

$$Z_1(s) = \frac{s(s+2)}{s^2+3s+1}, \quad Z_2(s) = \frac{s+1}{s^2+3s+1}$$

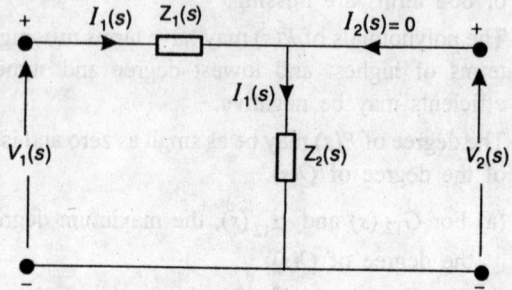

Fig. 13.12

$$Z_{11}(s) = \text{Input impedance at port } 1 = Z_1(s) + Z_2(s) = 1$$

$$I_1(s) = \frac{V_1(s)}{Z_{11}(s)} = V_1(s)$$

$$V_2(s) = I_1(s)Z_2(s) = V_1(s)Z_2(s)$$

$$\therefore \quad G_{12}(s) = \frac{V_2(s)}{V_1(s)} = Z_2(s) = \frac{s+1}{s^2+3s+1}$$

13.7 For the network shown in Fig. 13.13, determine the transfer function E_2/E_1, the driving point impedance Z_{11} and the driving point admittance Y_{11}. (Kuvempu University)

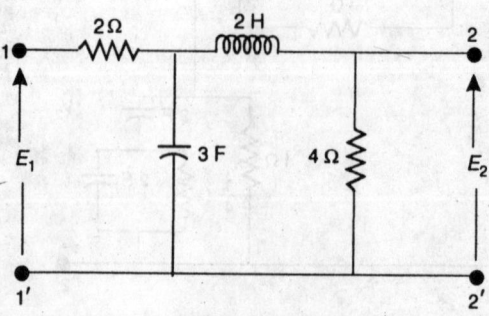

Fig. 13.13

Solution: The transformed circuit is as shown in Fig. 13.14.

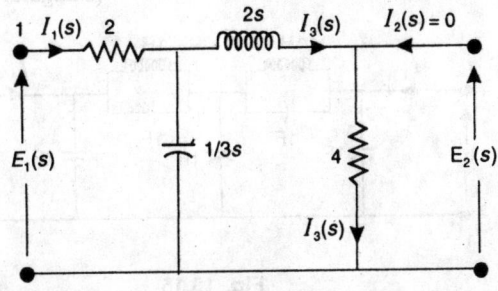

Fig. 13.14

As port 2 is open

$$I_2(s) = 0,$$

$$Z_{11}(s) = \frac{E_1(s)}{I_1(s)}$$

$$(4+2s)\,\|\,\frac{1}{3s} = \frac{2s+4}{6s^2+12s+1}$$

$$\therefore \quad Z_{11}(s) = \frac{2s+4}{6s^2+12s+1} + 2 = \frac{12s^2+26s+6}{6s^2+12s+1}$$

$$\therefore \quad Y_{11}(s) = \frac{1}{Z_{11}(s)} = \frac{6s^2+12s+1}{12s^2+26s+6}$$

$$I_3(s) = I_1(s) \times \frac{\dfrac{1}{3s}}{\dfrac{1}{3s}+(2s+4)} = I_1(s)\frac{1}{6s^2+12s+1}$$

$$E_2(s) = I_3(s)\,4 = \frac{4}{6s^2+12s+1}\,I_1(s)$$

$$E_1(s) = Z_{11}(s)I_1(s) = \frac{12s^2+26s+6}{6s^2+12s+1}\,I_1(s)$$

$$\therefore \quad \frac{E_2(s)}{E_1(s)} = \frac{4}{12s^2+26s+6}$$

13.8 For the network shown in Fig. 13.15, determine the voltage ratio transfer function. (Bangalore University)

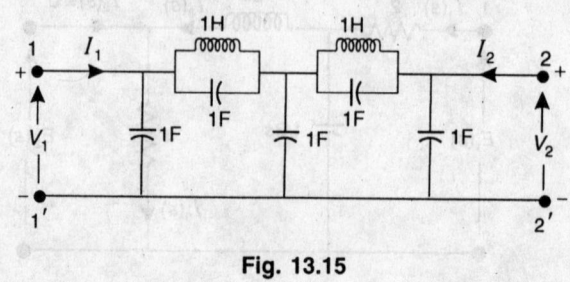

Fig. 13.15

Solution: As the second part is open, $I_2 = 0$. The transformed circuit is as shown in Fig. 13.16.

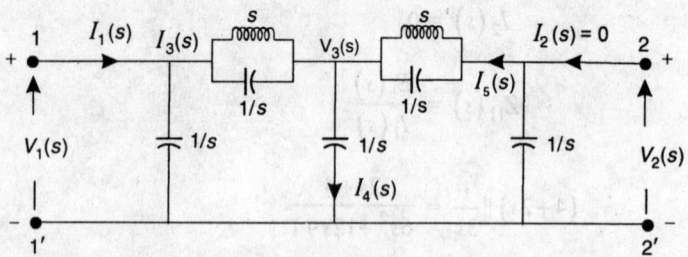

Fig. 13.16

At node $V_3(s)$

$$I_3(s) = I_4(s) - I_5(s)$$

i.e. $\left[V_1(s) - V_3(s)\right]\left[s + \dfrac{1}{s}\right] = V_3(s)s - \left[V_2(s) - V_3(s)\right]\left[s + \dfrac{1}{s}\right]$

i.e. $V_1(s)\left[s + \dfrac{1}{s}\right] = V_3(s)\left[3s + \dfrac{2}{s}\right] - V_2(s)\left[s + \dfrac{1}{s}\right]$ \hfill (1)

Also $\qquad 0 = V_2(s)s + \left[V_2(s) - V_3(s)\right]\left[s + \dfrac{1}{s}\right]$

i.e. $\qquad 0 = -V_3(s)\left[s + \dfrac{1}{s}\right] + V_2(s)\left[2s + \dfrac{1}{s}\right]$ \hfill (2)

Multiplying (1) by $\left[s + \dfrac{1}{s}\right]$ and (2) by $\left[3s + \dfrac{2}{s}\right]$ and cancelling

out $V_3(s)$, we get

$$G_{12}(s) = \frac{V_2(s)}{V_1(s)} = \frac{\left(s^2 + 1\right)^2}{5s^4 + 5s^2 - 1}$$

13.9 For the network shown in Fig. 13.17 (a), show that the input impedance has the form

$$Z(s) = \frac{K(s-z_1)}{(s-p_1)(s-p_2)}$$

and determine z_1, p_1 and p_2 in terms of R, L and C, if the pole-zero plot of $Z(s)$ is as shown in Fig. 13.17 (b), with $z(j0) = 1$. Find the values of R, L and C.

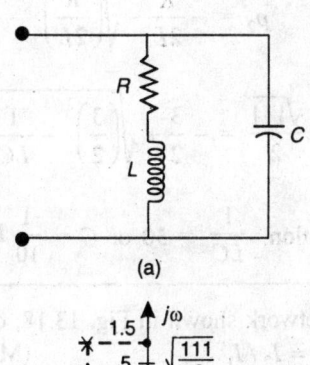

(a)

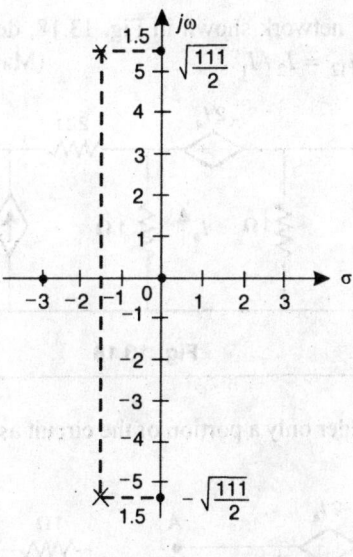

(b)

Fig. 13.17

Solution

$$Z(s) = \frac{1}{sC + \dfrac{1}{R+sL}} = \frac{R+sL}{s^2LC + sCR + 1} = \frac{1}{C}\frac{s + \dfrac{R}{L}}{s^2 + s\dfrac{R}{L} + \dfrac{1}{LC}}$$

$$z(j0) = 1 \qquad\qquad \therefore \ R = 1\,\Omega$$

$$z_1 = -\frac{R}{L} = -3 \qquad \therefore \ L = \frac{1}{3}\,\text{H}$$

$$p_1 = -\frac{R}{2L} + \sqrt{\left(\frac{R}{2L}\right)^2 - \frac{1}{LC}}$$

and

$$p_2 = -\frac{R}{2L} - \sqrt{\left(\frac{R}{2L}\right)^2 - \frac{1}{LC}}$$

i.e. $-1.5 + j\,\dfrac{\sqrt{111}}{2} = -\dfrac{3}{2} + \sqrt{\left(\dfrac{3}{2}\right)^2 - \dfrac{1}{LC}}$

on simplification, $\dfrac{1}{LC} = 30$ or $C = \dfrac{1}{10}\,\text{F}$

13.10   For the network shown in Fig. 13.18, determine the current ratio function $\alpha_{12} = I_2/I_1$. (Mangalore University)

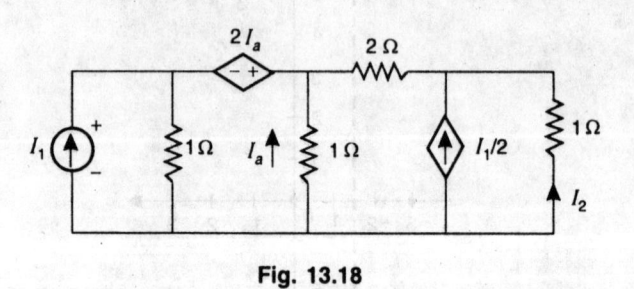

Fig. 13.18

Solution: Consider only a portion of the circuit as shown in Fig. 13.19.

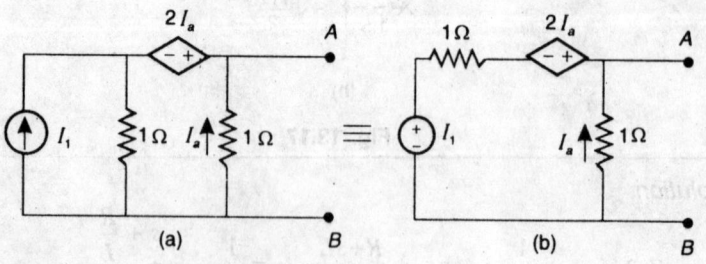

Fig. 13.19

$$\therefore \qquad I_a = \frac{-I_1 - 2I_a}{2}$$

$$\therefore \qquad I_a = -\frac{I_1}{4}$$

$$\therefore \qquad E_0 = -I_a = \frac{I_1}{4}$$

On short circuiting AB, $I_a = 0$ and $I_{sc} = I_1$

$$\therefore \qquad R_0 = \frac{E_0}{I_{sc}} = \frac{\frac{I_1}{4}}{I_1} = \frac{1}{4}\ \Omega$$

The circuit in Fig. 13.18, may be written as

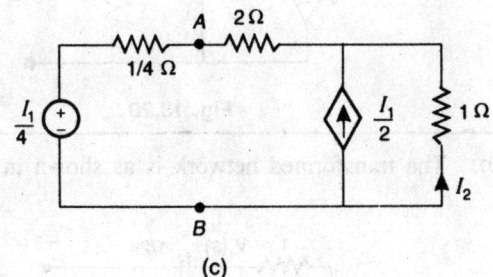

(c)

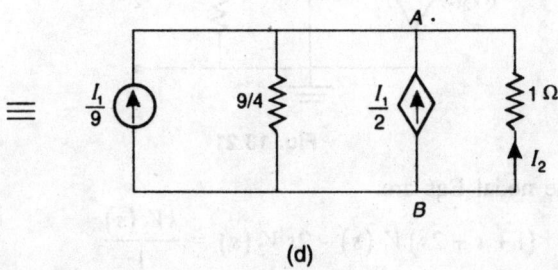

(d)

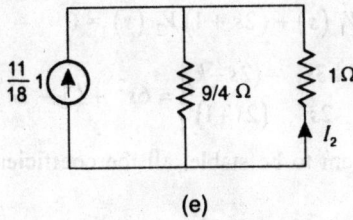

(e)

Fig. 13.19

$$\therefore \qquad I_2 = -\frac{11}{18} I_1 \times \frac{\frac{9}{4}}{\frac{9}{4} + 1} = -\frac{11}{26} I_1$$

$$\therefore \qquad \alpha_{12} = \frac{I_2}{I_1} = -\frac{11}{26}$$

13.11   For the network shown in Fig. 13.20, for what values of k, will the network be stable?

$R_1 = R_2 = 1\Omega, C_1 = 1F$ and $C_2 = 2F$.   (Gulbarga University)

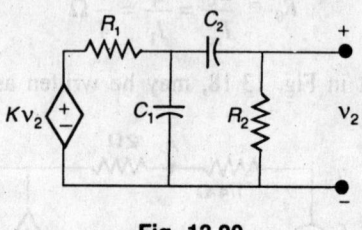

Fig. 13.20

Solution: The transformed network is as shown in Fig. 13.21.

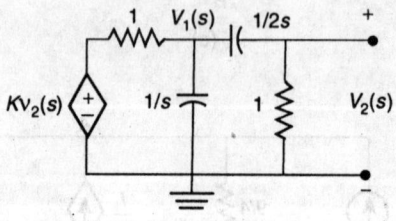

Fig. 13.21

The nodal Eqs are:

$$(1 + s + 2s)\, V_1(s) - 2s\, V_2(s) = \frac{kV_2(s)}{1}$$

i.e. $(1 + 3s)\, V_1(s) - (2s + k)\, V_2(s) = 0$ \hfill (1)

$$- 2s\, V_1(s) + (2s + 1)\, V_2(s) = 0 \hfill (2)$$

$$G(s) = \Delta = \begin{vmatrix} 1 + 3s & -(2s + k) \\ -2s & (2s + 1) \end{vmatrix} = 6s^2 + (5 - 2k)s + 1 = 0$$

For the system to be stable, all the coefficients of $G(s)$ must be positive.

$$\therefore \quad k < \frac{5}{2} \quad \text{for stability.}$$

13.12 Plot the pole–zero diagram of the system
$F(s) = \dfrac{10s}{(s+1)(s+3)}$. Obtain $f(t)$ using this diagram.

(Mysore University)

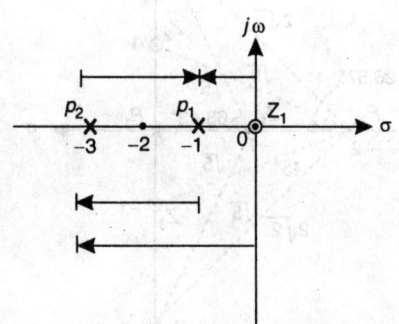

Fig. 13.22 Pole–zero diagram

Solution

$$F(s) = \frac{10s}{(s+1)(s+3)} = \frac{K_1}{s+1} + \frac{K_2}{s+3}$$

$$K_1 = \left(\text{residue of pole at } s = -1\right) = 10\,\frac{1\,\angle 180°}{2\,\angle 0°} = -5$$

$$K_2 = \left(\text{residue of pole at } s = -3\right) = 10\,\frac{3\angle 180°}{2\angle 180°} = 15$$

$$\therefore \qquad F(s) = \frac{-5}{s+3} + \frac{15}{s+3}$$

$$\therefore \qquad f(t) = -5\,e^{-t} + 15\,e^{-3t}$$

13.13 Plot the pole–zero diagram for the system

$$F(s) = \frac{(s+1)(s^2+1)}{s(s+2)(s^2+4)}.$$

Obtain $f(t)$ using pole–zero diagram. (Bangalore University)

Solution:

$$F(s) = \frac{(s+1)(s+j1)(s-j1)}{s(s+2)(s+j2)(s-j2)}$$

$$= \frac{K_1}{s} + \frac{K_2}{s+2} + \frac{K_3}{s+j2} + \frac{K_4}{s-j2}$$

The pole-zero plot is as shown in Fig. 13.23.

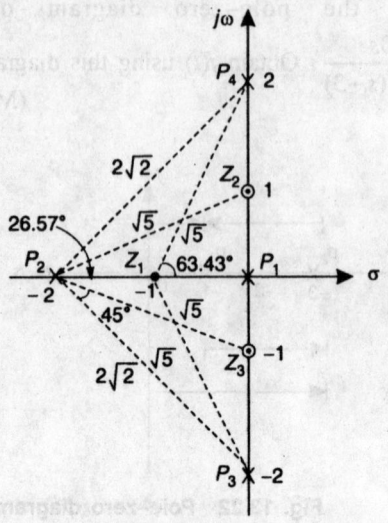

Fig. 13.23

$$K_1 = \text{residue of pole 1}$$

$$= \frac{1\angle 180° \times 1 \angle 90° \times 1 \angle -90°}{2 \angle 180° \times 2 \angle 90° \times 2 \angle -90°} = \frac{1}{8}$$

$$K_2 = \text{residue of pole 2}$$

$$= \frac{1\angle 0° \times \sqrt{5}\angle 26.57° \times \sqrt{5}\angle -26.57°}{2 \angle 0° \times 2\sqrt{2}\angle 45° \times 2\sqrt{2}\angle -45°} = \frac{5}{16}$$

$$K_3 = \text{residue of pole 3}$$

$$= \frac{\sqrt{5}\angle 116.57° \times 1 \angle -90° \times 3 \angle -90°}{2 \angle -90° \times 2\sqrt{2}\angle 45° \times 4 \angle -90°}$$

$$= 0.3 \angle -18.43° = 0.3\, e^{-j18.43°}$$

$$K_4 = \overset{*}{K_3} = 0.3\, e^{j18.43°}$$

$$\therefore \quad F(s) = \frac{1/8}{s} + \frac{5/16}{s+2} + \frac{0.3 e^{-j18.43°}}{(s+j2)} + \frac{0.3\, e^{j18.43°}}{(s-j2)}$$

$$\therefore \quad f(t) = \frac{1}{8} + \frac{5}{16} e^{-2t} + 0.3\, e^{-j2t}.e^{-j18.43°} + 0.3 e^{j2t}.e^{j18.43°}$$

$$= \frac{1}{8} + \frac{5}{16} e^{-2t} + 0.3\left[\frac{2\left\{e^{j(2t+18.43°)} + e^{-j(2t+18.43°)}\right\}}{2}\right]$$

$$= \frac{1}{8} + \frac{5}{16} e^{-2t} + 0.6 \cosh \left(2t + 18.43° \right)$$

NUMERICAL PROBLEMS

13.1 For the resistive two-port network shown in Fig. 13.24, determine the numerical values for (i) G_{12}, (ii) Z_{12}, (iii) Y_{12} and (iv) α_{12}.
(Kuvempu University)

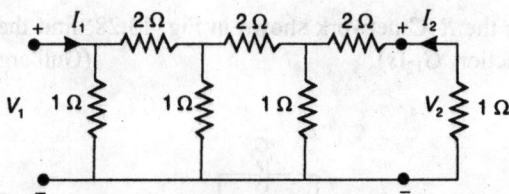

Fig. 13.24

13.2 For the *R–C* ladder network shown in Fig. 13.25, determine the voltage transfer function $V_0(s)/V_i(s)$. (Mysore University)

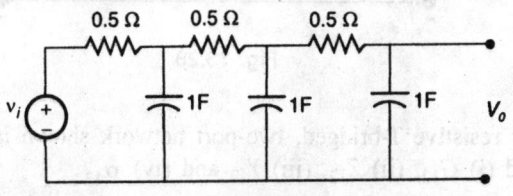

Fig. 13.25

13.3 Find the voltage transfer function $V_2(s)/V_1(s)$ for the network shown in Fig. 13.26. Also find $V_1(s)/I_1(s)$.
(Karnataka University)

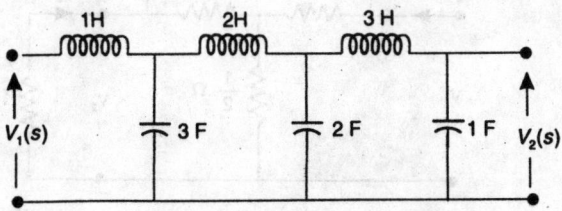

Fig. 13.26

13.4 Determine the driving point admittance $Y(s)$ for the network shown in Fig. 13.27, and plot its pole–zero diagram.
(Bangalore University)

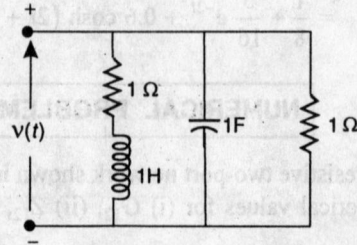

Fig. 13.27

13.5 For the R–C network shown in Fig. 13.28, find the voltage ratio function $G_{12}(s)$. (Gulbarga University)

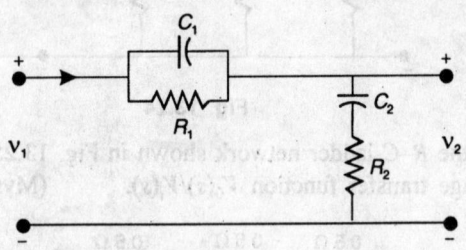

Fig. 13.28

13.6 For resistive T-bridged, two-port network shown in Fig. 13.29, find (i) G_{12}, (ii) Z_{12}, (iii) Y_{12} and (iv) α_{12}.

(Mangalore University)

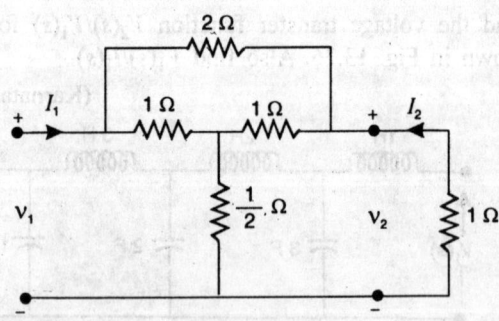

Fig. 13.29

13.7 The network given in Fig. 13.30, contains resistors and controlled sources. Compute $G_{12} = V_2/V_1$. (Bangalore University)

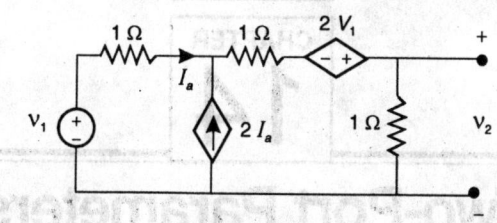

Fig. 13.30

13.8 For the network shown in Fig. 13.31, find (i) $G_{12}(s)$ and (ii) $\alpha_{12}(s)$.

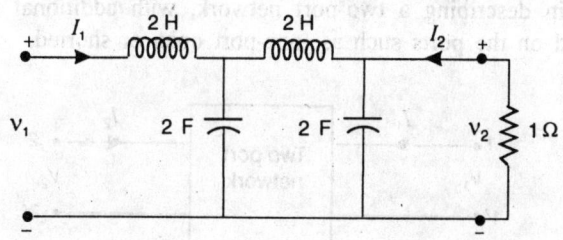

Fig. 13.31

13.9 Plot the pole–zero diagram for the function $I(s) = \dfrac{2s+3}{s^2+3s+2}$.

Obtain $i(t)$ using pole–zero diagram.   (Mysore University)

14

Two-Port Parameters

14.1 INTRODUCTION

Two-port parameters are a special class of network functions, which are useful in describing a two-port network, with additional restrictions imposed on the ports such as, one-port open or shorted.

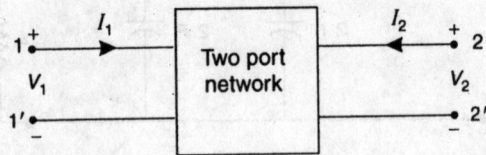

Fig. 14.1

Figure 14.1 represents a two-port network, which consists of two input terminals 1-1' and two output terminals 2-2'. Examples of two-port networks are: transformers, transmission lines, etc. The two-port network is represented by a box which is assumed to consist of a passive network of liner resistors, inductors, capacitors and dependent sources. Voltages V_1 and V_2 feed currents I_1 and I_2 to the two-ports of the network. The two-ports are called *input port* and *output port*. The knowledge of what a two-port network contains can be imparted by a set of network parameters, instead of the actual network, which expresses the relations between V_1, I_1, V_2 and I_2. Various parameters are defined for a two-port network, depending on the way in which, the relations betweeen the above four quantities are expressed. These parameters are called *two-port parameters*.

V_1, I_1, V_2 and I_2 are assumed to be transformed quantities. Specification of any two of them determines the remaining two. The dependence of the two variables on the other two may be described in a number of ways, depending on, which variables are chosen as independent variables. Six sets of parameters are defined for a two-port network.

The names given to these parameters indicate the dimensions, (Admittance, Impedance), lack of consistent dimensions (hybrid) and the principal application (transmission) of the parameters. They are:

1. Short-circuit admittance parameters.
2. Open-circuit impedance parameters.
3. Transmission parameters.
4. Inverse transmission parameters.
5. Hybrid parameters.
6. Inverse hybrid parameters.

14.2 SHORT-CIRCUIT ADMITTANCE PARAMETERS (*y* PARAMETERS)

The defining Eqs for these parameters are:

$$I_1 = y_{11} V_1 + y_{12} V_2 \tag{14.1}$$
$$I_2 = y_{21} V_1 + y_{22} V_2 \tag{14.2}$$

By putting V_1 or V_2 equal to zero in the above Eqs, we get

$$y_{11} = \frac{I_1}{V_1} \bigg|_{V_2 = 0} \tag{14.3}$$

$$y_{21} = \frac{I_2}{V_1} \bigg|_{V_2 = 0} \tag{14.4}$$

$$y_{12} = \frac{I_1}{V_2} \bigg|_{V_1 = 0} \tag{14.5}$$

$$y_{22} = \frac{I_2}{V_2} \bigg|_{V_1 = 0} \tag{14.6}$$

y_{11}, y_{12}, y_{21} and y_{22} are called short-circuit admittance parameters, as they are obtained by putting $V_1 = 0$ or $V_2 = 0$, i.e. by short-circuiting the two ports alternately.

For reciprocal or bilateral networks, $y_{12} = y_{21}$.

14.3 OPEN-CIRCUIT IMPEDANCE PARAMETERS (*z* PARAMETERS)

The defining Eqs for these parameters are:

$$V_1 = z_{11}I_1 + z_{12}I_2 \tag{14.7}$$
$$V_2 = z_{21}I_1 + z_{22}I_2 \tag{14.8}$$

By putting $I_1 = 0$ or $I_2 = 0$ in the above Eqs, we get

$$z_{11} = \frac{V_1}{I_1}\bigg|_{I_2=0} \qquad (14.9)$$

$$z_{21} = \frac{V_2}{I_2}\bigg|_{I_2=0} \qquad (14.10)$$

$$z_{22} = \frac{V_2}{I_2}\bigg|_{I_1=0} \qquad (14.11)$$

$$z_{12} = \frac{V_1}{I_2}\bigg|_{I_1=0} \qquad (14.12)$$

z_{11}, z_{12}, z_{21} and z_{22} are called open-circuit impedance parameters, as they are obtained by putting $I_1 = 0$ or $I_2 = 0$, i.e. by open-circuiting the two ports alternately.

For reciprocal or bilateral networks, $z_{12} = z_{21}$.

14.4 TRANSMISSION PARAMETERS (*T* PARAMETERS)

The transmission parameters give the relation between the voltage and current at one port to the voltage and current at the other port. The defining Eqs for these parameters are:

$$V_1 = AV_2 - BI_2 \qquad (14.13)$$

$$I_1 = CV_2 - DI_2 \qquad (14.14)$$

A, B, C and D are the transmission parameters.

By putting $I_2 = 0$ or $V_2 = 0$ in the above Eqs, we get

$$\frac{1}{A} = \frac{V_2}{V_1}\bigg|_{I_2=0} = \text{Open-circuit voltage gain} \qquad (14.15)$$

$$-\frac{1}{B} = \frac{I_2}{V_1}\bigg|_{V_2=0} = \text{Short-circuit transfer admittance} \qquad (14.16)$$

$$\frac{1}{C} = \frac{V_2}{I_1}\bigg|_{I_2=0} = \text{Open-circuit transfer impedance} \qquad (14.17)$$

$$-\frac{1}{D} = \frac{I_2}{I_1}\bigg|_{V_2=0} = \text{Short-circuit currrent gain} \qquad (14.18)$$

The transmission parameters are also called as *chain parameters*, as they are helpful in describing two-port networks, which are connected in cascade (chain arrangement).

14.5 INVERSE TRANSMISSION PARAMETERS (T' PARAMETERS)

The defining Eqs for these parameters are:

$$V_2 = A'V_1 - B'I_1 \tag{14.19}$$

$$I_2 = C'V_1 - D'I_1 \tag{14.20}$$

A', B', C' and D' are the inverse transmission parameters.

By putting $I_1 = 0$ or $V_1 = 0$, in the above Eqs, we get

$$A' = \left.\frac{V_2}{V_1}\right|_{I_1 = 0} = \text{Open-circuit voltage gain} \tag{14.21}$$

$$-B' = \left.\frac{V_2}{I_1}\right|_{V_1 = 0} = \text{Short-circuit transfer impedance} \tag{14.22}$$

$$C' = \left.\frac{I_2}{V_1}\right|_{I_1 = 0} = \text{Open-circuit transfer admittance} \tag{14.23}$$

$$-D' = \left.\frac{I_2}{I_1}\right|_{V_1 = 0} = \text{Short-circuit current gain} \tag{14.24}$$

14.6 HYBRID PARAMETERS (h PARAMETERS)

The defining Eqs for these parameters are:

$$V_1 = h_{11} I_1 + h_{12} V_2 \tag{14.25}$$

$$I_2 = h_{21} I_1 + h_{22} V_2 \tag{14.26}$$

By putting $I_1 = 0$ or $V_2 = 0$, in the above eqs, we get

$$h_{11} = \left.\frac{V_1}{I_1}\right|_{V_2 = 0} = \text{Short-circuit input impedance} \tag{14.27}$$

$$h_{21} = \left.\frac{I_2}{I_1}\right|_{V_2 = 0} = \text{Short-circuit current gain} \tag{14.28}$$

$$h_{12} = \left.\frac{V_1}{V_2}\right|_{I_1 = 0} = \text{Open-circuit voltage gain} \tag{14.29}$$

$$h_{22} = \frac{I_2}{V_2}\bigg|_{I_1 = 0} = \text{Open-circuit output admittance} \quad (14.30)$$

h_{11}, h_{12}, h_{21} and h_{22} are dimensionally mixed and hence, are called hybrid parameters. These parameters are useful in constructing models for transistors used in electronic circuits.

14.7 INVERSE HYBRID PARAMETERS (g PARAMETERS)

The defining Eqs for these parameters are:

$$I_1 = g_{11} V_1 + g_{12} I_2 \quad (14.31)$$
$$V_2 = g_{21} V_1 + g_{22} I_2 \quad (14.32)$$

By putting $I_2 = 0$ or $V_1 = 0$, in the above Eqs, we get

$$g_{11} = \frac{I_1}{V_1}\bigg|_{I_2 = 0} = \text{Open-circuit input admittance} \quad (14.33)$$

$$g_{21} = \frac{V_2}{V_1}\bigg|_{I_2 = 0} = \text{Open-circuit voltage ratio} \quad (14.34)$$

$$g_{12} = \frac{I_1}{I_2}\bigg|_{V_1 = 0} = \text{Short-circuit current ratio} \quad (14.35)$$

$$g_{22} = \frac{V_2}{I_2}\bigg|_{V_1 = 0} = \text{Short-circuit input impedance at port 2} \quad (14.36)$$

g_{11}, g_{12}, g_{21} and g_{22} are called inverse hybrid parameters.

14.8 RELATION BETWEEN y PARAMETERS AND OTHER TYPES OF PARAMETERS

(a) Relation between y and z Parameters

We know that
$$V_1 = z_{11} I_1 + z_{12} I_2 \quad (1)$$
$$V_2 = z_{21} I_1 + z_{22} I_2 \quad (2)$$

and
$$I_1 = y_{11} V_1 + y_{12} V_2 \quad (3)$$
$$I_2 = y_{21} V_1 + y_{22} V_2 \quad (4)$$

Substituting for I_2 from (2) in (1), we get

$$V_1 = z_{11} I_1 + z_{12}\left(\frac{V_2 - z_{21} I_1}{z_{22}}\right)$$

$$= I_1 \left(z_{11} - \frac{z_{12} \, z_{21}}{z_{22}} \right) + \frac{z_{12}}{z_{22}} \, V_2$$

$$= I_1 \left(\frac{z_{11} \, z_{22} - z_{12} \, z_{21}}{z_{22}} \right) + \frac{z_{12}}{z_{22}} \, V_2$$

$$\therefore \qquad I_1 = \frac{z_{22}}{z_{11} z_{22} - z_{12} z_{21}} V_1 - \frac{z_{12}}{z_{11} z_{22} - z_{12} z_{21}} V_2$$

Comparing this with Eq. (3), we get

$$y_{11} = \frac{z_{22}}{\Delta_z} \quad \text{and} \quad y_{12} = -\frac{z_{12}}{\Delta_z}$$

where $\quad \Delta_z = z_{11} z_{22} - z_{12} z_{21}$

Substituting for I_1 from (2) in (1), we get

$$V_1 = z_{11} \left(\frac{V_2 - z_{22} \, I_2}{z_{21}} \right) + z_{12} \, I_2$$

$$= I_2 \left(z_{12} - \frac{z_{12} \, z_{22}}{z_{21}} \right) + \frac{z_{11}}{z_{21}} \, V_2$$

$$\therefore \qquad I_2 = \frac{z_{21}}{z_{12} z_{21} - z_{11} z_{22}} V_1 - \frac{z_{11}}{z_{12} z_{21} - z_{11} z_{22}} V_2$$

Comparing this with Eq. (4), we get

$$y_{21} = -\frac{z_{21}}{\Delta_z} \quad \text{and} \quad y_{22} = \frac{z_{11}}{\Delta_z}$$

Similarly, by writing the Eqs, of other types of parameters with the Eqs of y parameters, and manipulating the Eqs, we can get the relations of y parameters with the other types of parameters. These relations are tabulated in Table 14.1.

Table 14.1. Relations between y parameters and other types of parameters

$[Y]$	$[Z]$		$[T]$		$[T']$		$[h]$		$[g]$	
$y_{11} \quad y_{12}$	$\dfrac{z_{22}}{\Delta_z}$	$-\dfrac{z_{12}}{\Delta_z}$	$\dfrac{D}{B}$	$-\dfrac{\Delta_r}{B}$	$\dfrac{A'}{B'}$	$-\dfrac{I}{B'}$	$\dfrac{I}{h_{11}}$	$\dfrac{h_{12}}{h_{21}}$	$\dfrac{\Delta_g}{g_{22}}$	$\dfrac{g_{12}}{g_{22}}$
$y_{21} \quad y_{22}$	$-\dfrac{z_{21}}{\Delta_z}$	$\dfrac{z_{11}}{\Delta_z}$	$-\dfrac{I}{B}$	$\dfrac{A}{B}$	$-\dfrac{\Delta_{T'}}{B'}$	$\dfrac{D'}{B'}$	$\dfrac{h_{21}}{h_{11}}$	$\dfrac{\Delta_h}{h_{11}}$	$-\dfrac{g_{21}}{g_{22}}$	$\dfrac{I}{g_{22}}$

where
$$\Delta_T = AD - BC,$$
$$\Delta'_T = A'D' - B'C',$$
$$\Delta_h = h_{11}h_{22} - h_{12}h_{21}$$

and
$$\Delta_g = g_{11}g_{22} - g_{12}g_{21}$$

14.9 RELATIONS BETWEEN z PARAMETERS AND OTHER TYPES OF PARAMETERS

(a) Relation between z and y Parameters

We know that

$$I_1 = y_{11}V_1 + y_{12}V_2 \tag{1}$$

$$I_2 = y_{21}V_1 + y_{22}V_2 \tag{2}$$

and
$$V_1 = z_{11}I_1 + z_{12}I_2 \tag{3}$$

$$V_2 = z_{21}I_1 + z_{22}I_2 \tag{4}$$

Substituting for V_2 from (2) in (1), we get

$$I_1 = y_{11}V_1 + y_{12}\left(\frac{I_2 - y_{21}V_1}{y_{22}}\right)$$

$$= V_1\left(y_{11} - \frac{y_{12}y_{21}}{y_{22}}\right) + \frac{y_{12}}{y_{22}}I_2$$

$$\therefore \qquad V_1 = \frac{y_{22}}{y_{11}y_{22} - y_{12}y_{21}}I_1 - \frac{y_{12}}{y_{11}y_{22} - y_{12}y_{21}}I_2$$

Comparing this Eq. with (3), we get

$$z_{11} = \frac{y_{22}}{\Delta_y} \quad \text{and} \quad z_{12} = -\frac{y_{12}}{\Delta_y}$$

Substituting for V_1 from (2) in (1), we get

$$I_1 = y_{11}\left(\frac{I_2 - y_{22}V_2}{y_{21}}\right) + y_{12}V_2$$

or
$$V_2 = -\frac{y_{21}}{y_{11}y_{22} - y_{12}y_{21}}I_1 + \frac{y_{11}}{y_{11}y_{22} - y_{12}y_{21}}I_2$$

Comparing this Eq. with (4), we get

$$z_{21} = -\frac{y_{21}}{\Delta_y} \quad \text{and} \quad z_{22} = \frac{y_{11}}{\Delta_y}$$

Similarly, by writing the eqs of other types of parameters with the Eqs of z parameters and manipulating the Eqs, we can get the relations of z parameters with the other types of parameters.

These relations are tabulated in Table 14.2.

Table 14.2. Relations between z parameters and other types of parameters

[z]		[y]		[T]		[T^1]		[h]		[g]	
z_{11}	z_{12}	$\dfrac{y_{22}}{\Delta_y}$	$-\dfrac{y_{12}}{\Delta_y}$	$\dfrac{A}{C}$	$\dfrac{\Delta_T}{C}$	$\dfrac{D'}{C'}$	$\dfrac{1}{C'}$	$\dfrac{\Delta_h}{h_{22}}$	$\dfrac{h_{12}}{h_{22}}$	$\dfrac{1}{g_{11}}$	$\dfrac{g_{12}}{g_{11}}$
z_{21}	z_{22}	$-\dfrac{y_{21}}{\Delta_y}$	$\dfrac{y_{11}}{\Delta_y}$	$\dfrac{1}{C}$	$\dfrac{D}{C}$	$\dfrac{\Delta_T}{C'}$	$\dfrac{A'}{C'}$	$-\dfrac{h_{21}}{h_{22}}$	$\dfrac{1}{h_{22}}$	$\dfrac{g_{21}}{g_{11}}$	$\dfrac{\Delta_g}{g_{11}}$

14.10 RELATIONS BETWEEN *T* PARAMETERS AND OTHER TYPES OF PARAMETERS

(a) Relation between *T* and *y* parameters

We know that
$$I_1 = y_{11} V_1 + y_{12} V_2 \qquad (1)$$
$$I_2 = y_{21} V_1 + y_{22} V_2 \qquad (2)$$
and
$$V_1 = AV_2 - BI_2 \qquad (3)$$
$$I_1 = CV_2 - DI_2 \qquad (4)$$

Equation (2) may be written as

$$V_1 = -\frac{y_{22}}{y_{21}} V_2 + \frac{1}{y_{21}} I_2$$

Comparing this Eq. with Eq. (3), we find that

$$A = -\frac{y_{22}}{y_{21}} \quad \text{and} \quad B = -\frac{1}{y_{21}}$$

Substituting for V_1 from (2) in (1), we get

$$I_1 = y_{11} \left(\frac{I_2 - y_{22} V_2}{y_{21}} \right) + y_{12} V_2$$

$$= \left(y_{12} - \frac{y_{22}}{y_{21}} \right) V_2 + \frac{y_{11}}{y_{21}} I_2$$

$$= -\frac{\Delta_y}{y_{21}} V_2 + \frac{y_{11}}{y_{21}} I_2$$

Comparing this Eq. with Eq. (4), we get

$$C = -\frac{\Delta_y}{y_{21}} \quad \text{and} \quad D = -\frac{y_{11}}{y_{21}}$$

Similarly, by writing the eqs of other types of parameters with the eqs of T parameters and manipulating the Eqs, we can get the relations of T parameters with the other types of parameters. These relations are tabulated in Table 14.3.

Table 14.3. Relations between T parameters and other types of parameters

[T]	[y]		[z]		[T']		[h]		[g]	
$A \quad B$	$-\dfrac{y_{22}}{y_{21}}$	$-\dfrac{1}{y_{21}}$	$\dfrac{z_{11}}{z_{21}}$	$\dfrac{\Delta_z}{z_{21}}$	$\dfrac{D'}{\Delta_{T'}}$	$\dfrac{B'}{\Delta_{T'}}$	$-\dfrac{\Delta_h}{h_{21}}$	$-\dfrac{h_{11}}{h_{21}}$	$\dfrac{1}{g_{21}}$	$\dfrac{g_{22}}{g_{21}}$
$C \quad D$	$-\dfrac{\Delta_y}{y_{21}}$	$-\dfrac{y_{11}}{y_{21}}$	$\dfrac{1}{z_{21}}$	$\dfrac{z_{22}}{z_{21}}$	$\dfrac{C'}{\Delta_{T'}}$	$\dfrac{A'}{\Delta_{T'}}$	$\dfrac{h_{22}}{h_{21}}$	$-\dfrac{1}{h_{21}}$	$\dfrac{g_{11}}{g_{21}}$	$\dfrac{\Delta_g}{g_{21}}$

14.11 RELATIONS BETWEEN h PARAMETERS AND OTHER TYPES OF PARAMETERS

(a) Relation between h and y Parameters

We know that

$$I_1 = y_{11} V_1 + y_{12} V_2 \tag{1}$$

$$I_2 = y_{21} V_1 + y_{22} V_2 \tag{2}$$

and

$$V_1 = h_{11} I_1 + h_{12} V_2 \tag{3}$$

$$I_2 = h_{21} I_1 + h_{22} V_2 \tag{4}$$

Equation (1) may be written as

$$V_1 = \frac{1}{y_{11}} I_1 - \frac{y_{12}}{y_{11}} V_2$$

Comparing this Eq. with Eq. (3), we find that

$$h_{11} = \frac{1}{y_{11}} \quad \text{and} \quad h_{12} = -\frac{y_{12}}{y_{11}}$$

Substituting for V_1 from (2) in (1), we get

$$I_1 = y_{11} \left(\frac{I_2 - y_{22} V_2}{y_{21}} \right) + y_{12} V_2$$

i.e. $\qquad I_1 = \dfrac{y_{11}}{y_{21}} I_2 - \dfrac{y_{11}\, y_{22} - y_{12}\, y_{21}}{y_{21}} V_2$

or $\qquad I_2 = \dfrac{y_{21}}{y_{11}} I_1 + \dfrac{\Delta_y}{y_{11}} V_2$

Comparing this Eq. with Eq. (4), we get

$$h_{21} = \frac{y_{21}}{y_{11}} \text{ and } h_{22} = \frac{\Delta_y}{y_{11}}$$

Similarly, by writing the Eqs of other types of parameters with the eqs of h parameters and manipulating the eqs, we canget the relations of h parameters with the other types of parameters.

These relations are tabulated in Table 14.4.

Table 14.4. Relations between h parameters and other types of parameters

$[h]$		$[y]$		$[z]$		$[T]$		$[T']$		$[g]$	
h_{11}	h_{12}	$\dfrac{1}{y_{11}}$	$-\dfrac{y_{12}}{y_{11}}$	$\dfrac{\Delta_z}{z_{22}}$	$\dfrac{z_{12}}{z_{22}}$	$\dfrac{B}{D}$	$\dfrac{\Delta_T}{D}$	$\dfrac{B'}{A'}$	$\dfrac{1}{A'}$	$\dfrac{g_{22}}{\Delta_g}$	$\dfrac{g_{12}}{\Delta_g}$
h_{21}	h_{22}	$\dfrac{y_{21}}{y_{11}}$	$\dfrac{\Delta_y}{y_{11}}$	$-\dfrac{z_{21}}{z_{22}}$	$\dfrac{1}{z_{22}}$	$-\dfrac{1}{D}$	$\dfrac{C}{D}$	$-\dfrac{\Delta_T}{A'}$	$\dfrac{C'}{A'}$	$-\dfrac{g_{21}}{\Delta_g}$	$\dfrac{g_{11}}{\Delta_g}$

14.12 RELATIONS BETWEEN *T'* PARAMETERS AND *g* PARAMETERS WITH OTHER TYPES OF PARAMETERS

Table 14.5 gives these relations.

Table 14.5. Relations between T' and g parameters with other types of parameters

$[T']$		$[y]$		$[z]$		$[T]$		$[h]$		$[g]$	
A'	B'	$-\dfrac{y_{11}}{y_{12}}$	$-\dfrac{1}{y_{12}}$	$\dfrac{z_{22}}{z_{12}}$	$\dfrac{\Delta_z}{z_{12}}$	$\dfrac{D}{\Delta_T}$	$\dfrac{B}{\Delta_T}$	$\dfrac{1}{h_{12}}$	$\dfrac{h_{11}}{h_{12}}$	$\dfrac{\Delta_g}{g_{12}}$	$-\dfrac{g_{22}}{g_{12}}$
C'	D'	$-\dfrac{\Delta_y}{y_{12}}$	$-\dfrac{y_{22}}{y_{12}}$	$\dfrac{1}{z_{12}}$	$\dfrac{z_{11}}{z_{12}}$	$\dfrac{C}{\Delta_T}$	$\dfrac{A}{\Delta_T}$	$\dfrac{h_{22}}{h_{12}}$	$\dfrac{\Delta_h}{h_{12}}$	$-\dfrac{g_{11}}{g_{12}}$	$\dfrac{1}{g_{12}}$

$[g]$		$[y]$		$[z]$		$[T]$		$[T']$		$[h]$	
g_{11}	g_{12}	$\dfrac{\Delta_y}{y_{22}}$	$\dfrac{y_{12}}{y_{22}}$	$\dfrac{1}{z_{11}}$	$-\dfrac{z_{12}}{z_{11}}$	$\dfrac{C}{A}$	$-\dfrac{\Delta_T}{A}$	$\dfrac{C'}{D'}$	$-\dfrac{1}{D'}$	$\dfrac{h_{22}}{\Delta_h}$	$\dfrac{h_{12}}{\Delta_h}$
g_{21}	g_{22}	$-\dfrac{y_{21}}{y_{22}}$	$\dfrac{1}{y_{22}}$	$\dfrac{z_{21}}{z_{11}}$	$\dfrac{\Delta_z}{z_{11}}$	$\dfrac{1}{A}$	$\dfrac{B}{A}$	$\dfrac{\Delta_T}{D'}$	$\dfrac{B'}{D'}$	$-\dfrac{h_{21}}{\Delta_h}$	$\dfrac{h_{11}}{\Delta_h}$

14.13   CASCADE CONNECTION OF TWO-PORT NETWORKS

The transmission parameters A, B, C and D are useful in describing two-port networks which are connected in cascade. Consider two networks N_a and N_b connected in cascade as shown in Fig.14.2.

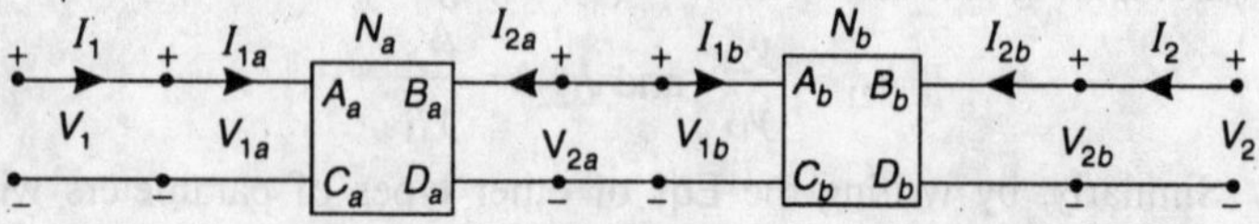

Fig. 14.2

Equations describing the network N_a are

$$\begin{bmatrix} V_{1a} \\ -I_{1a} \end{bmatrix} = \begin{bmatrix} A_a & B_a \\ C_a & D_a \end{bmatrix} \begin{bmatrix} V_{2a} \\ -I_{2a} \end{bmatrix} \tag{14.37}$$

Equations describing the network N_b are

$$\begin{bmatrix} V_{1b} \\ I_{1b} \end{bmatrix} = \begin{bmatrix} A_b & B_b \\ C_b & D_b \end{bmatrix} \begin{bmatrix} V_{2b} \\ -I_{2b} \end{bmatrix} \tag{14.38}$$

But $V_{1a} = V_1,$ $V_{2a} = V_{1b}$

             $I_{1a} = I_1,$ $I_{2a} = -I_{1b}$

and $V_2 = V_{2b},$ $I_2 = I_{2b}$

Substituting these relations Eq. (14.37), we get

$$\begin{bmatrix} V_1 \\ I_1 \end{bmatrix} = \begin{bmatrix} A_a & B_a \\ C_a & D_a \end{bmatrix} \begin{bmatrix} V_{1b} \\ I_{1b} \end{bmatrix}$$

$$= \begin{bmatrix} A_a & B_a \\ C_a & D_a \end{bmatrix} \begin{bmatrix} A_b & B_b \\ C_b & D_b \end{bmatrix} \begin{bmatrix} V_{2b} \\ -I_{2b} \end{bmatrix}$$

$$= \begin{bmatrix} A_a & B_a \\ C_a & D_a \end{bmatrix} \begin{bmatrix} A_b & B_b \\ C_b & D_b \end{bmatrix} \begin{bmatrix} V_2 \\ -I_2 \end{bmatrix}$$

$$= \begin{bmatrix} A & B \\ C & D \end{bmatrix} \begin{bmatrix} V_2 \\ -I_2 \end{bmatrix}$$

$$\therefore \quad \begin{bmatrix} A & B \\ C & D \end{bmatrix} = \begin{bmatrix} A_a & B_a \\ C_a & D_a \end{bmatrix} \begin{bmatrix} A_b & B_b \\ C_b & D_b \end{bmatrix} \tag{14.39}$$

The result obtained in Eq. (14.39), may be extended for any number of two-port networks connected in cascade.

If the given two-port network is complex, it can be viewed as being constructed from simpler two-port networks, whose ports are interconnected in certain ways to get the original network. There are a number of ways of interconnecting two-port networks.

Two-ports are said to be connected in cascade, if the output port of one is the input port for the second. This connection is also called as *Tandem-Connection*. The tandem connection is most conveniently studied by means of *ABCD* parameters.

It is much easier to design simple blocks and interconnect them, than to design a complex network. It is also much easier to shield smaller units, and thus reduce parasitic capacitances to ground.

14.14 SERIES CONNECTION OF TWO-PORTS

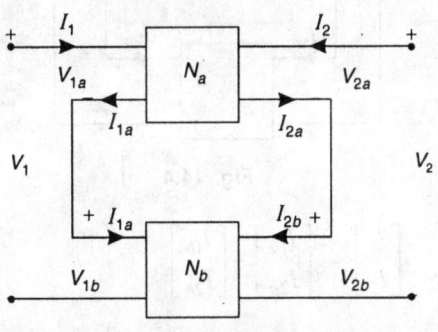

Fig. 14.3

Two two-port networks N_a and N_b are said to be connected in series, if the corresponding ports are connected in series as shown in Fig. 14.3. In this connection, the input and output currents at the corresponding ports are forced to be the same. The overall port voltages are equal to the sum of the corresponding port voltages of the individual two-ports.

$$\begin{bmatrix} V_1 \\ V_2 \end{bmatrix} = \begin{bmatrix} V_{1a} \\ V_{2a} \end{bmatrix} + \begin{bmatrix} V_{1b} \\ V_{2b} \end{bmatrix} = \begin{bmatrix} z_{11a} & z_{12a} \\ z_{21a} & z_{22a} \end{bmatrix} \begin{bmatrix} I_{1a} \\ I_{2a} \end{bmatrix}$$

$$+ \begin{bmatrix} z_{11b} & z_{12b} \\ z_{21b} & z_{22b} \end{bmatrix} \begin{bmatrix} I_{1b} \\ I_{2b} \end{bmatrix}$$

But, $I_{1a} = I_{1b} = I_1$ and $I_{2a} = I_{2b} = I_2$

$$\therefore \qquad \begin{bmatrix} V_1 \\ V_2 \end{bmatrix} = \begin{bmatrix} z_{11a} + z_{11b} & z_{12a} + z_{12b} \\ z_{21a} + z_{21b} & z_{22a} + z_{22b} \end{bmatrix} \begin{bmatrix} I_1 \\ I_2 \end{bmatrix} \qquad (14.40)$$

14.15 PARALLEL CONNECTION OF TWO-PORTS

Two two-port networks, N_a and N_b are said to be connected in parallel, if the corresponding ports are connected in parallel as shown in Fig.14.4. In this connection, the input and output voltages of the corresponding ports are forced to be the same. The overall port currents are equal to the sum of the corresponding port currents of the individual two-ports.

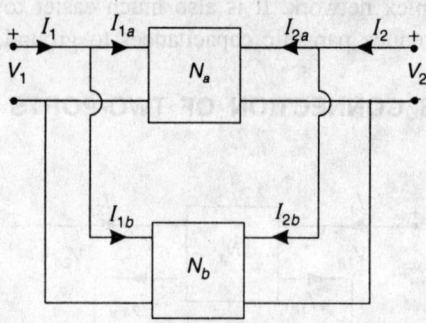

Fig. 14.4

$$\begin{bmatrix} I_1 \\ I_2 \end{bmatrix} = \begin{bmatrix} I_{1a} \\ I_{2a} \end{bmatrix} + \begin{bmatrix} I_{1b} \\ I_{2b} \end{bmatrix}$$

$$= \begin{bmatrix} y_{11a} & y_{12a} \\ y_{21a} & y_{22a} \end{bmatrix} \begin{bmatrix} V_{1a} \\ V_{2a} \end{bmatrix} + \begin{bmatrix} y_{11b} & y_{12b} \\ y_{21b} & y_{22b} \end{bmatrix} \begin{bmatrix} V_{1b} \\ V_{2b} \end{bmatrix}$$

But $\qquad V_{1a} = V_{1b} = V_1$ and $V_{2a} = V_{2b} = V_2 \qquad (14.41)$

$$\begin{bmatrix} I_1 \\ I_2 \end{bmatrix} = \begin{bmatrix} y_{11a} + y_{11b} & y_{12a} + y_{12b} \\ y_{21a} + y_{21b} & y_{22a} + y_{22b} \end{bmatrix} \begin{bmatrix} V_1 \\ V_2 \end{bmatrix}$$

14.16 PERMISSIBILITY OF INTERCONNECTION OF TWO-PORT NETWORKS

When two two-port networks are connected in parallel, the parameters characterising the individual two-ports and the overall two-port are the short-circuit admittance parameters. In such a parallel, connection of two two-ports, either the input ports are connected in parallel and the output ports are shorted, i.e. $V_2 = 0$, as shown in Fig.14.5 or the output ports are connected in parallel and the input ports are shorted, i.e. $V_1 = 0$, as shown in Fig. 14.6.

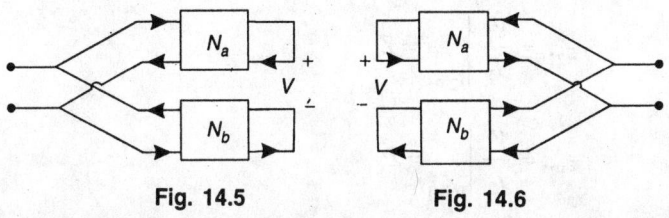

Fig. 14.5   **Fig. 14.6**

When such connections are made, if V is non-zero, there will be circulating currents as shown in Figs 14.5 and 14.6. This alters the port relationships of the individual two-port relations. When it is found that the parallel connection of two two-port networks cannot be made because of circulating currents, there is a way of stopping such currents and thus permitting the connection to be made. The approach is simply to put an isolating ideal 1 : 1 transformer at one of the ports as shown in Fig. 14.7.

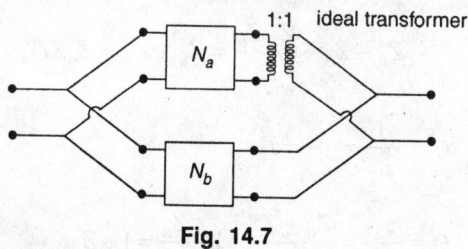

Fig. 14.7

14.17 *T* SECTION REPRESENTATION OF A TWO-PORT NETWORK

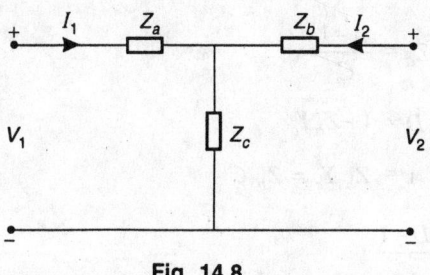

Fig. 14.8

Z_a, Z_b and Z_c are the three impedances connected as a T network. The impedance parameters for the network are given by

$$z_{11} = \frac{V_1}{I_1}\bigg|_{I_2=0} = Z_a + Z_c$$

$$z_{21} = \frac{V_2}{I_1}\bigg|_{I_2=0} = Z_c$$

$$z_{12} = \frac{V_1}{I_2}\bigg|_{I_1=0} = Z_c$$

and $$z_{22} = \frac{V_2}{I_2}\bigg|_{I_1=0} = Z_b + Z_c$$

But $$A = \frac{z_{11}}{z_{21}} = \frac{Z_a + Z_c}{Z_c} = 1 + \frac{Z_a}{Z_c} = 1 + Z_a Y_c \qquad (14.42)$$

$$B = \frac{\Delta_z}{z_{21}} = \frac{z_{11} z_{22} - z_{12} z_{21}}{z_{21}} = \frac{(Z_a + Z_c)(Z_b + Z_c) - Z_c^2}{Z_c}$$

$$= Z_a + Z_b + \frac{Z_a Z_b}{Z_c} = Z_a + Z_b + Z_a Z_b Y_c \qquad (14.43)$$

$$C = \frac{1}{z_{12}} = \frac{1}{Z_c} = Y_c \qquad (14.44)$$

$$D = \frac{z_{22}}{z_{21}} = \frac{Z_a + Z_c}{Z_c} = 1 + \frac{Z_a}{Z_c} = 1 + Z_a Y_c \qquad (14.45)$$

Conversely

$$A = 1 + Z_a Y_c \quad \text{or} \quad A - 1 = Z_a Y_c = Z_a C,$$

$$\therefore \qquad Z_a = \frac{A-1}{C} \qquad (14.46)$$

$$D = 1 + Z_b Y_c$$

$$\text{or} \quad D - 1 = Z_b Y_c = Z_b C$$

$$\therefore \quad Z_b = \frac{D-1}{C} \qquad (14.47)$$

and $$Z_c = \frac{1}{Y_c} = C \qquad (14.48)$$

To show that $AD - BC = 1$

$$AD - BC = \left(1 + Z_a\, Y_c\right)\left(1 + Z_b\, Y_c\right) - \left(Z_a + Z_b + Z_a\, Z_b\, Y_c\right) Y_c$$

$$= 1 + Z_a\, Y_c + Z_b\, Y_c + Z_a\, Z_b\, Y_c^2 - Z_a\, Y_c - Z_b\, Y_c - Z_a\, Z_b\, Y_c^2 = 1$$

14.8 Π REPRESENTATION OF A TWO-PORT NETWORK

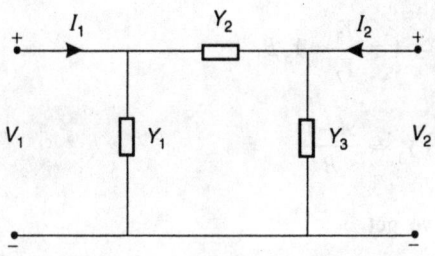

Fig. 14.9

Y_1, Y_2 and Y_3 are the three admittances connected as a π network. The admittance parameters for the network are given by

$$y_{11} = \left.\frac{V_1}{I_1}\right|_{V_2 = 0} = Y_1 + Y_2$$

$$y_{21} = \left.\frac{I_2}{V_1}\right|_{V_2 = 0} = -Y_2 \qquad \left[\because V_1 + \frac{I_2}{Y_2} = 0\right]$$

$$y_{12} = \left.\frac{I_1}{V_2}\right|_{V_1 = 0} = -Y_2 \qquad \left[\because V_2 + \frac{I_1}{Y_2} = 0\right]$$

$$y_{22} = \left.\frac{I_2}{V_2}\right|_{V_1 = 0} = Y_2 + Y_3$$

Conversely

$$A = -\frac{y_{22}}{y_{21}} = \frac{Y_2 + Y_3}{Y_2} = 1 + \frac{Y_3}{Y_2} \tag{14.49}$$

$$B = -\frac{1}{y_{21}} = \frac{1}{Y_2} = Z_2 \tag{14.50}$$

$$C = -\frac{\Delta_y}{y_{21}} = \frac{y_{11}\, y_{22} - y_{12}\, y_{21}}{y_{21}}$$

$$= \frac{(Y_1 + Y_2)(Y_2 + Y_3) - Y_2^2}{Y_2} = Y_1 + Y_3 + \frac{Y_1 Y_3}{Y_2} \qquad (14.51)$$

$$D = -\frac{y_{11}}{y_{21}} = \frac{Y_1 + Y_2}{Y_2} = 1 + \frac{Y_1}{Y_2} \qquad (14.52)$$

From (14.49), we get

$$A - 1 = \frac{Y_3}{Y_2} = Y_3 B,$$

$$\therefore \qquad Y_3 = \frac{A-1}{B} \qquad (14.53)$$

From (14.52), we get

$$D - 1 = \frac{Y_1}{Y_2} = Y_1 B,$$

$$\therefore \qquad Y_1 = \frac{D-1}{B} \qquad (14.54)$$

and $$\qquad Y_2 = \frac{1}{B} \qquad (14.55)$$

14.19 RECIPROCAL AND SYMMETRICAL NETWORKS

(a) Reciprocal network: Any two-port network in which, the ratio of the response to the excitation remains constant, when the positions of the excitation and response are interchanged, is called a *reciprocal network*. Consider a two-port network N as shown in Fig.14.10 (a), in which V is the excitation and I is the response. Let the excitation be shifted to the output terminals as shown in Fig.14.10 (b). Then, if the response at the input terminal remains I, then, the network is said to be reciprocal.

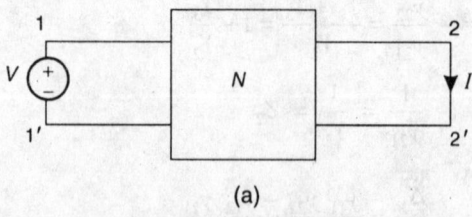

(a)

Fig. 14.10

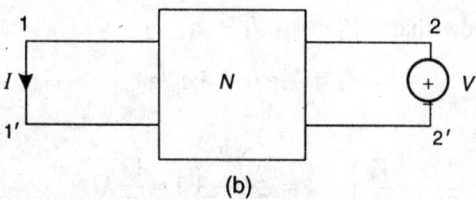

(b)

Fig. 14.10

(b) Symmetrical network: When a certain voltage is applied across the input terminals of a two-port network, a definite voltage appears across the open circuited output terminals and a definite current flows from the input terminals to the network. For the same output voltage, definite currents flow at the input and output ports, when the output port is short-circuited. Now, if the input and output ports are interchanged and the network exhibits the same characteristics as before, then the network is said to be *symmetrical*.

Table 14.6 gives the conditions of reciprocity and symmetry of a two-port network in terms of network parameters.

Table 14.6 Conditions for reciprocity and symmetry

Parameters	Condition for reciprocity	Condition for symmetry
y	$y_{12} = y_{21}$	$y_{11} = y_{22}$
z	$z_{12} = z_{21}$	$z_{11} = z_{22}$
h	$h_{12} = -h_{21}$	$h_{11}h_{22} - h_{12}h_{21} = 1$
g	$g_{11} = -g_{21}$	$g_{11}g_{22} - g_{12}g_{21} = 1$
A, B, C, D	$AD - BC = 1$	$A = D$
A', B', C', D'	$A'D' - B'C' = 1$	$A' = D'$

WORKED EXAMPLES

14.1 Find the y and z parameters for the network given in Fig. 14.11. (Kuvempu University)

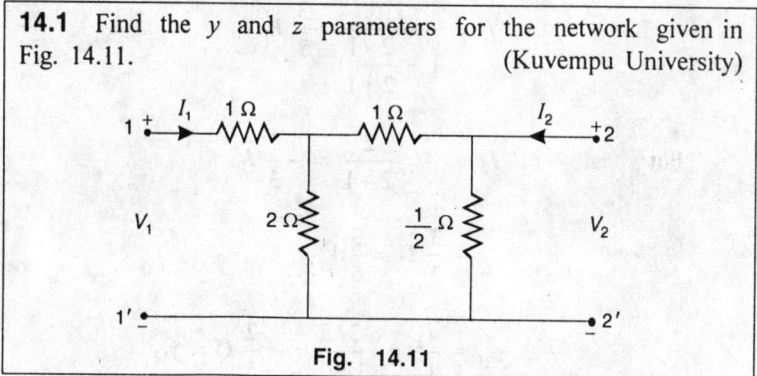

Fig. 14.11

Solution: *z parameters*

We know that $V_1 = z_{11} I_1 + z_{12} I_2$

$$V_2 = z_{21} I_1 + z_{22} I_2$$

$$\therefore \quad z_{11} = \frac{V_1}{I_1}\bigg|_{2=0} = \frac{\dfrac{3}{2} \times 2}{\dfrac{3}{2} + 2} + 1 = \frac{13}{7}\ \Omega$$

$$z_{21} = \frac{V_2}{I_1}\bigg|_{I_2=0} = \frac{1}{2}\left[\frac{I_1\ \dfrac{2}{2 + \dfrac{3}{2}}}{I_1}\right] = \frac{2}{7}\ \Omega = Z_{21}$$

$$z_{22} = \frac{V_2}{I_2}\bigg|_{I_1=0} = \frac{3 \times \dfrac{1}{2}}{3 + \dfrac{1}{2}} = \frac{3}{7}\ \Omega$$

y parameters

We know that $I_1 = y_{11} V_1 + y_{12} V_2$

$$I_2 = y_{21} V_1 + y_{22} V_2$$

$$\therefore \quad y_{11} = \frac{I_1}{V_1}\bigg|_{V_2=0} = \frac{1}{\dfrac{2 \times 1}{2+1} + 1} = \frac{3}{5}\ \mho$$

When $V_2 = 0,$

$$I_1 = \frac{V_1}{1 + \dfrac{2 \times 1}{2+1}} = \frac{3}{5} V_1$$

but $I_2 = -I_1 \dfrac{2}{2+1} = -\dfrac{2}{3} I_1$

$$\therefore \quad -\frac{3}{2} I_2 = \frac{3}{5} V_1$$

$$\therefore \quad y_{21} = \frac{I_2}{V_1} = \frac{-3/5}{3/2} = -\frac{2}{5}\ \mho = y_{12}$$

$$y_{22} = \frac{I_2}{V_2}\bigg|_{V_1 = 0} = \frac{1}{\dfrac{\left\{\dfrac{2\times 1}{2+1}+1\right\}\times \dfrac{1}{2}}{\left\{\dfrac{2\times 1}{2+1}+1\right\}+\dfrac{1}{2}}} = \frac{13}{5}\ \mho$$

14.2 The network given in Fig. 14.12 (a), contains a controlled current source. Find y and z parameters.   (Mysore University)

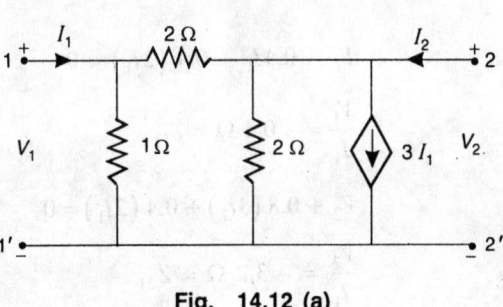

Fig. 14.12 (a)

Solution: z *parameters*

$$V_1 = z_{11}\ I_1 + z_{12}\ I_2$$

$$V_2 = z_{21}\ I_1 + z_{22}\ I_2$$

∴ $$z_{11} = \frac{V_1}{I_1}\bigg|_{I_2 = 0} \quad \text{and} \quad z_{21} = \frac{V_2}{I_1}\bigg|_{I_2 = 0}$$

When $I_2 = 0$, the resulting network is as shown in Fig.14.12 (b).

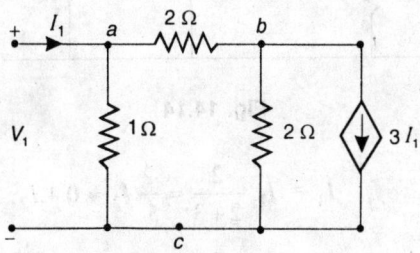

Fig. 14.12 (b)

Converting the delta network *abc* into star network, the resulting network is as shown in Fig.14.13, from which,

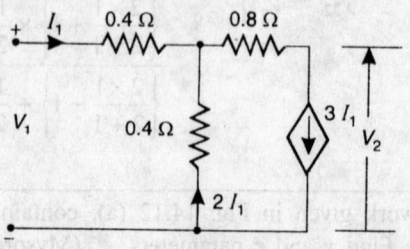

Fig. 14.13

$$V_1 - 0.4I_1 + 0.4(2I_1) = 0$$

or

$$\frac{V_1}{I_1} = -0.4\ \Omega = Z_{11}$$

Also

$$V_2 + 0.8(3I_1) + 0.4(2I_1) = 0$$

∴

$$\frac{V_2}{I_1} = -3.2\ \Omega = Z_{21}$$

$$Z_{12} = \frac{V_1}{I_2}\bigg|_{I_1 = 0}$$

and

$$Z_{22} = \frac{V_2}{I_2}\bigg|_{I_1 = 0}$$

When $I_1 = 0$, the resulting network is as shown in Fig.14.14.

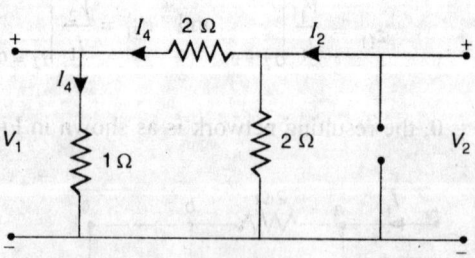

Fig. 14.14

$$I_4 = I_2\ \frac{2}{2+3} = \frac{2}{5}I_2 = 0.4\ I_2$$

$$V_1 = I_4 \times 1 = 0.4\ I_2$$

$$\therefore \qquad \frac{V_1}{I_2} = 0.4 \ \Omega = z_{12},$$

$$\frac{V_2}{I_2} = \frac{2 \times 3}{2 + 3} = 1.2 \ \Omega = z_{22}$$

y parameters

$$I_1 = y_{11} \ V_1 + y_{12} \ V_2$$

$$I_2 = y_{21} \ V_1 + y_{22} \ V_2$$

$$\therefore \qquad y_{11} = \left. \frac{I_1}{V_1} \right|_{V_2 = 0}$$

$$y_{21} = \left. \frac{I_2}{V_1} \right|_{V_2 = 0}$$

When $V_2 = 0$, the network in Fig. 14.12 (a), can be written as shown in Fig. 14.15.

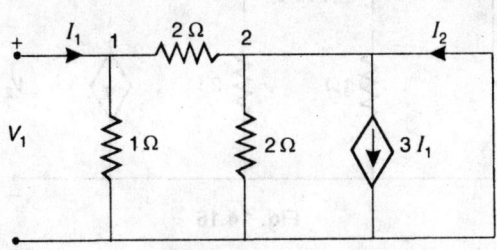

Fig. 14.15

At node 1, $\qquad I_1 - \dfrac{V_1}{1} - \dfrac{V_1 - V_2}{2} = 0$

But $\qquad V_2 = 0$

$$\therefore \qquad I_1 - V_1 - \frac{V_1}{2} = 0$$

i.e. $\qquad \dfrac{I_1}{V_1} = 1.5 \ \mho = y_{11}$

At node 2, $\qquad I_2 - 3I_1 - \dfrac{V_2}{2} + \dfrac{V_1 - V_2}{2} = 0$

But $\qquad V_2 = 0$

$$\therefore \qquad I_2 - 3I_1 + \frac{V_1}{2} = 0$$

i.e. $\qquad I_2 - 1.5\,V_1 + \dfrac{V_1}{2} = 0$

i.e. $\qquad \dfrac{I_2}{V_1} = 4\,\text{℧} = y_{21}$

$$y_{12} = \left.\frac{I_1}{V_2}\right|_{V_1 = 0}$$

and $\qquad y_{22} = \left.\dfrac{I_2}{V_2}\right|_{V_1 = 0}$

When $V_1 = 0$, the network in Fig. 14.12 (a) may be written as shown in Fig. 14.16

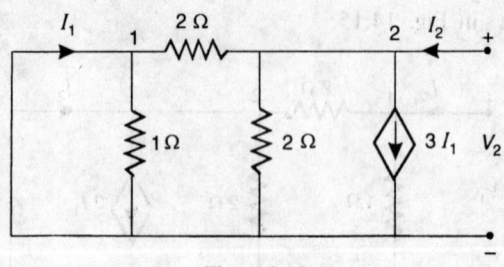

Fig. 14.16

At node 1, $\qquad I_1 - \dfrac{V_1}{1} - \dfrac{V_1 - V_2}{2} = 0$

But $\qquad V_1 = 0$

$$\therefore \qquad \frac{I_1}{V_2} = -0.5\,\text{℧} = y_{12}$$

At node 2, $\qquad I_1 - 3I_1 - \dfrac{V_2}{2} - \dfrac{V_2 - V_1}{2} = 0$

But $\qquad V_1 = 0$

$$\therefore \qquad \frac{I_2}{V_2} = -0.5\,\text{℧} = y_{22}$$

14.3 Find the z parameters for the network shown in Fig. 14.17, which contains a controlled voltage source. (Bangalore University).

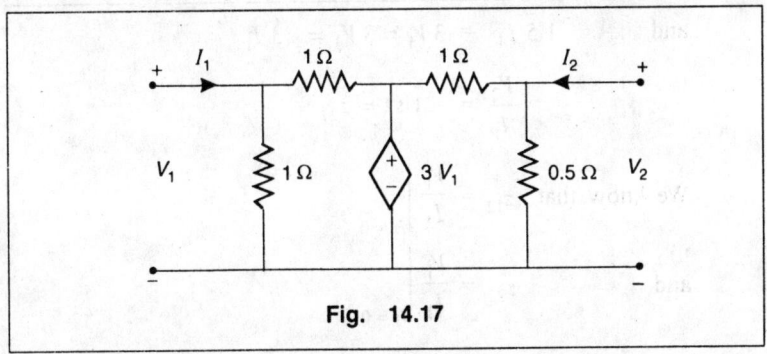

Fig. 14.17

Solution: We know that $z_{11} = \dfrac{V_1}{I_1}\bigg|_{I_2 = 0}$

and $\qquad\qquad z_{21} = \dfrac{V_2}{I_1}\bigg|_{I_2 = 0}$

When $I_2 = 0$, the network in Fig.14.17 may be written as shown in Fig. 14.18.

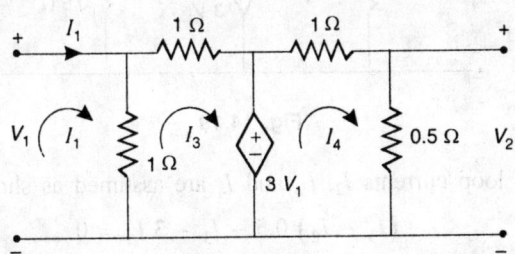

Fig. 14.18

$$V_1 = \left(I_1 - I_3\right) \times 1 \qquad\qquad (1)$$

and $\qquad\qquad -\left(I_3 - I_1\right) - I_3 - 3\,V_1 = 0$

or $\qquad\qquad I_3 = \dfrac{I_1 - 3\,V_1}{2} \qquad\qquad (2)$

Substituting (2) in (1), we get

$$V_1 = I_1 - \dfrac{I_1 - 3\,V_1}{2}$$

or $\qquad\qquad \dfrac{V_1}{I_1} = -1\,\Omega = z_{11}$

$$V_2 = 0.5\,I_4$$

and
$$1.5 I_4 = 3 V_1 = 3 V_2 = -3 I_1$$

$$\therefore \qquad \frac{V_2}{I_1} = -1 \, \Omega = z_{21}$$

We know that
$$z_{12} = \left. \frac{V_1}{I_2} \right|_{I_1 = 0}$$

and
$$z_{22} = \left. \frac{V_1}{I_2} \right|_{I_1 = 0}$$

When $I_1 = 0$, the network in Fig.14.17 may be written as shown in Fig. 14.19.

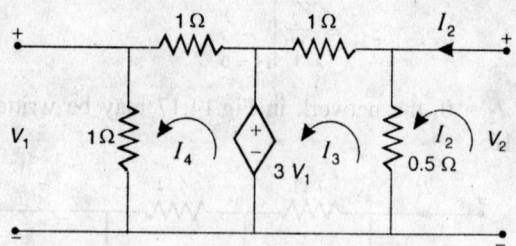

Fig. 14.19

The loop currents I_2, I_3, and I_4 are assumed as shown.

$$-\left(I_3 - I_2\right) 0.5 - I_3 - 3 V_1 = 0 \tag{1}$$

and
$$V_2 = \left(I_2 - I_3\right) 0.5$$

or
$$2 V_2 = I_2 - I_3 \tag{2}$$

$$V_1 = I_4 \tag{3}$$

and
$$1.5 I_4 = 3 V_1 \tag{4}$$

Equations (3) and (4) are satisfied only when, $V_1 = 0$ and $I_4 = 0$.

Equation (1) becomes, $I_2 = 3I_3$ and Eq. (2) becomess

$$2 V_2 = I_2 - \frac{I_2}{3}$$

or
$$V_2 = \frac{I_2}{3}$$

$$\therefore \qquad \frac{V_2}{I_2} = \frac{1}{3} \, \Omega = Z_{22}$$

$$z_{12} = \left. \frac{V_1}{I_2} \right|_{I_1 = 0} = 0$$

14.4 For the network shown in Fig. 14.20, find y and z parameters.

(Karnataka University)

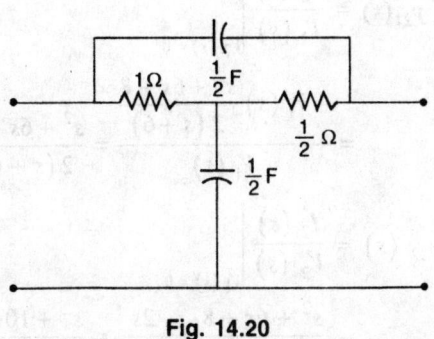

Fig. 14.20

Solution: The transformed network is as shown in Fig. 14.21.

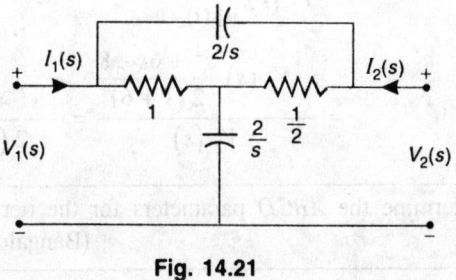

Fig. 14.21

Converting the star network into delta and simplifying, the network in Fig. 14.21 can be written as in Fig. 14.22.

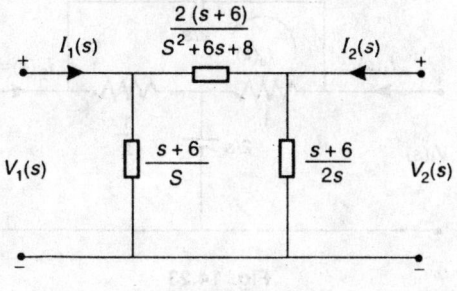

Fig. 14.22

$$y_{11}(s) = \frac{I_1(s)}{V_1(s)}\bigg|_{V_2(s)=0}$$

$$= \frac{s^2 + 6s + 8}{2(s+6)} + \frac{s}{s+6} = \frac{s^2 + 8s + 8}{2(s+6)}$$

$$y_{21}(s) = \frac{I_2(s)}{V_1(s)}\bigg|_{V_2(s)=0}$$

$$= \frac{-V_1(s)\dfrac{s^2 + 6s + .8}{2(s+6)}}{V_1(s)} = \frac{s^2 + 6s + 8}{2(s+6)}$$

$$y_{22}(s) = \frac{I_2(s)}{V_2(s)}\bigg|_{V_1(s)=0}$$

$$= \frac{s^2 + 6s + 8}{2(s+6)} + \frac{2s}{s+6} = \frac{s^2 + 10s + 8}{2(s+6)}$$

$$y_{12}(s) = \frac{I_1(s)}{V_2(s)}\bigg|_{V_1(s)=0}$$

$$= \frac{-V_2(s)\dfrac{s^2 + 6s + 8}{2(s+6)}}{V_2(s)} = -\frac{s^2 + 6s + 8}{2(s+6)}$$

14.5 Determine the *ABCD* parameters for the network shown in Fig. 14.20. (Bangalore University).

Solution: The transformed network for the network shown in Fig. 14.20, is as shown in Fig. 14.23.

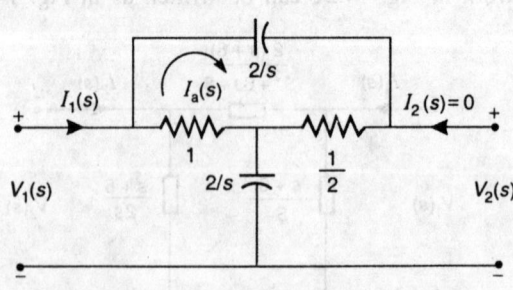

Fig. 14.23

We know that $V_1(s) = A V_2(s) - B I_2(s)$

$$I_1(s) = C V_2(s) - D I_2(s)$$

Assume the loop current $I_a(s)$ as shown.

(*i*) When $I_2(s) = 0$, $A = \dfrac{V_1(s)}{V_2(s)}\bigg|_{I_2(s)=0}$

$$V_1(s) - \{I_1(s) - I_a(s)\} 1 - \frac{2}{s} I_1(s) = 0$$

i.e. $V_1(s) = \left(1 + \dfrac{2}{s}\right) I_1(s) - I_a(s)$ (1)

Also $0 = \dfrac{2}{s} I_a(s) + \dfrac{1}{2} I_a(s) + 1\{ I_a(s) - I_1(s) \}$

$$= -I_1(s) + \left(1.5 + \frac{2}{s}\right) I_a(s)$$ (2)

Solving (1) and (2), we get

$$I_1(s) = \frac{s V_1(s)[1.5s + 2]}{0.5s^2 + 5s + 4} \quad \text{and} \quad I_a(s) = \frac{s^2 V_1(s)}{0.5s^2 + 5s + 4}$$

$$V_2(s) = \frac{1}{2} I_a(s) + \frac{2}{s} I_1(s)$$

$$= \frac{1}{2}\left[\frac{s^2 V_1(s)}{0.5s^2 + 5s + 4}\right] + \frac{2}{s}\left[\frac{s V_1(s)(1.5s + 2)}{0.5s^2 + 5s + 4}\right]$$

$$= V_1(s)\left[\frac{0.5s^2 + 3s + 4}{0.5s^2 + 5s + 4}\right]$$

$$\therefore \quad A = \frac{V_1(s)}{V_2(s)} = \frac{0.5s^2 + 5s + 4}{0.5s^2 + 3s + 4}$$

$$C = \frac{I_1(s)}{V_2(s)}\bigg|_{I_2(s)=0} = \frac{s(1.5s + 2)}{0.5s^2 + 3s + 4}$$

(*ii*) When $V_2(s) = 0$, $B = \dfrac{V_1(s)}{I_2(s)}$

Converting the star network into delta and simplifying, the network in Fig.14.23, may be written as in Fig. 14.24.

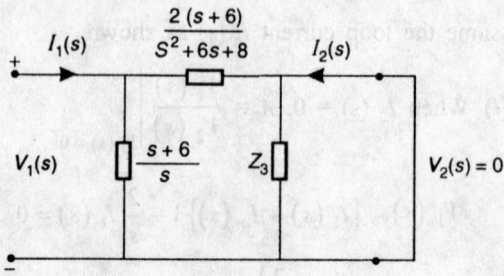

Fig. 14.24

As there is a short-circuit across z_3, it is neglected. Hence, there is no necessity to calculate z_3.

$$I_1(s) = V_1(s)\, Y(s) = V_1(s)\left[\frac{s}{s+6} + \frac{s^2 + 6s + 8}{2(s+6)}\right]$$

$$= V_1(s)\frac{s^2 + 8s + 8}{2(s+6)}$$

$$I_2(s) = -V_1(s)\left[\frac{s^2 + 6s + 8}{2(s+6)}\right]$$

$$\therefore \qquad B = -\frac{V_1(s)}{I_2(s)} = \frac{2(s+6)}{s^2 + 6s + 8}$$

$$D = -\frac{I_1(s)}{I_2(s)} = \frac{s^2 + 6s + 8}{s^2 + 8s + 8}$$

14.6 For the circuit shown in Fig. 14.25, determine the h parameters.
(Gulbarga University).

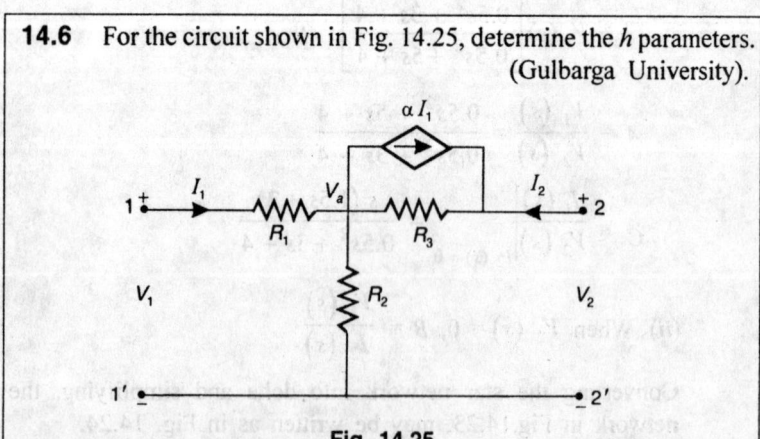

Fig. 14.25

Solution: We know that
$$V_1 = h_{11} I_1 + h_{12} V_2$$
$$I_2 = h_{21} I_1 + h_{22} V_2$$

(*i*) When $V_2 = 0$, port 2 is short circuited. Then

$$I_1 = \frac{V_a}{R_2} + \frac{V_a}{R_3} + \alpha I_1$$

i.e.
$$V_a = \frac{(1-\alpha) I_1 R_2 R_3}{R_2 + R_3}$$

$$V_1 - V_a = I_1 R_1$$

$\therefore \qquad V_a = V_1 - I_1 R_1$

$\therefore \qquad V_1 - I_1 R_1 = \dfrac{(1-\alpha) I_1 R_2 R_3}{R_2 + R_3}$

$$h_{11} = \frac{V_1}{I_1}\bigg|_{V_2 = 0} = R_1 + \frac{(1-a) R_2 R_3}{R_2 + R_3}$$

Also
$$I_2 + \alpha I_1 = \frac{V_a}{R_3} = \frac{(1-\alpha) I_1 R_2}{R_2 + R_3}$$

$$h_{21} = \frac{I_2}{I_1}\bigg|_{V_2 = 0} = \frac{(1-\alpha) R_2 R_3}{R_2 + R_3} - \alpha$$

(*ii*) When $I_1 = 0$, $V_a = V_1$

$\therefore \qquad \dfrac{V_1}{R_2} = \dfrac{V_2 - V_1}{R_3}$

i.e.
$$\frac{V_1}{V_2} = \frac{R_2}{R_2 + R_3} = h_{12}$$

Also
$$V_2 = I_2 (R_2 + R_3)$$

i.e.
$$\frac{I_2}{V_2} = \frac{1}{R_2 + R_3} = h_{22}$$

14.7 The network shown in Fig. 14.26 consists of a T network connected in parallel with a π network. Find the y parameters.

(Karnataka University)

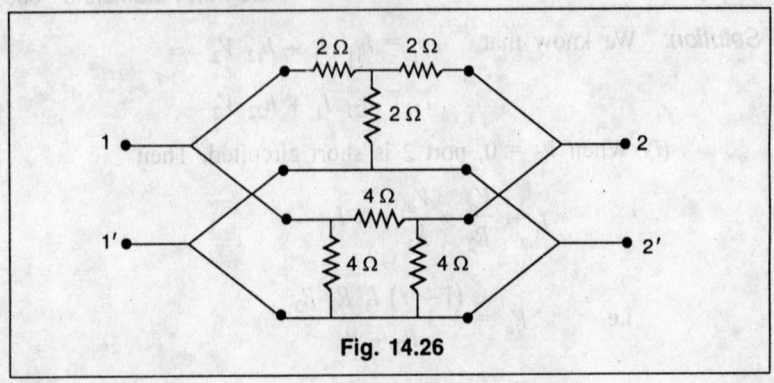

Fig. 14.26

Solution: Let the *a* network be as shown in Fig. 14.27. After $Y - \Delta$ conversion, the network is as shown in Fig. 14.28.

We get
$$y_{11a} = y_{22a} = \frac{1}{3} \mho$$
$$y_{12a} = y_{21a} = -6 \mho$$

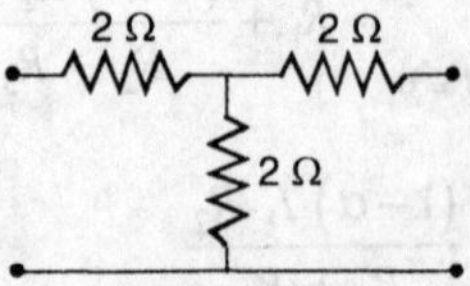

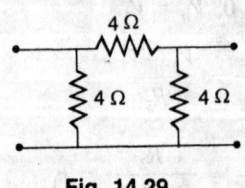

Fig. 14.27 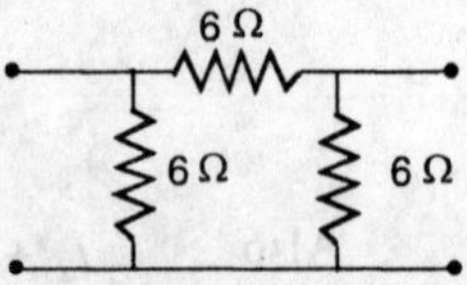   **Fig. 14.28**

Let the *b* network be as shown in Fig. 14.29.

Fig. 14.29

We get
$$y_{11b} = y_{22b} = \frac{1}{2} \mho$$
$$y_{12b} = y_{21b} = -4 \mho$$

The *y* parameters for the complete network are given by

$$y_{11} = y_{11a} + y_{11b} = \frac{1}{3} + \frac{1}{2} = \frac{5}{6} \mho = y_{22}$$

and
$$y_{12} = y_{12a} + y_{12b} = -6 - 4 = -10 \mho = y_{21}$$

14.8 The following results were obtained on measurements made on a two-port passive network (*i*) with port 2 short circuited, a voltage of 8 volts applied to port 1, results in $I_1 = 1.6$ A and $I_2 = -0.8$ A. (*ii*) with port 1 short circuited, a voltage of 10 V, applied to port 2, results in $I_1 = -1$ A and $I_2 = 1.2$ A. Find *y* parameters of the network and write the *y* parameter Eqs. Compute the voltage across 8 Ω resistor connected across port 2.   (Mangalore University)

Solution: We know that

$$I_1 = y_{11} V_1 + y_{12} V_2$$
$$I_2 = y_{21} V_1 + y_{22} V_2$$

(*i*) When $V_2 = 0$

$$y_{11} = \frac{I_1}{V_1} = \frac{1.6}{8} = 0.2 \; \mho$$

and

$$y_{21} = \frac{I_2}{V_1} = -\frac{0.8}{8} = -0.1 \; \mho$$

(*ii*) When $V_1 = 0$

$$y_{12} = \frac{I_1}{V_2} = -\frac{1}{10} = -0.1 \; \mho$$

and

$$y_{22} = \frac{I_2}{V_2} = \frac{1.2}{10} = 0.12 \; \mho$$

The *y* parameter Eqs are

$$I_1 = 0.2 V_1 - 0.1 V_2$$

$$I_2 = -0.1 V_1 + 0.12 V_2$$

When 8 Ω resistor is connected across port 2 and 8 volts is connected across port 1

$$V_1 = 8 \text{ V}$$

and

$$V_2 = -8 I_2 = -8 (-0.1 V_1 + 0.12 V_2)$$
$$= -8 (-0.1 \times 8 + 0.12 V_2)$$
$$= 6.4 - 0.96 V_2$$

∴

$$V_2 = 3.265 \text{ V}$$

14.9 The port currents of a two-port network are given by

$$I_1 = 2.5 V_1 - V_2$$
$$I_2 = -V_1 + 5 V_2$$

Find the equivalent π network.

Solution: From the given Eqs

$$y_{11} = 2.5 \, \mho$$

$$y_{12} = -1 \, \mho$$

$$y_{21} = -1 \, \mho$$

and

$$y_{22} = 5 \, \mho$$

We know that $Y_2 = -y_{21} = 1 \, \mho$

$$Y_1 + Y_2 = y_{11} = 2.5 \, \mho$$

∴ $$Y_1 = 1.5 \, \mho$$

$$Y_2 + Y_3 = y_{22} = 5 \, \mho$$

∴ $$y_3 = 4 \, \mho$$

The equivalent π network is as shown in Fig. 14.30.

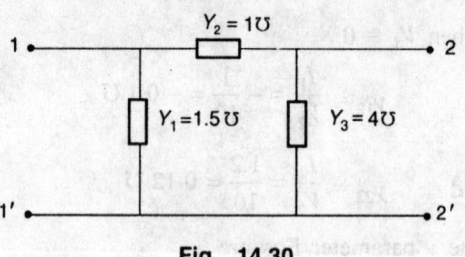

Fig. 14.30

14.10 The impedance parameters of a T network are given by $\begin{bmatrix} 50 & 25 \\ 25 & 100 \end{bmatrix}$. Find the parameters of the network.

(Karnataka University).

Solution: Given $z_{11} = 50 \, \Omega$

$$z_{12} = 25 \, \Omega = z_{21}$$

$$z_{22} = 100 \, \Omega$$

The network elements are:

$$Z_c = z_{12} = 25 \, \Omega$$

$$Z_a + Z_c = z_{11} = 50 \, \Omega$$

$$Z_a = 50 - 25 = 25 \, \Omega$$

$$Z_b + Z_c = z_{22} = 100 \,\Omega$$

$$\therefore \qquad Z_b = 100 - 25 = 75 \,\Omega$$

The network is as shown in Fig.14.31.

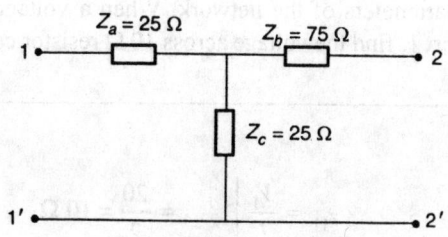

Fig. 14.31

14.11 A two-port network has the following short-circuit and open-circuit impedances. $Z_{oc1} = 900 \,\Omega$, $Z_{oc2} = 1000 \,\Omega$ and $Z_{sc1} = 650 \,\Omega$. Determine T section parameters to represent the two-port network.
(Mysore University)

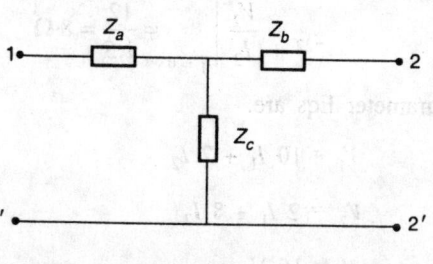

Fig. 14.32

Solution: Z_{oc1} = open-circuit impedance of the network, when the second port is open circuited = $Z_a + Z_c = 900 \,\Omega$

Similarly $\qquad Z_{oc2} = Z_b + Z_c = 1000 \,\Omega$

Z_{sc1} = short-circuit impedance of the network, when port 2 is

short-circuited = $Z_a + \dfrac{Z_b\, Z_c}{Z_b + Z_c} = 650 \,\Omega$

Solving the above eqs, we get $Z_c = 500 \,\Omega$

$$Z_a = 400 \,\Omega$$

and $\qquad\qquad Z_b = 500 \,\Omega$

14.12 The following results were obtained, when measurements were made on a two-port resistive network. (*i*) when port 2 is open circuited, a voltage of 20 V applied at port 1 results in $I_1 = 2$ A and $V_2 = 4$ volts, (*ii*) with port 1 open circuited, a voltage of 12 volts applied at port 2 results in $I_2 = 1.5$ A and $V_1 = 3$ volts. Find the *z* parameters of the network. When a voltage of 15 volts is applied to port 1, find the voltage across 10 Ω resistor connected across port 2.

Solution

$$z_{11} = \frac{V_1}{I_1}\bigg|_{I_2 = 0} = \frac{20}{2} = 10\ \Omega$$

$$z_{21} = \frac{V_2}{I_1}\bigg|_{I_2 = 0} = \frac{4}{2} = 2\ \Omega$$

$$z_{12} = \frac{V_1}{I_2}\bigg|_{I_1 = 0} = \frac{3}{1.5} = 2\ \Omega$$

$$z_{22} = \frac{V_2}{I_2}\bigg|_{I_1 = 0} = \frac{12}{1.5} = 8\ \Omega$$

The *z*-parameter Eqs are:

$$V_1 = 10\ I_1 + 2\ I_2$$

$$V_2 = 2\ I_1 + 8\ I_2$$

When $V_1 = 15$ V

$$V_{10\Omega} = -10\ I_2 = V_2$$

Substituting these values in the above Eqs and simplifying, we get

$$I_2 = -\frac{15}{7}\ \text{A}$$

$$V_{10\Omega} = 10 \times \frac{15}{7} = \frac{150}{7}\ \text{volts}$$

14.13 The network given in Fig. 14.33 represents two *T* networks connected in cascade. Find the transmission parameters of the network. Determine the input impedance, when the output terminals are shorted.

(Mangalore University).

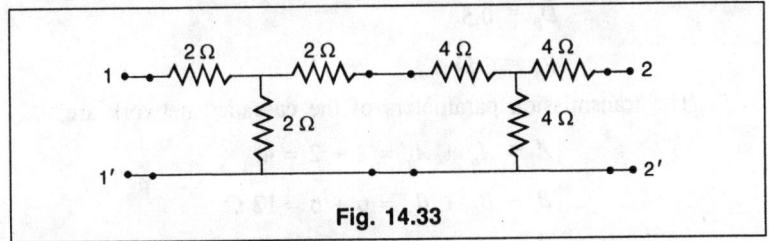

Fig. 14.33

Solution: Let the a network be as shown in Fig. 14.34.

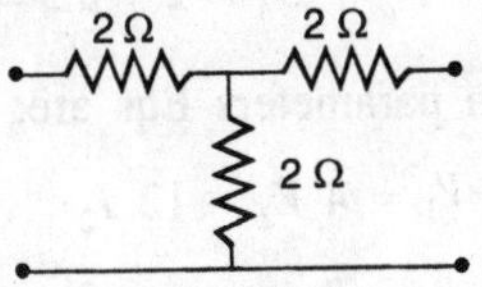

$$2\,\Omega \quad 2\,\Omega$$
$$2\,\Omega$$

Fig. 5.34

$$A_a = \left.\frac{V_1}{V_2}\right|_{I_2 = 0} = 2$$

$$C_a = \left.\frac{I_1}{V_2}\right|_{I_2 = 0} = \frac{1}{2}\,\mho$$

$$D_a = -\left.\frac{I_1}{I_2}\right|_{V_2 = 0} = 2$$

$$B_a = -\left.\frac{V_1}{I_2}\right|_{V_2 = 0} = 6\,\Omega$$

Let the b network be as shown in Fig. 14.35.

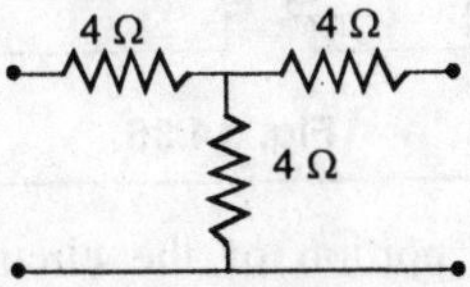

$$4\,\Omega \quad 4\,\Omega$$
$$4\,\Omega$$

Fig. 14.35

We can show that

$$A_b = 2$$

$$C_b = \frac{1}{4}\,\mho$$

$$D_b = 0.5$$

$$B_b = 6 \ \Omega$$

The transmission parameters of the cascaded network are:

$$A = A_a + A_b = 2 + 2 = 4$$

$$B = B_a + B_b = 6 + 6 = 12 \ \Omega$$

$$C = C_a + C_b = \frac{1}{2} + \frac{1}{4} = \frac{3}{4} \ \mho$$

$$D = D_a + D_b = 2 + 0.5 = 2.5$$

The transmission parameters Eqs are:

$$V_1 = 4 \ V_2 - 12 \ I_2$$

$$I_1 = \frac{3}{4} \ V_2 - 2.5 \ I_2$$

When $\qquad V_2 = 0$

$$Z_i = \frac{V_1}{I_1} = \frac{-12 \ I_2}{-2.5 \ I_2} = 4.8 \ \Omega$$

14.14 For the network shown in Fig. 14.36, find the z-parameters.
(Gulbarga University)

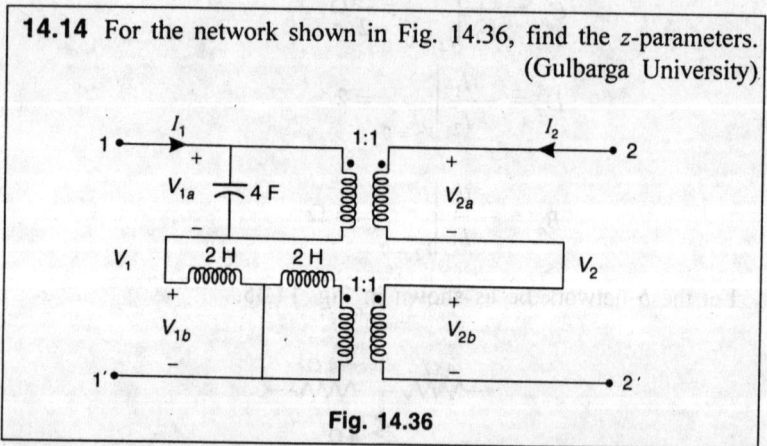

Fig. 14.36

Solution: For the top portion of the circuit, as the turns ratio is 1 : 1, the primary and secondary currents and voltages are equal.

$$V_{1a}(s) = \frac{1}{4s} \left[I_1(s) + I_2(s) \right] = \frac{1}{4s} \ I_1(s) + \frac{1}{4s} \ I_2(s) \qquad (1)$$

$$V_{2a}(s) = V_{1a}(s) = \frac{1}{4s} \ I_1(s) + \frac{1}{4s} \ I_2(s) \qquad (2)$$

For the bottom portion of the circuit

$$V_{1b}(s) = 2s\, I_1(s) + 0\, I_2(s) \tag{3}$$

$$V_{2b}(s) = 0\, I_1(s) + 2s\, I_2(s) \tag{4}$$

$$V_1(s) = V_{1a}(s) + V_{1b}(s)$$

$$= \left(2s + \frac{1}{4s}\right) I_1(s) + \frac{1}{4s} I_2(s)$$

$$V_2(s) = V_{2a}(s) + V_{2b}(s)$$

$$= \frac{1}{4s} I_1(s) + \left(2s + \frac{1}{4s}\right) I_2(s)$$

$$\therefore \qquad z_{11} = 2s + \frac{1}{4s}, \quad z_{12} = \frac{1}{4s}, \quad z_{21} = \frac{1}{4s}$$

and $\qquad z_{22} = 2s + \dfrac{1}{4s}$

As $\qquad z_{11} = z_{22}$, the network is symmetrical.

$\qquad z_{12} = z_{21}$, the network is reciprocal.

14.15 The Fig.14.37 shows two two-port networks connected in parallel. One two-port contains only a gyrator and the other is a resistive network containing a single controlled source. For this network, determine the y parameters. (Mysore University)

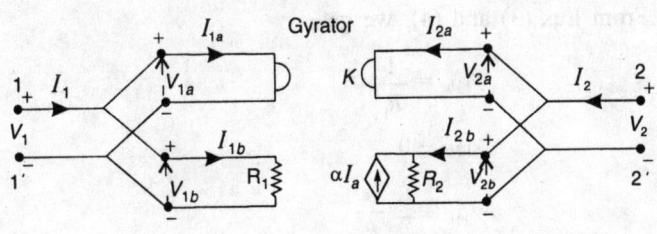

Fig. 14.37

Solution: For the gyrator network, we know that

$$V_1 = kI_2$$

i.e. $\qquad V_{1a} = kI_{2a}$

and $\qquad V_2 = -kI_1$

i.e. $\qquad V_{2a} = kI_{1a}$

i.e.
$$I_{1a} = 0\,V_{1a} - \left(\frac{1}{k}\right)V_{2a} = y_{11a}\,V_{1a} + y_{12a}\,V_{2a} \quad (1)$$

and
$$I_{2a} = \frac{1}{k}\,V_{1a} + 0\,V_{2a} = y_{21a}\,V_{1a} + y_{22a}\,V_{2a} \quad (2)$$

From the Eqs (1) and (2), we get
$$y_{11a} = 0$$
$$y_{12a} = -\frac{1}{k}$$
$$y_{21a} = \frac{1}{k}$$

and
$$y_{22a} = 0$$

For the lower part of the network
$$V_{1b} = I_a\,R_1 = I_{1b}\,R_1$$

i.e.
$$I_{1b} = \frac{1}{R_1}\,V_{1b} + 0\,V_{2b} = y_{11b}\,V_{1b} + y_{12b}\,V_{2b} \quad (3)$$

$$V_{2b} = \left(\alpha\,I_a + I_{2b}\right)R_2 = \left(\alpha\,I_{1b} + I_{2b}\right)R_2$$

i.e.
$$I_{2b}\,R_2 = V_{2b} - \alpha\,R_2\,I_{1b} = V_{2b} - \alpha\,R_2\,\frac{V_{1b}}{R_1}$$

i.e.
$$I_{2b} = -\frac{\alpha}{R_1}\,V_{1b} + \frac{1}{R_2}\,V_{2b} = y_{21b}\,V_{1b} + y_{22b}\,V_{2b} \quad (4)$$

From Eqs (3) and (4), we get
$$y_{11b} = \frac{1}{R_1}$$
$$y_{12b} = 0$$
$$y_{21b} = -\frac{\alpha}{R_1}$$

and
$$y_{22b} = \frac{1}{R_2}$$

For the entire network
$$y_{11} = y_{11a} + y_{11b} = \frac{1}{R_1}$$
$$y_{12} = y_{12a} + y_{12b} = -\frac{1}{k}$$

$$y_{21} = y_{21a} + y_{21b} = \frac{1}{k} - \frac{\alpha}{R_1}$$

and

$$y_{22} = y_{22a} + y_{22b} = \frac{1}{R_2}$$

NUMERICAL PROBLEMS

14.1 Obtain z parameters for the two-port network shown in Fig. 14.38.

(Bangalore University)

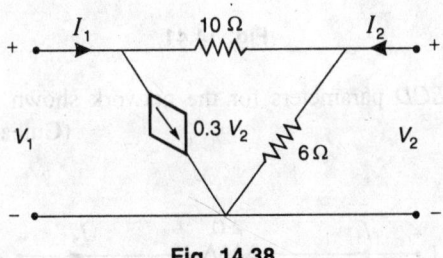

Fig. 14.38

14.2 Find the y parameters for the network shown in Fig. 14.39.

(Karnataka University)

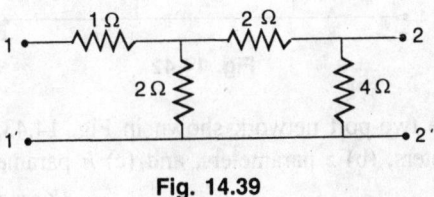

Fig. 14.39

14.3 Find z parameters for the network shown in Fig. 14.40.

(Mysore University)

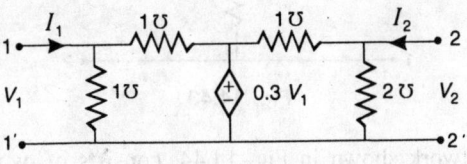

Fig. 14.40

14.4 Find y parameters for the network shown in Fig. 14.41

(Kuvempu University)

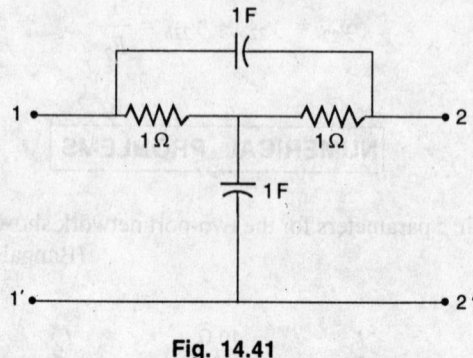

Fig. 14.41

14.5 Find *ABCD* parameters for the network shown in Fig. 14.42.

(Gulbarga University)

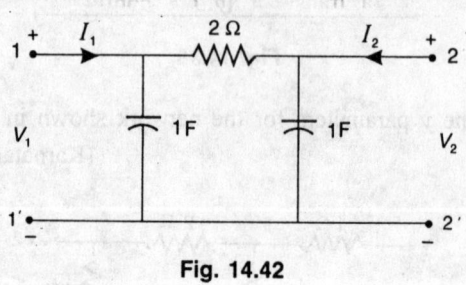

Fig. 14.42

14.6 For the two-port network shown in Fig. 14.43, find (a) *ABCD* parameters, (b) z parameters, and (c) h parameters.

(Karnataka University)

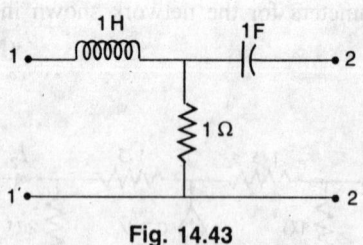

Fig. 14.43

14.7 The network shown in Fig. 14.44, consists of two networks in parallel. Find the y parameters. (Mysore University)

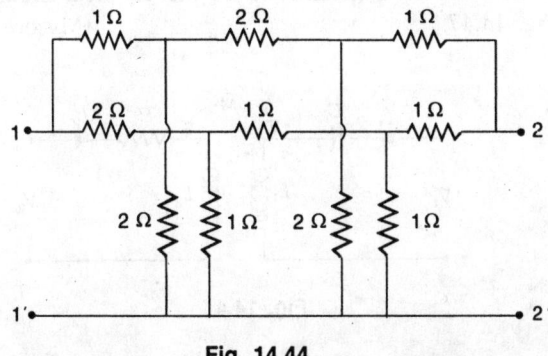

Fig. 14.44

14.8 The *h* parameters of a certain two-port network are $h_{11} = 1\Omega$, $h_{12} = 2$, $h_{21} = -2$ and $h_{22} = 1\mho$. Find (a) *z* parameters, (b) *y* parameters, and (c) *ABCD* parameters. Find whether the network is (*i*) reciprocal, (*ii*) symmetrical. (Bangalore University)

14.9 The model of a transistor in *CE* configuration is shown in Fig. 14.45. Determine its *h* parameters.

(Bangalore University)

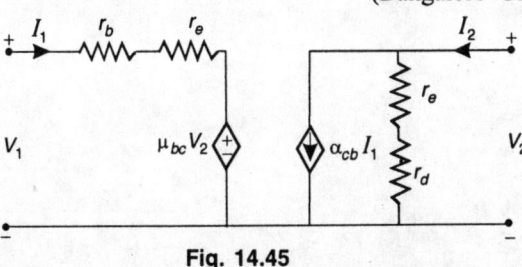

Fig. 14.45

14.10 Determine the *y* and *z* parameters for the circuit shown in Fig. 14.46. (Mangalore University)

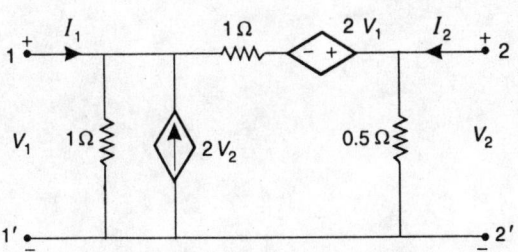

Fig. 14.46

14.11 Determine the z parameters of the coupled circuit shown in Fig. 14.47. (Mysore University)

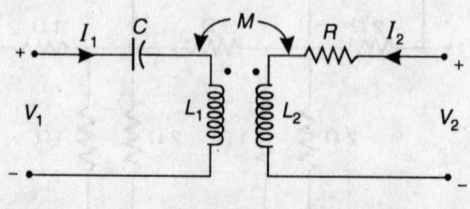

Fig. 14.47

15

Network Synthesis

15.1 INTRODUCTION

The solutions to problems on network theory are of two distinct types.
(1) Network analysis and (2) Network synthesis. Network analysis deals
with the problem of determining the response, when the network is
energised by a particular input. Network synthesis deals with the
realisation of the network for a particular response. In network analysis,
finding the response is simple and the response is generally specific. On
the otherhand, in network synthesis, realisation of the network is not so
simple and the network realised is not specific. There may be more than
one realisable networks, for a given response. Sometimes, it may not be
possible to realise a network for a given response. In this chapter, only
passive network synthesis of one port network is discussed, that too,
restricting to the synthesis of driving point immittance (impedance or
admittance) functions, containing only two elements.

The network described by an immittance function $F(s) = \dfrac{P(s)}{Q(s)}$ can
be realised only if, (1) $F(s)$ is a positive real function and (2) $P(s)$ and
$Q(s)$ are Hurwitz Polynomials.

The immittance function $F(s)$ may be either an impedance function
$Z(s)$ or an admittance function $Y(s)$.

15.2 POSITIVE REAL FUNCTION (PRF)

A passive network is one which does not contain any dependent or
controlled energy sources. Let the driving point impedance of a passive
network be given by

$$F(s) = \frac{P(s)}{Q(s)} = \frac{a_0 s^n + a_1 s^{n-1} + a_2 s^{n-2} + \cdots + a_{n-1} s + a_n}{b_0 s^m + b_1 s^{m-1} + b_2 s^{m-2} + \cdots + b_{m-1} s + b_m} \quad (15.1)$$

Where $a_0, a_1, a_2, \ldots a_{n-1}, a_n$ and $b_0, b_1, b_2, \ldots b_{m-1}, b_m$ are constant coefficients of the polynomials $P(s)$ and $Q(s)$ respectively.

If the function $F(s)$ as given in Eq. (15.1) is to be a positive real function, the following conditions are required to be satisfied.

1. The coefficients of both $P(s)$ and $Q(s)$, i.e. $a_0, a_1, a_2, \ldots a_{n-1}, a_n$, and $b_0, b_1, b_2, \ldots b_{m-1}, b_m$ must be real and positive.

2. The degrees of $P(s)$ and $Q(s)$ in s should not differ by more than one, i.e. $m - n \not> 1$ and $n - m \not> 1$.

3. Equation (15.1) may be factorised and written as

$$F(s) = H \frac{(s - z_1)(s - z_2) \cdots (s - z_n)}{(s - p_1)(s - p_2) \ldots (s - p_m)} \tag{15.2}$$

Where, $H = \dfrac{a_0}{b_0}$ is a constant known as the *scale factor*.

$z_1, z_2 \ldots z_n$ and $p_1, p_2, \ldots p_m$ are the complex frequencies, known as *zeros* and *poles* of the function $F(s)$, respectively. The real parts of the zeros and poles must be either $-$ve or 0. If the zero or pole of $F(s)$ is complex, then that zero or pole must have a complex conjugate.

4. If there are poles lying on the imaginary axis, they must be simple.

5. The function $F(s)$ should not have multiple zeros or poles at the origin. In other words, after factorisation of $Z(s)$, neither $P(s)$ nor $Q(s)$ must contain terms like s^2, s^3 ...., etc.

6. After the partial fraction expansion of $F(s)$, the residues of all poles must be real and positive.

WORKED EXAMPLE

15.1 Examine the following network functions and state whether they are positive real functions or not.

1. $$Z(s) = \frac{4s^4 - 8s^3 + 2s^2 + 5s + 5}{3s^3 + 5s^2 + 4}$$

As the coefficient of s^3 in the numerator of $Z(s)$ is $-$ve, it is not PRF.

2. $$Z(s) = \frac{4s^4 + 8s^3 + 2s^2 + 5s + 5}{3s^3 - 5s^2 + 4}$$

As the coefficient of s^2 in the denominator of $Z(s)$ is $-$ve, it is not PRF.

3. $$Y(s) = \frac{s^2 + 4s + 8}{s^4 + 8s^3 + 3s^2 + 9}$$

$m - n > 1$, Hence, $Y(s)$ is not PRF.

4. $$Y(s) = \frac{s^3 + 3s^2 + 7}{s + 8}$$

$n - m > 1$, Hence, $Y(s)$ is not PRF.

5. $$Z(s) = \frac{2s^2 + 9}{s\left(s^2 + 3\right)}$$

On partial fraction expansion, $Z(s)$ may be written as

$$Z(s) = \frac{3}{s} + \frac{1/2}{\left(s + j\sqrt{3}\right)} + \frac{-1/2}{\left(s - j\sqrt{3}\right)}$$

One of the residues of the poles is $-$ve. Hence, the function is not PRF.

6. $$Z(s) = \frac{3s^2 + 9s + 5}{s\left(s + 2\right)\left(s + 3\right)}$$

On partial fraction expansion, $Z(s)$ may be written as

$$Z(s) = \frac{5/6}{s} + \frac{1/2}{s + 2} + \frac{5/3}{s + 3}$$

The function $Z(s)$ satisfies all the conditions and hence, it is a PRF.

15.3 HURWITZ POLYNOMIALS

Let the polynomial $P(s)$ be written as

$$P(s) = a_0 s^n + a_1 s^{n-1} + a_2 s^{n-2} + \ldots + a_{n-1} s + a_n$$

$P(s)$ is said to be an Hurwitz polynomial, if the following conditions are satisfied.

1. All the coefficients $a_0, a_1, a_2, \ldots a_{n-1}, a_n$ must be real and positive.

2. There should not be any missing terms between the highest degree term, i.e. s^n and the lowest degree term, i.e. s^0 in the Eq. of the polynomial $P(s)$. But however, all the even degree terms or the odd degree terms may be absent.

3. The real parts of the roots of $P(s)$ must be –ve or 0.

4. If the polynomial $P(s)$ consists of both even degree and odd degree terms in s, then $P(s)$ can be written as

$$P(s) = O(s) + E(s) \tag{15.3}$$

Where, $O(s)$ is the group of odd degree terms and $E(s)$ is the group of even degree terms. If $P(s)$ is to be Hurwitz, then the roots of $O(s)$ and $E(s)$ must lie on $j\omega$ axis.

5. If $P(s)$ has to be Hurwitz, then all the quotients of the continued fraction expansion of $\dfrac{O(s)}{E(s)}$ or $\dfrac{E(s)}{O(s)}$ must be positive. (The group of the terms having highest degree is written as numerator).

WORKED EXAMPLE

15.2 Examine the following functions and say whether, they are Hurwitz polynomials or not.

(i) $\qquad P(s) = s^4 + 3s^3 + 2s^2 + 7s + 9$

$$\frac{E(s)}{O(s)} = \frac{s^4 + 2s^2 + 9}{3s^3 + 7s}$$

The continued fraction expansion of $\dfrac{E(s)}{O(s)}$ is as shown.

$$
3s^3 + 7s \overline{)s^4 + 2s^2 + 9} \left(\frac{1}{3}s \right.
$$

$$\underline{s^4 + \frac{7}{3}s^2}$$

$$-\frac{1}{3}s^2 + 9 \overline{)3s^3 + 7s} \left(-9s \right.$$

$$\underline{3s^3 - 81s}$$

$$88s \overline{)-\frac{1}{3}s^2 + 9} \left(-\frac{1}{88 \times 3}s \right.$$

$$\underline{-\frac{1}{3}s^2}$$

$$9 \overline{)88s} \left(\frac{88}{9}s \right.$$

$$\underline{88s}$$

$$0$$

The quotients obtained are; $\dfrac{1}{3} s, -9s, -\dfrac{1}{88 \times 3} s,$ and $\dfrac{88}{9} s.$

Two quotient terms are $-$ve. Hence, $P(s)$ is not Hurwitz.

From the above continued fraction expansion, $\dfrac{E(s)}{O(s)}$ can also be written as

$$\frac{E(s)}{O(s)} = \frac{1}{3} s + \cfrac{1}{-9s + \cfrac{1}{-\cfrac{s}{88 \times 3} + \cfrac{1}{\cfrac{88}{9} s}}}$$

(ii) $\qquad P(s) = s^4 + s^3 + 6s^2 + 3s + 4$

$$\frac{E(s)}{O(s)} = \frac{s^4 + 6s^2 + 4}{s^3 + 3s},$$

The continued fraction expansion of $P(s)$ is as shown.

$$
s^3 + 3s \overline{\smash{\big)}\ s^4 + 6s^2 + 4}\ \big(s \\
\underline{s^4 + 3s^2}
$$

$$
3s^2 + 4 \overline{\smash{\big)}\ s^3 + 3s}\ \left(\frac{1}{3} s\right) \\
\underline{s^3 + \frac{4}{3} s}
$$

$$
\frac{5}{3} s \overline{\smash{\big)}\ 3s^2 + 4}\ \left(\frac{9}{5} s\right) \\
\underline{3s^2}
$$

$$
4 \overline{\smash{\big)}\ \frac{5}{3} s}\ \left(\frac{5}{12} s\right) \\
\underline{\frac{5}{3} s} \\
0
$$

The quotients of the continued fraction expansion are

$$s, \frac{1}{3} s, \frac{9}{5} s, \text{ and } \frac{5}{12} s$$

All the quotients are positive. Hence, $P(s)$ is Hurwitz.

From the above continued fraction expansion, $\dfrac{E(s)}{O(s)}$ can also be written as

$$\frac{E(s)}{O(s)} = s + \cfrac{1}{\cfrac{1}{3}s + \cfrac{1}{\cfrac{9}{5}s + \cfrac{1}{\cfrac{5}{12}s}}}$$

(iii) $\qquad\qquad P(s) = 2s^4 + s^2 + 5$

$P(s)$ consists of only even terms. Odd terms are completely missing. Hence, $P(s)$ cannot be considered as a ratio of $\dfrac{E(s)}{O(s)}$. In such cases, the ratio $\dfrac{P(s)}{P'(s)}$ is considered.

$P'(s) = 8s^3 + 2s$. The continued fraction expansion of $\dfrac{P(s)}{P'(s)}$ is as shown below:

$$
8s^3 + 2s \overline{\smash{\big)}\, 2s^4 + s^2 + 5 \left(\frac{1}{4}s \right.}
$$
$$
\underline{2s^4 + \frac{1}{2}s^2}
$$
$$
\frac{1}{2}s^2 + 5 \overline{\smash{\big)}\, 8s^3 + 2s \left(16s \right.}
$$
$$
\underline{8s^3 + 80s}
$$
$$
-72s \overline{\smash{\big)}\, \frac{1}{2}s^2 + 5 \left(-\frac{1}{144}s \right.}
$$
$$
\underline{\frac{1}{2}s^2}
$$
$$
5 \overline{\smash{\big)}\, -72s \left(-\frac{72}{5}s \right.}
$$
$$
\underline{-72s}
$$
$$
0
$$

Two quotients of $\dfrac{P(s)}{P'(s)}$ are not positive. Hence, $P(s)$ is not Hurwitz.

(iv) $\qquad P(s) = s^6 + 2s^5 + 6s^4 + 10s^3 + 9s^2 + 8s + 4$

$$\frac{E(s)}{O(s)} = \frac{s^6 + 6s^4 + 9s^2 + 4}{2s^5 + 10s^3 + 8s}$$

The continued fraction expansion of $P(s)$ is as shown below.

$$2s^5 + 10s^3 + 8s \overline{\smash{\big)}\, s^6 + 6s^4 + 9s^2 + 4} \left(\frac{1}{2}s \right.$$
$$\underline{s^6 + 5s^4 + 4s^2}$$

$$s^4 + 5s^2 + 4 \overline{\smash{\big)}\, 2s^5 + 10s^3 + 8s} \left(2s \right.$$
$$\underline{2s^5 + 10s^3 + 8s}$$
$$0 + 0 + 0$$

In this particular problem, the continued fraction expansion of $P(s)$ has ended abruptly. For testing the polynomial $P(s)$ to be Hurwitz, it is necessary that the number of quotients in the continued fraction expansion should be equal to the highest degree of s in $P(s)$. In the given polynomial, the highest degree of s is 6, but the number of quotients obtained in the continued fraction expansion is only 2. Hence, the normal procedure explained earlier cannot be adopted to test whether $P(s)$ is Hurwitz or not. In such a case, the following procedure is adopted.

The given polynomial is written as a product of two functions. Each factor is tested for Hurwitz *property*. If both the functions are Hurwitz, then $P(s)$ is also Hurwitz.

Consider the same polynomial $P(s)$ given by

$$P(s) = s^6 + 2s^5 + 6s^4 + 10s^3 + 9s^2 + 8s + 4$$

Now $P(s)$ is written as a product of two functions.

$$P(s) = A(s) B(s)$$

One of the factors of $P(s)$, i.e. $A(s)$ is given as

$$A(s) = s^4 + 5s^2 + 4$$

$B(s)$ is obtained by

$$B(s) = \frac{P(s)}{A(s)}$$

$$
s^4 + 5s^2 + 4 \overline{\Big)} \begin{array}{l} s^6 + 2s^5 + 6s^4 + 10s^3 + 9s^2 + 8s + 4 \\ \underline{s^6 + 0 \;\; + 5s^4 + 0 \;\;\;\;\; + 4s^2} \end{array} \Big(s^2 + 2s + 1
$$

$$
\begin{array}{l}
2s^5 + s^4 + 10s^3 + 5s^2 + 8s + 4 \\
\underline{2s^5 + 0 \;\; + 10s^3 + 0 \;\; + 8s} \\
\quad\quad s^4 + 5s^2 + 4 \\
\quad\quad \underline{s^4 + 5s^2 + 4} \\
\quad\quad\quad 0 \;\; + 0 \;\;\; + 0
\end{array}
$$

$$
\therefore \quad\quad B(s) = s^2 + 2s + 1
$$

On testing, both $A(s)$ and $B(s)$ are found to be Hurwitz. Hence, $P(s)$ is also Hurwitz.

15.4 SYNTHESIS OF NETWORKS

By now, we are familiar with the properties of positive real functions and Hurwitz polynomials. The network can be realised only when the given function, which may be either a driving point impedance function $Z(s)$ or a driving point admittance function $Y(s)$, is a PRF and a Hurwitz polynomial. We restrict ourselves to the realisation of one-port networks containing only two elements. During the process of network synthesis, it is possible to realise the network elements and the way in which they are connected to form the network. The realised network for the given immittance function is not unique. For the same function, different networks may be realised, because of the different methods used for the realisation of the networks.

An immittance function $Z(s)$ or $Y(s)$ must have equal number of zeros and poles. The poles at the origin or infinity are called *external poles*, and the other finite poles are called *internal poles*.

Total number of poles = number of internal poles + number of *external* poles, i.e. $p = p_i + p_e$.

III^{ly} Total number of zeros = number of internal zeros + number of external zeros, i.e. $z = z_i + z_e$.

By observing the given immittance function, it is possible to identify the internal and external poles and zeros. The external poles and zeros decide the frequency response characteristic of the network.

The fundamental step involved in the synthesis of a driving point immittance function is to break up the given function into a sum of simpler functions.

For example, the driving point impedance function $Z(s)$ is written as a sum of simpler functions $Z_1(s)$, $Z_2(s)$, $Z_3(s)$ ....., etc. However, all the above functions must be positive real.

WORKED EXAMPLES

15.3 Synthesize the network described by the positive real impedance function $Z(s)$.

$$Z(s) = \frac{s^2 + 5s + 40}{s(s+8)} = \frac{s^2}{s(s+8)} + \frac{5(s+8)}{s(s+8)}$$

$$= \frac{5}{s} + \frac{s}{s+8} = Z_1(s) + Z_2(s)$$

Solution: $Z_1(s) = \dfrac{5}{s}$ can be realised as a capacitor of $\dfrac{1}{5}$F in series with $Z_2(s)$. $Z_2(s)$ can be realised as follows:

$$Y_2(s) = \frac{1}{Z_2(s)} = \frac{s+8}{s} = 1 + \frac{8}{s} = Y_a(s) + Y_b(s)$$

$Y_2(s)$ represents a parallel combination of $Y_a(s)$ and $Y_b(s)$.

$Y_a(s)$ represents a conductance of $1\,\mho$ or 1Ω resistance.

$Y_b(s)$ represents an inductance of $\dfrac{1}{8}$H.

Hence, the network described by the function

$$Z(s) = \frac{s^2 + 5s + 40}{s(s+8)} \quad \text{is as shown in Fig. 15.1}$$

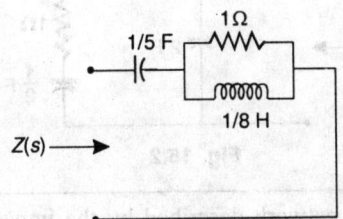

Fig. 15.1

15.4 Realise the network for the following positive real impedance function.

$$Z(s) = \frac{s+2}{s(2s+5)}$$

$$Y(s) = \frac{1}{Z(s)} = \frac{s(2s+5)}{s+2} = \frac{2s^2 + 5s}{s+2}$$

Solution: By observing the function $Y(s)$, it is found that there is a zero at the origin and is corresponding pole is at ∞.

We can remove the pole at ∞ by dividing the numerator by the denominator.

$$
s + 2 \overline{\smash{\big)}\, 2s^2 + 5s}\, \big((2s
$$
$$
\underline{2s^2 + 4s}
$$
$$
s
$$

$\therefore$ $$Y(s) = 2s + \frac{s}{s+2} = Y_1(s) + Y_2(s)$$

$Y_1(s) = 2s$ can be realised as a capacitor of 2 F in parallel with $Y_2(s)$.

$$Y_2(s) = \frac{s}{s+2}$$

or $$Z_2(s) = \frac{s+2}{s} = 1 + \frac{2}{s} = Z_a(s) + Z_b(s)$$

$Z_a(s)$ can be realised as a resistance of 1 Ω in series with $Z_b(s)$, which can be realised as a capacitance of $\frac{1}{2}$ F. The network described by the given function $Z(s)$ is as shown in Fig. 15.2.

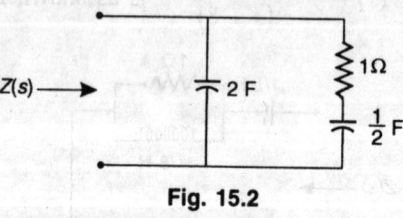

$Z(s)\longrightarrow$ 2 F   1Ω $\frac{1}{2}$ F

Fig. 15.2

15.5 Synthesize the network described by the impedance function

$$Z(s) = \frac{3s + 5}{3s + 3}.$$

Solution: On inspection of the function $Z(s)$, we find that, there are no poles or zeros on the imaginary axis. We have to remove the real positive constant by dividing the numerator by the denominator. On doing this, $Z(s)$ can be written as

$$Z(s) = 1 + \frac{2}{3s + 3} = Z_1(s) + Z_2(s)$$

$Z_1(s)$ can be realised as 1 Ω in series with $Z_2(s)$.

$$Z_2(s) = \frac{2}{3s + 3}$$

or $$Y_2(s) = \frac{3s + 3}{2} = \frac{3}{2}s + \frac{3}{2} = Y_a(s) + Y_b(s)$$

$Y_a(s) = \dfrac{3}{2}s$ is realised as a capacitance of $\dfrac{3}{2}$ F and $Y_b(s)$ is

realised as a conductance of $\dfrac{3}{2}\mho$ or a resistance of $\dfrac{2}{3}\,\Omega$. Hence, the realised network is as shown in Fig. 15.3.

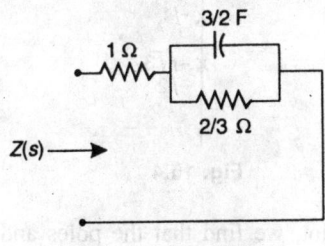

Fig. 15.3

15.5 PROPERTIES OF *L-C* IMMITTANCE FUNCTIONS

The various properties of the *L-C* immittance functions are:

1. $Z_{LC}(s)$ or $Y_{LC}(s)$ is the ratio of even to odd or odd to even polynomials.
2. The poles and zeros are simple and lie on $j\omega$ axis.
3. The poles and zeros must alternate or interlace on $j\omega$ axis.
4. The highest powers of the numerator and the denominator polynomials must differ by one. The lowest powers also must differ by one.
5. There must be either a pole or a zero at the origin and infinity.

$$\boxed{\textbf{WORKED EXAMPLES}}$$

15.6 Check whether the following functions are *L-C* immittance functions.

(i) $$Z(s) = \frac{K\,s\left(s^2 + 4\right)}{\left(s^2 + 1\right)\left(s^2 + 3\right)}$$

Solution: The zeros of $Z(s)$ are : $s = 0, s = \pm j2$

The poles of $Z(s)$ are : $s = \pm j1, s = \pm j\sqrt{3}$

The pole zero plot is as shown in Fig. 15.4.

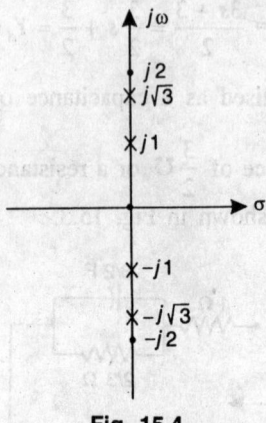

Fig. 15.4

From pole zero plot, we find that the poles and zeros does not alternate or interlace on the $j\omega$ axis. Hence, the given function is not an *L-C* impedance function.

(ii) $$Z(s) = \frac{s^4 + 6s^2 + 5}{3s^4 + 6s^2} = \frac{\left(s^2 + 5\right)\left(s^2 + 1\right)}{3s^2\left(s^2 + 2\right)}$$

By observing the nature of poles and zeros, we find that, some zeros and poles does not lie on $j\omega$ axis and there are multiple poles at the origin. Hence, $Z(s)$ is not an *L-C* function.

(iii) $$Z(s) = \frac{k\left(s^2 + 1\right)\left(s^2 + 9\right)}{\left(s^2 + 2\right)\left(s^2 + 10\right)}$$

The function $Z(s)$ is not the ratio of odd to even or even to odd polynomials in s. Hence, it is not an *L-C* function.

(iv) $$Z(s) = \frac{(s + 2)^2}{s^5 + 3s^3 + 25}$$

There are multiple zeros at $s = \pm j2$ and hence, $Z(s)$ is not an *L-C* function.

$$(v) \qquad Z(s) = \frac{2\left(s^2 + 1\right)\left(s^2 + 9\right)}{s\left(s^2 + 4\right)}$$

The pole zero diagram of the above function is as shown in Fig. 15.5.

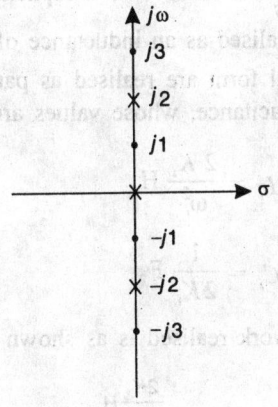

Fig. 15.5

The function $Z(s)$ satisfies all the conditions. Hence, it is an *L-C* function.

15.6 SYNTHESIS OF *L-C* IMMITTANCE FUNCTIONS

For the synthesis of an *L-C* immittance function, the basic requirement is that, it should be a PRF with all the poles and zeros on $j\omega$ axis only. Hence, the partial fraction expansion of an *L-C* immittance function in its general form is as given in equation (15.4).

$$F(s) = \frac{K_0}{s} + \frac{2K_2 s}{s^2 + \omega_2^2} = + \cdots + \frac{2K_i s}{s^2 + \omega_i^2} + \cdots + K_\infty s \qquad (15.4)$$

The synthesis of $F(s)$ is accomplished directly from its partial fraction expansion, by realising the network elements from the individual terms in the expansion. The network described by $F(s)$ given in equation (15.4) can be realised in the following forms.

1. **Foster forms:** (a) First Foster Form (series network realisation) and (b) Second Foster Form (parallel network realisation).
2. **Cauer forms:** (Ladder network realisation) (a) First Cauer Form and (b) Second Cauer Form.

15.7 FIRST FOSTER FORM

If the given function $F(s)$ is an impedance function $Z(s)$, then it can be realised in First Foster Form.

Let, $$Z(s) = \frac{K_0}{s} + \frac{2K_2\, s}{s^2 + \omega_2^2} + \cdots + \frac{2K_i\, s}{s^2 + \omega_i^2} + \cdots + K_\infty\, s \qquad (15.5)$$

The first term $\dfrac{K_0}{s}$ is realised as a capacitance of $C_0 = \dfrac{1}{K_0}$ F. The last term $K_\infty s$ is realised as an inductance of $L_\infty = K_\infty$ H. The middle terms of the general form are realised as parallel combinations of an inductance and capacitance, whose values are given by

$$L_i = \frac{2K_i}{\omega_i^2}\ \text{H}$$

and

$$C_i = \frac{1}{2K_i}\ \text{F}$$

The general network realised is as shown in Fig. 15.6.

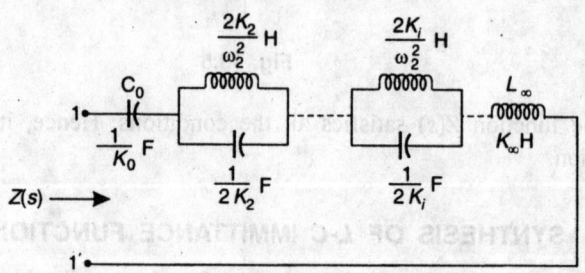

Fig. 15.6

If $Z(s)$ has no pole at the origin, then C_0 is not present. Similarly, if there is no zero at ∞, then L_∞ is not present.

WORKED EXAMPLES

15.7 Realise the following L-C impedance functions in the First Foster Form.

(i) $$Z(s) = \frac{2\left(s^2 + 1\right)\left(s^2 + 9\right)}{s\left(s^2 + 4\right)}$$

On partial fraction expansion, we get

$$Z(s) = 2s + \frac{A}{s} + \frac{Bs + C}{s^2 + 4} = 2s + \frac{\frac{9}{2}}{s} + \frac{\frac{15}{2}s}{s^2 + 4}$$

The realised network is as shown in Fig. 15.7.

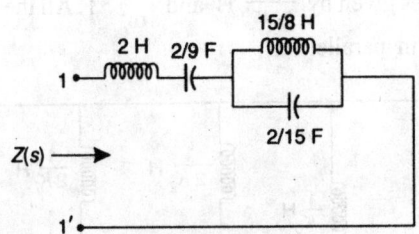

Fig. 15.7

(ii)
$$Z(s) = \frac{s^3 + 2s}{\left(s^2 + 1\right)\left(s^2 + 3\right)}$$

On partial fraction expansion, we get

$$Z(s) = \frac{As + B}{s^2 + 1} + \frac{Cs + D}{s^2 + 3} = \frac{\frac{1}{2}s}{s^2 + 1} + \frac{\frac{1}{2}s}{s^2 + 3}$$

The realised network is as shown in Fig. 15.8.

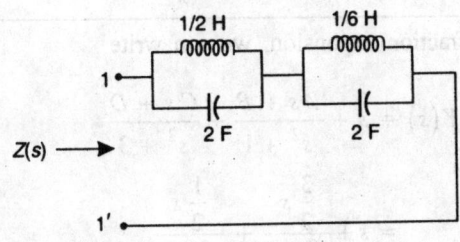

Fig. 15.8

15.8 SECOND FOSTER FORM

If the given function is an admittance function $Y(s)$, given by the Eq. (15.6).

$$Y(s) = \frac{K_0}{s} + \frac{2 K_2 s}{s^2 + \omega_2^2} + \cdots + \frac{2 K_i s}{s^2 + \omega_i^2} + \cdots + K_\infty s \qquad (15.6)$$

then, the realised network is as shown in Fig. 15.9.

The first element $\dfrac{K_0}{s}$ represents an inductance of $\dfrac{1}{K_0}$ H. The last element represents a capacitance of K_∞ F.

The elements in between represent a series combinations of inductances and capacitances given by $\dfrac{1}{2 K_i}$ H and $\dfrac{2 K_i}{\omega_i^2}$ F. All the realised elements are connected in parallel.

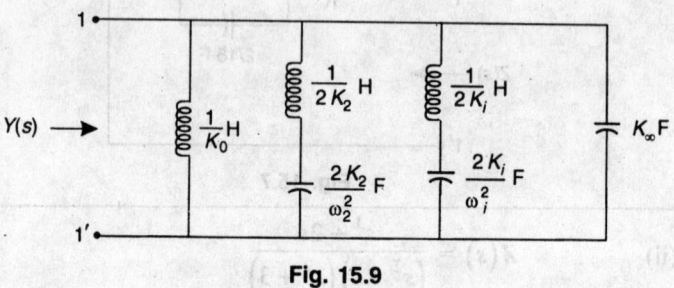

Fig. 15.9

WORKED EXAMPLE

15.8 Realise the following *L-C* admittance functions in the Second Foster form.

(i) $$Y(s) = \dfrac{s \, (s^2 + 2) \, (s^2 + 4)}{(s^2 + 1) \, (s^2 + 3)}$$

On partial fraction expansion, we can write

$$Y(s) = s + \frac{As + B}{s^2 + 1} + \frac{Cs + D}{s^2 + 3}$$

$$= s + \frac{\dfrac{3}{2} s}{s^2 + 1} + \frac{\dfrac{1}{2} s}{s^2 + 3}$$

The realised network is as shown in Fig. 15.10.

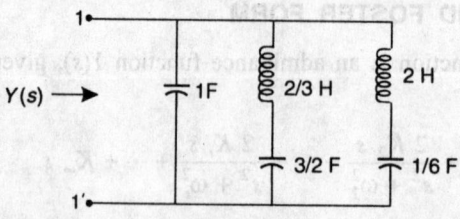

Fig. 15.10

(ii) $$Y(s) = \frac{\left(s^2 + 1\right)\left(s^2 + 3\right)}{s\left(s^2 + 2\right)}$$

On partial fraction expansion, we can write

$$Y(s) = s + \frac{A}{s} + \frac{Bs + C}{s^2 + 2}$$

$$= s + \frac{\dfrac{3}{2}}{s} + \frac{\dfrac{1}{2}s}{s^2 + 2}$$

The realised network is as shown in Fig. 15.11.

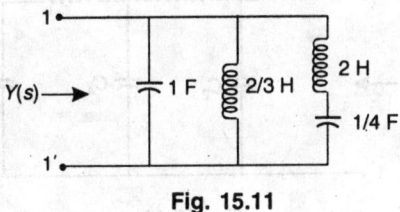

Fig. 15.11

15.9 FIRST CAUER FORM

One important property of an L-C immittance function is that the numerator and the denominator polynomials of the function always differ in degree by one. Hence, there is always a pole or a zero at ∞. By successive removal of the pole or a zero at ∞ leads to an L-C ladder network, which was first suggested by W. Cauer in 1927.

Consider an L-C immittance function, whose numerator is of degree $2n$ and the denominator of degree $(2n - 1)$. Hence, there is a pole at $s = \infty$. This pole can be removed by removing an inductor, say $L_1 s$, so that the remainder function $Z_2(s)$ is still an L-C function.

$$Z_2(s) = Z(s) - L_1 s$$

The degree of the denominator of $Z_2(s)$ is $(2n - 1)$ and hence, the degree of the numerator of $Z_2(s)$ has to be $(2n - 2)$. Therefore, we find that, $Z_2(s)$ has a zero at $s = \infty$.

$Y_2(s) = \dfrac{1}{Z_2(s)}$ will have a pole at $s = \infty$. This pole can be removed by removing a capacitor say $C_2 s$, so that the remainder function $Y_3(s)$ may be written as

$$Y_3(s) = Y_2(s) - C_2 s$$

Further, $Y_3(s)$ has a zero at $s = \infty$.

Now, $Z_3(s) = \dfrac{1}{Y_3(s)}$ has a pole at $s = \infty$, which can be removed

by removing an inductance as explained earlier. This process of successively removing a pole at $s = \infty$, by inverting the remainder and dividing is continued till the remainder is zero. Each time we remove a pole, either an inductor is removed or a capacitor is removed depending upon whether, the function is an impedance function or an admittance function. Thus, the final structure of the network synthesised is a ladder network, whose series arms are inductors and the shunt arms are capacitors as shown in Fig. 15.12.

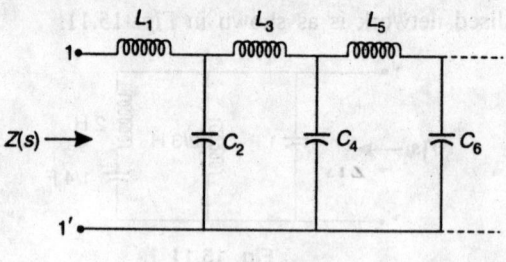

Fig. 15.12

Since, a pole at $s = \infty$ is always removed by inverting the remainder and dividing, an *L-C* ladder network can be synthesised by continued fraction expansion of $Z(s)$. The driving point impedance $Z(s)$ of a ladder network shown in Fig. 15.12 may be written as

$$Z(s) = Z_1(S) + \cfrac{1}{Y_2(s) + \cfrac{1}{Z_3(s) + \cfrac{1}{Y_4(s) + \cfrac{1}{Z_5(s) + \cdots}}}} \qquad (15.7)$$

If the given impedance function has no pole at $s = \infty$, it has a zero

at $s = \infty$. In such a case, $Z(s)$ is written as $Y(s) = \dfrac{1}{Z(s)}$ and the,

procedure explained above is followed to synthesise the ladder network. The first element of the ladder network depends on the nature of $Z(s)$ at $s = \infty$, is an inductor (series arm). If $Z(s)$ has a pole at $s = \infty$, the first element. If $Z(s)$ has a zero at $s = \infty$, the first element of the ladder network is a capacitor (shunt arm).

WORKED EXAMPLES

15.9 Realise the following *L-C* impedance function in the First Cauer Form.

$$Z(s) = \frac{2\,s^5 + 12\,s^3 + 16\,s}{s^4 + 4\,s^2 + 3}$$

Solution: $Z(s)$ has a pole at $s = \infty$. This pole can be removed by dividing the numerator by the denominator. $Z(s)$ can be written as

$$Z(s) = 2\,s + \frac{4\,s^3 + 10\,s}{s^4 + 4\,s^2 + 3}$$

By removing the pole at $s = \infty$, we realise an inductance of 2H. The remainder function $Z_2(s)$ can be written as

$$Z_2(s) = Z(s) - 2\,s = \frac{4\,s^3 + 10\,s}{s^4 + 4\,s^2 + 3}$$

$Z_2(s)$ has a zero $s = \infty$. Invert $Z_2(s)$ and write $Y_2(s) = \dfrac{1}{Z_2(s)}$

$$Y_2(s) = \frac{s^4 + 4\,s^2 + 3}{4\,s^3 + 10\,s} = \frac{1}{4}\,s + \frac{\dfrac{3}{2}\,s^2 + 3}{4\,s^3 + 10\,s}$$

$Y_2(s)$ has a pole at $s = \infty$. By removing this pole at $s = \infty$, we realise a capacitance of $\dfrac{1}{4}$ F and $Y_3(s)$ can be written as

$$Y_3(s) = Y_2(s) - \frac{1}{4}\,s = \frac{\dfrac{3}{2}\,s^2 + 3}{4\,s^3 + 10\,s}$$

$$Z_3(s) = \frac{4\,s^3 + 10\,s}{\dfrac{3}{2}\,s^2 + 3} = \frac{8}{3}\,s + \frac{2\,s}{\dfrac{3}{2}\,s^2 + 3}$$

The third element realised is an inductance of $\dfrac{8}{3}$ H.

$$Z_4(s) = Z_3(s) - \frac{8}{3}\,s = \frac{2\,s}{\dfrac{3}{2}\,s^2 + 3}$$

$$Y_4(s) = \frac{\frac{3}{2}s^2 + 3}{2s} = \frac{3}{4}s + \frac{3}{2s}$$

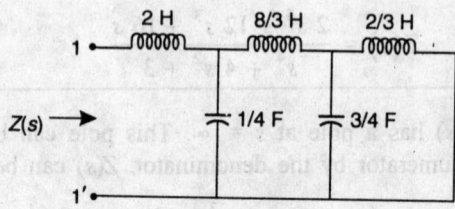

Fig. 15.13

The last two elements realised are (i) a capacitance of $\frac{3}{4}$ F, and (ii) an inductance of $\frac{2}{3}$ H, which are in parallel. Hence, the complete network realised can be written as shown in Fig. 15.13.

Since we always remove a pole at $s = \infty$, by inverting the remainder and dividing, we can synthesis an L-C ladder network by a continued fraction expansion, which is an easier method as shown below.

$$s^4 + 4s^2 + 3 \overline{) \, 2s^5 + 12s^3 + 16s \,} \left(2s \to Z \right.$$
$$\underline{2s^5 + 8s^3 + 6s}$$

$$\qquad 4s^3 + 10s \overline{) \, s^4 + 4s^2 + 3 \,} \left(\frac{1}{4}s \to Y \right.$$
$$\qquad \underline{s^4 + \frac{5}{2}s^2}$$

$$\qquad\qquad \frac{3}{2}s^2 + 3 \overline{) \, 4s^3 + 10s \,} \left(\frac{8}{3}s \to Z \right.$$
$$\qquad\qquad \underline{4s^3 + 8s}$$

$$\qquad\qquad\qquad 2s \overline{) \, \frac{3}{2}s^2 + 3 \,} \left(\frac{3}{4}s \to Y \right.$$
$$\qquad\qquad\qquad \underline{\frac{3}{2}s^2}$$

$$\qquad\qquad\qquad\qquad 3 \overline{) \, 2s \,} \left(\frac{2}{3}s \to Z \right.$$
$$\qquad\qquad\qquad\qquad \underline{2s}$$
$$\qquad\qquad\qquad\qquad \overline{00}$$

The quotients of the continued fraction expansion give the elements of the ladder network, as shown in Fig. 15.13.

It is important to note that, if the given function $F(s)$ is an impedance, then the first quotient in the continued fraction expansion must necessarily be an impedance. If the given function is an admittance, then the first quotient in the continued fraction expansion must necessarily be an admittance.

15.10 Realise the given *L-C* admittance function in the First Cauer Form.

$$Y(s) = \frac{s^6 + 9 s^4 + 23 s^2 + 15}{s^5 + 6 s^3 + 8 s}$$

Solution: The continued fraction expansion is as follows:

$$s^5 + 6s^3 + 8s \overline{\smash{\big)}\ s^6 + 9s^4 + 23s^2 + 15}\left(s \to Y\right.$$
$$\underline{s^6 + 6s^4 + 8s^2}$$

$$3s^4 + 15s^2 + 15 \overline{\smash{\big)}\ s^5 + 6s^3 + 8s}\left(\frac{1}{3}s \to Z\right.$$
$$\underline{s^5 + 5s^3 + 5s}$$

$$s^3 + 3s \overline{\smash{\big)}\ 3s^4 + 15s^2 + 15}\left(3s \to Y\right.$$
$$\underline{3s^4 + 9s^2}$$

$$6s^2 + 15 \overline{\smash{\big)}\ s^3 + 3s}\left(\frac{1}{6}s \to Z\right.$$
$$\underline{s^3 + \frac{15}{6}s}$$

$$\frac{1}{2}s \overline{\smash{\big)}\ 6s^2 + 15}\left(12s \to Y\right.$$
$$\underline{6s^2}$$

$$15 \overline{\smash{\big)}\ \frac{1}{2}s}\left(\frac{1}{30}s \to Z\right.$$
$$\underline{\frac{1}{2}s}$$
$$0$$

The realised network is as shown in Fig. 15.14.

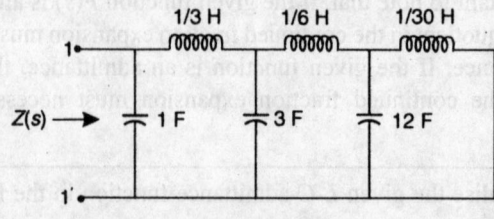

Fig. 15.14

15.10 SECOND CAUER FORM

Following the procedure adopted to synthesise an immittance function in the First Cauer Form, by successively removing the pole at $s = 0$, it is possible to realise the ladder structure of the network in an alternate form known as Second Cauer Form, by arranging the given immittance function in the ascending order of s, both in the numerator and in the denominator.

Let $Z(s)$ which is arranged in the ascending order of s, both in the numerator and the denominator be given by equation (15.8).

$$Z(s) = \frac{a_0 + a_2 s^2 + \cdots + a_{n-2} s^{n-2} + a_n s^n}{b_1 s + b_3 s^3 + \cdots + b_{m-2} s^{m-2} + b_m s^m} \tag{15.8}$$

Where, $n - m = \pm 1$.

The continued fraction expansion of $Z(s)$ in equation (15.8), may be written as in equation (15.9).

$$Z(s) = \frac{1}{C_1 s} + \cfrac{1}{\cfrac{1}{L_2 s} + \cfrac{1}{\cfrac{1}{C_3 s} + \cfrac{1}{\cfrac{1}{L_4 s} + \cdots}}} \tag{15.9}$$

Since the function considered is an impedance, the first term in the continued fraction expansion is an impedance representing the series arm. The second term must be an admittance representing the shunt arm, alternating as impedance and admittance for the remaining arms. The general form of the realised ladder network is as shown in Fig. 15.15.

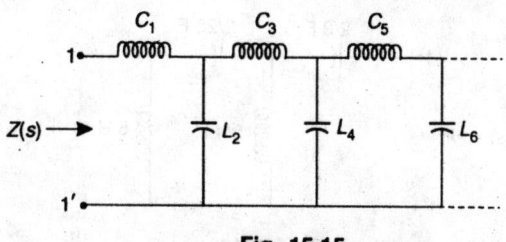

Fig. 15.15

WORKED EXAMPLES

15.11 Synthesize the given *L-C* impedance function in the Second Cauer Form.

$$Z(s) = \frac{(s^2 + 1)(s^2 + 3)}{s(s^2 + 2)} = \frac{s^4 + 4s^2 + 3}{s^3 + 2s}$$

Solution: Rewrite $Z(s)$ in the ascending order of s.

$$\therefore \qquad Z(s) = \frac{3 + 4s^2 + s^4}{2s + s^3}$$

The continued fraction expansion of $Z(s)$ is as shown below:

$$2s + s^3 \overline{)\, 3 + 4s^2 + s^4 \,} \left(\frac{3}{2s} \to Z \right.$$
$$\underline{3 + \frac{3}{2}s^2}$$
$$\frac{5}{2}s^2 + s^4 \overline{)\, 2s + s^3 \,} \left(\frac{4}{5s} \to Y \right.$$
$$\underline{2s + \frac{4}{5}s^3}$$
$$\frac{1}{5}s^3 \overline{)\, \frac{5}{2}s^2 + s^4 \,} \left(\frac{25}{2s} \to Z \right.$$
$$\underline{\frac{5}{2}s^2}$$
$$s^4 \overline{)\, \frac{1}{5}s^3 \,} \left(\frac{1}{5s} \to Y \right.$$
$$\underline{\frac{1}{5}s^3}$$
$$0$$

The realised network is as shown in Fig. 15.16.

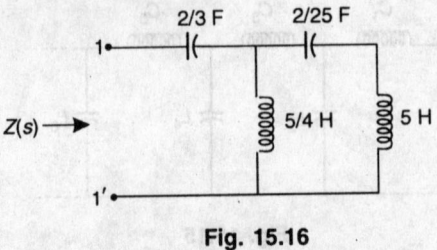

Fig. 15.16

It is interesting to note that, in both Foster and Cauer Forms of the realisation of the network, the number of elements is one greater than the number of internal frequencies, which are defined as the poles and zeros of the function, excluding those at $s = 0$ and $s = \infty$.

In other words, the total number of elements is equal to the degree of the highest power of s in $F(s)$. The realised networks in Foster Forms or Cauer Forms contain the minimum number of elements for the given L–C immittance function. Hence, they are called as *cononical forms*. For any given L–C immittance function, the network can be realised in two Foster forms and two Cauer Forms. All the four types of networks realised are equivalent. One form may be preferred over the other because of practical considerations, elements size, compensation for parasitic effects, etc. It is therefore not possible to conclude that any one particular form is always the best.

15.11. PROPERTIES OF *R-C* IMPEDANCE FUNCTIONS AND *R-L* ADMITTANCE FUNCTIONS

By observing the Series Foster Form of an L-C impedance network as shown in Fig. 15.6, we can obtain the Series Form of Foster realisation of an R-C impedance network, by simply replacing the inductances by resistances, as shown in Fig. 15.17.

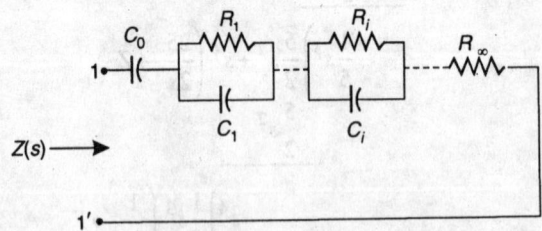

Fig. 15.17

For the R-C impedance network shown in Fig. 15.17, the general equation for the driving point impedance $Z(s)$ may be written as

$$Z(s) = \frac{K_0}{s} + \frac{K_1}{s + \sigma_1} + \cdots + \frac{K_i}{s + \sigma_i} + \cdots + K_\infty\, s \quad (15.10)$$

Where $C_0 = \dfrac{1}{K_0}$ F, $R_\infty = K_\infty\ \Omega$, $C_1 = \dfrac{1}{K_1}$ F,

$$R_1 = \frac{K_1}{\sigma_1}\ \Omega,\ \ C_i = \frac{1}{K_i}\ \text{F},\ \ R_i = \frac{K_i}{\sigma_i}\ \Omega$$

The properties of an *R-C* impedance function or an *R-L* admittance function are:

1. All poles and zeros are simple and are located on the negative real axis.
2. Poles and zeros alternate or interlace.
3. The singularity nearest to (or at) the origin is a pole.
4. The singularity nearest to (or at) infinity is a zero.
5. The residues of the poles of $Z(s)$ are real and positive.
6. The slope $\dfrac{dZ}{d\sigma}$ is negative.
7. $Z(0) > Z(\infty)$.

WORKED EXAMPLE

15.12 Test whether the following functions represent *R-C* driving point impedance functions.

---

1. $$Z(s) = \frac{(s + 2)(s + 5)}{(s + 3)(s + 4)}$$

The singularity nearest to the origin, i.e. $s = -2$ is a zero. Hence, it is not an *R-C* impedance function.

---

2. $$Z(s) = \frac{(s + 3)(s + 5)}{(s + 2)}$$

The poles and zeros do not alternate. Hence, it is not an *R-C* impedance function.

---

3. $$Z(s) = \frac{(s + 1)(s - 2)(s + 3)}{s\,(s + 4)(s + 5)}$$

There is a zero at the origin. Hence, $Z(s)$ is not an *R-C* impedance function.

$$4. \qquad Z(s) = \frac{(s+1)(s+5)(s+9)}{s(s+2)(s+7)}$$

This function satisfies all the conditions and hence, it is an R-C impedance function.

If the given function $F(s)$ is a driving point admittance function given by Eq. (15.11), it can be realised as an R-L admittance function.

$$Y(s) = \frac{K_0}{s} + \frac{K_1}{s+\sigma_1} + \cdots + \frac{K_i}{s+\sigma_i} + \cdots + K_\infty \qquad (15.11)$$

Then, the network elements are realised as

$$L_0 = \frac{1}{K_0} \text{ H,} \qquad\qquad R_1 = \frac{\sigma_1}{K_1} \Omega,$$

$$L_1 = \frac{1}{K_1} \text{ H,} \qquad\qquad R_i = \frac{\sigma_i}{K_i} \Omega,$$

$$L_i = \frac{1}{K_i} \text{ H,} \qquad \text{and} \qquad R_\infty = \frac{1}{K_\infty}$$

The realised network is as shown in Fig 15.18.

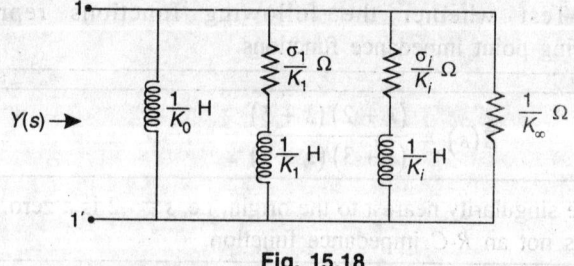

Fig. 15.18

15.12. SYNTHESIS OF R-C IMPEDANCE OR R-L ADMITTANCE FUNCTIONS

The R-C impedance networks or R-L admittance networks can be realised in all the four forms, i.e. two Foster Forms and two Cauer Forms.

(a) Foster Forms: The First Foster Form of R-C impedance function is obtained by the partial fraction expansion of $Z_{RC}(s)$, which is in the form of Eq. (15.10) and the realised network is as shown in Fig. 15.17.

The Second Foster Form of R-L admittance function is obtained by the partial fraction expansion of $Y_{RL}(s)$ which is in the form of equation (15.11) and the realised network is as shown in Fig. 15.18.

It is important to note that for an R-C impedance function or an R-L admittance function, the degree of the numerator polynomial can not be greater than the degree of the denominator polynomial. When the numerator and the denominator are of the same degree, then the numerator is divided by the denominator and the process of the realisation of the network is carried on.

WORKED EXAMPLES

15.13 Synthesize the given R-C impedance function in the Series Foster Form.

$$Z(s) = \frac{3(s + 2)(s + 4)}{s(s + 3)}$$

Solution: In the above example, the degree of the numerator is the same as the degree of the denominator. Hence, the numerator is divided by the denominator and the partial fraction expansion of the remainder function is obtained. After doing this, $Z(s)$ can be written as

$$Z(s) = 3 + \frac{9s + 24}{s(s + 3)} = 3 + \frac{8}{s} + \frac{1}{s + 3}$$

The realised network is as shown in Fig. 15.19.

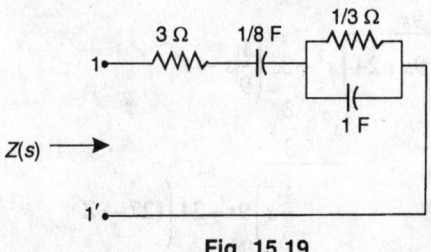

Fig. 15.19

15.14 Synthesize the given R-L admittance function in the Parallel Foster Form.

$$Y(s) = \frac{3(s + 2)(s + 4)}{s(s + 3)}$$

Solution: $Y(s) = 3 + \dfrac{8}{s} + \dfrac{1}{s+3}$

The realised network is as shown in Fig. 15.20.

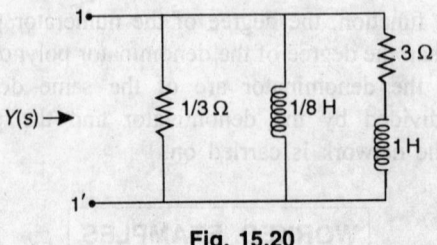

Fig. 15.20

(b) Cauer Forms: The Cauer Forms of networks for the given *R-C* impedance functions or *R-L* admittance functions can be obtained by continued fraction expansion method.

WORKED EXAMPLES

15.15 Let $Z(s) = \dfrac{3(s+2)(s+4)}{s(s+3)}$

Solution: The continued fraction expansion of $Z(s)$ as shown below:

$$Z(s) = \frac{3s^2 + 18s + 24}{s^2 + 3s}$$

$$
s^2+3s\ \overline{\big)\ 3s^2+18s+24\ \big(}\ (3 \to Z
$$
$$
\underline{3s^2 + 9s}
$$
$$
9s+24\ \overline{\big)\ s^2+3s\ \big(}\ \tfrac{1}{9}s \to Y
$$
$$
\underline{s^2 + \tfrac{8}{3}s}
$$
$$
\tfrac{1}{3}s\ \overline{\big)\ 9s+24\ \big(}\ (27 \to Z
$$
$$
\underline{9s}
$$
$$
24\ \overline{\big)\ \tfrac{1}{3}s\ \big(}\ \tfrac{1}{72}s \to Y
$$
$$
\underline{\tfrac{1}{3}s}
$$
$$
00
$$

The realised network in the First Cauer Form is as shown in Fig. 15.21.

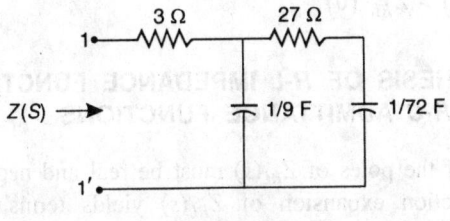

Fig. 15.21

15.16 Let $Y(s) = \dfrac{3(s+2)(s+4)}{s(s+3)}$

Solution: The continued fraction expansion of the given $Y(s)$ is the same as that of $Z(s)$ given in the previous example. The only difference is that the quotients in the continued fraction expansion are Y in places of Z and Z in places of Y. The realised network for $Y(s)$ is as shown in Fig. 15.22.

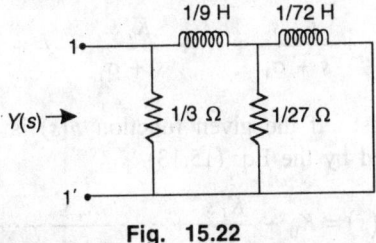

Fig. 15.22

15.13 PROPERTIES OF *R-L* IMPEDANCE FUNCTIONS AND *R-C* ADMITTANCE FUNCTIONS

The properties of an *R-L* impedance function and an *R-C* admittance function are the same.

They are:

1. All poles and zeros are simple and are located on the negative real axis.
2. Poles and zeros alternate.
3. The singularity nearest to (or at) the origin is a zero.
4. The singularity nearest to (or at) infinity is a pole.
5. The residues of the poles of $Z_{RL}(s)$ are real and negative.

   However, the residues of $\dfrac{Z_{RL}(s)}{s}$ are real and positive.

6. The slope $\dfrac{d}{d\sigma} Z_{RL}(\sigma)$ is negative.

7. $Z_{RL}(\infty) > Z_{RL}(0)$.

15.14 SYNTHESIS OF *R-L* IMPEDANCE FUNCTIONS AND *R-C* ADMITTANCE FUNCTIONS

The residues of the poles of $Z_{RL}(s)$ must be real and negative. Hence, the partial fraction expansion of $Z_{RL}(s)$ yields terms of the form $-\dfrac{K_i}{s + \sigma_i}$. But this term does not represent an *R-L* impedance at all. To obtain the Foster Form of an *R-L* impedance function $Z_{RL}(s)$, $\dfrac{Z_{RL}(s)}{s}$ is expanded into partial fractions, which yields positive residues for poles.

The general form of an immittance function, which represents an *R-L* impedance function or an *R-C* admittance function is as shown in Eq. (15.11). After obtaining the partial fractions of $\dfrac{F(s)}{s}$, it is multiplied by s, which results in the equation (15.12), which is realisable.

$$F(s) = K_0 + \frac{K_1 s}{s + \sigma_1} + \cdots + \frac{K_i s}{s + \sigma_i} + \cdots + K_\infty s \qquad (15.12)$$

(a) Foster Forms: If the given function $F(s)$ is an *R-L* impedance function described by the Eq. (15.13).

$$F(s) = Z_{RL}(s) = K_0 + \frac{K_1 s}{s + \sigma_1} + \cdots + \frac{K_i s}{s + \sigma_i} + \cdots + K_\infty s \qquad (15.13)$$

Then, the Series Foster Form of the network realised is as shown in Fig. 15.23.

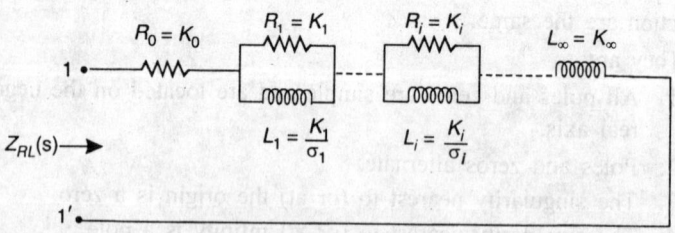

Fig. 15.23

If the given function $F(s)$ is an *R-C* admittance function described by the Eq. 15.14.

$$F(s) = Y_{RC}(s) = K_0 + \frac{K_1 s}{s + \sigma_1} + \cdots + \frac{K_i s}{s + \sigma_i} + \cdots + K_\infty s$$

$$(15.14)$$

Then, the Parallel Foster Form of the network realised is as shown in Fig. 15.24.

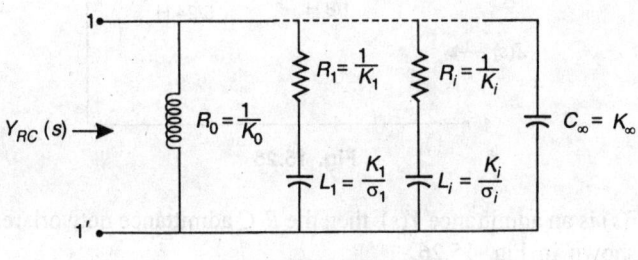

Fig. 15.24

WORKED EXAMPLE

15.17 Let $\quad F(s) = \dfrac{2(s+1)(s+3)}{(s+2)(s+6)}$

Solution: $F(s)$ represents an R-L impedance or an R-C admittance function. After partial fraction expansion, $F(s)$ can be written as

$$F(s) = 2 - \frac{\dfrac{1}{2}}{s+2} - \frac{\dfrac{15}{2}}{s+6}$$

From the above equation, we find that the residues of poles are negative. Hence, the partial fraction expansion of $\dfrac{F(s)}{s}$ is obtained.

$$\frac{F(s)}{s} = \frac{2(s+1)(s+3)}{s(s+2)(s+6)} = \frac{\dfrac{1}{2}}{s} + \frac{\dfrac{1}{4}}{s+2} + \frac{\dfrac{5}{4}}{s+6}$$

Multiplying the above equation by s, we get

$$F(s) = \frac{1}{2} + \frac{\dfrac{1}{4}s}{s+2} + \frac{\dfrac{5}{4}s}{s+6}$$

If $F(s)$ is an impedance $Z(s)$, then the Series Foster Form of the R-L impedance network realised is as shown in Fig. 15.25.

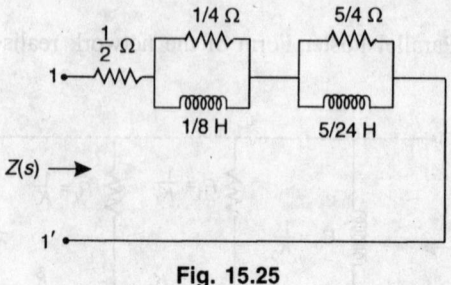

Fig. 15.25

If $F(s)$ is an admittance $Y(s)$, then the R-C admittance network realised is as shown in Fig. 15.26.

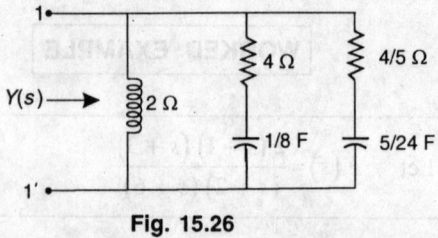

Fig. 15.26

(b) Cauer Forms: The Cauer Forms or Ladder Forms of the networks for the given immittance function $F(s)$ are obtained by arranging the numerator and denominator polynominals of $F(s)$ in the ascending order of s and by obtaining the continued fraction expansion of $F(s)$. The quotients of the continued fraction expansion of $F(s)$ give the elements of the network.

WORKED EXAMPLES

15.18 Let $F(s) = \dfrac{2\,(s+1)\,(s+3)}{(s+2)\,(s+6)}$

Solution: $F(s)$ is written in the ascending order of s.

$$F(s) = \frac{6 + 8\,s + 2\,s^2}{12 + 8\,s + s^2}$$

The continued fraction expansion of $F(s)$ is as given below.

$$12+8s+s^2 \overline{\smash{\big)}\ 6+8s+2s^2} \left(\frac{1}{2} \to Z\right.$$
$$\underline{6+4s+\frac{1}{2}s^2}$$
$$4s+\frac{3}{2}s^2 \overline{\smash{\big)}\ 12+8s+s^2} \left(\frac{3}{s} \to Y\right.$$
$$\underline{12+\frac{9}{2}s}$$
$$\frac{7}{2}s+s^2 \overline{\smash{\big)}\ 4s+\frac{3}{2}s^2} \left(\frac{8}{7} \to Z\right.$$
$$\underline{4s+\frac{8}{7}s^2}$$
$$\frac{5}{14}s^2 \overline{\smash{\big)}\ \frac{7}{2}s+s^2} \left(\frac{49}{5s} \to Y\right.$$
$$\underline{\frac{7}{2}s}$$
$$s^2 \overline{\smash{\big)}\ \frac{5}{14}s^2} \left(\frac{5}{14} \to Z\right.$$
$$\underline{\frac{5}{14}s^2}$$
$$\overline{00}$$

If the given function $F(s)$ is an R-L impedance function, then the quotients of continued fraction expansion of $F(s)$ represent the impedance and admittance functions alternately and the realised R-L impedance network is as shown in Fig. 15.27.

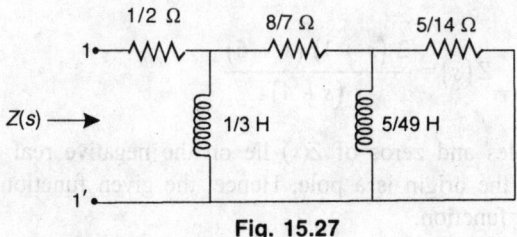

Fig. 15.27

If the given function $F(s)$ is an R-C admittance function, then the quotients of continued fraction expansion of $F(s)$ represent the admittance and impedance functions alternately and the realised R-C admittance network is as shown in Fig. 15.28.

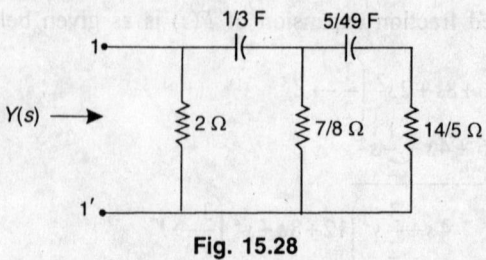

Fig. 15.28

15.19 An impedance function has the pole-zero configuration as shown in Fig. 15.29. If $Z(-3) = 4$, synthesize the impedance function in a Foster Form and a Cauer Form. (Bangalore University)

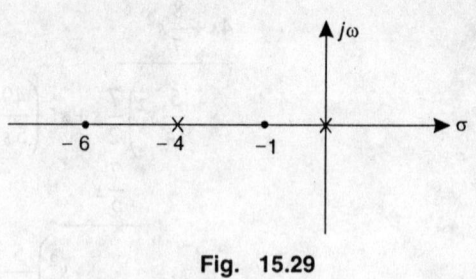

Fig. 15.29

Solution: There are two zeros at $s = -1$ and $s = -6$. There are two poles at $s = 0$ and $s = -4$.

If H is the scale factor, the impedance function may be written as

$$Z(s) = H \frac{(s+1)(s+6)}{s(s+4)}$$

Substituting $s = -3$ and $Z(s) = 4$ in the above equation, we get $H = 2$.

$$\therefore \quad Z(s) = \frac{2(s+1)(s+6)}{s(s+4)}$$

All the poles and zeros of $Z(s)$ lie on the negative real axis. The singularity at the origin is a pole. Hence, the given function is an R-C impedance function.

(i) First Foster Form

As the numerator and the denominator of $Z(s)$ are of the same degree, the numerator is divided by the denominator and $Z(s)$ is re-written as:

$$Z(s) = 2 + \frac{6s + 12}{s(s+4)} = 2 + \frac{A}{s} + \frac{B}{s+4} = 2 + \frac{3}{s} + \frac{3}{s+4}$$

The realised *R-C* impedance network is as shown in Fig. 15.30.

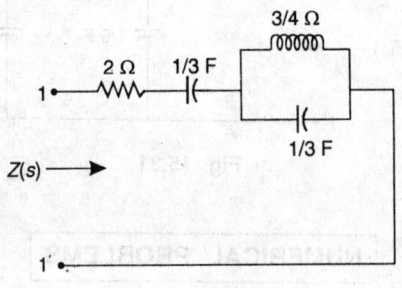

Fig. 15.30

(ii) First Cauer Form

$$Z(s) = \frac{2(s+1)(s+6)}{s(s+4)} = \frac{2s^2 + 14s + 12}{s^2 + 4s}$$

The continued fraction expansion of $Z(s)$ is as follows:

$$s^2 + 4s \overline{\smash{\big)}\ 2s^2 + 14s + 12} \left(2 \to Z \right.$$
$$\underline{2s^2 + 8s}$$

$$6s + 12 \overline{\smash{\big)}\ s^2 + 4s} \left(\frac{1}{6}s \to Y \right.$$
$$\underline{s^2 + 2s}$$

$$2s \overline{\smash{\big)}\ 6s + 12} \left(3 \to Z \right.$$
$$\underline{6s}$$

$$12 \overline{\smash{\big)}\ 2s} \left(\frac{1}{6}s \to Y \right.$$
$$\underline{2s}$$
$$0$$

The realised network in the First Cauer Form is as shown in Fig. 15.31.

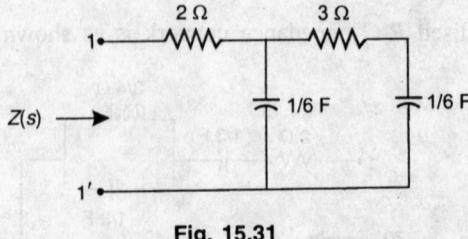

Fig. 15.31

NUMERICAL PROBLEMS

15.1 Test whether the following functions are Positive Real.

(i) $F(s) = \dfrac{s\left(s^2 + 2\right)}{\left(s^2 + 1\right)\left(s^2 + 3\right)}$

(ii) $F(s) = \dfrac{2s^2 + 5}{s\left(s^2 + 1\right)}$

(iii) $F(s) = \dfrac{(s + 2)(s + 4)}{(s + 1)(s + 3)}$

(iv) $F(s) = \dfrac{s + 4}{s^2 + 2s + 1}$

(v) $F(s) = \dfrac{s^2 + s + 6}{s^2 + s + 1}$

(vi) $F(s) = \dfrac{3s^2 + 5}{s\left(s^2 + 1\right)}$

(vii) $F(s) = \dfrac{s^3 + 5s^2 + 11s + 7}{s^3 + 5s^2 + 2s + 10}$

15.2 Test whether the following Polynomials are Hurwitz.

(i) $F(s) = s^3 + 6s^2 + 2s + 1$

(ii) $F(s) = s^3 + 7s^2 + 15s + 9$

(iii) $F(s) = s^7 + 3s^5 + 2s^3 + s$

(iv) $F(s) = s^6 + 2s^5 + 14s^4 + 26s^3 + 49s^2 + 72s + 36$

(v) $F(s) = s^3 + 2s^2 + 3s + 6$

(vi) $F(s) = s^4 + s^3 + 5s^2 + 3s + 4$

15.3 Synthesis the following driving point impedance function into one Foster Form of network.

$$Z(s) = \frac{s\left(s^2 + 2\right)}{\left(s^2 + 1\right)\left(s^2 + 3\right)}$$

15.4 Synthesis the following driving point impedance function in Cauer Form.

$$Z(s) = \frac{s\left(s^2 + 2\right)}{\left(s^2 + 1\right)\left(s^2 + 3\right)}$$

15.5 Synthesis the following driving point function in Second Foster Form.

$$Z(s) = \frac{(s + 5)(s + 7)}{(s + 1)(s + 6)(s + 8)}$$

15.6 Synthesis the following driving point admittance function in the Second Foster Form.

$$Y(s) = \frac{\left(s^2 + 1\right)\left(s^2 + 4\right)}{s\left(s^2 + 2\right)}$$

15.7 Realise the given function in a Foster Form.

$$Z(s) = \frac{s^5 + 105\, s^3 + 945\, s}{15\, s^4 + 360\, s^2 + 945}$$

15.8 Realise the given function by continued fraction expansion.

$$Z(s) = \frac{(s + 2)(s + 5)}{(s + 1)(s + 3)}$$

15.9 Realise the given *R-C* driving point impedance.

$$Z(s) = \frac{s^2 + 7\, s + 10}{s^2 + 4\, s + 3}$$

15.10 Realise the First and Second Foster Forms of the network for the given impedance function.

$$Z(s) = \frac{s^3 + 2\, s}{s^4 + 4\, s^2 + 3}$$

15.11 Realise the given *L-C* function in Cauer-I and Cauer-II Forms.

$$Z(s) = \frac{s^4 + 10\, s^2 + 9}{s^3 + 4\, s}$$

15.12 Synthesis the given driving point impedance function in the Second Cauer Form.

$$Z(s) = \frac{(s + 2)\,(s + 4)}{(s + 1)\,(s + 3)}$$

15.13 Of the three pole zero diagrams given in Figs 6.32 (*a*), (*b*) and (*c*), pick the one that represents an *R-C* admittance function and synthesis it in Parallel Foster Form.

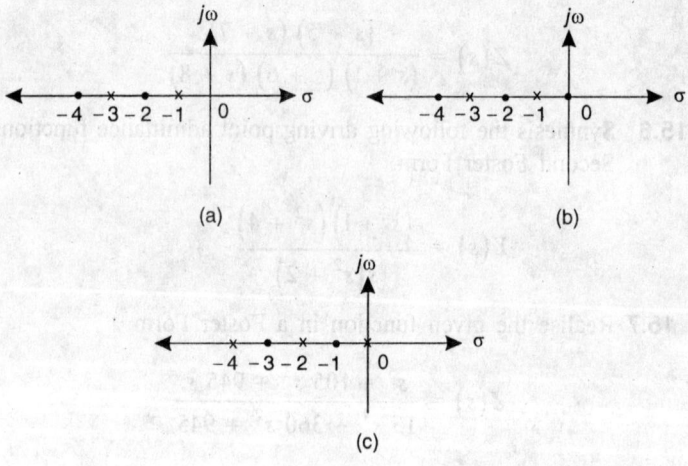

(a)

(b)

(c)

Fig. 15.32

15.14 A driving point immittance function has the pole zero configuration as shown in Fig. 15.33. If $F(-7) = 48/5$, synthesize the immittance function in Two Foster Forms and Two Cauer Forms.

(Bangalore University)

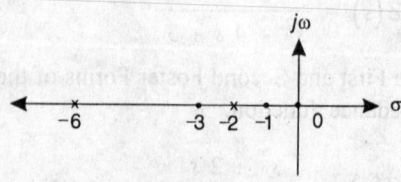

Fig. 15.33

Answers to Numerical Problems

ELECTRIC CIRCUITS

Chapter 1: Basic Circuit Concepts (D.C. Circuits)

1.1 30.25 Ω, 9.68 Ω   **1.2** 5 Ω

1.3 6 A, 4 A   **1.4** 5.75 A, 1.49 A,1.67 W

1.5 6.52 A, 4.35 A, 408.73 W   **1.6** 146.486 V, 1.35 A

1.7 85.71 V, 114.29 V   **1.8** 10,000 Ω

1.9 6 Ω, 2 A   **1.10** 3.57 A, 12.745 W

1.11 150 V, 40 V   **1.12** 5 A, 2.5 V

1.13 – 9.488 V   **1.14** 3.83 A, 2.33 A, 1.5 A

1.15 4.4 A, 0.758 A, – 0.49 A, 4.668 A

1.16 29.41 mA   **1.17** 51.65 A, 38.68 A, 12.96 A

1.18 5.5 A, 2.5 A   **1.19** 10.62 µA.

1.20 2.56 A, 1.83 A, – 3.128 A   **1.21** 18.67 Ω

1.22 23.74 Ω   **1.23** 4.23 Ω

1.24 0.471 A   **1.25** 4 A

1.26 5.44 Ω, 1.589 Ω

Chapter 2: Mesh Current and Node Voltage Analysis

2.1 $1.58\angle34.7°\,\Omega, 3.162\angle34.7°\,\Omega$ and $3.162\angle-55.26°\,\Omega$

2.2 $\infty, 7.07\angle45°\,\Omega$ and $7.07\angle45°\,\Omega$

2.3 $1.58\angle27.56°\,\Omega$

2.4 −16.41 V

2.5 8.57 A

2.6 14.53 V, 255.4 W

2.7 $V_A = -5.55$ V, $V_B = 4.44$ V, $V_C = -0.406$ V

2.8 -0.692 A, 2.308 A and -0.1538 A

2.9 3.55 W

2.10 $V_1 = 6.18$ V, $V_2 = -8.82$ V and $V_3 = -0.5$ V

2.11 2.635 W, 53.13 W

2.12 $6.23\angle223.76°$ V

2.13 201.8 W

2.14 $7.4 \times 10^{-3}\sin(\omega t - 187.15°)V$

2.15 $34.42\angle23.55°$ V

2.16 1.8 A

2.17 $I_R = 46.18\angle53.13°$ A, $I_Y = 46.18\angle-66.87°$ A and
$I_B = 46.18\angle173.13°$ A

2.18 $I_R = 69.28\angle-60°$ A, $I_Y = 69.28\angle180°$ A and $I_B = 69.28\angle60°$ A

2.19

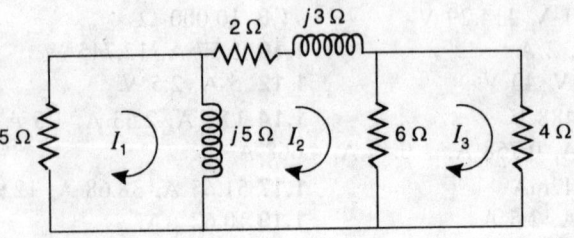

2.20 $69.125\angle156.1°$ V

2.21 $I_1 = 5.77\angle-30°$ A, $I_2 = 11.54\angle150°$ A and $I_3 = 5.77\angle-30°$ A

2.22

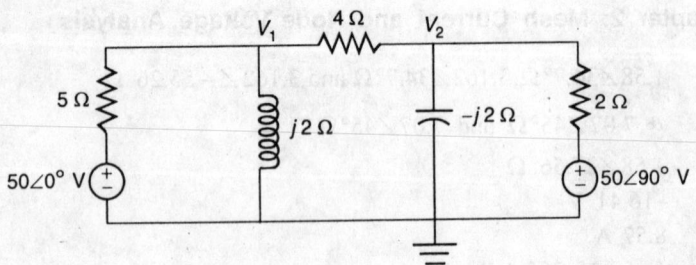

Chapter 3: Network Theorems

3.1 4.38 A, 5.17 A, 2.78 A and 12.33 A

3.2 6 V

3.3 0.1 A

3.4 61.09∠56.56° V

3.5 −7.5 V

3.6 4 A

3.7 83.018 V

3.8 0.676∠ − 2.37° A

3.9 $E_0 = 2.38 \text{V}, R_0 = \Omega, 1.5 \text{A}$

3.10 $I_{SC} = 1.718 \text{A}, R_0 = 5.3 \Omega, I_1 = 0.88 \text{A}$

3.11 $E_0 = 2.38 \text{V}, R_0 = 2 \Omega$

3.12 $I_{SC} = 6.44 \angle 39.23° \text{A}, Z_0 = 3.676 \angle 36.03° \Omega$

3.13 $E_0 = 9.74 \angle -83.6° \text{V}, Z_0 = 5.124 \angle 79.56° \Omega$ and
$I = 1.11 \angle -67.33° \text{A}$

3.14 $I_{SC} = 4.146 \angle -105.7° \text{A}, Z_0 = 5.66 \angle 54.34° \Omega$

3.15 $E_0 = 8 \text{V}, R_0 = 10 \text{k}\Omega$

3.16 $I_{SC} = 5 \text{A}, R_0 = 25 \Omega$

3.17 $E_0 = 2.7 \angle 26.04° \text{V}, Z_0 = j31 \Omega$

3.18 6.477 Ω, 0.296 W

3.19 10 Ω, 10.7 mW

3.20 11.18 Ω, 236.1 W

3.21 $R_L = 5.05 \Omega, 3382.36 \text{W}$

3.22 $Z_L = 1.788 \angle -10.3° \Omega, 1136 \text{W}$

3.23 $Z_L = (6.27 - j3.62) \Omega, 224.46 \text{W}$

3.24 0.187∠ − 26.28° A

3.25 0.48∠114.3° A

3.26 43.76∠61.35° V

3.27 5.246∠19.37° A

3.28 3.384∠12.6° A

3.29 2.263∠ − 130° A

3.30 39.25 W

Chapter 4: Resonance

4.1   (a) 41.1 Hz (b) 5 Ω (c) 1187.6 V (d) 5.16 (e) 7.97 Hz

4.2   (a) 37.33 Hz, 45.29 Hz (b) 7.96 Hz (c) 5.16 (d) 41.10 Hz, 1193.3 V (e) 41.51 Hz, 1193.3 V

4.3   (a) 10 µF (b) 2311.47 V (c) 10

4.4   (a) 67.62 mH (b) 1.06 (c) 200 V, 212.33 V, 212.33 V (d) 47.17 Hz

4.5 $R = 15.33\,\Omega, L = 244.11$ mH and $C = 41.5$µF

4.6   (a) 50.36 Hz (b) 2.11 (c) 38.43 Hz, 62.6 Hz (deviation method) (d) 23.88 Hz (e) 13.33 A (f) 9.42 A (g) 200 V, 422 V and 422 V

4.7   101.36 Hz, 98.49 Hz

4.8   421.34 Hz, 578.66 Hz

4.9   250 Hz, 0.239 H, 1.6 µF

4.10   (a) 76.59 Ω, 0.169 mH (b) 6.93

4.11   (a) 0.916 A (b) 0.0796 lagging (c) 288.21 V, 58.29 V

4.12   16.2 Hz, 1000 Ω

4.13   1.5678 H, 5 A, 0.21572 A

4.14   92 µF, 2.1176 A

4.15   205.57 Hz, 28.75 A

4.16   17.85 mH or 5.76 mH

4.17   18.3 µF or 94.4 µF

4.18 $R_L = R_C = 10\Omega$

4.19   21.4 Ω

4.20   2.44 Ω

4.21   (a) 73.1 µH, 813 pF (b) 6,44,953.02 Hz, 6,60,876.57 Hz (c) 50

4.22   0.255 H, 3.1 µF

4.23 $R_L = R_C = 5\Omega$

4.24   (a) 60 Hz (b) 0.747$\angle 51.5°$ A, 0.623 leading

  (c) 1.13$\angle -55.67°$ A, 0.564 lagging

4.25   There is no value of C for which resonance occurs.

4.26 9.1 µF or 62.6 µF

4.27 (a) 148.57 mH, 10.67 mH (b) $0.345 \angle -12.93°$ A,
8.345 $\angle -12.93°$ A

4.28 124.03 Ω, 50.57 µH, 0.408

4.29 (i) 79.62 Hz (ii) 10 (iii) 7.962 Hz

4.30 (i) 159 Hz (ii) 0.0002499 (iii) 4 (iv) 39.75 Hz, 140.13 Hz, 180.11 Hz

4.31 9696.96 rad/sec

4.32 25 Ω, 8 µF

4.33 26.5 µF

4.34 266.68 rad/sec, 730.27 rad/sec

4.35 40.82 Ω, 8.8 µF, 476.73 rad/sec

4.36 54.05 Hz, 56.3 Hz, 58.64 Hz

4.37 45.4 Hz

Chapter 5: Network Topology

5.1 Cut-set schedule:

tree-branch voltages and 'f' cut-sets	Branches						
	1	2	3	4	5	6	7
v_1, A (1, 5)	+1	0	0	0	+1	0	0
v_2, B (2, 5, 6)	0	+1	0	0	+1	−1	0
v_3, C (3, 6, 7)	0	0	+1	0	0	−1	+1
v_4, D (4, 7)	0	0	0	+1	0	0	−1

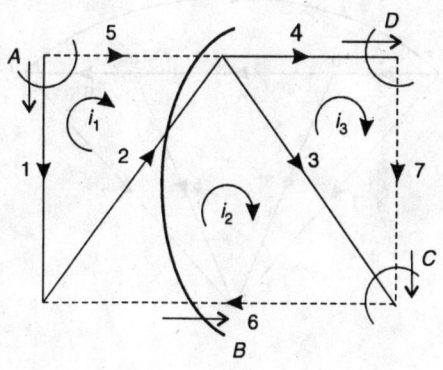

5.2

$$
\mathbf{B} = \begin{array}{c} \\ i_1 \\ i_2 \\ i_3 \end{array}
\begin{array}{cccccccc}
1 & 2 & 3 & 4 & 5 & 6 & 7 \\
\left[\begin{array}{ccccccc}
-1 & +1 & 0 & 0 & 0 & 0 & 0 \\
-1 & 0 & -1 & -1 & +1 & 0 & +1 \\
0 & 0 & 0 & 0 & -1 & +1 & 0
\end{array}\right]
\end{array}
$$

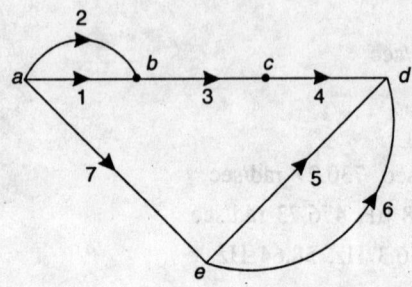

5.3

$$
\mathbf{Q} = \begin{array}{c} \\ v_1 \\ v_2 \\ v_3 \\ v_4 \end{array}
\begin{array}{cccccccc}
1 & 2 & 3 & 4 & 5 & 6 & 7 \\
\left[\begin{array}{ccccccc}
+1 & +1 & 0 & 0 & 0 & 0 & +1 \\
0 & 0 & +1 & 0 & 0 & 0 & +1 \\
0 & 0 & 0 & +1 & 0 & 0 & +1 \\
0 & 0 & 0 & 0 & +1 & +1 & -1
\end{array}\right]
\end{array}
$$

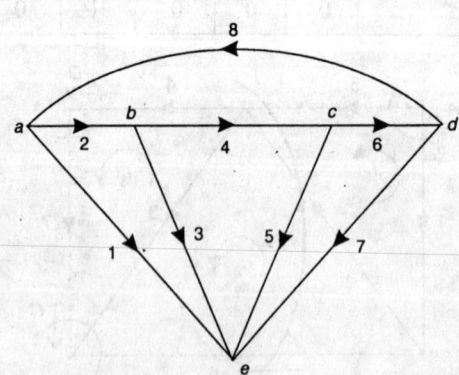

5.4

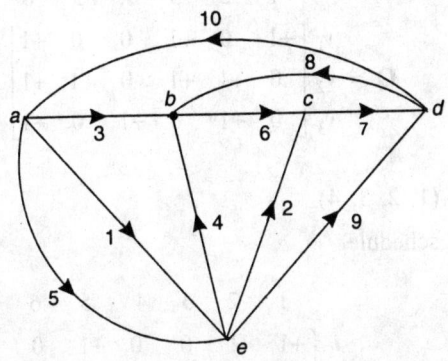

5.5 Cut-set matrix

$$
\begin{array}{c}
 \begin{array}{ccccc} 1 & 2 & 3 & 4 & 5 \end{array} \\
\mathbf{Q} = \begin{array}{c} v_2 \\ v_3 \\ v_4 \end{array}
\left[\begin{array}{ccccc}
+1 & +1 & 0 & 0 & 0 \\
0 & 0 & +1 & 0 & +1 \\
0 & +1 & 0 & +1 & -1
\end{array}\right]
\end{array}
$$

Branch admittance matrix

$$
\begin{bmatrix}
1/R_1 & 0 & 0 & 0 & 0 \\
0 & 1/C_2 s & 0 & 0 & 0 \\
0 & 0 & 1/R_3 & 0 & 0 \\
0 & 0 & 0 & 1/R_4 & 0 \\
0 & 0 & 0 & 0 & 1/C_5 s
\end{bmatrix}
$$

5.6 (i) i_1 (2, 4, 5), i_2 (1, 3, 5), i_3 (1, 4, 5, 6)

(ii) v_1 (1, 3, 6), v_2 (2, 3, 5, 6), v_3 (2, 4, 6)

$$
\begin{array}{c}
\phantom{\mathbf{B} = i_1} \begin{array}{cccccc} 1 & 2 & 3 & 4 & 5 & 6 \end{array} \\
\mathbf{B} = \begin{array}{c} i_1 \\ i_2 \\ i_3 \end{array}
\left[\begin{array}{cccccc}
0 & +1 & 0 & -1 & -1 & 0 \\
-1 & 0 & +1 & 0 & -1 & 0 \\
-1 & 0 & 0 & -1 & -1 & +1
\end{array}\right]
\end{array}
$$

$$\begin{array}{c} \\ v_1 \\ \mathbf{Q} = v_2 \\ v_3 \end{array} \begin{array}{cccccc} 1 & 2 & 3 & 4 & 5 & 6 \\ \begin{bmatrix} +1 & 0 & +1 & 0 & 0 & +1 \\ 0 & +1 & +1 & 0 & +1 & +1 \\ 0 & +1 & 0 & +1 & 0 & +1 \end{bmatrix} \end{array}$$

5.7 For tree (1, 2, 3, 4)

(i) Tie-set schedule

$$\begin{array}{c} \\ i_1 \\ i_2 \\ \mathbf{B} = i_3 \\ i_4 \end{array} \begin{array}{cccccccc} 1 & 2 & 3 & 4 & 5 & 6 & 7 & 8 \\ \begin{bmatrix} +1 & -1 & 0 & 0 & +1 & 0 & 0 & 0 \\ 0 & +1 & -1 & 0 & 0 & +1 & 0 & 0 \\ 0 & 0 & +1 & -1 & 0 & 0 & +1 & 0 \\ -1 & 0 & 0 & +1 & 0 & 0 & 0 & +1 \end{bmatrix} \end{array}$$

(ii) Cut-set schedule

$$\begin{array}{c} \\ v_1 \\ v_2 \\ \mathbf{Q} = v_3 \\ v_4 \end{array} \begin{array}{cccccccc} 1 & 2 & 3 & 4 & 5 & 6 & 7 & 8 \\ \begin{bmatrix} +1 & 0 & 0 & 0 & -1 & 0 & 0 & +1 \\ 0 & +1 & 0 & 0 & +1 & -1 & 0 & 0 \\ 0 & 0 & +1 & 0 & 0 & +1 & -1 & 0 \\ 0 & 0 & 0 & +1 & 0 & 0 & +1 & -1 \end{bmatrix} \end{array}$$

5.8 For tree (4, 5, 6)

(i) Tie-set schedule

$$\begin{array}{c} \\ i_1 \\ \mathbf{B} = i_2 \\ i_3 \end{array} \begin{array}{cccccc} 1 & 2 & 3 & 4 & 5 & 6 \\ \begin{bmatrix} +1 & 0 & 0 & -1 & 0 & -1 \\ 0 & +1 & 0 & -1 & +1 & 0 \\ 0 & 0 & +1 & 0 & -1 & -1 \end{bmatrix} \end{array}$$

(ii) Cut-set schedule

$$\begin{array}{c} \\ v_4 \\ \mathbf{Q} = v_5 \\ v_6 \end{array} \begin{array}{cccccc} 1 & 2 & 3 & 4 & 5 & 6 \\ \begin{bmatrix} +1 & +1 & 0 & +1 & 0 & 0 \\ 0 & -1 & +1 & 0 & +1 & 0 \\ -1 & 0 & -1 & 0 & 0 & -1 \end{bmatrix} \end{array}$$

5.9 Oriented graph

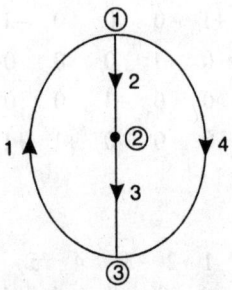

Node admittance matrix

$$
\begin{bmatrix}
\dfrac{1}{R_1} + \dfrac{1}{R_2} + C_4 s & -\dfrac{1}{R_2} & -\dfrac{1}{R_1} \\[2mm]
-\dfrac{1}{R_2} & \dfrac{1}{R_2} + \dfrac{1}{L_3 s} & -\dfrac{1}{L_3 s} \\[2mm]
-\dfrac{1}{R_1} & -\dfrac{1}{L_3 s} & \dfrac{1}{L_3 s} + \dfrac{1}{R_1} + C_4 s
\end{bmatrix}
$$

Loop impedance matrix

$$
\begin{bmatrix}
R_1 + R_2 + L_3 s & -(R_2 + L_3 s) \\[2mm]
-(R_2 + L_3 s) & R_2 + L_3 s + \dfrac{1}{C_4 s}
\end{bmatrix}.
$$

5.10 Oriented graph

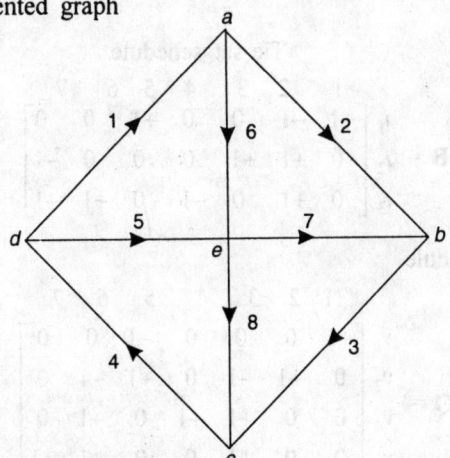

Tie-set schedule

$$\mathbf{B} = \begin{array}{c} \\ i_1 \\ i_2 \\ i_3 \\ i_4 \end{array} \begin{array}{cccccccc} 1 & 2 & 3 & 4 & 5 & 6 & 7 & 8 \\ \left[\begin{array}{cccccccc} +1 & 0 & 0 & 0 & -1 & +1 & 0 & 0 \\ 0 & +1 & 0 & 0 & 0 & -1 & -1 & 0 \\ 0 & 0 & +1 & 0 & 0 & 0 & +1 & -1 \\ 0 & 0 & 0 & +1 & +1 & 0 & 0 & +1 \end{array} \right] \end{array}$$

Cut-set schedule

$$\mathbf{Q} = \begin{array}{c} \\ v_5 \\ v_6 \\ v_7 \\ v_8 \end{array} \begin{array}{cccccccc} 1 & 2 & 3 & 4 & 5 & 6 & 7 & 8 \\ \left[\begin{array}{cccccccc} +1 & 0 & 0 & -1 & +1 & 0 & 0 & 0 \\ -1 & +1 & 0 & 0 & 0 & +1 & 0 & 0 \\ 0 & +1 & -1 & 0 & 0 & 0 & +1 & 0 \\ 0 & 0 & +1 & -1 & 0 & 0 & 0 & +1 \end{array} \right] \end{array}$$

5.11 Oriented graph

Tie-set schedule

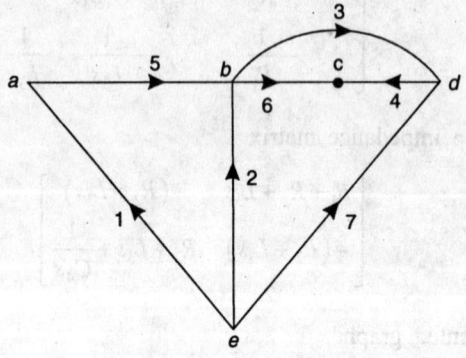

Tie-set schedule

$$\mathbf{B} = \begin{array}{c} \\ i_1 \\ i_2 \\ i_3 \end{array} \begin{array}{ccccccc} 1 & 2 & 3 & 4 & 5 & 6 & 7 \\ \left[\begin{array}{ccccccc} +1 & -1 & 0 & 0 & +1 & 0 & 0 \\ 0 & +1 & +1 & 0 & 0 & 0 & -1 \\ 0 & +1 & 0 & -1 & 0 & +1 & -1 \end{array} \right] \end{array}$$

Cut-set schedule

$$\mathbf{Q} = \begin{array}{c} \\ v_1 \\ v_2 \\ v_4 \\ v_7 \end{array} \begin{array}{ccccccc} 1 & 2 & 3 & 4 & 5 & 6 & 7 \\ \left[\begin{array}{ccccccc} +1 & 0 & 0 & 0 & -1 & 0 & 0 \\ 0 & +1 & -1 & 0 & +1 & -1 & 0 \\ 0 & 0 & +1 & +1 & 0 & +1 & 0 \\ 0 & 0 & +1 & 0 & 0 & +1 & +1 \end{array} \right] \end{array}$$

5.12 Tie-set schedule

loop currents and tie-sets	Branches							
	1	2	3	4	5	6	7	8
i_1 (1, 2, 3, 4)	+1	+1	+1	+1	0	0	0	0
i_2 (2, 3, 4, 5)	0	+1	+1	+1	+1	0	0	0
i_3 (2, 6)	0	−1	0	0	0	+1	0	0
i_4 (3, 7)	0	0	−1	0	0	0	+1	0
i_5 (4, 8)	0	0	0	−1	0	0	0	+1

Cut-set schedule

Tree branch voltages and f cut-sets	Branches							
	1	2	3	4	5	6	7	8
v_1 (1, 2, 5, 6)	−1	+1	0	0	−1	+1	0	0
v_2 (1, 3, 5, 7)	−1	0	+1	0	−1	0	+1	0
v_3 (1, 4, 5, 8)	−1	0	0	+1	−1	0	0	+1

5.13

$$
e_b = \begin{bmatrix} 0.3623 \\ -1.6 \\ 0.76 \\ 0.477 \\ -1.237 \\ -0.84 \end{bmatrix} \text{volts} \qquad i_b = \begin{bmatrix} 0.3623 \\ -0.3 \\ 0.25 \\ 0.12 \\ -0.247 \\ -0.14 \end{bmatrix} \text{amps}
$$

5.14 Tie-set schedule

Loop currents and tie-sets	Branches					
	1	2	3	4	5	6
i_1 (1, 2, 3)	−1	+1	+1	0	0	0
i_2 (3, 4, 5)	0	0	−1	+1	+1	0
i_3 (1, 4, 6)	+1	0	0	−1	0	+1

$$i_b = \begin{bmatrix} 0.695 \\ 0.428 \\ -0.053 \\ -0.64 \\ +0.4813 \\ 1.12 \end{bmatrix} A \qquad e_b = \begin{bmatrix} 0.695 \\ 0.856 \\ -0.159 \\ -2.56 \\ 2.4 \\ 6.74 \end{bmatrix} V$$

5.15 Tie-set schedule

Loop currents and tie-sets	Branches					
	1	2	3	4	5	6
i_1 (1, 2, 4)	−1	−1	0	+1	0	0
i_2 (2, 3, 5)	0	−1	−1	0	+1	0
i_3 (1, 3, 6)	−1	0	+1	0	0	+1

$$i_b = \begin{bmatrix} -0.434 \\ -0.106 \\ -0.062 \\ 0.186 \\ -0.08 \\ -0.1426 \end{bmatrix} amps$$

5.16 $A = \begin{bmatrix} +1 & +1 & +1 & +1 & 0 & 0 \\ -1 & -1 & 0 & 0 & +1 & +1 \end{bmatrix}$, $\begin{bmatrix} 3 & -1.5 \\ -1.5 & 3 \end{bmatrix}\begin{bmatrix} v_a \\ v_b \end{bmatrix} = \begin{bmatrix} 8 \\ 2 \end{bmatrix}$

$$e_b = \begin{bmatrix} 1.33 \\ 1.33 \\ 4.0 \\ 4.0 \\ 2.667 \\ 2.667 \end{bmatrix} volts \qquad i_b = \begin{bmatrix} 1.33 \\ 2.667 \\ 2.0 \\ -6.0 \\ 1.33 \\ 2.667 \end{bmatrix} amps$$

5.17 Tie-set schedule

Loop currents and tie-sets	Branches					
	1	2	3	4	5	6
i_1 (1, 2, 4)	+1	−1	0	+1	0	0
i_2 (1, 3, 5)	−1	0	+1	0	+1	0
i_3 (2, 3, 6)	0	+1	−1	0	0	+1

$$i_b = \begin{bmatrix} -0.286 \\ 0.286 \\ 0.0 \\ -0.857 \\ -0.57 \\ -0.57 \end{bmatrix} \text{A} \qquad e_b = \begin{bmatrix} -2.857 \\ +2.857 \\ 0.0 \\ 5.715 \\ -2.857 \\ -2.857 \end{bmatrix} \text{V}$$

5.18 Cut-set schedule

Tree-branch voltages and f cut-sets	Branches					
	1	2	3	4	5	6
v_1 (1, 4, 5)	+1	0	0	−1	+1	0
v_2 (2, 4, 6)	0	+1	0	+1	0	−1
v_3 (3, 5, 6)	0	0	+1	0	−1	+1

5.19

$$\mathbf{Q} = \begin{matrix} \\ v_1 \\ v_2 \end{matrix} \begin{matrix} 1 & 2 & 3 & 4 \\ \begin{bmatrix} +1 & 0 & +1 & 0 \\ 0 & +1 & -1 & -1 \end{bmatrix} \end{matrix}$$

Equilibrium equations are:

$$\begin{bmatrix} 3 & -1 \\ -1 & 4 \end{bmatrix} \begin{bmatrix} v_1 \\ v_2 \end{bmatrix} = \begin{bmatrix} 5 \\ 20 \end{bmatrix}$$

$$e_b = \begin{bmatrix} 3.636 \\ 5.91 \\ -2.274 \\ -5.91 \end{bmatrix} \text{volts} \qquad i_b = \begin{bmatrix} 2.272 \\ 5.91 \\ -2.274 \\ 8.18 \end{bmatrix} \text{amps}$$

5.20

$$\mathbf{B} = \begin{array}{c} \\ i_1 \\ i_2 \end{array} \begin{array}{cccc} 1 & 2 & 3 & 4 \\ \left[\begin{array}{cccc} -1 & +1 & +1 & 0 \\ 0 & +1 & 0 & +1 \end{array} \right] \end{array}$$

Equilibrium equations are:

$$\begin{bmatrix} 2.5 & 1 \\ 1 & 1.5 \end{bmatrix} \begin{bmatrix} i_1 \\ i_2 \end{bmatrix} = \begin{bmatrix} 2.5 \\ 10 \end{bmatrix}$$

$$i_b = \begin{bmatrix} 2.272 \\ 5.91 \\ -2.274 \\ 8.18 \end{bmatrix} \text{amps} \qquad e_b = \begin{bmatrix} 3.636 \\ 5.91 \\ -2.274 \\ -5.91 \end{bmatrix} \text{volts}$$

5.21 $(R_2 + R_3)i_1 - R_3 i_2 - R_2 i_3 = v_1(0) - e_3$ (1)

$$-R_3 i_1 + \left\{ R_3 + j\omega(L_4 + L_5) - j\frac{j}{\omega C_4} \right\} i_2 - \left\{ j\omega L_4 - j\frac{1}{\omega C_4} \right\} i_3 = e_3 \quad (2)$$

$$-R_2 i_1 - \left(j\omega L_4 - \frac{j}{\omega C_4} \right) i_2 + \left[R_2 + R_6 + j\omega(L_4 + L_6) \right.$$
$$\left. - j\left\{ \frac{1}{\omega C_4} + \frac{1}{\omega C_6} \right\} \right] i_3 = e_2 \qquad (3)$$

5.22

$$\mathbf{B} = \begin{array}{c} \\ i_1 \\ i_2 \\ i_3 \end{array} \begin{array}{ccccccc} 1 & 2 & 3 & 4 & 5 & 6 & 7 \\ \left[\begin{array}{ccccccc} +1 & 0 & 0 & +1 & +1 & 0 & 0 \\ -1 & +1 & -1 & 0 & 0 & 0 & 0 \\ -1 & 0 & -1 & 0 & -1 & +1 & +1 \end{array} \right] \end{array}$$

Equilibrium equations are:

$$\begin{bmatrix} 1.2285 & -1 & -1.0625 \\ -1 & 1.83 & 1.33 \\ -1.0625 & 1.33 & 1.9925 \end{bmatrix} \begin{bmatrix} i_1 \\ i_2 \\ i_3 \end{bmatrix} = \begin{bmatrix} 0.625 \\ 0 \\ -8.625 \end{bmatrix}$$

$$v_b = \begin{bmatrix} 0.895 \\ 2.4625 \\ 1.5625 \\ -0.637 \\ -0.26 \\ 2.67 \\ -0.466 \end{bmatrix} \text{ volts} \qquad \text{and} \qquad i_b = \begin{bmatrix} 0.895 \\ 4.925 \\ 4.735 \\ -3.84 \\ 5.82 \\ -9.66 \\ -9.66 \end{bmatrix} \text{ amps}$$

5.23

$$\mathbf{B} = \begin{array}{c} \\ i_1 \\ i_2 \\ i_3 \end{array} \begin{array}{cccccc} 1 & 2 & 3 & 4 & 5 & 6 \\ \begin{bmatrix} +1 & +1 & 0 & +1 & 0 & 0 \\ 0 & -1 & -1 & 0 & +1 & 0 \\ -1 & 0 & +1 & 0 & 0 & +1 \end{bmatrix} \end{array}$$

Equilibrium equations are:

$$\begin{bmatrix} 25 & -10 & -5 \\ -10 & 25 & -5 \\ -5 & -5 & 12 \end{bmatrix} \begin{bmatrix} i_1 \\ i_2 \\ i_3 \end{bmatrix} = \begin{bmatrix} -80 \\ 0 \\ 80 \end{bmatrix}$$

$$v_b = \begin{bmatrix} 40.43 \\ -22.87 \\ 28.135 \\ -17.6 \\ 5.27 \\ 12.3 \end{bmatrix} \text{ volts and } i_b = \begin{bmatrix} -7.914 \\ -2.287 \\ 5.627 \\ -1.76 \\ 0.527 \\ 6.154 \end{bmatrix} \text{ amps}$$

5.24

$$\mathbf{Q} = \begin{array}{c} \\ v_1 \\ v_2 \\ v_3 \end{array} \begin{array}{cccccc} 1 & 2 & 3 & 4 & 5 & 6 \\ \begin{bmatrix} +1 & 0 & 0 & -1 & 0 & +1 \\ 0 & +1 & 0 & 0 & +1 & -1 \\ 0 & 0 & +1 & +1 & -1 & 0 \end{bmatrix} \end{array}$$

Equilibrium equations are:

$$\begin{bmatrix} 0.4 & -0.1 & -0.1 \\ -0.1 & +0.4 & -0.1 \\ -0.1 & -0.1 & +0.4 \end{bmatrix} \begin{bmatrix} v_1 \\ v_2 \\ v_3 \end{bmatrix} = \begin{bmatrix} 0.5 \\ 0.0 \\ -0.5 \end{bmatrix}$$

$$v_b = \begin{bmatrix} 1.0 \\ 0.0 \\ -1.0 \\ -2.0 \\ 1.0 \\ 1.0 \end{bmatrix} \text{ volts} \qquad \text{and} \qquad i_b = \begin{bmatrix} 0.2 \\ 0.0 \\ -0.2 \\ 0.3 \\ 0.1 \\ 0.1 \end{bmatrix} \text{ amps}$$

5.25
$$\begin{bmatrix} 8 & 5 & -5 \\ 5 & 6 & -3 \\ -5 & -3 & 7 \end{bmatrix} \begin{bmatrix} i_1 \\ i_2 \\ i_3 \end{bmatrix} = \begin{bmatrix} -1 \\ -1 \\ -1 \end{bmatrix}$$

$$i_b = \begin{bmatrix} -0.3146 \\ 0.426 \\ -0.4157 \\ -0.112 \\ -0.011 \\ 0.1011 \end{bmatrix} \text{ amps} \qquad \text{and} \qquad v_b = \begin{bmatrix} -0.3146 \\ 4.852 \\ 5.168 \\ 4.888 \\ -0.033 \\ 5.202 \end{bmatrix} \text{ volts}$$

5.26

$$\mathbf{B} = \begin{array}{c} \\ i_1 \\ i_2 \\ i_3 \end{array} \begin{array}{c} \begin{array}{cccccc} 1 & 2 & 3 & 4 & 5 & 6 \end{array} \\ \begin{bmatrix} +1 & 0 & 0 & -1 & 0 & -1 \\ 0 & +1 & 0 & 0 & +1 & +1 \\ 0 & 0 & +1 & -1 & +1 & 0 \end{bmatrix} \end{array}$$

$$i_b = \begin{bmatrix} -0.55 \\ 0.866 \\ -0.916 \\ 1.466 \\ -0.05 \\ 1.416 \end{bmatrix} \text{amps} \quad \text{and} \quad v_b = \begin{bmatrix} 8.7 \\ -2.536 \\ 6.17 \\ 5.864 \\ -0.3 \\ 2.832 \end{bmatrix} \text{volts}$$

5.27 With tree branches as 3 and 4 (6 is not a branch).

$$\mathbf{B} = \begin{matrix} & \begin{matrix} 1 & 2 & 3 & 4 & 5 \end{matrix} \\ \begin{matrix} i_1 \\ i_2 \\ i_3 \end{matrix} & \begin{bmatrix} 0 & 0 & +1 & -1 & +1 \\ 0 & +1 & +1 & -1 & 0 \\ 0 & 0 & +1 & 0 & +1 \end{bmatrix} \end{matrix}$$

$$\mathbf{Q} = \begin{matrix} & \begin{matrix} 1 & 2 & 3 & 4 & 5 \end{matrix} \\ \begin{matrix} v_3 \\ v_4 \end{matrix} & \begin{bmatrix} -1 & -1 & +1 & 0 & -1 \\ +1 & +1 & 0 & +1 & 0 \end{bmatrix} \end{matrix}$$

5.28

$$\mathbf{B} = \begin{matrix} & \begin{matrix} 1 & 2 & 3 & 4 \end{matrix} \\ \begin{matrix} i_1 \\ i_2 \end{matrix} & \begin{bmatrix} +1 & 0 & +1 & -1 \\ 0 & +1 & 0 & +1 \end{bmatrix} \end{matrix}$$

Equilibrium equations are:

$$\begin{bmatrix} 8 & -4 \\ -4 & 6 \end{bmatrix} \begin{bmatrix} i_1 \\ i_2 \end{bmatrix} = \begin{bmatrix} -15 \\ 10 \end{bmatrix}$$

$$i_b = \begin{bmatrix} -1.5625 \\ 0.625 \\ -1.5625 \\ 2.1875 \end{bmatrix} \text{amps} \quad v_b = \begin{bmatrix} -1.5625 \\ -8.75 \\ 10.3125 \\ 8.75 \end{bmatrix} \text{volts}$$

5.29

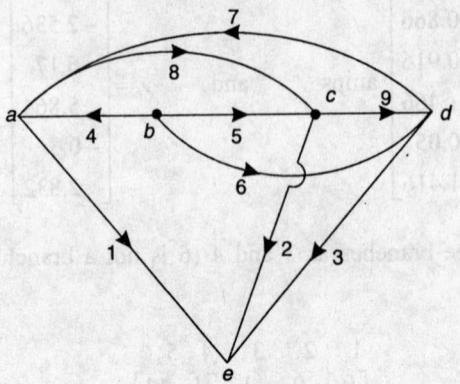

5.30

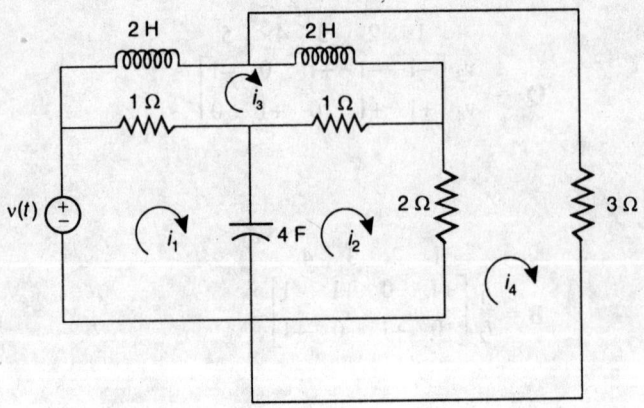

The loop equations are:

$$1(i_1 - i_3) + \frac{1}{4}\int(i_1 - i_2)\,dt = v(t)$$

$$\frac{1}{4}\int(i_2 - i_1)\,dt + 1(i_2 - i_3) + 2(i_2 - i_4) = 0$$

$$1(i_3 - i_1) + 2\frac{di_3}{dt} + 2\frac{d}{dt}(i_3 - i_4) + 1(i_3 - i_2) = 0$$

$$3i_4 + 2(i_4 - i_2) + 2\frac{d}{dt}(i_4 - i_3) = 0$$

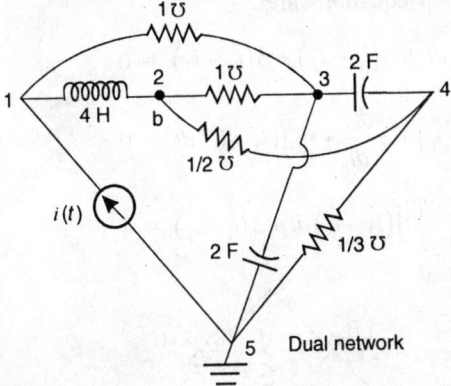

Dual network

The nodal equations are:

$$1(v_1 - v_3) + \frac{1}{4}\int(v_1 - v_2)dt = i(t)$$

$$\frac{1}{4}\int(v_2 - v_1)dt + 1(v_2 - v_3) + \frac{1}{2}(v_2 - v_4) = 0$$

$$1(v_3 - v_1) + 2\frac{dv_3}{dt} + 2\frac{d}{dt}(v_3 - v_4) + 1(v_3 - v_2) = 0$$

$$\frac{1}{3}v_4 + \frac{1}{2}(v_4 - v_2) + 2\frac{d}{dt}(v_4 - v_3) = 0$$

5.31

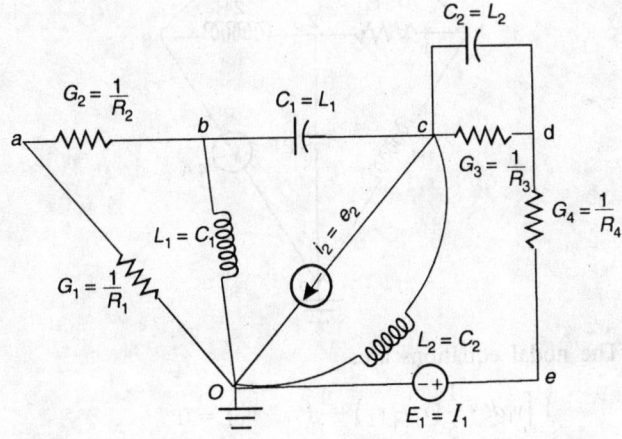

5.32 The loop equations are:

$$1\int i_1 \, dt + 5(i_1 - i_2) + 3(i_1 - i_3) = 0$$

$$5(i_2 - i_1) + 2\frac{di_2}{dt} + \frac{1}{2}\int(i_2 - i_3)\, dt = 0$$

$$\frac{1}{2}\int(i_2 - i_3)\, dt + 3(i_1 - i_3) = 4$$

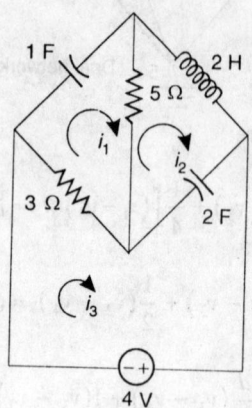

Dual network

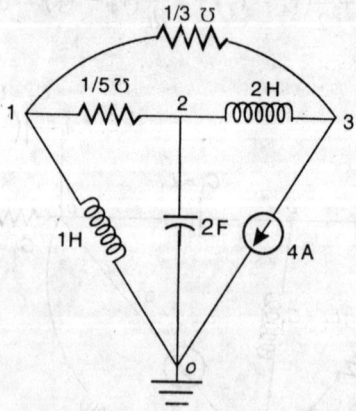

The nodal equations are:

$$1\int v_1 dt + \frac{1}{5}(v_1 - v_2) + \frac{1}{3}(v_1 - v_3) = 0$$

$$\frac{1}{5}(v_2 - v_1) + 2\frac{dv_2}{dt} + \frac{1}{2}\int(v_2 - v_3)\, dt = 0$$

$$\frac{1}{2}\int(v_2 - v_3)\,dt + \frac{1}{3}(v_1 - v_3) = 4$$

5.33

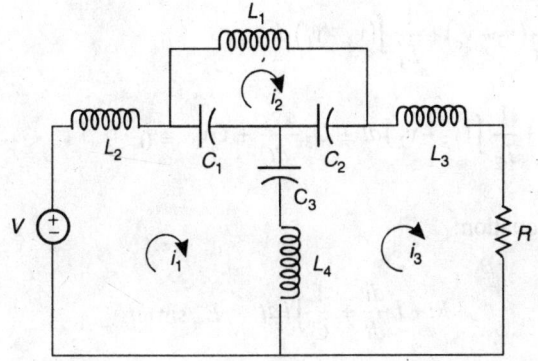

The loop equations are:

$$L_2\frac{di_1}{dt} + \frac{1}{C_1}\int(i_1 - i_2)\,dt + \frac{1}{C_3}\int(i_1 - i_3)\,dt + L_4\frac{d}{dt}(i_1 - i_3) = V \quad (1)$$

$$L_1\frac{di_2}{dt} + \frac{1}{C_2}\int(i_2 - i_3)\,dt + \frac{1}{C_1}\int(i_2 - i_1)\,dt = 0 \quad (2)$$

$$L_4\frac{d}{dt}(i_3 - i_1) + \frac{1}{C_3}\int(i_3 - i_1)\,dt + \frac{1}{C_2}\int(i_3 - i_2)\,dt + L_3\frac{di_3}{dt} + Ri_3 = 0 \quad (3)$$

Dual network

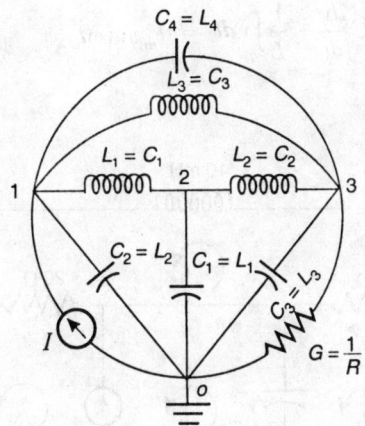

The nodal equations are:

$$C_2\frac{dv_1}{dt} + \frac{1}{L_1}\int(v_1 - v_2)\,dt + \frac{1}{L_3}\int(v_1 - v_3)\,dt + C_4\frac{d}{dt}(v_1 - v_3) = I \quad (4)$$

$$C_1 \frac{dv_2}{dt} + \frac{1}{L_2} \int (v_2 - v_3) dt + \frac{1}{L_1} \int (v_2 - v_1) dt = 0 \qquad (5)$$

$$C_4 \frac{d}{dt} (v_3 - v_1) + \frac{1}{L_3} \int (v_3 - v_1) dt$$

$$+ \frac{1}{L_2} \int (v_3 - v_2) dt + C_3 \frac{dv_3}{dt} + Gv_3 = 0 \qquad (6)$$

5.34 Loop equation:

$$Ri + L\frac{di}{dt} + \frac{1}{C} \int i \, dt = E_m \sin \omega t \qquad (1)$$

Dual network:

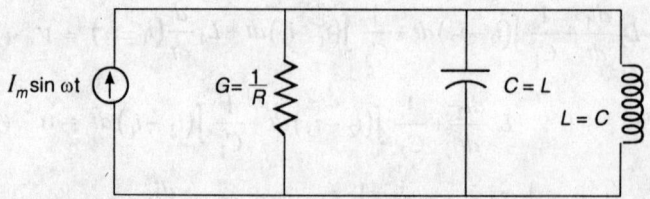

Nodal equation:

$$GV + C\frac{dv}{dt} + \frac{1}{L} \int v \, dt = I_m \sin \omega t \qquad (2)$$

5.35

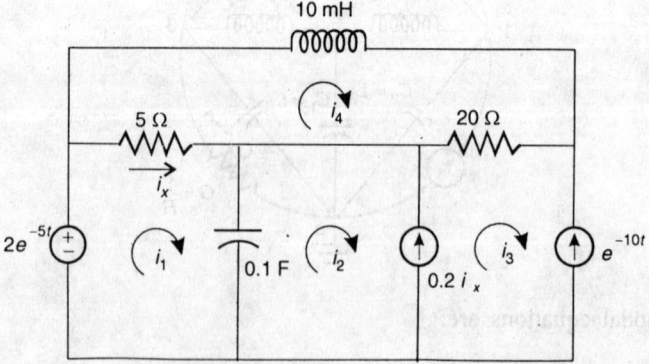

The loop equations are:

$$i_3 - i_2 = 0.2i_x \tag{1}$$

$$i_3 = -e^{-10t} \tag{2}$$

$$i_x = i_1 - i_4 \tag{3}$$

$$5(i_1 - i_4) + \frac{1}{0.1}\int(i_1 - i_2)dt = 2e^{-5t} \tag{4}$$

$$5(i_4 - i_1) + 10 \times 10^{-3}\frac{di_4}{dt} + 20(i_4 - i_3) = 0 \tag{5}$$

The dual graph is

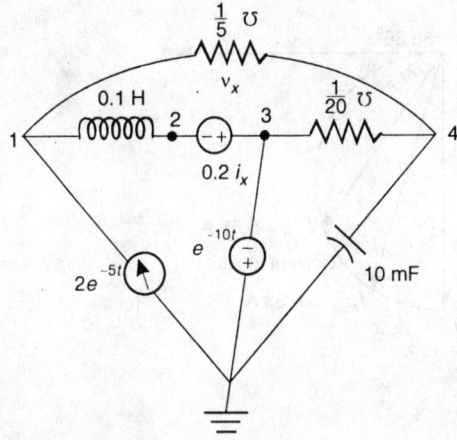

The nodal equations are:

$$v_3 - v_2 = 0.2v_x \tag{6}$$

$$v_3 = -e^{-10t} \tag{7}$$

$$v_x = v_1 - v_4 \tag{8}$$

$$\frac{1}{5}(v_1 - v_4) + \frac{1}{0.1}\int(v_1 - v_2)dt = 2e^{-5t} \tag{9}$$

$$\frac{1}{5}(v_4 - v_1) + 10 \times 10^{-3}\frac{dv_4}{dt} + \frac{1}{20}(v_4 - v_3) = 0 \tag{10}$$

5.36

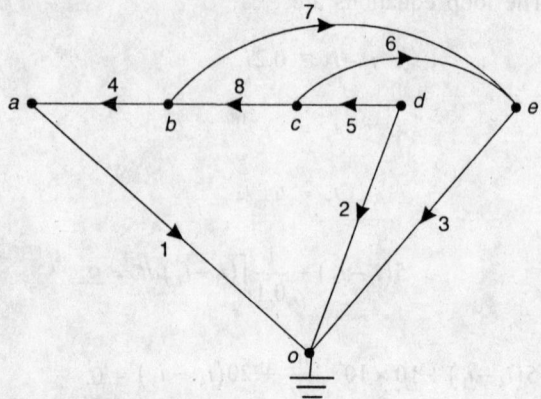

Chapter 6: Locus Diagrams

6.1

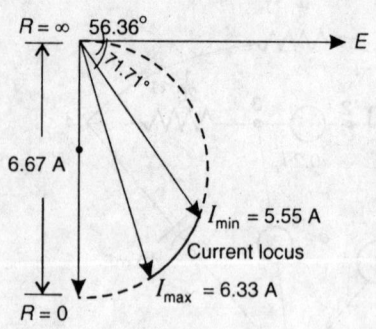

6.2

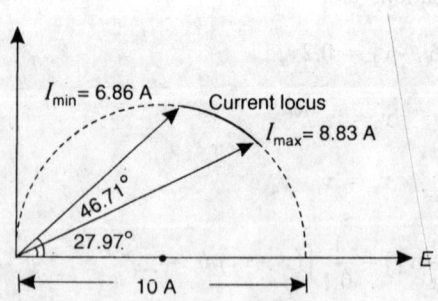

6.3

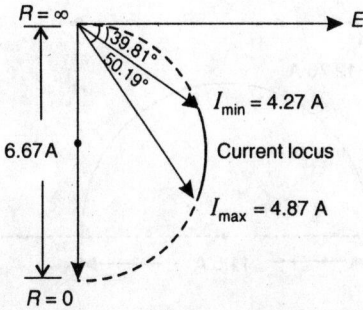

6.4

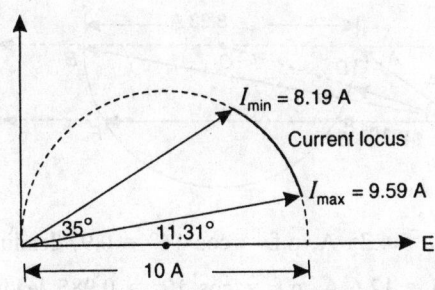

6.5

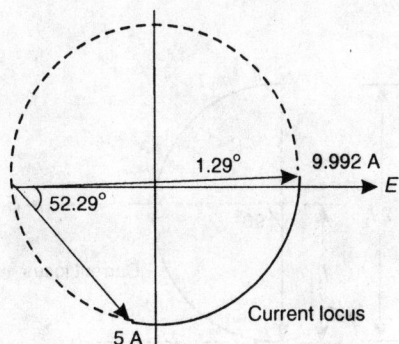

6.6

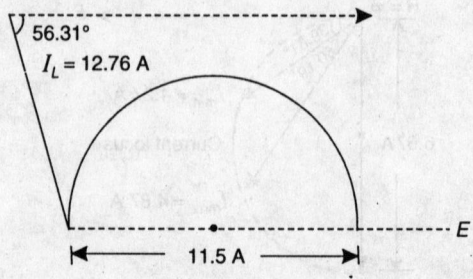

The circuit p.f. will never be unity.

6.7

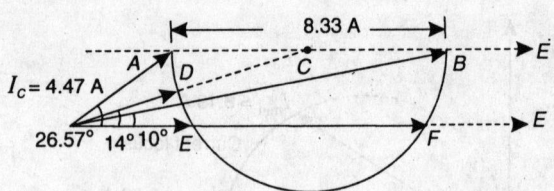

(i) $I_{min} = OD = 4.25$ A, p.f. $= \cos 14° = 0.97$ leading.

(ii) $I_{max} = OB = 12.6$ A, p.f. $= \cos 10° = 0.985$ leading.

(iii) u.p.f. currents are, $OE = 4.7$ A and $OF = 11.9$ A.

Corresponding inductive reactances are $46.08\,\Omega$ and $2.2\,\Omega$.

6.8

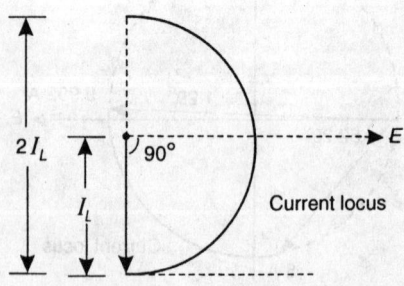

6.9

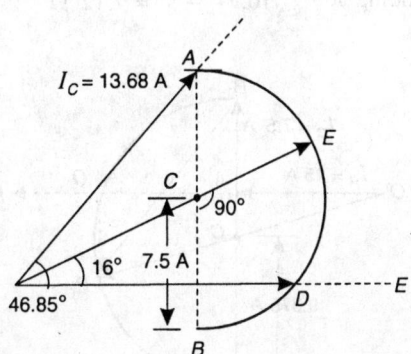

$I_C = 13.68$ A

90°

C

16° 7.5 A

46.85°

(i) u.p.f. current $OD = 16$ A
(ii) minimum p.f. $= \cos 46.85° = 0.684$ leading
     Corresponding current $= 13.68$ A
(iii) $I_{max} = OE = 17$ A
     Corresponding p.f. $= \cos 16° = 0.96$ leading.

6.10 Maximum p.f. $= \cos 25° = 0.906$
Corresponding current $= OD = 5.8$ A

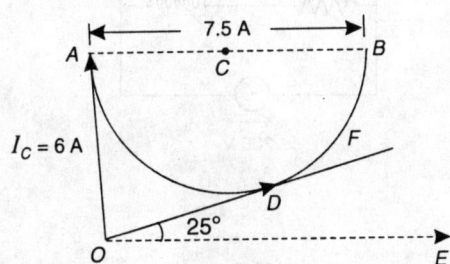

7.5 A

A C B

$I_C = 6$ A F

D

25°

O E

6.11 $I_m = OP = 25$ A,

In phase component of $I_m = OQ = 24.2$ A

$P_m = 150 \times 24.2$

$= 3630$ W, p.f. $\cos 8° = 0.99$

Corresponding current $I_L = BP = 14$ A

Corresponding $Z = \dfrac{150}{14} = 10.71 \ \Omega$

Corresponding $R = \sqrt{10.71^2 - 8^2} = 7.12 \; \Omega$

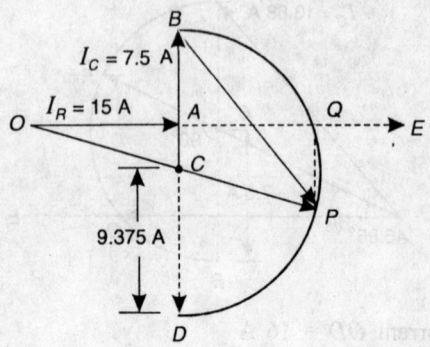

6.12

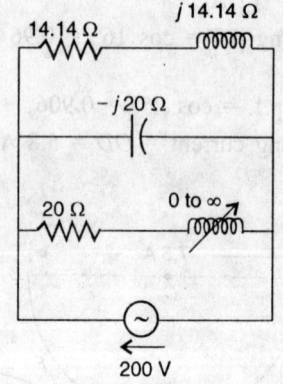

6.13

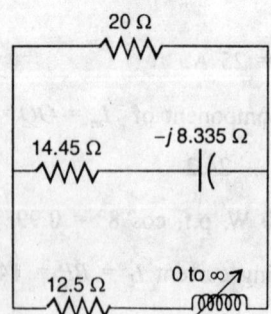

6.14

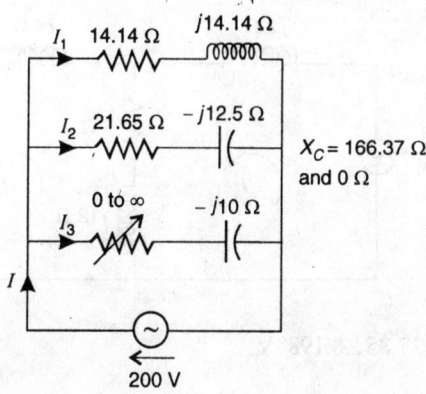

Chapter 7: Coupled Circuits

7.1 137.36 mH, 12.64 mH, 25 mH

7.2 290 μH, 440 μH

7.3 150 mH

7.4 $L_1 = L_2 = 2$mH, $K = 0.25$

7.5 19.2 mH

7.6 16.282 $\angle$ 32.166° volts

7.7

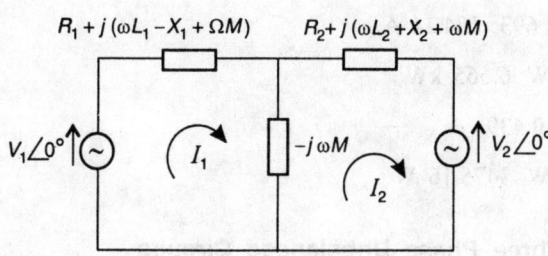

7.8 $E_0 = 4.519\angle9.49°$ volts

$Z_0 = 11.573\angle32.47°$ Ω

7.9 0.0398

7.10 0.67$\angle$21.376° A

7.11

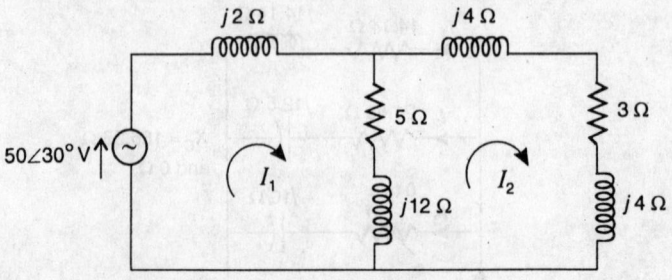

7.12 $E_0 = 27.85\angle 8.198° \text{ V}$

$Z_0 = 21.42\angle -247.2° \ \Omega$

$I = 1.15\angle -96.12° \text{ A}$

Chapter 8: Three Phase Balanced Circuits

8.1 8 A, 13.86 A, 7682 W

8.2 12.41 A, 4617.07 W, 4402.27 W, 214.795 W

8.3 9.24 A, 9.24 A, 5121.33 W, 0.8 lagging

8.4 3000 W, 0.866, When p.f. is less than 0.5

8.5 5 Ω , 12.47 mH

8.6 (10.39 + j33.046) Ω

8.7 50 kW, 0.693, 104.14 A

8.8 13.335 kW, 6.665 kW

8.9 4400 W, 0.439

8.10 8783.17 W, 3475.16 W

Chapter 9: Three Phase Unbalanced Circuits

9.1 (a) *RYB* sequence: $60 \angle -30° \text{A}$, $64.5 \angle 161.93°$ A, $13.7 \angle 46.92°$ A, 26.118 kW

   (b) *RBY* sequence: 34.64 $\angle 0°$ A, 48.08 $\angle 116.3°$ A,

   45.115 $\angle -107.2°$ A, 26.118 kW

9.2 (i) 12 kW, 17.263 kW, (ii) 4 kW, 25.24 kW

9.3 (i) 4.657 ∠137.6° A, 4.73 ∠42.17° A, 6.32 ∠−90.63° A

(ii) 200.1 Ω

9.4 342.8 ∠−123.74° V

9.5 18.5 kW, 20.5 kW

9.6 1:1.732 ∠150°:1∠300°

9.7 $(8.66 − j 5)$ Ω

9.8 10.67 kW, 19.184 kW, $W_1 + W_2$ = 29.854 kW

Actual power = 28.524 kW

9.9 3.3 kW, 24.3 kW

9.10 − 0.428 kW

9.11 − 0.346 kW

9.12 (a) 18.8 ∠− 40° A, 6.86 ∠170.6° A, 13.28 ∠125.05° A

(b) 265.8 ∠5° V, 137.2 ∠170.6° A, 375.5 ∠80.05° A

(c) 8 kW

9.13 4618.8 W, 5334.7 W

9.14 (a) 80.8 ∠32.15° V

(b) 206 ∠−50.3° V, 311.7 ∠−149.4° V

9.15 75.25 ∠−79.96° A, 141.68 ∠111° A, 56.1 ∠−59.2° A,

13.46 ∠132.35° A

9.16 309.8 ∠−193.3° V, 705.11 ∠−5.8° V, 429.85 ∠−76.34° V,
293.26 ∠69.75° V, 31.53 ∠−69.23° A, 14.33 ∠−76.3° A,
45.82 ∠108.4° A

9.17 2.957 ∠153.7° Ω

9.18 247.12 W

9.19 −1.69 kW

9.20 23.15%, 86%, 77.46%

NETWORK ANALYSIS

Chapter 10: Non-Sinusoidal Waveforms

10.1 $\quad i = \dfrac{I_m}{\pi}\left[1 - \dfrac{\pi}{2}\sin\theta - \dfrac{2}{3}\cos 2\theta - \dfrac{2}{15}\cos 4\theta \cdots\right]$ amps

10.2 $\quad v = \dfrac{100}{\pi}\left[1 + \dfrac{\pi}{2}\cos\theta + \dfrac{2}{3}\cos 2\theta - \dfrac{2}{15}\cos 4\theta + \dfrac{2}{35}\cos 6\theta + \cdots\right]$ volts

10.3 $\quad i = \dfrac{I}{4} + \dfrac{I}{\pi}\left[\sin\theta - \dfrac{1}{2}\sin 2\theta + \dfrac{1}{3}\sin 3\theta + \cdots\right]$

$$-2I\left[\dfrac{1}{\pi^2}\cos\theta + \dfrac{1}{(3\pi)^2}\cos 3\theta + \cdots\right] \text{A}$$

10.4 $\quad v = \dfrac{V}{2} - \dfrac{V}{\pi}\left[\sin\theta + \dfrac{1}{2}\sin 2\theta + \dfrac{1}{3}\sin 3\theta \right.$

$$\left. + \dfrac{1}{5}\sin 5\theta + \cdots + \dfrac{1}{n}\sin n\theta\right] \text{V}$$

10.5 $\quad v = \dfrac{800}{\pi}\left[\sin\theta + \dfrac{1}{3}\sin 3\theta + \dfrac{1}{5}\sin 5\theta + \cdots + \dfrac{1}{n}\sin n\theta\right]$ V

10.6 $\quad i = \dfrac{5}{2} + \dfrac{30}{\pi}\left[\sin\theta + \dfrac{1}{3}\sin 3\theta + \dfrac{1}{5}\sin 5\theta + \cdots + \dfrac{1}{n}\sin n\theta\right]$ A

10.7 $\quad v = \dfrac{4\,V}{\pi^2}\left[\cos\theta + \dfrac{1}{9}\cos 3\theta + \dfrac{1}{25}\cos 5\theta + \ldots\right]$

$$-\dfrac{2\,V}{\pi}\left[\sin\theta + \dfrac{1}{3}\sin 3\theta + \dfrac{1}{5}\sin 5\theta + \cdots\right] \text{V}$$

10.8 $\quad v = \dfrac{V}{2} + \dfrac{4\,V}{\pi^2}\cos\theta + \dfrac{4\,V}{(3\pi)^2}\cos 3\theta + \dfrac{4\,V}{(5\pi)^2}\cos 5\theta +$

$$\ldots + \dfrac{4\,V}{(n\pi)^2}\cos n\theta\right] \text{V, When } n \text{ is odd.}$$

10.9 $v = \dfrac{V}{6} + \dfrac{2V}{\pi}\left[\dfrac{1}{2}\cos\theta + \dfrac{\sqrt{3}}{4}\cos 2\theta\right.$

$$\left. + \dfrac{1}{3}\cos 3\theta + \dfrac{1}{10}\cos 5\theta + \cdots\right]V$$

10.10 $v = \dfrac{V}{2\pi} + \dfrac{V}{2\pi}\cos\theta + \displaystyle\sum_{n=2}^{\infty}\dfrac{V\left[n\sin\dfrac{n\pi}{2}-1\right]}{\pi\left(n^2-1\right)}\cos n\theta \text{ V}$

10.11 $v = \dfrac{2V}{\pi}\left[1 - \dfrac{2}{3}\cos 2\theta - \dfrac{2}{15}\cos 4\theta - \dfrac{2}{35}\cos 6\theta - \cdots\right]V$

10.12 $v = \dfrac{200}{\pi}\left[1 + \dfrac{2}{3}\cos 2\theta - \dfrac{2}{15}\cos 4\theta + \dfrac{2}{35}\cos 6\theta - \cdots\right]V$

10.13 $v = \dfrac{4}{\pi}\left[\sin\theta + \dfrac{1}{3}\sin 3\theta + \dfrac{1}{5}\sin 5\theta + \cdots + \dfrac{1}{n}\sin n\theta\right]V$

10.14 $v = \dfrac{4V}{\pi}\left[\cos\theta - \dfrac{1}{3}\cos 3\theta + \dfrac{1}{5}\cos 5\theta - \dfrac{1}{7}\cos 7\theta + \cdots\right]V$

10.15 $v = \dfrac{80}{\pi^2}\left[\sin\dfrac{\pi}{2}t - \dfrac{1}{9}\sin\dfrac{3\pi}{2}t + \dfrac{1}{25}\sin\dfrac{5\pi}{2}t + \cdots\right]V$

10.16 $e = 45 + 1204\sin\left(1000\,t + 86.63°\right)$

$\qquad\qquad + 256.13\sin\left(3000t + 129.44°\right)$ volts, 45296.93 watts.

10.17 $i = 16\sin\left(377t + 36.87°\right) + 5.32\sin\left(1131t + 43.76°\right)$ amps,

$\qquad\qquad\qquad\qquad\qquad\qquad$ 11.92 amps, 0.835.

10.18 $i = 9.83\sin\left(314t - 38.13°\right) + 0.977\sin\left(942t - 6.99°\right)$

$\qquad\qquad\qquad + 0.247\sin\left(1570t + 73.87°\right)$ A,

$\qquad\qquad$ 180.83 V, 990.217 W, 0.784 lagging.

10.19 $i = 2.58\sin\left(314t + 71.8°\right)$

$\qquad\qquad + 0.765\sin\left(942t + 17.12°\right)$ amps, 87.89 W, 0.311 leading.

10.20 $i = 34.99\sin(\omega t + 14.04°) + 18.116\sin(3\omega t + 62.783°)$

$$- 7.152\sin(5\omega t - 28.776°) \text{ A, } 3140.96 \text{ W, } 0.977.$$

10.21 (a)   81 V, 16.2 A, 918.54 W, 0.7

(b)   85.15 V, 44.86 A, 2183.924 W, 0.572.

(c)   81 V, 16.2 A, 425.82 W, 0.325.

(d)   218.66 V, 25.087 A, 1374.33 W, 0.251.

10.22 48.085 V, 195.024 W

10.23 $i = 10 + 3.53\sin(500\, t - 28.1°)$ A

Chapter 11: Initial Conditions

11.1 $i_{R_1}(0+) = \dfrac{V}{R_1 + R_3},$ $\quad i_{R_2}(0+) = 0,$

$$i_{R_3}(0+) = \dfrac{V}{R_1 + R_3}, \qquad i_L(0+) = 0, \ i_C(0+) = 0$$

$$i_{R_1}(\infty) = \dfrac{V(R_2 + R_3)}{\Sigma R_1\, R_2}$$

$$i_{R_2}(\infty) = i_L(\infty) = \dfrac{V\,R_3}{\Sigma\, R_1\, R_2}$$

$$i_{R_3}(\infty) = \dfrac{V\,R_2}{\Sigma\, R_1\, R_2}, \qquad i_C(\infty) = 0$$

11.2 $\dfrac{di_1}{dt}(0+) = 0$ A/ sec, $\qquad \dfrac{di_2}{dt}(0+) = \dfrac{13}{3}$ A/ sec,

$$\dfrac{di_3}{dt}(0+) = 0 \text{ A/ sec}, \qquad \dfrac{dv_c}{dt}(0+) = 0 \text{ V/ sec},$$

$$\dfrac{dv_L}{dt}(0+) = -\dfrac{13}{3} \text{ V/sec}$$

11.3 $R = 10^4\ \Omega,$ $\qquad\qquad C = 2.5\ \mu\text{ F},$

$$i(t) = 0.01e^{-40\,t}$$

11.4 $i(0+) = 2$ A, $\qquad\qquad \dfrac{di}{dt}(0+) = -20$ A/sec,

$\dfrac{d^2i}{dt^2}(0+) = -1.9988$ A / sec^2

11.5 $i(0+) = 0$ A, $\qquad\qquad \dfrac{di}{dt}(0+) = 10$ A/sec,

$\dfrac{d^2i}{dt^2}(0+) = -100$ A/sec^2

11.6 $v_1(0+) = 0$ V, $\qquad\qquad v_2(0+) = 0$ V,

$\dfrac{dv_1}{dt}(0+) = 2$ V/sec, $\qquad\qquad \dfrac{dv_2}{dt}(0+) = 0$ V/sec

11.7 $v(0+) = 100$ V, $\qquad\qquad \dfrac{dv}{dt}(0+) = -10^4$ V/sec,

$\dfrac{d^2v}{dt^2}(0+) = 10^6$ V/sec,

11.8 $v_a(0-) = 5$ V, $v_a(0+) = 3$ V

11.9 $\dfrac{d^2i_1}{dt^2}(0+) = 3.78125 \times 10^{-5}$ A/sec^2,

$\dfrac{d^2i_2}{dt^2} = 1.40625 \times 10^{-5}$ A/sec^2

11.10 $v_1(0+) = V_0, \dfrac{dv_1}{dt}(0+) = 0$ V/sec,

$\dfrac{d^2v_1}{dt^2}(0+) = 0$ V / sec^2

11.11 $i(0+) = 2$ A, $\qquad\qquad \dfrac{di}{dt}(0+) = -20$ A/sec,

$\dfrac{d^2i}{dt^2}(0+) = -0.5 \times 10^6$ A/sec^2

11.12 (i) $v_1(0+) = 0$ V, $\qquad\qquad\qquad$ $v_2(0+) = 0$ V

$\qquad$ (ii) $v_1(\infty) = 0$ V, $\qquad\qquad\qquad$ $v_2(\infty) = 5$ V

$\qquad$ (iii) $\dfrac{dv_1}{dt}(0+) = 0$ V/sec,

$\qquad\qquad$ $\dfrac{dv_2}{dt}(0+) = 2.5 \times 10^5$ V/sec

$\qquad$ (iv) $\dfrac{d^2 v_2}{dt^2}(0+) = 2.5 \times 10^5$ V/sec

11.13 $\qquad$ $v_1(0+) = 100$ V, $\qquad\qquad$ $v_2(0+) = 0$ V

$\qquad\qquad$ $\dfrac{dv_1}{dt}(0+) = -1000$ V/sec, $\qquad$ $\dfrac{dv_2}{dt}(0+) = 0$ V/sec

11.14 $\qquad$ $i_1(0+) = 2$ A, $\qquad\qquad\qquad$ $i_2(0+) = 1.33$ A,

$\qquad\qquad$ $\dfrac{di}{dt}(0+) = -66.99 \times 10^3$ A/sec,

$\qquad\qquad$ $\dfrac{di_2}{dt}(0+) = 3.365$ A/sec,

11.15 $\qquad$ $v_i(0+) = \dfrac{VR_2}{R_1 + R_3} = v_2(0+)$

$\qquad\qquad$ $v_1(\infty) = \dfrac{V_1 R_2}{R_1 + R_2} = v_2(\infty)$

11.16 $\qquad$ $i_1(0+) = 3.33$ A, $\qquad\qquad$ $i_2(0+) = 1.667$ A,

$\qquad\qquad$ $\dfrac{di_1}{dt}(0+) = 33.4$ A/sec,

$\qquad\qquad$ $\dfrac{di_2}{dt}(0+) = -83.35 \times 10^3$ A/sec,

$\qquad\qquad$ $\dfrac{di_1}{dt}(\infty) = 0$ A/sec

11.17 $\qquad$ $v_a(0+) = V$ volts, $\qquad\qquad$ $v_{c_1}(\infty) = V$ volts

11.18 $\qquad$ $i(0+) = 0.1$ A, $\qquad\qquad$ $\dfrac{di}{dt}(0+) = -100$ A/sec,

$$\frac{d^2i}{dt^2}(0+) = -9 \times 10^5 \text{ A/sec}^2$$

11.19 $\quad \dfrac{di_1}{dt}(0+) = 0 \text{ A/sec}$ $\qquad\qquad \dfrac{di_2}{dt}(0+) = \dfrac{V}{L} \text{ A/sec,}$

$$\frac{d^2i_2}{dt^2}(0+) = -\frac{V(R_1 + R_2)}{L} \text{ A/sec}$$

11.20 $\quad \dfrac{d^2v_C}{dt^2}(0+) = 0 \text{ V/sec}^2,$ $\qquad\qquad \dfrac{d^3v_C}{dt^3}(0+) = -4096 \text{ V/sec}^3,$

Chapter 12: Laplace Transformation

12.1 (i) $\quad \dfrac{60}{s^4} - \dfrac{5s}{s^2 + 9} + \dfrac{40}{s^2 + 25}$

(ii) $\quad \dfrac{1}{4}\left[\dfrac{9}{(s+3)^2 + 9} - \dfrac{27}{(s+3)^2 + 81}\right]$

(iii) $\quad \dfrac{1}{4}\left[\dfrac{s-3}{(s-3)^2 + 36} - \dfrac{3(s-3)}{(s-3)^2 + 4}\right]$

(iv) $\quad \dfrac{-2(s+a)\left[(s+a)^2 + \omega^2\right] + 4(s+a)\left[\omega^4 - (s+a)^4\right]}{\left[(s+a)^2 + \omega^2\right]^4}$

(v) $\quad \left[\dfrac{2s(s\cos\theta - \omega\sin\theta) - \left(s^2 + \omega^2\right)\cos\theta}{\left(s^2 + \omega^2\right)^2}\right]$

12.2 (i) $\quad e^{-5t} + e^{-t}$

(ii) $\quad e^{-2jt} + e^{2jt} - e^{-\sqrt{5}jt} - e^{\sqrt{5}jt}$

(iii) $\quad (0.5 - j1)e^{-(1-j2)t} + (0.5 + j1)e^{-(1+j2)t}$

(iv) $\quad e^{-t} - e^{-2t} - te^{-2t}$

(v) $\quad \dfrac{1}{8}e^{-2t} + \dfrac{3+j4}{8+j8}e^{-j2t} + \dfrac{-3+j4}{-8-j8}e^{j2t}$

(vi) $\dfrac{2 + j\sqrt{2}}{j\,2\sqrt{2}} e^{\left(2 + j2\sqrt{2}\right)t} + \dfrac{2 - j\sqrt{2}}{-j\,2\sqrt{2}} e^{-\left(-2 + j\sqrt{2}\right)t}$

(vii) $\dfrac{1}{2} e^{t} - e^{2t} + 2.5\, e^{3t}$

(viii) $-\dfrac{1}{3} + t + \dfrac{1 + j2}{j3} e^{-j3t} + \dfrac{1 - j2}{-j3} e^{j3t}$

12.3 (i) $1 + e^{-(2 - j2)t} + e^{-(2 + j2)t}$

(ii) $2e^{t} - 3t\, e^{t} + 3t^{2}$

(iii) $\dfrac{1}{2} e^{-j2t} + \dfrac{1}{2} e^{j2t} - \dfrac{1}{j2\sqrt{3}} e^{-\left(1 - j\sqrt{3}\right)t} + \dfrac{1}{j2\sqrt{3}} e^{-\left(1 + j\sqrt{3}\right)t}$

(iv) $-e^{-t} + e^{-3t} + 6t\, e^{-3t}$

(v) $10e^{-t} + 6e^{-2t} - 16e^{-\frac{3}{2}t}$

12.4 (i) $0, -\dfrac{1}{a^{2}}$ (ii) $1, 2.5$

(iii) $2, 0$ (iv) $0, 0.4$

(v) $\infty, 0$ (vi) $\sin\theta, 0$

(vii) $8, 0$

12.5 (i) $\dfrac{1}{2a} + \sin at$

(ii) $0.0125 \cos 3t + 0.0375 \cos 5t$

(iii) $-\dfrac{1}{a^{2}} t + \dfrac{1}{a^{3}} \sin h\, at$

(iv) $\dfrac{1}{4} - \dfrac{1}{2} j \cos 2t - \dfrac{1}{4} \sin 2t$

(v) $\dfrac{5}{4}\left[-1 + t + e^{-2t} + t\, e^{-2t}\right]$

12.6 $2e^{-2t} - e^{-t}$

12.7 $4te^{-2t} - 2e^{-t} + 2e^{-2t}$

12.8 $e^{-t} + e^{-2t}$

12.9 $-10e^{-2t} + 30e^{-3t} - 20e^{-4t}$

12.10 $1.5 + \dfrac{6}{5} e^{-\frac{1}{3}t} - \dfrac{3}{10} e^{-2t}$

12.11 $2.5e^{-37.5t}$

12.12 $2 - e^{-0.5t}$

12.13 $e^{-t} \left[\sin t + \cos t \right]$

12.14 $\dfrac{1}{2} \cos 2t + \dfrac{1}{4} \sin 2t + 0.75\, e^{-2t} - 2t\, e^{-2t}$

12.15 $-10 - 0.034\, e^{-17.8t} + 14.03\, e^{2.81t}$

12.16 $E_0(s) = \dfrac{2\left(s^2 + 40s + 500\right)}{s\left(s + 20\right)}$, $\quad Z_0(s) = \dfrac{\left(s^2 + 30s + 100\right)}{s + 20}$

$$i(t) = 2\left[\dfrac{5}{3} - e^{-10t} + \dfrac{1}{3} e^{-30t} \right]$$

12.17 $\dfrac{100}{9} - \dfrac{100}{9} e^{-\frac{5}{3}t} - \dfrac{100}{9} u\,(t - a) + \dfrac{100}{9} u\,(t - a)\, e^{-\frac{5}{3}(t - a)}$

12.18 $10e^{-5t}\, u(t) - 10e^{-5(t - a)}\, u\,(t - a)$

12.19 (i) $-\dfrac{1}{8} u\,(t - 5) + \dfrac{1}{4} r\,(t - 5) + \dfrac{1}{8} u\,(t - 5)\, e^{-2(t - 5)}$

(ii) $0.5\, e^{-2(t - 5)}\, u\,(t - 5)$

(iii) $0.25\, u\,(t - 5) - 0.25\, e^{-2(t - 5)}\, u\,(t - 5)$

(iv) $0.5\delta(t - 5) - e^{-2(t - 5)} u(t - 5)$

12.20 (i) $0.2u(t-2) - 0.2 e^{-1.25(t-2)} u(t-2)$

   (ii) $0.25 e^{-1.25(t-2)} u(t-2)$

   (iii) $0.25 \delta(t-2) - 0.3125 e^{-1.25(t-2)} u(t-2)$

   (iv) $0.25 \delta'(t-2) - 0.3125 \delta(t-2)$
   $$+ 0.39 e^{-1.25(t-2)} u(t-2)$$

12.21 $v_L(t) = \dfrac{3}{4} \delta(t) - 12 e^{-12t} + 36t\, e^{-12t}$

12.22 $Z(t) = 2\left[\delta(t) + 2u(t)\right], \quad v_s(t) = 10 u(t)$

12.23 (i) $C = \dfrac{1}{2}$ F   (ii) $L = \dfrac{3}{5}$ H

   (iii) $C = \dfrac{1}{2}$ F   (iv) $R = 1.5\ \Omega$

   (v) $L = \dfrac{1}{9}$ H

12.24 $I(s) = \dfrac{1 - e^{-\frac{T}{2}s}}{s\left(1 - e^{-Ts}\right)}$

12.25 $F(s) = \dfrac{1}{s}\left[1 - 3e^{-s} + 4e^{-2s} - 4e^{-4s} + 2e^{-5s}\right]$

12.26 $\dfrac{2}{s^2}\left[1 - 3e^{-s} + 5e^{1.5s} - 6e^{-2s} + 6e^{-3s}\right]$

12.27 $\dfrac{1 - e^{-as} - e^{-3as} + e^{-4as}}{as^2\left(1 - e^{-4as}\right)}$

12.28 $\dfrac{1}{as^2} - \dfrac{e^{-as}}{s\left(1 - e^{-as}\right)}$

Chapter 13: Network Functions

13.1   (i) $\dfrac{1}{41}$ (ii) $\dfrac{1}{56}\,\Omega$

       (iii) $-\dfrac{1}{41}\,\mho$ (iv) $-\dfrac{1}{56}$

13.2 $\dfrac{1}{0.125s^3 + 1.25s^2 + 3s + 1}$

13.3 $\dfrac{\left(s^2 + 1\right)^2}{5s^4 + 5s^2 + 1}$

13.4 $\dfrac{(s+1+j1)(s+1-j1)}{s+1}$

13.5 $\dfrac{s^2 + \dfrac{R_1 C_1 + R_2 C_2}{R_1 R_2 C_1 C_2}\,s + \dfrac{1}{R_1 R_2 C_1 C_2}}{s^2 + \dfrac{R_1 C_1 + R_1 C_2 + R_2 C_2}{R_1 R_2 C_1 C_2}\,s + \dfrac{1}{R_1 R_2 C_1 C_2}}$

13.6   (i) $\dfrac{1}{3}$ (ii) $\dfrac{1}{3}\,\Omega$

       (iii) $-\dfrac{1}{3}\,\mho$ (iv) $-\dfrac{1}{3}$

13.7 $-\dfrac{3}{7}$

13.8   (i) $\dfrac{s^4}{s^4 + 2s^3 + 3s^2 + 1}$

       (ii) $\dfrac{1}{8s^3 + 4s^2 + 4s + 1}$

13.9 $e^{-t} + e^{-2t}$

Chapter 14: Two-Port Parameters

14.1 $z_{11} = 5.71 \ \Omega,$ $z_{12} = -4.3 \ \Omega,$

$z_{21} = 2.14 \ \Omega,$ $z_{22} = 2.14 \ \Omega.$

14.2 $y_{11} = 0.5 \ \mho, \ y_{12} = y_{21} = -0.25 \ \mho,$

$y_{22} = 0.625 \ \mho$

14.3 $z_{11} = z_{21} = -1 \ \Omega,$ $z_{12} = 0 \ \Omega, \ z_{22} = \dfrac{1}{3} \Omega$

14.4 $y_{11}(s) = y_{22}(s) = \dfrac{s^2 + 3s + 1}{s + 2}$

$y_{12}(s) = y_{21}(s) = -\dfrac{s^2 + 2s + 1}{s + 2}$

14.5 $A = \dfrac{1}{2s + 1},$ $B = 2,$

$C = 2s(s + 1),$ $D = 2s + 1$

14.6 (a) $A = s + 1, \ B = \dfrac{s^2 + s + 1}{s},$

$C = 1, \ D = \dfrac{s + 1}{s}$

(b) $z_{11} = s + 1, z_{12} = z_{21} = 1, z_{22} = \dfrac{s + 1}{s}$

(c) $h_{11} = \dfrac{s^2 + s + 1}{s + 1}, h_{12} = \dfrac{s}{s + 1},$

$h_{21} = -\dfrac{1}{3}, h_{22} = \dfrac{1}{3}$

14.7 $y_{11} = \dfrac{166}{195} \ \mho, y_{12} = y_{21} = -\dfrac{41}{195} \ \mho,$

$y_{22} = \dfrac{211}{195} \ \mho$

14.8 (a) $z_{11} = 5 \ \Omega,$ $z_{12} = 1 \Omega$

$z_{21} = 2 \ \Omega,$ $z_{22} = 1 \Omega$

(b) $y_{11} = \dfrac{1}{3} \, \mho$, $\qquad\qquad$ $y_{12} = -\dfrac{1}{3} \, \mho$

$\qquad\qquad y_{21} = -\dfrac{2}{3} \, \mho$, $\qquad\qquad$ $y_{22} = \dfrac{5}{3} \, \mho$

(c) $A = \dfrac{5}{2}$, $\qquad\qquad\qquad$ $B = \dfrac{1}{2} \, \Omega$

$\qquad\qquad C = \dfrac{1}{2} \, \mho$ $\qquad\qquad\qquad$ $D = \dfrac{1}{2}$

The network is neither reciprocal nor symmetrical.

14.9 $\quad h_{11} = r_b + r_e$, $\qquad\qquad$ $h_{12} = \mu_{bc}$,

$\qquad\quad h_{21} = \alpha_{cb}$, $\qquad\qquad\qquad$ $h_{22} = \dfrac{1}{r_e + r_d}$

14.10 $\quad y_{11} = 4 \, \mho$ $\qquad\qquad\qquad$ $y_{12} = y_{21} = -3 \, \mho$,

$\qquad\quad y_{22} = 3 \, \mho$ $\qquad\qquad\qquad$ $z_{11} = 1 \, \Omega$

$\qquad\quad z_{12} = z_{21} = 1 \, \Omega$, $\qquad\qquad$ $z_{22} = \dfrac{4}{3} \, \Omega$

14.11 $z_{11}(s) = \dfrac{1}{sC} + sL$, $\qquad\qquad$ $z_{12}(s) = z_{21}(s) = sM$,

$\qquad\quad z_{22}(s) = R + sL_2$

Chapter 15: Network Synthesis

15.1 (i) P.R.F. $\qquad\qquad\qquad$ (ii) Not P.R.F.

$\qquad$ (iii) P.R.F. $\qquad\qquad\qquad$ (iv) Not P.R.F.

$\qquad$ (v) Not P.R.F. $\qquad\qquad\quad$ (vi) Not P.R.F.

$\qquad$ (vii) Not P.R.F.

15.2 (i) Hurwitz $\qquad\qquad\qquad$ (ii) Hurwitz

$\qquad$ (iii) Not Hurwitz $\qquad\qquad$ (iv) Hurwitz

$\qquad$ (v) Hurwitz $\qquad\qquad\qquad$ (vi) Hurwitz

First Foster From

15.3

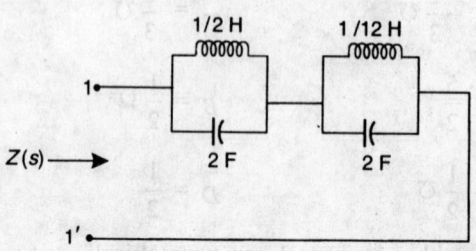

$Z(s) \longrightarrow$

1/2 H   1/12 H

2 F   2 F

15.4

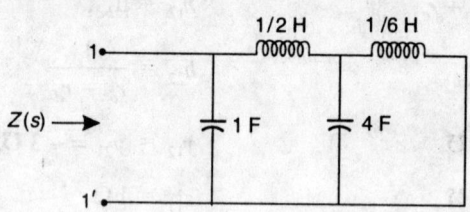

$Z(s) \longrightarrow$

1/2 H   1/6 H

1 F   4 F

15.5

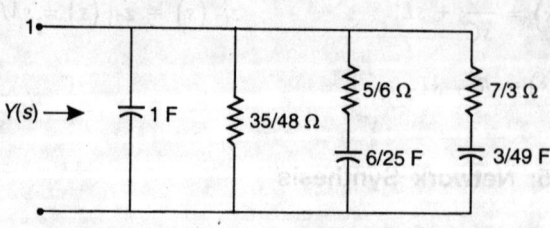

$Y(s) \longrightarrow$

1 F   35/48 Ω   5/6 Ω   7/3 Ω

6/25 F   3/49 F

15.6

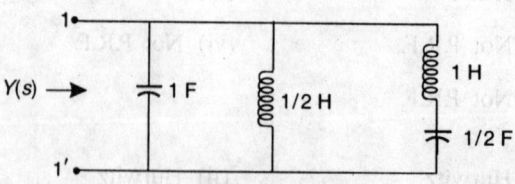

$Y(s) \longrightarrow$

1 F   1/2 H   1 H

1/2 F

15.7

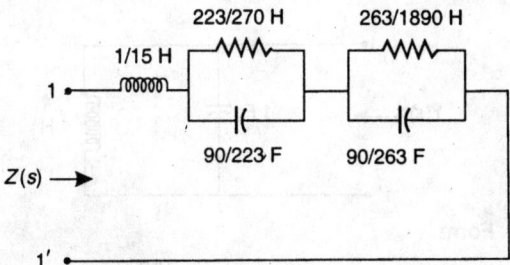

15.8

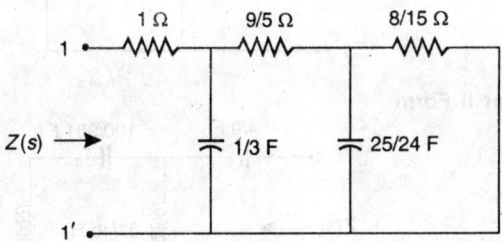

15.9

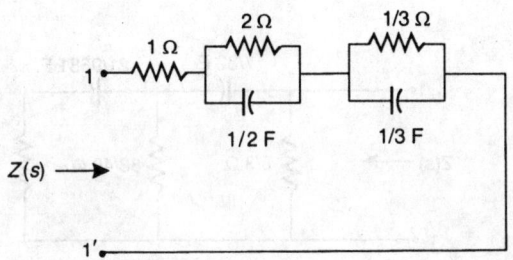

15.10 First Foster Form

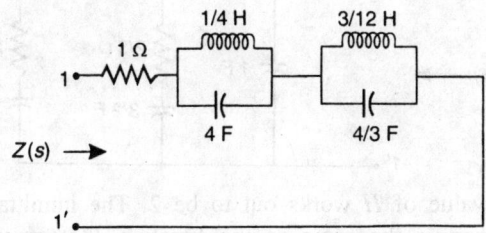

Second Foster Form

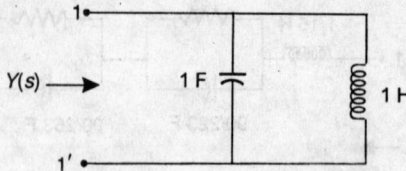

15.11 Cauer I Form

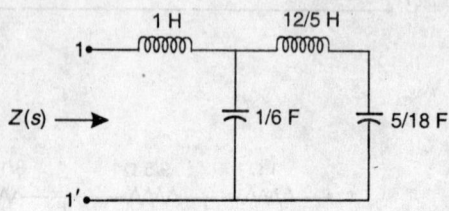

Cauer II Form

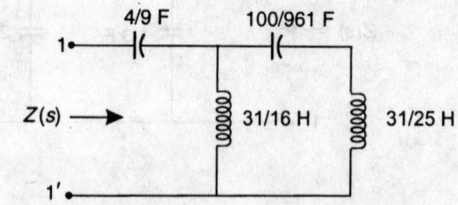

15.12

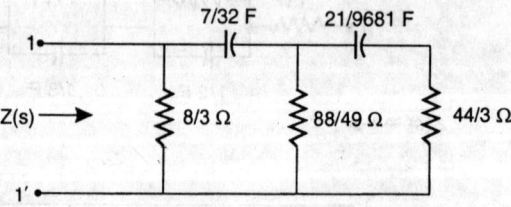

15.13 Figure 15.32 (b) represents an *R-C* admittance function and the realised *R-C* network is as shown.

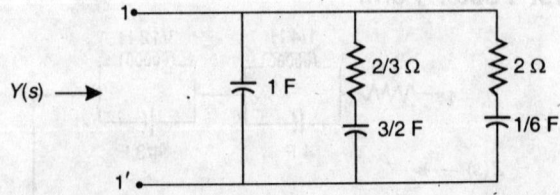

15.14 The value of *H* works out to be 2. The immittance function obtained is the same as the function *F(s)* given in worked example 15.16. The network is realised in all the Four Forms.

Index

Reader's Notes